P9-CJF-961

REAL-TIME COMPUTING

With Applications to Data Acquisition and Control

Edited by

Duncan A. Mellichamp
Department of Chemical and Nuclear Engineering
University of California, Santa Barbara
Santa Barbara, California

Van Nostrand Reinhold Electrical/Computer Science and Engineering Series

VNR VAN NOSTRAND REINHOLD COMPANY
NEW YORK CINCINNATI TORONTO LONDON MELBOURNE

Copyright © 1983 by Van Nostrand Reinhold Company Inc.

Library of Congress Catalog Card Number: 82-8663
ISBN: 0-442-21372-7

All rights reserved. No part of this work covered by the copyright hereon may be reproduced or used in any form or by any means—graphic, electronic, or mechanical, including photocopying, recording, taping, or information storage and retrieval systems—without permission of the publisher.

Manufactured in the United States of America

Published by Van Nostrand Reinhold Company Inc.
135 West 50th Street, New York, N.Y. 10020

Van Nostrand Reinhold
480 Latrobe Street
Melbourne, Victoria 3000, Australia

Van Nostrand Reinhold Company Limited
Molly Millars Lane
Wokingham, Berkshire, England

15 14 13 12 11 10 9 8 7 6 5 4 3

Library of Congress Cataloging in Publication Data

Main entry under title:

Real-time computing.

   (Van Nostrand Reinhold electrical/computer science
and engineering series)
   Includes bibliographical references and index.
   1. Process control--Data processing. 2. Real-time
data processing. 3. Automatic data collection systems.
I. Mellichamp, Duncan A. II. Series.
TS156.8.R4 1982   001.64   82-8663
ISBN 0-442-21372-7   AACR2

004.33
R288
1983

Contributors

Dr. David E. Clough
Department of Chemical Engineering
University of Colorado
Boulder, Colorado

Dr. Thomas F. Edgar
Department of Chemical Engineering
University of Texas
Austin, Texas

Dr. George P. Engelberg
Canadian National
Montreal, Quebec, Canada

Dr. D. Grant Fisher
Department of Chemical Engineering
University of Alberta
Edmonton, Alberta, Canada

Dr. James A. Howard
Department of Electrical and Computer
  Engineering
University of California
Santa Barbara, California

Dr. William R. Hughes
Department of Electrical Engineering
University of Colorado
Boulder, Colorado

Dr. Duncan A. Mellichamp
Department of Chemical and Nuclear
  Engineering
University of California
Santa Barbara, California

Dr. Walter G. Rudd
Computer Science Department
Louisiana State University
Baton Rouge, Louisiana

Dr. Cecil L. Smith
Cecil L. Smith, Inc.
Baton Rouge, Louisiana

Dr. James Wm. White
Modular Mining Systems, Inc.
Tucson, Arizona

Dr. Joseph D. Wright
Xerox Research Centre of Canada
Mississauga, Ontario, Canada

Mills College Library
Withdrawn

MILLS COLLEGE
LIBRARY

Van Nostrand Reinhold
Electrical/Computer Science and Engineering Series
Sanjit Mitra, Series Editor

HANDBOOK OF ELECTRONIC DESIGN AND ANALYSIS PROCEDURES USING PROGRAMMABLE CALCULATORS, by Bruce K. Murdock

COMPILER DESIGN AND CONSTRUCTION, by Arthur B. Pyster

SINUSOIDAL ANALYSIS AND MODELING OF WEAKLY NONLINEAR CIRCUITS, by Donald D. Weiner and John F. Spina

APPLIED MULTIDIMENSIONAL SYSTEMS THEORY, by N. K. Bose

MICROWAVE SEMICONDUCTOR ENGINEERING, by Joseph F. White

INTRODUCTION TO QUARTZ CRYSTAL UNIT DESIGN, by Virgil E. Bottom

REAL-TIME COMPUTING, edited by Duncan A. Mellichamp

LIGHT TRANSMISSION OPTICS, Second Edition, by Deitrich Marcuse

HARDWARE AND SOFTWARE CONCEPTS IN VLSI, edited by Guy Rabbat

MODELING AND IDENTIFICATION OF DYNAMIC SYSTEMS, by N. K. Sinha and B. Kuszta

COMPUTER METHODS FOR CIRCUIT ANALYSIS AND DESIGN, by Jiri Vlach and Kishore Singhal

Preface

With the advent of the first minicomputers in the mid to late 1960s, many university departments, industrial companies, and research laboratories suddenly found that they could afford to install digital computers for purposes of data acquisition and control. The explosion in applications which followed was complicated by a shortage of people with training in the field. Education of new people was, in turn, hampered by the extreme lack of suitable textual materials.

Over ten years ago the CACHE* Real-Time Task Force was formed to assist in this sudden swing to computer automation. A group of us, all with industrial computer control or hybrid computing backgrounds, took on the task of putting together appropriate teaching materials. To make sure the fundamental computer areas were suitably covered, we enlisted the help of several knowledgeable electrical engineers and computer scientists.

The project turned out to be bigger than anyone had imagined. Covering such a broad field, from both fundamental and applied points of view, was not made easier by the plethora of machine architectures and hardware designs nor by the sudden changes which occurred as hardware and software rapidly developed and became more complex. Nevertheless, the eight volumes of the CACHE Monograph Series in Real-Time Computing, published between 1977 and 1979, obviously met many of the requirements, as much favorable feedback has indicated subsequently.

The present volume came directly from the monographs. It retains all of the features which made them so useful — the introductory material followed by more detailed and advanced coverage, the many figures, tables, and examples which put the text into perspective, the applications examples which range from simple to moderately complex, etc. In addition to a complete updating of the materials, a number of the contributions have been substantially revised to increase the cohesiveness of this single volume. Although there always will be some unevenness in style and treatment in a contributed book, we have attempted to maintain a common "vocabulary" and approach.

This book now seems to accomplish our original goal, i.e., to give the real-time user, in one volume, all of the material that is required to develop and/or use computer data acquisition and control systems. Hence, it contains introductory material from all important real-time sub-fields and extended treatments of important topics such as process/computer interfacing, multitask programming, and applications implementation.

Even with the attempt to give the new or experienced real-time user the most important material of six to eight separate computer engineering and process control references, there have to be some gaps. Hence, we have restricted ourselves somewhat to process-oriented applications, our main area of expertise, and to single computer applications. Networks, though briefly discussed, would require a separate volume, as would any extended discussion of mini- and microcomputer achitectures. Also, we have limited our discussion of high-level real-time languages to BASIC and FORTRAN, those most widely used at the present time. Perhaps a future volume will discuss PASCAL and ADA; both of these languages have tremendous future potential for real-time applications as a re-

---

*CACHE (Computer Aids in Chemical Engineering Education), formerly a committee formed under the auspices of the National Academy of Engineering and, now, a not-for-profit corporation.

sult of features designed *into* them rather than patched onto them as with BASIC and FORTRAN. In any case, we have furnished a large number of references and/or supplemental reading sources which can be used for extended study.

I would like to thank each of the contributors personally for being willing to suffer through many revisions of their work. The discipline required in a group effort of this sort is considerable; and, by projects's end, several of my colleagues dreaded to answer the phone or open their mail. I hope that now can change.

I also would like to thank Darcy Radcliffe, who typed most of the photocopy for the monographs and drew the figures, and Renee Cella and Carina Billigmeier, who typed the revised manuscript. These women did far more than "process the words"; their professional concern and comments on style and usage have been greatly appreciated.

As any teacher will know, I owe a particular debt to my students. It seems appropriate to mention here those former students who played a major part in the development of the real-time facilities and program at Santa Barbara; their influence on this volume was exerted through our many discussions, arguments, design sessions, etc. Jay Bayne, Larry Nelson, John Balster, George Engelberg, Gerry Brown, and Bill Hughes (electrical and computer engineering graduates) and Michel Idier, Ed McNeal, Ferhan Kayihan, Dennis Marston, and Tom Moore (chemical engineering graduates) all have made their significant individual contributions.

Finally, the editor of such volumes usually gives credit to the contributors for all of the good features in the book and graciously accepts the blame for any deficiencies. Why not?

Duncan A. Mellichamp
Santa Barbara

Contents

Part I
An Introduction to Real-Time Computing

1

Digital Computing and Real-Time Digital Computing

Duncan A. Mellichamp
Department of Chemical and Nuclear Engineering
University of California, Santa Barbara

1.0 SOME PRELIMINARY REMARKS

"Real-time computing" is one of those general terms which has come to mean different things to different people. We will devote considerable effort in this introductory chapter to the development of our own definition of real-time computing; however, to begin with, let us simply agree that the term will imply the use of a computer in conjunction with some external "process" (or processes). The object of this interconnection will be to obtain information from the process (*monitor* its operation through measurement of important variables) and, perhaps, to manipulate it in some desired fashion (*control* the way in which it operates based on the information previously acquired). For the computer to be able to accomplish one or both of these objectives, its operations have to be carefully sequenced in time; we might say that it must operate in "real time".

In our treatment of real-time computing, we will be dealing with a field of digital computer applications that has been around practically since the modern digital computer was invented. The first stored program computers were large, cumbersome, hard-to-use devices which could be justified and built only to carry out long, tedious or complex arithmetic calculations. Still, it was not long after their introduction before the idea of using a computer to monitor or control some physical process became obvious. Since the early computers were encumbered with so many deficiencies, at least by today's standards, and were designed primarily

for calculation purposes, not for operations in conjunction with external processes, it was no simple task to implement a computer system for process monitoring or control. Almost all of these early applications were very high-cost, defense-oriented projects that could not have been justified using ordinary economic criteria. Some early applications were, for example, in ship fire-control systems, radar networks, and ballistic missile navigation systems.

Vast improvements have been made in computer architecture, hardware, software, and reliability, and computer system costs have been reduced by many orders of magnitude since these first exploratory applications. Still, the central theme in real-time computing is the connection of a process to a digital computer. Today the process might be as simple as a single laboratory instrument that makes use of the computer to interpret raw results, or it might be as complex as an entire petroleum refinery. Almost all modern industrial facilities utilize computers to monitor and control operating processes. Computer-operated equipment test stands are used to evaluate new gas turbines, electronic components, or perhaps even new computers themselves. Manufacturing processes such as paper machines, chemical plants, or steel rolling mills represent good practical examples of present-day processes. Doubtless the most sophisticated and glamorous processes supervised and controlled by computers to date have been the various launch, orbiting, and landing vehicles used in space projects such as the Apollo project to land on the moon or the

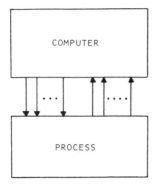

Figure 1.1. A real-time system.

space shuttle project. In these cases whole networks of expensive computers were used to keep track of all the elements in the space flight system.

Despite the tremendous range of properties (scale, configuration, purpose, etc.) embodied in all of these processes, any real-time system can be depicted very simply as in Figure 1.1. Here the interconnection of the computer and the process (we might use the term "coupling" to describe this physical connection for the time being) is graphically indicated by some arrows representing flow of information.

Furthermore, there are a number of common features we can focus on when considering the connection of a process to a computer. The primary, unifying idea is that of "real time". All of these processes operate with their own inherent time scales, and, whether the process time scale is measured in milliseconds, days, or years, operations involving the computer must be keyed to operations of the process and not the other way around. To accomplish this interconnection or coupling of computer and process we must design and program the computer in such a way that it can keep track of the passage of "real time" independently of its own internal operations. Real time, actually our own familiar system of seconds (or fractions of seconds), minutes, hours, etc., becomes the standard of reference by which the computer initiates actions and records results concerning the coupled external process.

The concept of "real time" also implies the ability of the computer to respond to stimuli from the process in a timely fashion, i.e., sufficiently fast to accommodate the needs of the process. For example, if some emergency condition arises in the process and is signaled to the computer, the computer must be capable of reacting to the process's requirements fast enough to handle the emergency. Clearly, for simple processes operating on long time scales (hours or days), the design of a real-time computer system will likely not involve any critical timing problems, and practically any digital computer could be selected for the job. For complex processes (involving the transfer back and forth of large amounts of information) or for processes operating on short time scales (milliseconds or seconds), the idea of real-time response becomes a much more critical one, requiring careful attention in the selection of the computer and in designing the total real-time system.

A further common feature of process/computer systems involves the physical means of connecting the process to the computer. Digital computers have their own internal languages, using digital "words" to represent information, with "bits" of information composing the words themselves. By way of contrast, data (information) from typical processes seldom are obtained entirely in digital format: voltage is the deflection of a needle on a meter dial; temperature, the physical height of a column of mercury in an enclosed glass tube; chemical concentration, the area under the output curve of a chromatograph, etc. We would say that information of this type is in "analog" form. Fortunately, by means of the combined steps of measurement, transduction, sampling, and conversion, we can change most process data, including the analog variety, into a form usable by a digital computer. And, by something like a reversal of these procedures, internal computer information can be converted to a form that can be used to manipulate or control the connected process. The measurement, transduction, sampling, and conversion equipment needed to connect a physical process to a computer is called the "computer/process interface"; our approach will be to generalize discussions of these interfaces as much as possible, emphasizing the elements common to most physical processes, in order that too much effort is not expended dealing with special or exceptional cases.

Despite all our efforts to generalize real-time concepts, it must be admitted that at the present time almost every real-time system is a unique creation. Hence the real-time computer user—despite his central concern with his process and what he wants to do with it—is forced to pay some attention to the structure of the computer system, to the computer/process interface, and to programming fundamentals if he wishes to design and to make best use of the computer/process system. In many ways the real-time field has characteristics of the data processing field 20 to 25 years ago. At that time of relatively inefficient, slow computers with their primitive programming languages, it was necessary for a potential user to be familiar with many details of the system hardware, perhaps to know the machine language itself, in order to be able to program the computer and certainly to minimize the size and complexity of his program. Now of course, with large, fast computers and standardized high-level languages such as FORTRAN, BASIC, or PASCAL, present-day *non*-real-time computer users don't even need to pay attention to the type of computer their program will run on. Concern about system hardware has been reduced to an occasional consideration of machine accuracy (which is related to internal number representation and format).

Real-time systems computers have come a long way from the first-generation machines which were, for the most part as we have noted, unique designs or special adaptations of existing data processing computers. The second and third generations of real-time processors included some specifically intended for on-line applications, e.g., the Ramo Wooldridge RW-400 which was designed for defense applications and subsequently was used in several industrial process control projects. Then later the IBM 1800, a modified version of the IBM 1130, the CDC 1700, and several other commercial machines were widely used in industrial applications. The DEC PDP-8 which appeared in 1965 was the first minicomputer to obtain widespread usage in real-time, particularly laboratory, applications. Today we are in the midst of an explosion of technology, and there are several trends—the rapid development of inexpensive, fast computer hardware (generally based on a minicomputer or on dedicated microcomputers), the introduction of standard high-level languages (with necessary extensions for real-time applications), plus the wider availability of sophisticated executive systems to monitor computer operations—all of which are reducing the requirements for specialized knowledge. Computer hardware and software systems now on the market require the real-time user to understand only a few additional programming commands whose use is relatively straightforward. This sort of progress is to be expected and necessary if widespread application of real-time concepts ever is to occur. We should note, however, that the ordinary batch or time-shared computer user who understands the fundamentals of what he is doing always has been able to compute more effectively and efficiently than the casual user. Similarly, a person attempting real-time applications will always make most efficient use of his system (or choose the most efficient system for his purpose) if he has a true understanding of the basics of computing in general, and real-time computing in particular. Furthermore, despite major efforts to simplify and generalize the interconnection of process and real-time computer, the interface between them will always present particular problems which require detailed and specific information for their proper solution. Hence each real-time system, as distinguished from the real-time processor only, likely will remain unique.

This book represents an attempt to bring together all of the fundamentals of real-time computing, illustrating the concepts wherever possible by means of examples. It has been written with the real-time user in mind rather than the systems designer. Hence, it is more introductory in scope, and certain simplifications have been introduced to aid in developing an understanding of the most important features of real-time computing. In this regard we feel we have not taken any unnecessary liberties. Where possible, references dealing with particular topics have been included for supplemental reading. However, the computer science literature is large and growing rapidly, and our list of references is by no means intended to be all-inclusive.

1.1 SOME WORKING DEFINITIONS

Before beginning a detailed discussion of real-time computing we want to define the term as precisely as possible, at least to build a working understanding of what we mean by real-time computing. To do this effectively requires that we first discuss some of the characteristics of digital computing in general and then take a rather philosophical look at where real-time computing fits in among all the other variants of digital computer applications.

1.1.1 The Computer: A Serial Processor

The most fundamental concept to be aware of at the outset is that the digital computer is, in essence, a serial device. By this we mean that, although a certain amount of overlap can be designed into its internal structure and operations, the computer basically can perform only one operation at a time. This characteristic turns out to be one that demands considerable attention from systems designers and users, no matter whether the system is to be used substantially for solving problems (data processing) or in operating external processes. Now even though the serial nature of the machine is limiting in this sense, further difficulties arise because the machine also must communicate with the outside world through external devices or peripheral equipment (peripherals). In virtually every case these external devices operate at speeds several orders of magnitude slower than the computer's central processor, usually because they contain slow mechanical components. For example, a typing device might be able to output results at speeds of ten to a hundred characters per second, or a printer at speeds of hundreds to thousands of characters per second. However, a typical computer might be capable of outputting results at speeds measured in millions of characters per second.

Now unless we are willing to underutilize a processor's capabilities drastically whenever input or output operations must be performed

(and input/output operations make up a substantial portion of a typical real-time computer's work-load), there has to be some way to avoid this dilemma. A second potential problem is tied to real-time system requirements that we mentioned earlier—the computer must be able to respond to changing process conditions in a reasonably short period of time. It may be difficult or impossible to program a simple sequential machine in such a way that it can monitor a changing process and do other things as well. The danger is that important process changes may be missed while the computer is involved in some other, extended program sequence.

1.1.2 Interrupts and Interrupt Processing

The approach taken to overcome these problems has been to design processors so that they can respond the "external interrupts". We will pay considerable attention later on to the hardware and software details necessary to make "interrupt processing" both possible and efficient. At this point we only need to describe the concept briefly and mention how it can be utilized.

Suppose that it is possible for an external (peripheral) device to break into the normal sequence of a processor's operations, i.e., to cause the processor to stop whatever it is doing and then, by executing some preplanned steps, take over the job of "servicing the interrupt". By this phrase we mean "to determine which device caused the interrupt, to find out what needs or requirements the device has that caused it to initiate the interrupt, and to respond to its needs, e.g., by furnishing it information or taking information from it". Suppose further that, after completing the interrupt servicing operation, the processor is able to resume operations at the point where it originally was interrupted as if nothing had happened. With this single simple feature (which admittedly has vast implications with respect to the complexity of systems hardware and software required to implement and use it), the sequential limitations of the processor can be overcome.

For example, if a printing device has been

designed so that it interrupts the processor *after* it has completed some output operation, say after printing a character (or a line of characters), then the computer need only transfer information (data) to an electronic (buffer) storage area in the printing device itself, doing this essentially at processor speeds, and then start the mechanical operations. The processor subsequently can turn its attention to something else, perhaps to performing some additional calculations, since it will be interrupted when the device has printed the information and hence will be informed that it can proceed to print out another character (or line). In this way the processor does not sit "idle" waiting for the printing device to complete its mechanical operation.

Some feeling for the way in which this simple interrupt concept has been elaborated in practice will be developed in later chapters. For present purposes, however, it is only necessary to realize that modern computers intended for general real-time applications will have extensive interrupt handling capabilities.

1.1.3 Two Classifications of Computer Applications

"In the beginning", when computers first appeared on the scene, classification of computer applications was quite simple. There was only one type of application, namely, computation. Preparation of a computer to solve a particular problem required intense, specialized efforts—to set-up the computer, to program it, and to prepare any relevant data for processing. Simply entering the program and data was itself a chore. Later, as computers evolved for the purpose of solving mathematical problems or of performing some administrative or business functions on a more routine basis, the concept of *batch computing* evolved. Typically a batch of programs and associated data was prepared to read into such a system from punched cards or magnetic tape; the computer performed its calculations; the results were read out onto magnetic tape, printed paper, or punched cards; and the next programs and data were entered.

Another class of computer applications evolved at about the same time. This category,

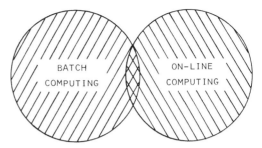

Figure 1.2. Two major classes of digital computer applications (some years ago).

which some people have referred to as *on-line computing*, involved the use of a computer with external (perhaps multiple) user-oriented terminals, with processes as we described above, or with other computers (i.e., in a form of computer network). There was relatively little similarity either in configuration or in function between the two types of systems, batch and on-line, as they had very different objectives. (Such a lack of overlap is shown schematically in Figure 1.2.) Batch facilities had the common goal of running computational jobs as rapidly as possible. The idea of "throughput" or the number of typical programs that could be run in a given length of time became an important concept. On the other side, on-line systems were designed to perform specific external functions not strictly related to performing calculations rapidly, and the idea of "throughput" made little sense. At this stage of computer design the interrupt concept was relatively little developed, meaning, for example, that multiuser (time-shared) systems had to rely on "polling" extensively (i.e., checking each terminal relatively often to determine whether the user had typed-in a character). Early "real-time" systems involving a process or processes had to utilize a precisely-designed sequence of program steps, executed repetitively, in order that the computer's operations might be keyed to real or process time.

In the last 15 to 20 years everything—computer architecture, hardware and peripherals, and software—has become more sophisticated and complex. Elaborate interrupt structures now are available so that it is no longer necessary for the computer to use primitive polling techniques to determine if external devices (peripherals or processes) need to be serviced. Internal hardware items have taken over many of the functions that formerly had to be programmed; and medium-speed bulk storage devices such as magnetic disks allow for much flexibility in terms of the inputting, outputting, and storage of programs as well as data. Information now can be transferred into and out of the system without tying up the computer's central processor; and operating systems (i.e., the internal program elements designed to control the scheduling and sequencing of input, output, and calculation operations) now are available for even very small and inexpensive computers with sophisticated features of the sort available only for large computers just a few years ago.

As a result, the old idea of pure batch processing has practically disappeared in the sense that most data processing systems now utilize what formerly would have been considered on-line concepts to increase throughput. Today a "batch" processor may utilize its sophisticated hardware and operating system to read multiple computing jobs into its bulk storage units while it simultaneously is performing calculations and, perhaps, writing out results from other jobs. Individual jobs may be scheduled and executed using a complicated algorithm to maximize throughput by overlapping operations of the central processor and all of its peripheral devices. Information supplied as part of each computing job's entry data becomes the basis for executing immediately or deferring. Hence,

we have come around to the situation shown in Figure 1.3 where almost all computing could be considered to be on-line.

1.1.4 Internally- and Externally-Controlled Interrupts

Some authors (see, e.g., Ref. 1) have noted the distinction between interrupts that are controlled completely by a particular processor (internal) and interrupts that are under the control of devices, peripherals, or processes not completely within the jurisdiction of the processor (external). An attempt can be made to base a classification scheme on the nature of system interrupts. Hence the classical batch system might be defined as one in which only internal interrupts exist. In this case the processor only has to deal with its own peripherals, none of which can cause a spontaneous interrupt to occur. (This means that the computer could never get an interrupt from its line printer if it had not first loaded the printer buffer and started the printing operation.)

Such is not the case with an on-line system which ordinarily must contend with an environment in which interrupts can occur at any time. These could be caused by external devices such as other computers or user terminals wishing to communicate or by a process requiring attention.

Note that the present-day batch systems have at least a control console which subjects the computer to external interrupts from the operator, and many so-called batch systems have remote job entry stations and remote terminals as well. Hence, even following this more formalized classification scheme, we can see that a true batch system will seldom exist, and that most computer systems will be on-line both in terms of their external configuration and also how they function internally.

1.2 A DEFINITION OF REAL-TIME COMPUTING

We have seen that just about every computer application involves on-line concepts, and some people would be inclined to use the term "real-time" wherever we have said "on-line". Now

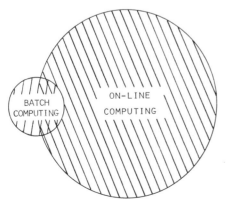

Figure 1.3. Two major classes of digital computer applications (today).

this book purports to deal with real-time computing, but we have no aspirations to try to
cover all aspects of digital computer applications. Hence we need to arrive at a definition
of real-time computing which more reasonably
limits our area of interest. Numerous definitions have appeared in the literature [1–4] but
there certainly is no consensus definition. Also,
from a report of the IEEE Computer Society
Model Curricula Subcommittee [5] it is clear
that at least some people view the field as a
very theoretically-oriented area of computer
science. Our interest in writing this book is
almost entirely from a very particular standpoint—that of the active user, developer,
teacher. We see real-time computing primarily
as an applications area; so, in an attempt to define the field for our own purposes and to show
where it fits into the larger category of on-line
computing, we need to make some further dis-
tinctions. Based on system hardware configuration alone we might arrive at the following
three subclassifications of on-line computing:

Clock-Based or Time-Based Computing—The
   computer utilizes a device (or devices) to
   key its operation to time in the real-world
   sense, i.e., the computer is able to maintain
   an accurate measure of physical time or to
   measure the passage of a particular interval
   of time in terms of usual time units (seconds,
   minutes, etc.).
Sensor-Based Computing—The computer utilizes an interface, i.e., external devices (generalized analog and/or digital sensors), to measure the conditions or state of a physical process. The computer may only monitor the
   process without influencing it, in which case
   the relation might be termed "passive", or the
   computer may manipulate certain process inputs, in which case there would be an "active"
   coupling with the process.
Interactive Computing—The computer is connected to external devices, processes, individuals (terminals), or other computers. If
   the computer communicates actively with the
   external units, the response of the on-line
   computer may need to be within certain time
   constraints.

Now the distinction between the second and

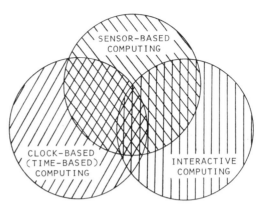

Figure 1.4. A subdivision of the on-line computing
category based on system configuration.

third categories may not be totally clear at this
point, but the intention here is to attempt to
distinguish between those external "processes"
that involve information that may be largely
analog in nature (and will require the type of
generalized interface equipment described in
Chapter 2 in order to be connected to the
computer) and those "processes" that are completely digital (generate and utilize only digital
information and will be connected to the computer by means of more specific hardware interface equipment).

Figure 1.4 indicates that these three categories overlap considerably; however, we can
see, by looking at several specific examples
listed below, that there are some clear distinc-
tions. Consider:

(1) A time-shared computing system involv-
    ing multiple terminals all connected to a
    single computer.
(2) A laboratory instrument that is moni-
    tored and controlled by a computer.
(3) An airlines reservation system consisting
    of many terminals connected to one
    centrally-located computer.
(4) A computerized burglar alarm which
    monitors all of the doors and windows in
    a house and turns on lights, rings alarms,
    and dials the police when an intruder is
    detected.
(5) A computer that monitors the passage of
    time via its own real-time clock and re-
    lates its other computational activities to
    that time.
(6) A computer connected to a second com-

puter with which it communicates from time to time.

With reference to Figure 1.4, Examples (1), (3), and (6) all fall in the area of interactive computing, Examples (2) and (4) fall in the area of sensor-based computing, and Example (5) falls in the area of clock-based computing. Example (5) also happens to describe the operations of a typical batch facility where jobs are logged in and logged out using the system clock to determine appropriate billing charges (another example of the demise of batch computing as a pure concept).

Using this system of classification we would be most interested in systems that fall into the sensor-based area. Since most real-time systems utilize a clock as well, we will likely be dealing with systems that lie in the union of the sensor-based and clock-based areas as shown in Figure 1.5. Assuming that his set of restrictions or conditions is acceptable, we now can define real-time computing as follows:

> Real-time processing involves the interconnection of a process with a computer utilizing analog/digital and digital/analog interfaces and/or generalized digital (binary) data interfaces. Data acquisition by the computer must be keyed to the time scale of the process. If the computer is to influence the process as well, then its own response must be timely, resulting in an appropriate process response.

This definition is not rigorous but descriptive. What we are saying is that we will not be particularly interested in remote batch or time-shared systems, although many principles would apply

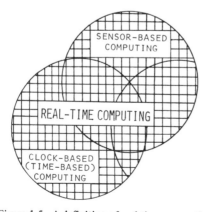

Figure 1.5. A definition of real-time computing.

to these systems as well as to those we will be studying. By excluding them we do not wish to imply that they are not real-time systems. They simply are not systems that we primarily are interested in here.

1.3 ORGANIZATION OF THE BOOK

The remaining two chapters of Part I are intended to give a broad overview of real-time system structures (hardware and architecture) and of real-time programming philosophy (software). Part II deals with a description of some typical processes and representative examples of real-time applications, with an introduction to the measurement of process variables, to transduction, transmission, and any subsequent signal processing such as the analog filtering of noise. Part III covers an introduction to digital representation and manipulation of information and to simple digital logic and hardware.

Part IV covers computer architecture, standard peripheral devices, and data communications and contains an important chapter on computer/process interfacing. General information dealing with software and programming—assembly language, systems software, multitask programming and operating systems—is covered in Part V; and the important area of programming for real-time applications is covered in Part VI. There we have stressed real-time versions of BASIC and FORTRAN and have included material on general, table-driven applications software of the type frequently used for data acquisition and control in the process industries. Part VII deals with the management of real-time facilities, particularly the smaller systems which might typically be used for pilot plant scale operations, in a laboratory, or for university teaching and research.

It has been our intention to make these materials as nonmathematical as possible. There often is, however, a need for detailed mathematical analysis of the computer/process system. Hence Part VIII has been included to cover: (1) an introduction to process analysis (with an emphasis on the analysis of process dynamics) and how process characteristics affect such important details as data sampling rates, and (2) a brief introduction to the topic of control of processes, in particular some materials on the digital computer implementation of process control and signal processing algorithms.

For potential real-time system users—we hope that this volume will furnish the wherewithal to get started; for present users—we believe it will fill in some remaining knowledge gaps; for those who would just like to know what real-time computing is all about—we recommend the introductory chapters as a way to get a good start in learning the working concepts and, more importantly, a basic understanding of this rapidly developing and growing field.

1.4 REFERENCES

1. Saffer, S. I. and Mishelevich, D. J., "A Definition of Real-Time Computing", *Comm. ACM, 18,* 544, Sept. 1975.
2. Martin, J., *Programming Real-Time Computer Systems,* Prentice-Hall, Englewood Cliffs, N.J., 1965.
3. Martin, J., *Design of Real-Time Computer Systems,* Prentice-Hall, Englewood Cliffs, N.J., 1967.
4. Sackman, H., *Computer, Systems Science and Evolving Society,* Wiley, New York, 1967.
5. Mulder, M. C., "Model Curricula for Four-Year Computer Science and Engineering Programs: Bridging the Tar Pit", *Computer, 8,* 28, Dec. 1975.

2

The Structure of Real-Time Systems

Duncan A. Mellichamp
Department of Chemical and Nuclear Engineering
University of California, Santa Barbara

2.0 INTRODUCTION

To begin with we would like to develop a general understanding of real-time system structures. By the word structure we mean the system hardware or physical components and, also, what these components are intended to do and how they are interconnected. Real-time systems, as we have defined them, typically consist of a process whose operations must be monitored or controlled, an associated real-time computer, the equipment necessary to operate the computer, and the equipment required to marry the computer to the process. We also want to discuss some general ideas concerning how all these devices communicate with each other, since these concepts dictate to some extent how the system is structured. Finally, we need to cover the general characteristics or properties of present-day real-time computers such as size, cost, and speed in order to give some idea of the potential applications of these systems and how this potential has expanded tremendously with the recent changes in capability and cost of computers appropriate for real-time applications. In later chapters we go back over many of these same points, and in much greater detail, but for now we wish to restrict ourselves to an overview of the subject.

First, we comment again on the *process*, the central reason for a real-time system. Despite its central importance, the process will not be the focus of this book. Chapter 4, however, is intended to discuss the process side of real-time systems, for the most part by describing a number of typical real-time applications examples. It is assumed that most people interested in real-time digital computing already will have a process in mind for potential application and will be knowledgeable concerning its nature and normal operating characteristics and their own reasons for wishing to connect it to an operating computer.

2.1 THE PROCESS

The concept of "process" will be used in a very general sense. Specifically, a process might be a laboratory instrument, a pilot plant, a manufacturing plant, a machine tool, a satellite or its launch rocket. It might be a computer itself or a network of terminals although, as we discussed in Chapter 1, such on-line systems will be of only passing interest to us here. We will take the point of view that all of the processes we are interested in exhibit certain common features; in general they all may be depicted in the form shown in Figure 2.1.

The basic idea illustrated in this figure is that the process always will have some combination of outputs to monitor and, often, inputs to manipulate or control. This will be true whether the process involves sequential operations (as with a machine tool), parallel continuous operations (a chemical process), or some combination of the two. It will also be the case regardless of the process size, configuration, location, or any of a number of other important process characteristics. Also, we will assume that process inputs and outputs can be manipulated into one of several forms which are compatible with a digital computer or, at least, compatible with a generalized process/computer interface. This interface will be a primary emphasis in our subsequent discussion.

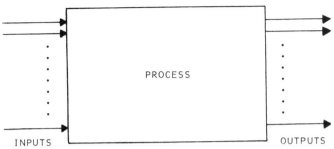

Figure 2.1. Schematic depiction of a process.

As a consequence of this situation we will be able to take the relatively detached point of view that, even though the box marked "Process" is important—its properties may perhaps be critical in a specific application—discussion of processes best can be dealt with by example in our introductory study of real-time systems. This is not the same as saying that the process can be treated lightly when the time comes to consider a specific application; indeed the characteristics of the process should always be considered first in assessing a potential application. Often the characteristics of the process dictate key requirements for the computer system and the process/computer interface, and for the programming language or software which can or should be used. These important points are discussed in Chapters 4 and 19.

2.2 THE DIGITAL COMPUTER

Almost any digital computer can be used in a real-time application; however, all computers do not adapt to this purpose equally easily. For the moment, any general-purpose, stored program computer can be considered as a candidate for real-time applications, and we may visualize such a computer to be composed of four subunits as shown in Figure 2.2.

The Arithmetic Unit contains all hardware necessary to carry out arithmetic and logic commands, e.g., to add two numbers, subtract two numbers, check to see if one number is larger than a second number, etc. The Arithmetic Unit contains the "accumulators" or working registers of the computer in which numbers are temporarily stored before they are used in computations. Some older or small computers contain only one general-purpose accumulator, in which case a second number to be added in, subtracted, or otherwise used in calculation would have to be obtained directly from memory. Large scientific computers may contain as many as 16, 64, or 256 accumulators. Such a difference may have a significant effect on how easily the computer is programmed (more registers generally implies greater ease), but it does not affect our categorization of computer function.

All units in the computer are constantly under the supervision of the Control Unit. This part of the computer is responsible for reading a program statement from memory, interpreting it, causing the appropriate action to take

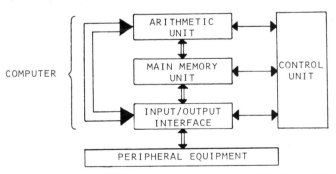

Figure 2.2. Schematic diagram of a general-purpose, digital computer.

place (e.g., add two numbers in the Arithmetic Unit), select the next consecutive program statement for reading, etc. The Control Unit is capable of carrying-out "branching" within a program, i.e., a change in the normal sequence of operations, when appropriate. This capability is responsible for much of the versatility of the general-purpose computer.

The Memory Unit in a digital computer is used for the storage of data and of the computer program itself. Normally the Control Unit causes a sequence of program statements stored in consecutive memory locations to be executed. These program statements may request operations to be performed in the Arithmetic Unit, data to be moved into or out of the Memory Unit, the normal sequence of operations to be altered (as noted above), or many other possibilities available to the computer. We speak of the set of operations which the computer performs, at least at the lowest or "machine language level", as the instruction set. Bigger computers have richer instruction sets (meaning more flexibility), but all computers have a certain minimum set of instructions from which all operation sequences must be derived. The main Memory Unit these days most often is constructed of solid state (integrated circuit) elements which serve to store digital information. These storage elements are so designed that it is possible to access (fetch or store) the contents of a memory location in one microsecond or less. Up until recent years, memory units were based on a matrix of ferrite or magnetic core elements. Solid state memory has the advantage of higher speed than core memory but the accompanying disadvantage that stored information is lost if a power interruption occurs. Hence, back-up battery power supply often is included. Main memory often is referred to by the name "Fast Memory" or sometimes as "Core Memory" in contradistinction to "Bulk Memory" or "Slow Memory", which usually is a separate, peripheral device that will be discussed later.

The final unit in this simplified visualization of the general-purpose digital computer is the Input/Output Interface. The I/O Interface (a shorter way of referring to it) is necessary for the computer to communicate with the "outside world", i.e., all of its peripheral equipment.

The interface in most computers consists of a set of bi-directional data lines and control lines, usually referred to as buses, and the logic necessary to detect and respond to external "events". These events usually take the form of a request for some kind of action on the part of the computer which then would have to interrupt its normal processing. The ability to respond to an external "interrupt" is a requirement for any computer, but it is in fact the very basis of the real-time computer design as we discussed at some length in the introduction. It is this capability that allows the real-time computer to keep track of the passage of time independently of its normal operations, and to watch multiple processes, each with a different set of demands which must be serviced by the computer. Obviously, the I/O interface also is necessary for reading a program into the computer and for transferring results out of the computer. As with the other three computer units, I/O Interface operations ordinarily are coordinated by the Control Unit.

We sometimes will refer to a computer using the generic term "processor". The term "central processing unit" (CPU) ordinarily is used more specifically to mean a combination of the Arithmetic and Control Units, i.e., the Memory and I/O Units are not included.

2.3 COMPUTER-RELATED PERIPHERAL EQUIPMENT

A complete description of real-time computing systems must include some discussion of computer-related peripheral equipment at this point. Computer-related in this context is used in opposition to process-related peripheral equipment; often the terms "data processing" or "information processing" are used in reference to such peripheral units. Figure 2.3 illustrates schematically a representative real-time computer system where the computer-related auxiliary devices have been located along the right-hand side of the drawing. The assortment of equipment shown here by no means is fixed or standard. Often a terminal or electric typewriter is available for communication with the system operator. Some means of reading in programs is necessary, such as the card reader or high-speed paper tape reader. Ordinarily,

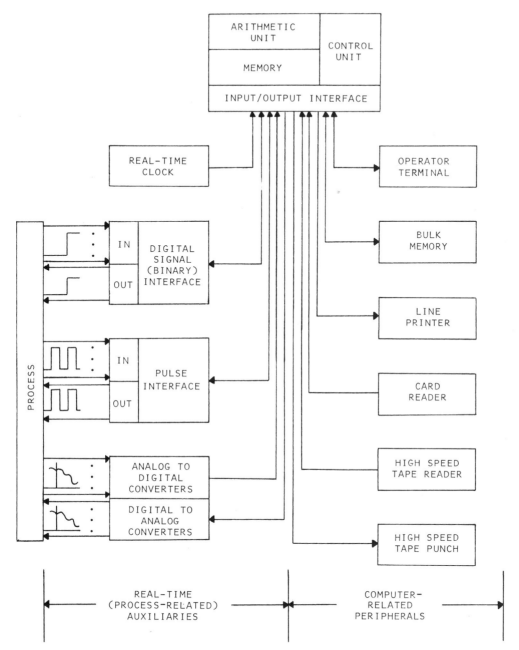

Figure 2.3. A representative real-time system showing the real-time peripheral devices.

computed results are desired, and these might be dumped out using a line printer or paper tape punch. Also, for reasons of convenience, some means of bulk storage, usually magnetic disk or magnetic tape units, is practically a necessity. This especially is true whenever general scientific or business computing is being carried out, or whenever program development and debugging are involved.

The smallest (so-called minimal) computer systems in *general* use at present consist of a microcomputer with video terminal and cassette tape unit (or diskette) as the only peripheral devices. In this case all user programs must be typed into the machine using the terminal; system software or programs such as the tape editor, interpreter, etc., are read in from tape or are stored in "read-only" memory. In a minimal system, program results can be output only on the terminal screen; and developed programs are stored using the attached tape unit. The above description of operations with a minimal system is not intended to be any sort of criticism. Many such systems are performing useful service at the present time; and often there are advantages in having a smaller, dedicated system rather than a large, general-purpose one.

Finally, the system referred to above as *minimal* really is the minimum useful computer system only for general-purpose computing. For some real-time applications, the computer-related peripherals may be dispensed with entirely. Such a stripped-down system is utilized, for example, where the computer is intended strictly for controlling a process. In these situations the computer's operating program usually is quite simple and is contained in permanent (nonerasable) form in a "read-only" memory unit. Microcomputer-based instrumentation (control) systems, of the type now widely replacing analog equipment in industry, would fall into this category.

2.4 THE REAL-TIME CLOCK

The real-time clock, shown in Figure 2.3, is the most important real-time auxiliary or peripheral device. The name—real-time clock—is something of a misnomer. The device is not a clock; it does not tell time; and it has nothing to do with actual or real-time. In fact, the real-time clock in its various forms ordinarily is no more than an interval timer connected to the computer I/O Interface so that the computer can be informed each time a particular interval of time has gone by. The computer must be programmed to use this information correctly. For example, if the clock interval is one second, a counter within the computer might be set to zero at the start of an experiment and incremented by one each time the clock interrupts the processor so as to give a continuous estimate of time in seconds from the start of the experiment. By resetting a counter to zero exactly at midnight, one could easily program a routine to give the time on a 24-hour basis.

Ordinarily the clock interval is based on the AC line supply (60 Hz or 60 interrupts per sec-

ond) or is selectable under program control at one of a number of fixed interrupt intervals, e.g., 1, 0.1, 0.01, or 0.001 second. The real-time clock actually is intended to be used to time long intervals during an experiment with high precision or to initiate some computer action periodically (on an integer multiple of the clock interval). As we will see later, operating the real-time clock at too high an interrupt frequency increases the operating system overhead, i.e., the fraction of time the system software spends just counting interrupts from the real-time clock rather than in performing useful operations. The timing interval of the real-time clock always should be chosen carefully, usually no shorter than the smallest data sampling interval that might be anticipated.

Before leaving the real-time clock we should mention that some computers are equipped with one or more hardware *interval* timers. This is a device that functions similarly to the clock we have described but is somewhat more general in design and flexible in its use. Usually an interval timer consists of an external register which can be set to some initial value under computer control. The register counts down to zero at a rate based on its own internal clock and causes an interrupt to occur in the computer at the end of the pre-set period. Thus the interval timer is nothing more than a real-time clock that can be set to operate at any desired frequency (a frequency that is integrally related to that of the timer's internal clock).

As a final point, although the real-time clock is a key peripheral device in most real-time computing systems, it is also a common peripheral in most batch-oriented or time-shared facilities as we have noted already. In the former case its use is desirable in order to have the computer log programs in and out using time-of-day (true time) and to charge users on the basis of the amount of time a program takes to run on the computer. Time-shared facilities have precisely the same needs with the additional requirement that each user should get the feeling that his terminal is the only one connected to the computer (central processor). Ordinarily a user expects to experience delays whenever his program is performing a large amount of calculating; but he does not want delays to interrupt the communications he makes with the computer from the terminal. Hence, to achieve these results in a time-shared system, no one program is allowed to run longer than a fraction of a second. If further processing is required at the end of this fixed length of time (measured by a real-time clock or interval timer), then the computer turns to another user, servicing all waiting users in a sort of "round-robin" fashion. This technique is known as *time slicing* and commonly is used on time-shared systems. In a sense, a time-shared computer is a type of a real-time system where the user terminals represent a set of processes that make random demands on the computer.

2.5 PROCESS-RELATED PERIPHERAL EQUIPMENT

Any real-time computer system of the sort we are interested in will possess those peripheral units depicted in Figure 2.3 as "Real-Time Auxiliaries" in addition to the real-time clock. These other peripheral devices are entirely devoted to the job of inputting information from the process to the computer or outputting it from the computer to the process. They are necessary because, as noted previously, digital computers utilize information only in a very specific form—one not ordinarily compatible with most processes. The computer Input/Output Interface itself is too specialized to be used for general process information transfer. Further, it is not desirable to design a specialized interface for each process that we would like to connect to the computer unless we foresee the possibility of connecting many similar processes to the same type of computer. In this case the design expense of the specialized interface might be justified.

Chapter 5 deals with the general problem of measuring and manipulating process variables. These quantities can be grouped generally into four major areas:

(1) Digital (or binary) quantities; e.g., a valve is open or closed, a switch is on or off, etc.

(2) Generalized digital quantities; e.g., many existing laboratory instruments normally make results available in BCD (binary coded decimal) format for operation of their own separate printing device or for input to a computer.

(3) Pulses or pulse trains; e.g., turbine liquid flowmeters often output a train of pulses with pulse frequency directly related to turbine speed; some stepping motors used to open and close devices such as valves require a pulse train as input.

(4) Analog quantities; e.g., thermocouples or strain gauges yield outputs measured in millivolts; operational amplifiers often have outputs in the −10 to +10 VDC range; industrial plant instrumentation often works on current signals in the range 10–50 ma or 4–20 ma. All of these quantities have the unifying property that they are continuous variables.

As a result of our ability to concentrate on only four classes of process signals, it is possible to generalize the process/computer interfacing problem considerably. Now we require a relatively small number of devices, all of which can be considered together as a general extension of the computer I/O Interface. A large, general-purpose real-time system probably would contain devices to handle all four forms of input/output described above. A smaller system might be oriented around a single type of process data such as BCD output from a laboratory instrument. At this point it will pay us to consider the specific nature of the devices required to handle process I/O in greater detail.

2.5.1 Digital Signal Interface

The Digital (Binary) Signal Interface ordinarily is designed around the computer data-handling structure itself. This means that a computer with memory elements and accumulators of a certain size would have output and input registers of the same size. We will discuss these ideas at length in Chapter 10; for our present needs it is sufficient to note that the smallest amount of digital information is a "bit". A bit is a binary number, i.e., can have only the values 1 or 0, corresponding to logical values— "True" or "False"—or switch con-

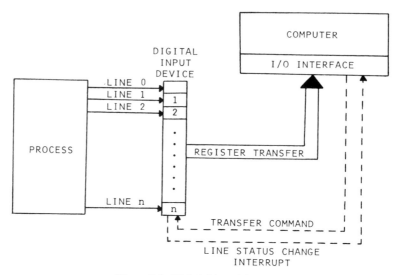

Figure 2.4. Digital (binary) input device.

ditions—"Open" or "Closed", etc. Digital computers invariably handle and store information in the form of binary numbers—an array of bits represents a "word" of computer information; and the computer memory and arithmetic units often handle data words of a specific length, say 8, 16, or 32 bits. Hence it makes sense to transfer an entire word into or out of the computer at one time. For this reason the Digital Input Device often consists simply of one or several registers; each register receives a number of binary output lines from the process and is capable of passing the status of all the lines to the computer on command. Figure 2.4 illustrates the idea. At this point it is not necessary to go into detail as to how binary information is stored and transmitted, but, ordinarily, electronic components which are used to construct computers and to interface peripheral devices utilize voltage levels (0 and +5 volts in the most widely-used system) to represent binary 0 and binary 1. Hence one common design for the Digital Input Device would accept process outputs of the same form. More commonly used in the process industries (and available from computer vendors) is a "contact-closure" device which outputs a 0 or 1 to the computer depending on whether the corresponding process output relay is open or closed (the device detects a small current flow).

The Digital Input Device described above transfers information to the computer only on command. In some cases it is necessary that the computer be informed whenever a process output changes status, e.g., whenever an out-of-limits alarm used to monitor some process variable turns on. It is relatively easy to design an input device that notifies the computer of a line status change, usually via an interrupt in computer operations through the I/O Interface. This feature is shown schematically in Figure 2.4. A device with this capability can be called a Digital Input Sensing Device.

Finally, it should be noted that process information in the second category above, generalized digital information, can be processed into the computer via the same type device. For example, as we shall see later, one digit (0–9) of binary coded decimal information is represented by four bits of binary information. Consequently a 16-bit register in a Digital Input Device could be used to transmit 4 digits of results (accuracy of 1 part in 10,000) from a particular laboratory instrument.

The Digital (or Binary) Output Device is a similar but simpler application of the same ideas. As shown in Figure 2.5, the device is nothing more than a register that holds binary information being transferred from the computer to the process. The register may operate a set of process relays contained within the device; or actual output lines representing the binary information (perhaps 0 and +5 volt signals) may be passed to the process as shown in

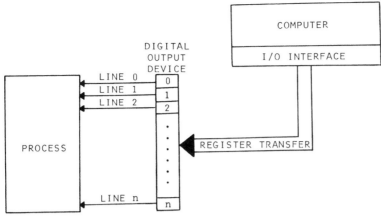

Figure 2.5. Digital (binary) output device.

Figure 2.5. Again, this same device can be used to represent digital quantities, such as BCD information to be displayed in a set of Nixie lights. Summarizing, we have shown that both binary and generalized digital forms of process information (Categories 1 and 2, above) can be transmitted to and from the computer by a single pair of devices.

2.5.2 Pulse Interface

The Input Pulse Interface ordinarily consists of a pulse counter in the form of a register for each process line. Any counter can be reset to zero under program control, and, after a specified length of time, the results (usually a binary or BCD count) can be transferred to the computer using techniques described above.

Figure 2.6 shows schematically the interface for a single pulse input line.

The Output Pulse Interface consists (conceptually) of a device to generate a continuous train of pulses of proper size and period. The device can be turned on and off by the computer. By turning the generator on for a certain length of time, any desired number of pulses can be sent to the process. Or a register in the device could be loaded with a number specifying the number of pulses that should be output. Figure 2.7 gives a schematic representation of the interface equipment required for a single pulse output line.

We have not considered at any length what form these pulses might have. Ordinarily we would be considering equipment that generates or utilizes pulses of a generally rectangular

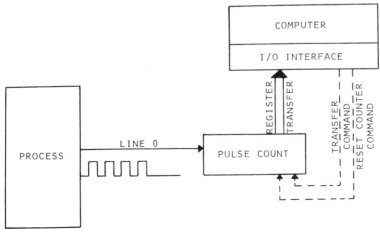

Figure 2.6. Input pulse device.

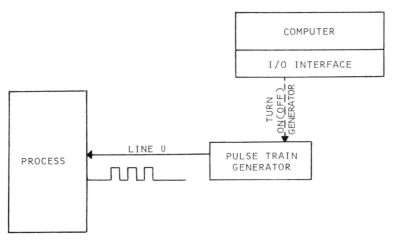

Figure 2.7. Output pulse device.

nature such as, for example, might be produced by a relay opening and closing. For this situation it should be clear that digital input and output devices might be used instead of interfacing equipment specifically designed for pulses. To output a pulse train on a particular digital (binary) output line would only require that the computer turn a particular bit on and off at appropriate times. Similarly, a Digital Input Interface with sensing capability could notify the computer via an interrupt each time a particular input pulse line turns on. Most computer systems can handle a small number of pulse lines, particularly if the pulse frequency is low, in this fashion. For many lines or higher frequency inputs and outputs, separate pulse generating or counting units should be used to save the computer these relatively trivial but time-consuming chores.

2.5.3 Analog Signal Interface

The final pair of process interfacing capabilities required of a real-time computing system involves the inputting and outputting of analog variables or signals. An analog variable is, simply, a measurable quantity or signal that can take on any value within a certain range, i.e., is not restricted to a predetermined fixed set of discrete values within that range. The pressure in a tank as measured by the height of mercury in a manometer and the temperature of a chemical reactor as measured by the output of a thermocouple represent examples of analog signals. Now a digital computer is not well-oriented to the task of storing or generating continuous signals. In the first place, its memory consists of a number of discontinuous storage locations—not very useful for storing continuous information. In the second place, a computer works sequentially in fixed operating steps; so even if a particular quantity could be expressed in the form of an algebraic expression (as a continuous function of time, for example), the computer could do no more than generate a sequence of values which would have to represent the continuous function.

As a result of these limitations we are forced to sample analog input signals periodically or at discrete points in time in order to obtain a sequence of values that can be used by the computer. Similarly, the computer can, at best, generate only a sequence of values to output to the process as an approximation of a true continuous or analog signal. Figure 2.8a illustrates this procedure with an analog signal that is sampled periodically (every Δt time units). Figure 2.8b shows the step-shaped approximation to the original signal that must now represent the form of the original signal, either as it would be stored in the computer or as it would be regenerated by the computer as an output to the process. In these figures the sampling period has been chosen so long that the step approximation in Figure 2.8b has been exaggerated. If sampling is done frequently enough, it may be possible to represent the continuous signal almost exactly. Often it

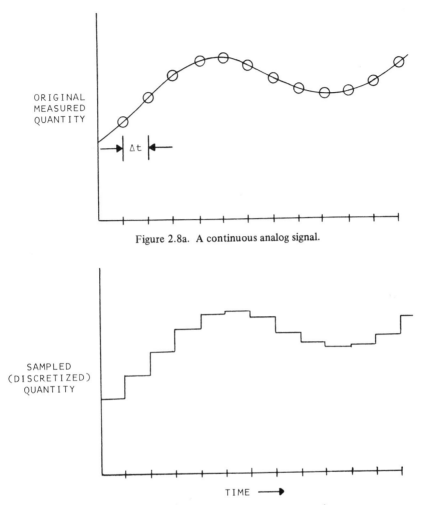

ORIGINAL
MEASURED
QUANTITY

Δt

Figure 2.8a. A continuous analog signal.

SAMPLED
(DISCRETIZED)
QUANTITY

TIME

Figure 2.8b. Its sampled (discretized) representation.

is not necessary to sample this often, and it may actually be wasteful to do so; for example, in logging process analog variables we might need to acquire only as much data as would be necessary to reconstruct the signal graphically. The important details to be considered in a sampling operation are covered in Chapter 19 at some length. Essentially we are quantizing a continuous quantity, and we will assume for the present that we can always find some sampling rate that satisfies our own particular requirements. The computer hardware device that performs this quantization operation on input signals is called a Sample and Hold unit.

Given that our system can operate satisfactorily with sampled data, there is still the problem of converting the sampled, analog input signal into a digital form which can be stored in the computer or to convert digital information in computer representation back to the analog, step-shaped form for output to the process. The Analog to Digital Converter is the device which is used for digitizing input information. In Chapter 10, we will consider in detail how this device operates. Here it will be sufficient to say that an analog signal, a voltage, can be converted to a digital (usually binary) representation for direct input to the computer. The conversion ordinarily is carried out electronically and at great speed, but the net result is to convert the signal, say a voltage in the range 0 to 10 volts, to a number. For example, this 10-volt range might be represented (digitally) by numbers in the range 0000 to 1000. In this

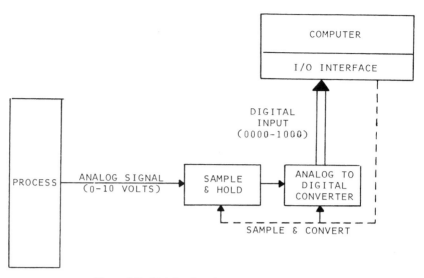

Figure 2.9. High-level analog to digital conversion.

example, we obtain an accuracy in the digital representation of one part in 1000 compared to the original analog voltage. Figure 2.9 shows schematically how an analog to digital converter would be connected to a computer. The example and figure use a decimal representation of the data. Ordinarily, a binary representation is used, and both positive and negative inputs can be handled. These discrepancies from our introductory explanation are not important at this point in our discussions.

As we have described it, a real-time computer would have to have a single ADC (analog to digital converter) for each analog input line. ADC prices have gone down considerably in the last few years, but this approach still would be far more expensive than necessary. Most analog input devices permit a Multiplexer to switch selectively a number of analog input lines into a single ADC, and this is the preferred procedure. As we mentioned, all analog signals must be transduced to a voltage signal (normally on the order of 5 to 10 volts full scale in magnitude) for conversion by a standard ADC. At these levels, the switching device can be electronic (usually field effect transistor switches) rather than mechanical. However, many process analog quantities to be input to the computer are low-level (millivolt) signals from thermocouples or strain gauges; hence the use of "low-level" multiplexers is widespread. These devices must be mechanical in nature, usually consist of an array of mercury-wetted reed relays, and ordinarily feed a programmable gain amplifier to boost the signal level up to the range of the Sample/Hold and ADC. Figure 2.10 illustrates a very general system involving both high- and low-level signals. With this system the computer can choose to convert any of a number of high- or low-level signals, amplify the signal if necessary, and convert to digital representation using only a single actual analog to digital converter.

The Digital to Analog Converter (DAC) is, by comparison, much simpler and cheaper than the ADC. As a consequence, each output line from the computer to the process usually has its own dedicated converter. It is possible to construct a multiplexer for outputting of analog signals, but this is done infrequently because a latching or holding device would be needed for each analog output line, anyway. A DAC performs both the conversion and latching functions. Figure 2.11 illustrates a typical configuration of DACs and their coupling to the computer. In this representation the digital output is made available to all DACs and the proper DAC (Channel) is selected (caused to convert the data) by means of a SELECT command. Most DACs are designed to hold a previous analog output (latch) until commanded to convert again; hence a DAC output will have the typical step appearance we have seen with other sampled signals.

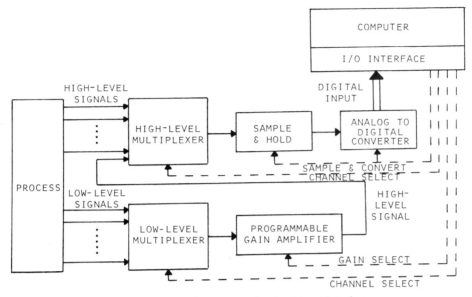

Figure 2.10. A general-purpose analog signal input interface.

The processes of analog to digital and digital to analog conversion take little time with modern equipment. Conversion time on the order of 5 to 20 microseconds might be considered typical, although these times are very much a function of the required conversion accuracy. The ordinary real-time system should have conversion accuracies on the order of 0.1% (of full scale, i.e., 0.01 volt on a 10-volt full scale converter) or better.

In summarizing this description of process interfacing equipment we note that it is not possible to "plug the process into the computer" in the sense that a lamp can be plugged into a 110 VAC main. On the other hand, the very specialized equipment required to inter-

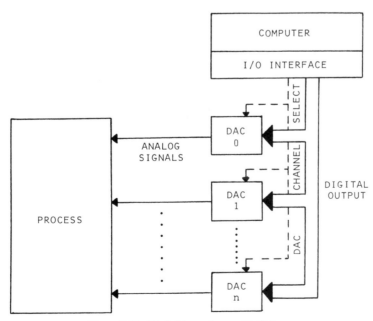

Figure 2.11. Digital to analog conversion.

face the process and computer ordinarily can be ordered "off-the shelf". Although it is not necessarily true, most computers suitable for real-time applications will reflect such capability through the ready availability of real-time peripherals from their manufacturers.

2.6 INFORMATION TRANSFER AMONG THE COMPUTER AND PERIPHERAL DEVICES

As we have seen, the real-time computer is required to communicate and interact with a number of peripheral devices in order to carry out its process-related assignments. In a complex real-time system, i.e., one with many different computer- and process-related peripheral devices, more attention must be paid to the interconnection of computer and devices. For example, where large amounts of process data are acquired by the computer and stored out

on a bulk storage device, it may not be satisfactory for the computer to have to oversee the taking and transfer into storage of each individual piece of data. Fortunately, there are more efficient alternatives.

A simplified schematic layout of one typical general-purpose computer is shown in Figure 2.12. In this diagram the flow of information within the system has been indicated by means of buses, shown in the diagram by double lines with arrows on the ends. Buses physically are no more than wires, cable connections, or internal printed circuits over which binary information is passed from one part of the system to another, usually in a parallel format. Hence in this so-called multiple bus system, there is one bus that transfers information from the accumulators or registers in the Arithmetic Unit to locations in memory (and in the reverse direction). A bus also is used for the transfer of addressing information and for the control

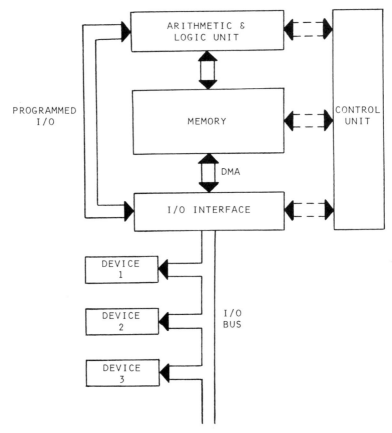

Figure 2.12. Multi-bus computer system.

of information flow. For example, it might be desired to transfer information *from* a specific register in the Arithmetic Unit *to* a specific location in memory. In this case a bus connecting the Control Unit with the Memory Unit might be used to direct the flow of information from register to memory.

The computer I/O Interface is connected to all peripheral devices by means of the I/O bus. Actually the I/O bus consists of a set of buses, one for data transfer, a second to specify addresses (device numbers and registers within the device controllers themselves), and a third to control the flow of data. We need not be concerned with details at this point; the major idea is that the computer communicates with all peripheral devices over the I/O bus.

Most computers are designed to permit the transfer of data between the computer and its peripheral devices in two ways. One way is known as *programmed transfer*; in this case information actually is brought into the Arithmetic Unit, one piece at a time, under program control. The information then can be utilized immediately or stored in memory, again under program control. Information transfer in this fashion would be via the bus labeled "Programmed I/O". A second mode of I/O transfer is oriented toward the transfer of large amounts or blocks of data, as when reading information from a tape or disk memory unit. In this situation the data cannot be moved through the Arithmetic Unit to the Memory Unit efficiently because of the large number of processor operations which would be required to effect the transfer. Hence it makes sense to transfer the data directly to memory, bypassing the Arithmetic Unit entirely. This form of transfer occurs over the "Direct Memory Access" (DMA) bus shown in Figure 2.12. Direct memory access permits the exchange of large amounts of data between the computer and a device with relatively little programming effort, in short amounts of time, and with minimal disturbance to operations under way in the Arithmetic Unit.

In recent years several manufacturers have introduced computers which are strictly laid out around a single (or perhaps dual) bus. As shown in Figure 2.13, all elements in these machines are connected to the primary bus—the individual accumulators, adder, multiplier, etc. (in the Arithmetic Unit), the memory elements, and all peripheral devices. The Control Unit in such a machine is the master device on the bus; in a sense it is a Bus Control Unit, supervising the flow of information from accumulator to memory, from device to accumulator, from device directly to memory, etc. It is the Control Unit's responsibility to schedule bus operations and to settle disputes among devices on the bus which might wish to use it at the same time.

Because information flow on a single bus must be bi-directional, there has been some incentive to develop a two-bus system similar to that shown in Figure 2.13. With a source bus and a destination bus the flow of information is generally in one direction, and consequently the bottlenecking character of a single bus is overcome to some extent.

Our major interest in these introductory discussions concerns the effect of system structure on potential applications of the system. From this point of view it is interesting to note that the unified bus systems, while appealing to computer designers perhaps, offer no significant general advantages over the more usual multi-bus systems. Of course there are differences in capabilities; for example, the unified bus system often offers a "cleaner" instruction set, since every memory location, accumulator, and register in the peripheral devices can be referenced (addressed) with the same form of instruction. As to disadvantages, the buses in the unified bus system and their associated electronics are more complicated by comparison to the more specialized buses in multi-bus systems. While these arguments for and against a particular computer architecture may be important to systems people, they make little difference to the end user. Regardless of the computer bus configuration, each peripheral device requires its own individual controller, which we noted earlier is a special-purpose interface mating the device to the computer and to its particular I/O bus. The manufacturer of any peripheral device will see to it that appropriate controllers are available to connect his device to any of a number of different computers. In this way it is possible for one particular line

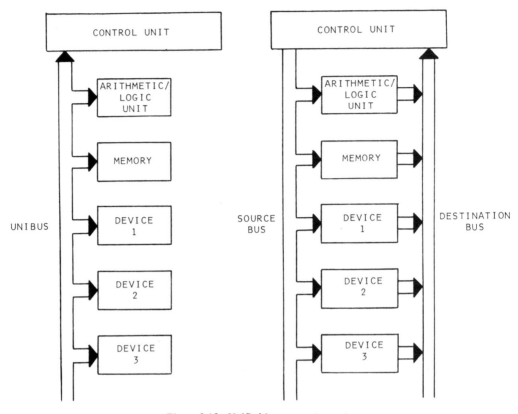

Figure 2.13. Unified bus computer systems.

printer, for example, to work perfectly well with computers from many different manufacturers (assuming, of course, that appropriate controllers have been designed and are available).

From the programming point of view, it is generally true that any computer system with sufficient I/O capability incorporated in its design can be used for real-time applications. It may be necessary for the designer of the executive or operating system software (usually the computer manufacturer) to overcome specific deficiencies in machine architecture by using longer, more complicated programs to handle the required peripheral devices.

Both hardware and software differences can be completely transparent, i.e., not visible, to the user. In particular, if a system is purchased off-the-shelf for a specific application and if the software (system programs) supplied by the manufacturer is appropriate, then the user need never know the details, either of the system structure or system programs, in order to implement his application. Similarly, the later addition of another peripheral device might involve no more effort than plugging in the device controller, a cable, and the device AC power plug.

Real-time applications seldom are this straightforward; often a user wishes to carry out operations involving specialized, nonstandard equipment. In these cases it is necessary for a user to know the details of his system structure quite intimately; hence we will want to spend considerably more time in Chapters 7 and 8 developing a better understanding of digital systems hardware and in Chapter 10 developing a particular emphasis on interfacing.

2.7 DISTRIBUTED (MULTI-PROCESSOR) SYSTEMS

It is not necessary that real-time computing systems be constrained to contain only a single computer. In fact many applications can best be handled by multiple processors. In central-

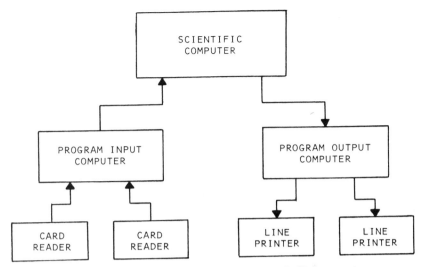

Figure 2.14. Scientific computer with input and output (buffer) computers.

ized computing facilities it is quite common to encounter distributed, i.e., multiple computer, systems. For example, a single scientific computer might use two smaller computers to buffer program input and output so as to free the scientific machine of tasks that waste its powerful computational ability. Such a system is depicted in Figure 2.14.

A multiple computer configuration most generally is referred to as a "network"; for example, a generalized computer network is shown in Figure 2.15. Networks often are used to permit computers to share a common set of data, to provide a wider range of services to users by making available different computers, or to distribute the computing load more evenly over a group of computers.

A special form of computer network, often used in real-time applications, is the computer "hierarchy". This structure has a characteristic pyramidal shape, with the function of an individual computer largely determined by its position or level in the hierarchy. One reason that this structure is natural for computer/process operations is that real-time applications often involve a large number of relatively simple tasks which do not require the power of a large computer. However, a more powerful computer may be required occasionally. Hence it often is useful to use a set of small, dedicated com-

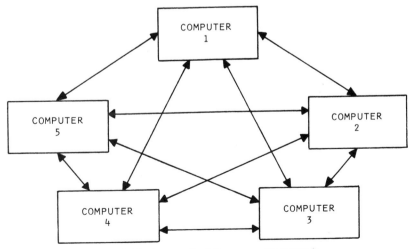

Figure 2.15. A generalized five-computer network.

puters to handle tasks with the following characteristics: (1) a required rapid response to process demands for service, (2) the taking of large quantities of process data, (3) simple calculations as, for example, the implementation of simple control algorithms, (4) little interaction with components other than the process, in particular, little interaction with human operators. In such a situation the dedicated machine may serve the specific purpose of data compressor. A useful design approach then is to connect several dedicated machines to a single supervisory computer as shown in Figure 2.16. The on-line supervisory computer can have real-time constraints of the sort described earlier; it may have to respond relatively quickly to requests for information from lower-level computers, or accept information passed up from lower-level computers but not necessarily in real-time; or it may serve simply as the device where computer programs are developed and tested before loading into a lower-level machine. In any case, intermachine communication probably will need to be handled by

some special device or devices. The Communications Processor shown in Figure 2.16 also might be a small, special-purpose computer with the sole responsibility of managing communications among the other computers.

The example hierarchy described here clearly is a very simple one. Computer hierarchies with three levels of computers presently exist, and it is not difficult to imagine the tremendous variety of systems that might be configured with three levels. As minicomputers entered the real-time computing picture some years back, the trend toward multi-processor systems, including hierarchies, increased greatly; the present situation with very inexpensive microcomputers flooding the market has accelerated this development even more. The hardware interfaces for multiple computer networks, the software interface drivers, the communication protocols, and scheduling strategies make up a complex subarea of real-time computing and deserve a completely separate treatment. Nevertheless, many of the principles that apply in the data acquisition and control area (what

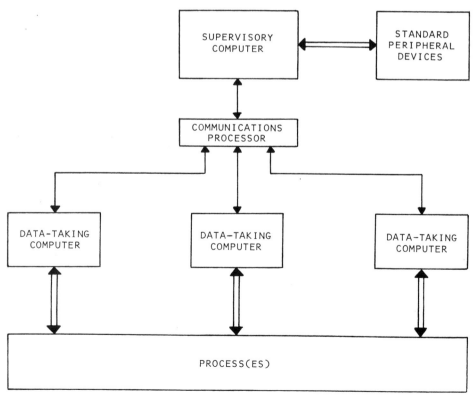

Figure 2.16. A two-level computer hierarchy.

we have called sensor-based computing) apply as well in the design and operation of networks. Our discussion of hardware and software and the detailed treatment of communication protocols in Chapter 9 constitute a good introduction to networking distributed computer systems.

2.8 REAL-TIME SYSTEM COSTS AND PERFORMANCE CHARACTERISTICS

Computer hardware costs and performance capabilities have been undergoing radical changes for the last 20 years, and there appears to be no reason to believe that hardware costs will not continue to drop, or that systems will not continue to become faster and more sophisticated. Only a few years ago someone suggested that a minicomputer could be defined as a processor with four thousand words of memory costing less than $20,000. Today a microprocessor-based computer with 32 thousand words of memory, capable of running high-level languages such as BASIC, can be obtained for one-hundredth as much. Solid state design and manufacturing technologies continue to develop, and economies of scale in production and marketing continue to be favorable; hence we can anticipate further drops in hardware prices.

On the subject of performance, we can say that the spectrum of real-time applications systems is a very broad one. At the small-scale end of the spectrum are the dedicated computer systems (often microcomputer-based) used for simple data taking, analysis, and, perhaps, control. Read-only memory (ROM) may well be adequate for storage of the operating computer program, and accuracy/speed tradeoffs may not be of overriding importance; hence an eight-bit microprocessor architecture may be adequate. In the last year or so, even while these devices have been getting more and more sophisticated, microprocessors have become amazingly inexpensive; costs on the order of several dollars per unit now are common. Hence the idea of a dedicated real-time computer to monitor and control truck braking systems for safety purposes, or automobile carburetion (fuel injection) and engine timing systems for economy and regulation of pollutants, now seems quite obvious. To penetrate mass markets such as the

automotive one, the cost of the entire real-time system including all measuring devices and actuators has had to be reduced drastically; hence a key area of commercial development in the next several years will be concerned with process-related peripherals and with process input and output devices. Many small-scale systems of this sort will find applications in the home and in our transportation and recreation equipment. The rapid spread of microprocessor-based calculators and video games should be evidence of an impending explosion in the real-time applications area.

At the other end of the real-time spectrum are, of course, the large industrial, aerospace, and defense systems. Commercial real-time systems costing from one hundred thousand to more than a million dollars are not uncommon. These systems may be used to monitor a hundred to a thousand process variables, perhaps control hundreds as well; often they are built in the form of a network of computers (a multiprocessor system) and utilize large numbers of peripheral devices.

Thus, real-time systems easily span a price range of three to four decades from smallest to largest systems. However, for a general-purpose real-time system—one that might be used in a laboratory, pilot plant facility, or for university teaching and research as contrasted to both the very small and very large special-purpose systems described above—there does appear to be something like a standard-sized, real-time "unit". If not a standard, at least it is so commonly encountered that we should discuss it here. That system typically is built around a 16-bit word computer with multiple registers or accumulators (4 to 16) and with 32K words of internal memory that can be accessed in less than one microsecond. There are several reasons why this machine has become so common: for one thing, as we shall discuss later, a general data acquisition computer needs at least 10 to 12 bits to represent process variable values adequately in digital format. Several early minicomputers utilized a 12-bit word internal structure, but, because textual (alphanumeric) information conveniently occupies one byte (8 bits) of memory per character, a 16-bit machine can efficiently store a full alphanumeric information set at two characters per word. Further, a 16-bit machine has (potentially) a more diverse set of arithmetic and logical instructions than a 12-bit machine and, with the extra 4 bits, gives greater mathematical accuracy in processing converted analog (process) data. Hence, by a fairly wide margin, the 16-bit machine has become something of a de facto standard. Most recent extensions of small computer architecture have retained a "byte-oriented" structure, i.e., word size of 24 or 32 bits.

We will be discussing all of these considerations in Chapters 6 through 10. In Chapter 8 we also will discuss how a 16-bit machine can access up to 32K (in some cases 64K) words of internal memory efficiently; consequently many of these machines are designed to be expanded up to 32K words of memory without having to use memory extension hardware. Because extra memory always translates into greater computational capabilities and convenience, general-purpose real-time computers often will be configured with the maximum possible memory. A minicomputer of this size can support a full range of peripheral devices and utilize high-level programming software such as FORTRAN. With the addition of a bulk storage unit (usually a disk) such a small computer system can run under the control of a disk-based, real-time executive with many of the conveniences of a very large computer facility.

Finally, concerning costs, the major cost of any general-purpose, real-time facility almost invariably will stem from its peripheral devices and for connecting (wiring in) of processes. Depending on the specific application, it is easily possible to spend five times the acquisition costs of the computer in purchasing the auxiliary devices: disks, magnetic tape units, printer, console typer, analog to digital converter, etc. It also is the case that most operating and maintenance costs will be incurred in running the system's electromechanical devices, i.e., the peripherals, and the measuring and transducing elements.

2.9 SUMMARY

In this chapter we have attempted to describe the structure of typical real-time systems, and to generalize concerning the real-time computer and its peripheral devices wherever possible. In doing this, we noted that a certain amount of standardization of real-time peripherals has occurred over the years, and it now is possible to discuss a generalized process interface consisting (usually) of a digital signal input/output device and an analog signal input/output device. We also have shown that there presently is a strong natural tendency for users to develop general-purpose systems involving computers of a particular size and performance level. It might be interesting to speculate briefly about future trends in real-time computing hardware, also.

The encouraging technological and economic developments discussed in Section 2.8 point in the direction of increased application of small systems, often in the form of networks, for data acquisition and control. This distributed computing approach is likely because, on the one hand, a single low-cost, microprocessor-based computer obviously will be appropriate for data taking, for analysis, and for controlling from one to ten process variables without elaborate auxiliary equipment. The availability of relatively powerful computers, designed for such a specialized application and with suitable analog I/O unit (ADC/DAC package) and terminal, means that it makes more sense to buy, program, and dedicate such a system to a single application than to try to expand an existing, even if larger and more general, real-time facility. On the other hand, users are not likely to want to give up the convenience of high-level programming languages and the convenience of an array of input/output peripherals which can be used for program development. Hence, the network or distributed system, which combines the best of the two worlds, will emerge both in industrial settings and in the academic environment.

This approach will be seen as a good way to take advantage of small-computer technology (low costs). Simultaneously, it avoids many problems of large, single-computer systems (nonmodularity and consequent difficulty of expansion, decreased reliability, etc.), and yet does not give up the significant advantages of large systems. Such an arrangement likely will exploit the interconnection of multiple small computer systems with a larger, general-purpose real-time system on which program input and development, mass storage, data analysis, and report generation facilities will be concentrated. The small systems we speak of here will be based either on microcomputers or minicomputers; the larger system likely will be designed around an expanded mini- or midicomputer with the powerful executive software necessary to service the hierarchy of smaller machines and to drive its own peripherals.

2.10 SUPPLEMENTARY READING

1. Brignell, J. E. and Rhodes, G. M., *Laboratory On-Line Computing*, Wiley, New York, 1975.
2. Harrison, T. J. (Ed.), *Minicomputers in Industrial Control*, ISA Press, Pittsburgh, 1978.
3. Johnson, C. D., *Process Control Instrumentation Technology*, Wiley, New York, 1977.
4. Skrokov, M. R. (Ed.), *Mini- and Microcomputer Control in Industrial Processes*, Van Nostrand Reinhold Co., New York, 1980.

2.11 EXERCISES

1. You have decided to hook up your backyard irrigation system to a microcomputer capable of performing real-time operations. A set of solenoid-operated valves is available to turn on the water in eight separate watering lines. You can purchase two types of elements to sense ground moisture:
   i) A continuous-measuring device whose output varies between 1 and 5 VDC as moisture content varies from 0 to 100%.
   ii) A binary output device that switches on when ground moisture drops below some value set at the factory.

   The binary output devices are considerably cheaper than the analog output units. To measure all watered plots, you need eight measuring devices of one kind or the other.
   a. For each of the two types of measuring devices, what real-time peripheral equipment would be required?
   b. Can you think of any good reason(s) which might justify the more expensive interfacing equipment? (Yes or No by themselves are not complete answers.)
   c. Would there be any need for a real-time clock in a system like this? (Consider the rate of ground surface drying as a function of time-of-day.) Explain why or why not.

2. A mechanical engineering student has the idea to use a real-time computer designed around a microprocessor to automate the scoring of a bowling alley. In order to become wealthy from this idea he has to sell a lot of units which means that he has to keep costs down.

   The student has access to a line of cheap cathode ray tube terminals (CRTs) that can be used to notify the computer that a game has begun and how many people are playing.
   a. Is a real-time clock necessary? Why (not)?
   b. What sort of instrumentation would be required? Describe briefly.
   c. Based on your selection of instrumentation, what process-related peripheral devices would be needed? ADC? DAC? etc.

3. Normally a bottling operation—such as might be used for soft drinks, beer, etc.—is mechanically designed for high-speed operation. In this case the machine often is designed so that bottles continue to move as they are filled; during this process multiple filling spouts move along with the bottles, each in a different stage of filling.

   To make the problem easier, consider a California wine maker who wants to use a real-time computer to automate his bottling line. In this case the wine is quite expensive, and the single filling spout is in a fixed position. Assume that a solenoid valve can be used to open or close the filling line.
   a. How would you instrument the filling process? (State all assumptions involving how the bottle line will be advanced. Don't worry about how bottles are placed on the line or removed from it.)
   b. What process-related peripherals are required?
   c. Is a real-time clock required?
   d. If you used any analog input or output devices in part (a), justify their use. Can you think of an alternative approach that uses only digital I/O?

4. The USSR chess federation wants a Japanese semiconductor manufacturer to design and build a microcomputer-based chess board for training purposes. Since 20 million units will be needed for the junior division alone and hard currency for imports is scarce, a simple digital method is required to determine where on the board the chess pieces are located.

a. What is an obvious way to instrument the 8 by 8 square chess board so that occupied squares can be determined?

b. How is the computer to know that a particular piece is being (has been) played?

c. What I/O interface equipment will be required? For a microcomputer with an internal 8-bit data representation, what else can you say about the I/O units?

d. If you choose a very simple instrumentation scheme, it will not be able to distinguish one chess piece from another. (Sixty-four tiny scales to distinguish King from Knight, etc., by weight is not likely to be cheap enough!) How can the computer logically maintain an internal representation of the distribution of chess pieces using your method?

3

An Overview of Real-Time Programming

Duncan A. Mellichamp
Department of Chemical and Nuclear Engineering
University of California, Santa Barbara

3.0 INTRODUCTION

Before discussing real-time programming it will be useful to review the basic steps involved in writing and executing programs in *non*-real-time applications. For this purpose it will be sufficient to consider the use of a higher language, such as FORTRAN. Most real-time users will have had considerable computing experience with FORTRAN in the batch or time-shared computing modes.

The actual steps involved in developing and running a user program ordinarily would consist of writing down a sequence of program statements which carry out the necessary mathematical and logical operations. These statements then are converted to a form suitable for reading into the computer, usually a deck of punched cards with one program statement per card. The FORTRAN statements in the program then must be broken down into a set of machine operations which can be executed by the computer; this conversion process is something of a "translation", and ordinarily it is accomplished by means of a special-purpose computer program known as a "compiler". The resulting machine language program then is "executed" or run.

All of these operations are expedited greatly by use of another program, called the "Operating System" or "Executive System". All programs are brought into the central processor and executed under supervision of the operating system which has the responsibility for coordinating operations of the computer and all of its peripheral devices. For example, the operating system might bring the FORTRAN compiler into the computer from a bulk memory device such as a disk, then read in a user program from a card reader for compilation. The machine language program resulting from compilation of the user program would be stored out on the disk, and, on completion of the compilation process, the machine language version of the user program would be brought back into the processor and executed.

Even in this last stage of the sequence it would be the operating system's responsibility to coordinate input/output operations with appropriate peripheral devices such as the line printer or card reader. With a typical high-level language such as FORTRAN, the program statements used for these communications would be a "PRINT_____" or a "READ_____". In the former case, the user's PRINT statement must be converted into the correct sequence of many hundreds of machine language operations necessary to transfer information to the line printer. The operating system coordinates the actual transfers at run time.

All of the details of these communications between the central processor and its peripheral devices are not known to the individual user-programmer; in fact, one of the major advantages of high-level languages is that I/O operations are practically "transparent" to the user in the sense that the user does not see all of the details. It should be the goal of user-oriented real-time systems to make communications operations involving the *process* just as transparent, and, sometimes, this can be done. More often, however, process communications cannot be made as straightforward. With the exception of one or two high-level languages developed for applications with specific processes such as machine tools, it is not yet possible to communicate directly with the process. Computer operations such as "READ TEMPERATURE" or "OPEN VALVE" have not and probably will not become extensions of our most widely used high-level scientific languages such as FORTRAN and BASIC.

Hence, the programmer actually is forced to communicate with the process through the generalized process interface equipment we discussed in Chapter 2, i.e., with relays, analog to digital converters, etc. In some cases it is possible to design a real-time programming language which at least makes communication with the generalized process interface transparent; for example, a statement such as "READ ANALOGVOLTAGE" or equivalent "CALL ANIO" might be made available. In other cases, the user might require an intimate understanding of both the computer and the interface to be able to write his program. This will always be true when a high-level real-time language is not available for a particular system or when the user wishes to incorporate his own specialized interfacing equipment which is not supported by the manufacturer's high-level language. In any event, the user should understand the basic ideas involved in real-time programming, all of which revolve around the computer's ability to interact with external devices. Hence it will be useful to take time to develop the fundamental ideas in this important area before proceeding to discuss the more general characteristics of operating sytems and programming languages. In Parts III through VI we will return to many of these same hardware and software topics but, of course, in much greater detail than in these introductory chapters.

3.1 TECHNIQUES FOR COMPUTER/DEVICE COMMUNICATIONS

Each peripheral device connected to the computer has its own specialized controller which is designed to permit the connection of the computer I/O Unit and the device's own electronic input, output, and status signals. In most cases, the relationship between computer and device is a master/slave one, i.e., the computer is responsible for initiating activity in the device, for servicing the device's continuing requirements, and for shutting the device down. In the case of a line printer, for example, the computer might initialize the device by advancing the paper to a new page, service the printer by supplying alphabetic or numeric (alphanumeric) information as rapidly as the printer prints it, and shut it down by clearing the finished page from the printing device.

Now all of these operations involve the transfer of information between the computer and the device controller: the transfer of control signals from the I/O Unit (start, stop, etc.) or the transfer of information from the device controller concerning the status and requirements of the device (busy, ready, service requested, etc.). In subsequent chapters it will be important to go into detail as to how these transfers take place and how all of these operations can be coordinated; for the moment we need only note that all of the above operations can be considered as input/output transfers.

Since ordinarily we consider the computer to be the master device, it is logical to imagine it initiating a transfer to or from the peripheral device as part of a user program. Figure 3.1a illustrates, via a portion of a program flow chart, how the simplest mode of transfer works. In this case we speak of an *unconditional transfer* because the computer carries out the operation whenever it executes that particular portion of the program. For simple and very fast devices such as the digital to analog converter this procedure may be completely appropriate. However, one disadvantage with this type of transfer is that a slower or more complicated device may be in no position to supply the required information to the computer, in the case of an input transfer, or to accept the information, in the case of an output transfer. As an example, a typewriter that outputs information from the computer one character at a time may be in the process of typing a character at the moment the computer supplies the subsequent character. This mismatch in timing may result

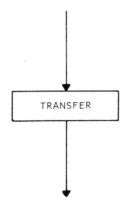

Figure 3.1a. Unconditional transfer.

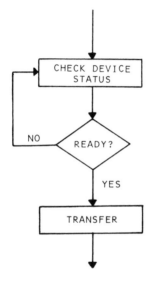

Figure 3.1b. Conditional transfer.

ever, it is important to note the disadvantage in this method. The computer must wait for the device to indicate that a transfer may be made via an appropriate status ("ready" perhaps), and, in the meantime, it can do no more than sit in the waiting loop. This may not be a problem in some cases; for example, some analog to digital converters can convert input signals at a rate of 100 kHz (100,000 conversions per second), meaning that a computer might only need to "waste" several microseconds waiting after initiating a conversion and before transferring in the results. Another example, one demonstrating a situation that is not so acceptable, would be a 30 character per second typewriter which might require a wait of up to one-thirtieth second for each character in a string of text typed-out. An alternate procedure utilizing conditional transfer is shown in Figure 3.1c. In this case the computer does not hang-up in a wait loop but can proceed to other jobs if a transfer is not possible. The inherent problem in this procedure is that the programmer must arrange his program so that all conditional transfer operations to be carried out are placed in a single section of program code which will be regularly accessed.

All of these problems are eliminated if the computer has, within its I/O structure, the capability of responding to an interrupt from any external device that requires attention. In this case, an interrupt initiated by any device causes the computer to branch to some other pre-

in a loss of the transferred information. One way to handle this problem would be to include some temporary or buffer storage areas in the device itself; however, if the buffers become momentarily filled, it still might be possible to lose transferred information.

A slightly more complicated yet more useful approach is to use the computer to determine first what the device status is (ready, busy, etc.) via one transfer command and, only when the device is in a position to accept or supply information, to carry out the second (information) transfer. Such a technique is shown in Figure 3.1b and can be called a *conditional transfer*.

The conditional transfer has all elements necessary for a complete matching of computer and device. By properly conditioning the transfer, there can be no loss of information; how-

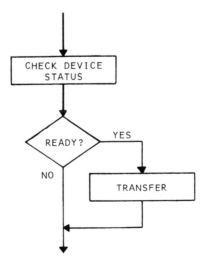

Figure 3.1c. Conditional transfer.

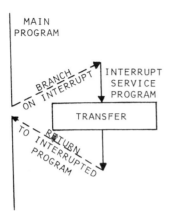

Figure 3.1d. Program interrupt initiated transfer.

determined portion of the computer program where the necessary operations required to service the device are carried out. Operations then can return to the "interrupted" program. A *program interrupt initiated transfer* is shown in Figure 3.1d.

Hardware interrupt capability is the foundation on which real-time systems operate. For example, as described in Chapter 2, the system real-time clock actually consists of nothing more than a hardware device that interrupts computer operations on a fixed frequency, perhaps ten times a second. It is the function of computer software (programs) to utilize the interrupt and I/O capability of a particular system in an efficient manner. For example, a real-time system must coordinate communications not only with computer-related peripherals and the real-time clock as do most data processing systems; also it must handle the requirements of connected processes which communicate demands of a more randomly occurring nature through the generalized process interface. These requirements lead us to a discussion of the software systems which have been developed to coordinate the complicated communications between a computer and its many external devices.

3.2 EXECUTIVE OR OPERATING SYSTEMS

3.2.1 Structure

All computers with interrupt capability also possess the ability to shut out interrupts when-

ever necessary or desired. Hence whenever the computer will permit itself to be interrupted, and a device requests an interrupt, the computer must cease its regular operations at the time of interruption, but it must make sure that it can return to them after taking care of the interrupting device's requirements. Ordinarily, these functions are accomplished by having the computer branch automatically on interrupt to a separate section of the operating system known as the Interrupt Handler. The Interrupt Handler then must store out temporarily all information necessary for later resumption of operations, i.e., the next statement to be executed in the interrupted program, the contents of all working registers or accumulators, etc. It then must determine which device caused the interrupt and branch to a subroutine to service that particular device. Finally, after servicing, the Interrupt Handler must restore the machine to its state at the time the interrupt occurred and return execution to the interrupted program. Table 3.1 summarizes the sequence of operations which must be carried out by this simplified interrupt handler.

For this basic interrupt handler to function correctly we have to agree that a second device request for service cannot actually interrupt the Interrupt Handler itself. Since the computer can allow itself to be interrupted or not (we say—"enable interrupts" or "disable interrupts") via appropriate program statements, one way to resolve this problem would be to *disable* subsequent interrupts as soon as the Interrupt Handler begins execution and only to *enable* interrupts as a last assignment of the Handler before returning to normal execution. A device requesting service could only interrupt at this point.

Such a procedure is not very efficient because some devices require rapid service (attention) while others are inherently less critical. For example, it would not make good sense to have the computer tied up in servicing one of its own devices, perhaps accepting a message someone is typing-in from a terminal, while at the same time leaving some more important device, such as a process alarm input, locked out until the terminal input operation is concluded. Hence we need to take a closer look at the way execu-

Table 3.1. Configuration of a Simplified Interrupt Handler

- Save state of interrupted program
- Determine which device caused interrupt
- Service interrupting device
- Restore state of interrupted program
- Open processor to further interrupts
- Return to interrupted program

tive systems are designed to eliminate this potential problem.

3.2.2 Operations Sequencing

Any computer, even the smallest, will possess certain hardware capabilities which establish a set of priorities for its external devices. For example, a very simple machine might be designed to have four priority levels—0, 1, 2, 3. The highest priority level (0) might be used for critical, process-related I/O; the next level (1) for high-speed computer-related I/O devices such as a disk that transfers large amounts of information in a short time. The next-to-the-lowest level (2) could be used for the real-time clock, and the bottom level (3) used for slow I/O devices such as an operator terminal. A very simple diagram of this structure is shown in Figure 3.2.

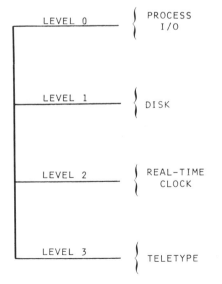

Figure 3.2. Interrupt priority levels in a simple computer.

In this case the hardware might be designed so that a device on one level, once it interrupts the computer, locks out all devices on lower levels until it has been serviced completely; however, an interrupt from a device on a higher level would be permitted by the hardware. For example, if the real-time clock has interrupted, and the computer is in the process of updating all portions of the program that are time-critical, then the terminal would not be permitted to intrude until these operations were completed. On the other hand, an interrupt caused by a process alarm (level 0) would be permitted; presumably something has happened that should be given immediate attention.

The operating system must be designed to handle such situations. For a simple hardware system, such as this one with four interrupt levels, the operating system would only need to have four storage areas allocated to preserve all of the information necessary for restoring operations on any of the interrupted levels. For example, a user program might be in the process of performing a routine (noncritical) set of calculations when the clock interrupts. If, while the operating system is servicing the clock, an interrupt comes in from the process I/O device, then clock servicing must be dropped. After servicing the process I/O device the operating system will return to where it left off in servicing the clock, and then back to where it was in the user calculations. The sequence of operations might be as shown in Figure 3.3. A similar analysis of all possible combinations of events would show that it would be impossible to have to store out more than four sets of data concerning the status of interrupted programs. Hence we can associate a storage area in the computer with the program used to service the device at each interrupting level (0, 1, 2, 3).

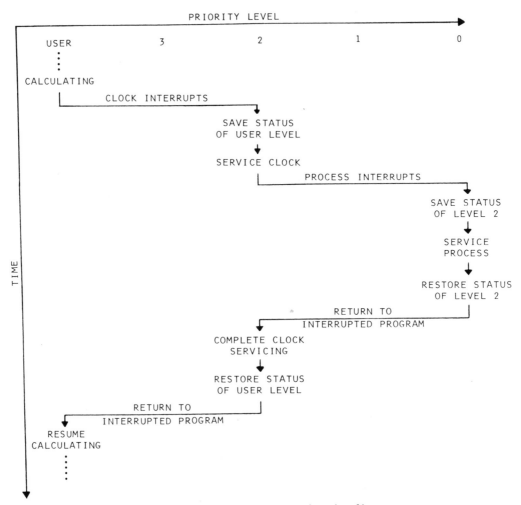

Figure 3.3. Operating system sequencing of a series of interrupts.

This situation would also hold for the case where several devices are associated with each hardware priority level. For example, if several terminals are connected to Level 3 and one of them causes an interrupt, the others would be locked out until servicing was completed. (Presumably one terminal is no more important than any other in this scheme).

3.2.3 Interrupt Handling

More sophisticated and, consequently, more complicated arrangements of system hardware than those described above are used generally. For example, most real-time computing systems have the capability of masking out selected external devices under control of the processor, of determining many more than four levels of interrupt priority via hardware, and of support-

ing on the order of one hundred separate external devices.

All of these capabilities imply that a much more complex handler must be available to take full advantage of the system architecture. Such a routine is shown in Figure 3.4 where virtually all of the chores associated with servicing external devices are performed by the Interrupt Handler and the individual device service subroutines, i.e., by software. The structure of this interrupt handler is designed to permit a device service routine to be interrupted itself, if the second device requiring service is not locked out by the *mask*. In this case the handler is called again and must save the status of the interrupted program—as shown here, the service routine for Device n. The status data cannot overwrite the previous status data; rather they

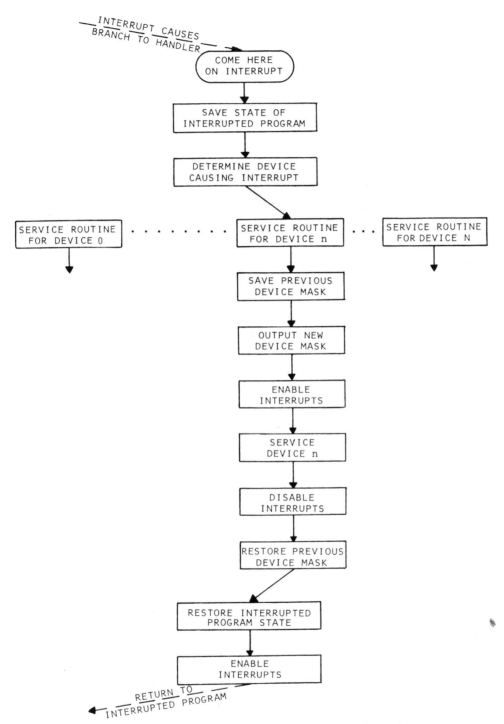

Figure 3.4. An interrupt handler with multiple device service routines.

are appended or, as is usually the case, placed on a "stack", i.e., stored in a set of successive memory locations. The final operations performed by the Interrupt Handler (sometimes called the interrupt dismissal routine) involve restoring the status data of the interrupted program and returning to it. The dismissal routine must be able to work its way back through the stack. In systems where the status data are saved and restored by software (the Interrupt Handler) the stack consists simply of a set of preallocated memory locations. The computer may be designed with a set of storage locations especially constructed in the I/O hardware which automatically save the machine state whenever an interrupt occurs and restore the state on a single machine command. Or the hardware may be designed to stack interrupts in a particular section of regular memory. Such a capability costs extra, of course, but materially reduces the overhead of the operating system; in particular, it reduces the length (number of instructions) of the Interrupt Handler and the amount of time to service and dismiss a device interrupt.

3.2.4 Multitasking

One of the advantages of being able to stack interrupts is that it then becomes feasible to interleave the operations of many or all external devices. For example, while the system is pushing data out to one device, say a terminal or console typer, it might also be bringing data in from some other device, say an analog to digital converter. It is not necessary that the system sit in a wait loop or periodically check flags if a set of program routines is available that coordinates all input/output operations. Statements in a user program to output a line of information on a terminal ("TYPE____") or read data from a card ("READ____") then become simply calls to the I/O routines; and, if the routines are designed appropriately, once a TYPE operation has been initiated, the computer can proceed to subsequent program statements. It then would return to the I/O program only when an interrupt occurs indicating that another character has been typed on the terminal and that it is time to transfer the next one.

Since many real-time user program chores involve input/output or the initiation of jobs on a periodic or real-time clock basis, it is natural to extend these ideas to the concept of a user program subdivided into multiple tasks and a Task Scheduler to take over the problem of scheduling them. Scheduling could be based both on a set of user-assigned priorities and on the tasks' "readiness" to be executed. Chapter 13 is devoted to a full description of multitasking, but the underlying philosophy of multitasking needs to be presented here. Briefly multitasking is an attempt to get around the inherent limitation of digital computers we discussed in Chapter 1, namely, the fact that only one program can be executing in the central processor at any given time. Hence, the user subdivides his program into sections of code that logically must be executed together. We refer to these program subdivisions as *Tasks* and assign a priority level to each task. In a sense, all tasks compete for an opportunity to take over the central processor (be executed), and it is the responsibility of the Task Scheduler to determine which task can be executed at any given time.

For example, a program may be designed to sample a set of analog input signals from a process on a strictly periodic basis (say every 10 seconds), compute whether the process is operating normally, and output a console typer message if not, then output a "report" on the line printer every 10 minutes which constitutes a summary of operating conditions and which takes about 30 seconds to print out. It is clear, in such a case, that the appropriate sequence of computer operations cannot be predicted in advance (because of randomly occurring console typer messages), and, in any event, there is no way to output results to the line printer over a period of 30 seconds without interfering with the other operations. However, the total program can be divided into three tasks:

(1) Input analog signals every 10 seconds and determine whether process is operating normally,

(2) Output error message on typer if not,

(3) Output report of operations on line printer every 10 minutes.

Now if a priority is assigned to each task, for example:

Task	Priority
1	High
2	Middle
3	Low

it becomes the Task Scheduler's job to make sure that each task will tie up the processor only when it needs to run and can run. (We must assume that the Task Scheduler will take over the processor whenever an interrupt occurs or whenever an I/O sequence is initiated or completed.)

If, for example, the real-time clock interrupts and the scheduler determines that analog data must be read-in and, also, that the report must be sent to the line printer, it is clear that the input task (1) will take over. However, at any time the input task initiates a conversion and is waiting for it to be completed, the report output task (3) will be brought in for execution. When the analog to digital converter interrupts (signifying that it has finished its conversion) the Task Scheduler will cause the input task to continue execution at the point where it stopped. Similarly, if all data have been brought in and an error condition has been found to exist, the input task will be dismissed, and the error output task (2) will take over. Again, while the computer is waiting for the typer to output each successive character in the error message, it will be possible for the report output task (3) to continue. To an observer not familiar with internal operations, it would appear as if the analog to digital converter, typer, and line printer all were being operated simultaneously by the computer. Multitasking software (an executive or operating system) makes possible a much more efficient utilization of the processor and, except for dedicated and fairly simple systems, is a practical necessity for real-time operations.

3.2.5 Background/Foreground Operations

An extension of priority multitasking, referred to as background/foreground, is important in single-computer operations where the computer must be supervising the process continuously. Such a programming system permits *non*-real-time jobs to be executed on a *lowest* priority basis (background) with any real-time opera-

tions taking precedence (foreground). In this case, when foreground operations are desired, the processor is turned over to the usual Task Scheduler which schedules tasks based on ordinary rules of priority. Whenever, however, there is no task remaining for the Foreground Task Scheduler, the background is permitted to run under the software which supervises normal queueing (or scheduling) assignments. In this manner, such operations as program development and debugging, or ordinary batch computing can be performed "at the same time" as time-critical or process-critical functions. It will, of course, be necessary for the foreground multitask scheduler to take over from the background job whenever an external device allocated to foreground tasks happens to interrupt. Since the background job presumably is not a critical function, these temporary interruptions are no more than an inconvenience.

One important potential problem in any operation involving two programs or two users is that of memory protection. Practically speaking, there must be some hardware method available to permit only one user (or user program) to write into selected portions of the machine memory whenever program development at the assembly language level is being done in background. If such capability—generally known as "memory mapping" or "memory protection"—is not available, then it is simply a question of time before some user program under development will operate incorrectly and knock out the foreground system. This problem *should not* occur with high-level language program development in the background because (presumably) the high-level compiler has been programmed to assign variable and program statement storage with foreground operations in mind; and the operating system should do the same with dynamic memory allocation on any stacks that are used during run time.

3.2.6 Multiprogramming

One of the methods used to make a single computer perform as if it were several computers has been termed multiprogramming. In this procedure—normally used only with large, high-capability systems but, sometimes with true real-time applications as well—the processor works on several different jobs at the same time. The usual method for multiprogramming allocates or divides up all available main memory among several user programs. The processor then executes these programs, in turn, using its own criteria for scheduling operations. For example, the criteria may be designed to minimize wasted time due to I/O waits (leave a program and proceed to the next whenever an I/O sequence has been initiated) or be designed to give on-line, time-sharing users the illusion that each is the only system user (by time-slicing or giving only a fixed fraction of a second to each user before proceeding to the next). Another procedure that also might be called multiprogramming involves swapping user programs into and out of bulk storage on some preprogrammed basis. Again, for example, with time-shared users the basis might be a time-slicing arrangement where no user can have a program executing longer than some fixed length of time before another user program takes over the processor. Some real-time computing systems extend these techniques to multiple real-time users—for example, to several users, each with a program operating in a separate *and* protected partition in main memory.

3.2.7 Multiprocessing

A final procedure which need be mentioned again only briefly here is the growing (particularly in the real-time applications area) use of multiple processors or computers. Advantages of multiple computer operations are that a single executive system can be designed and used with each computer; memory protection hardware no longer is necessary because independent programs are not running in the same memory unit at the same time; and job functions can be split among the individual processors on some rational basis—say real-time operations on one machine, program development on a second. Operating systems must be able to communicate with each other, and usually this is accomplished by means of a hardware buffer between computers, a common (shared) bulk storage unit, or an interprocessor bus (data highway).

3.3 PROGRAMMING LANGUAGES

Up to this point we have not concerned ourselves with a lengthy consideration of the many different kinds of programming languages and the appropriateness of one or the other to real-time applications. Our few specific references to program statements have been to "READ", "PRINT", "TYPE" statement, i.e., the sort seen by most FORTRAN programmers. These are examples of high-level language statements, and it is an unfortunate fact of life that computers cannot use such statements directly but must work with a more elementary (low-level) language, one that is strictly a function of the computer architecture and, therefore, specific and nongeneral.

Each type of computer then has its own internal language, usually referred to as machine language. Machine language is composed of the elemental instructions which tell the computer's Control Unit to add two accumulators, to bring the contents of some memory location into an accumulator, to transfer the contents of an accumulator to a particular register in a particular external device, etc. Whenever the computer is operating, it actually is executing a sequence of machine instructions; hence any high-level language ultimately must be reducible to machine language operations. Since (as we will see in Chapter 8) all machine operations consist of a set of binary operations, this implies that machine language consists of a set of binary (base 2) numbers. In the early days of computers, programmers actually had to work in machine language. Fortunately, this is not the case today, since somewhere along the way methods were developed to let the computer itself do the dirty work of converting the desired sequence of mathematical operations in a high-level program into the equivalent but much longer sets of executable machine language statements.

In this introductory chapter we will concentrate on the general properties of programming languages and how each language fits into the real-time picture. Parts V and VI cover the topic of real-time software in detail.

3.3.1 Assembly Language

The first step away from machine language is assembly language. In fact, assembly language usually is a direct (meaning statement-for-statement) equivalent of machine language with the chief advantage that it is considerably more verbally-oriented than the binary number sequences of machine language. To begin with, each machine operation is given a mnemonic code which stands for some operation such as "load accumulator". Secondly, in assembly language it is possible to refer to a variable stored in a particular location by name (or label) and to refer to a particular program statement by label. Further, external devices can be referred to by label rather than by number. This all means that a program segment used to transfer a number stored in memory to a register in an external device might look as follows:

LDA 1, NUMB

DOA 1, DEV20

instead of the machine language equivalent:

0011000100100100

1110000000010000

Clearly, it is easier to write programs in assembly as compared to machine language. The computer itself is used to convert assembly language (as described in Chapters 8 and 11) to machine language by means of a program called the "Assembler" (which is described in Chapter 12).

Systems software—Assemblers, Executive or Operating Systems, etc.—might originally be written in assembly language. Fortunately, because assembly language is a tedious mechanism for programming (except in comparison to machine language), higher-level languages ordinarily are available from the computer manufacturer. Unfortunately, if the user wishes to use any nonstandard device in his system (i.e., one not supplied by the manufacturer or supported by his software), there usually is no recourse but to program the Device Handler (Start, Stop, and Service routines) in assembly language and to append it to regular systems software. Of course, for exotic applications,

for extremely high-speed operations, or for software to be able to run in a very small amount of memory, it may be necessary that the user do all of his programming in assembly language. For those reasons, and because a machine's assembly language mirrors its internal architecture and operating capabilities exactly and therefore permits the hardware to be evaluated easily, a real-time user should plan to familiarize himself with assembly coding.

3.3.2 High-Level Languages with Real-Time Extensions

Most readers are familiar with one or another of the widely-used scientific programming languages such as FORTRAN or BASIC. These two languages are good examples of the two major types of high-level computer languages. FORTRAN is a compiler-based language, meaning that a FORTRAN source program must be translated in a series of steps involving the computer itself before execution can take place: (1) the FORTRAN program must be compiled, i.e., read into a computer where another program (the compiler) converts each high-level statement into a correct sequence of assembly language statements; (2) the assembly language program is then converted to machine language in the computer using the machine's assembler; and (3) the machine language program is loaded and executed under supervision of the operating system. Many computer compilers are designed to carry out both steps 1 and 2, i.e., to create directly the machine language version which then is loaded.

BASIC, by way of contrast, usually is an interpreter-based language although compiler versions of BASIC do exist. This means that the sequence of statements constituting a BASIC program is read into the computer along with the BASIC interpreter program and the operating system. The interpreter treats the BASIC program source statements just as a set of data. In executing the program, it proceeds to look at each statement, interpret it as to specific function(s), and call appropriate subroutines to carry out the functions. Hence, a BASIC program does not become an executing program;

each time a statement is "executed", it must be interpreted as if it were the first time.

The operating characteristics of a high-level language depend significantly on whether it is compiler- or interpreter-based. BASIC language programs run much slower than FORTRAN because of the extra time required to interpret. On the other hand, a BASIC program can be modified, on-line, simply by typing-in any desired changes and rerunning the program; whereas a FORTRAN program must have the changes edited-in, then be recompiled, reassembled, and reloaded before rerunning. What this means is that BASIC has advantages (assuming that it is fast enough) for situations where programs are being developed continuously, as in a teaching laboratory or pilot plant operation; FORTRAN has advantages for well-defined applications where a fixed program will be used for long periods of time once developed or for applications where speed is required.

In general both programming languages require more memory for execution than does an equivalent program written in assembly language by a proficient programmer (sometimes as much as four times the memory is required), and the assembly language program will also execute faster than the program generated by the FORTRAN compiler. Despite the clear-cut advantages of assembly coded programs, few users will want to forgo using the high-level languages for one very simple reason—programming ease. Several other reasons tip the balance decidedly in favor of high-level languages, viz., ease of documenting completed programs and the relative simplicity of modifying, expanding, and restructuring a high-level program at a later date.

Both BASIC and FORTRAN exist in a number of standardized forms for batch processing purposes (e.g., FORTRAN IV, FORTRAN 77, WATFOR, and WATFIV). BASIC does not have a well-defined or widely-accepted real-time variant; however, the Instrument Society of America is attempting to specify an extended FORTRAN Language for process control applications. In general, BASIC is available from computer manufacturers in single-task versions although multitasking versions do exist. FORTRAN, on the other hand, is commonly avail-able in a multitasking variation (of course, an operating system to support multitasking must also be available). In this book we have chosen to emphasize the single-task (connected program) nature of BASIC and its particular suitability for teaching real-time computing at the introductory level (Chapter 14). We have chosen to treat FORTRAN as a multitasking system (Chapter 15, where we have assumed that real-time extensions follow ISA standards insofar as these have been spelled out).

The real-time extensions needed for either BASIC or FORTRAN consist of a set of subroutines to input (and output) data from the process, both analog and digital data. Further, a set of time-keeping routines is necessary, along with routines to handle process-generated interrupts (such as alarms), and (for FORTRAN) a set of multitasking routines. In the next section we furnish a pair of examples—one in BASIC and one in FORTRAN—simply to give a feel for the appearance of a real-time program before we return to fundamentals.

3.4 TWO EXAMPLES OF REAL-TIME PROGRAMS

Almost any real-time computing system will contain at least a terminal (or console typer) and, of course, a real-time clock. Both examples described below show how the computer can be used to keep track of the passage of time. In addition, the FORTRAN example shows the extra capability obtained through multitasking—in this case, the ability to use the terminal simultaneously as an input device, to "reset" the timing routines whenever desired.

3.4.1 A Real-Time BASIC Example

Let us finish up this chapter by looking at representative real-time exercises. As the simplest example we might consider the assignment of making a real-time computer and terminal into a "digital clock". In terms of underused resources, this system would have to represent the ultimate in digital clocks; nevertheless the problem offers some interesting insights into programming practices and the complications that enter the picture whenever any unusual secondary requirements must be handled.

For this example we would like to program the computer to type-out the time every 15 seconds. Our clock also should be based on a 24-hour day and, otherwise, give the time in the usual "hours, minutes, seconds". We will ignore the presence of an operating system "clock", even if one exists, and use one of the timer subroutines defined below to update our own set of software registers. These then will be used to maintain clock time, which will be printed out periodically on the terminal.

Note that some versions of BASIC can be modified to perform real-time computing jobs by means of a set of user-supplied subroutines for input, output, and time-keeping chores. For example, a rather primitive system might use subroutine calls of the form

$$\begin{pmatrix} \text{Stmt.} \\ \text{No.} \end{pmatrix} \text{CALL} \begin{pmatrix} \text{Subr.} \\ \text{Name} \end{pmatrix},$$
$$\begin{pmatrix} \text{Parameter} \\ 1 \end{pmatrix}, \ldots \begin{pmatrix} \text{Parameter} \\ N \end{pmatrix}$$

where each subroutine name would specify a particular routine written in assembly language (which, after assembly, would have to be loaded along with the BASIC function library and, of course, the interpreter and the operating system). A complete set of such subroutines is described in Chapter 14; for purposes of this example we require only four routines: (1) to turn on the real-time clock and to link an update routine for a set of software timers to the operating system interrupt service routine (so the timers will be advanced at each clock interrupt), (2) to initialize a particular software timer with a desired timing period, (3) to check the software timer for completion of the elapsed timing period, and (4) to test a switch connected to one line of the Digital Input Device, noting whether it is "on" or "off".

We then might have the following four calls:

(1) Turn clock on and link software timers
    CALL CLON (no parameters)
(2) Initialize (start) one of the software timers
    CALL TMRST, Parameter 1, Parameter 2
    Parameter1 is the software timer number (0-7)
    Parameter2 is the number of real-time

clock "ticks" (the length of the desired timing period)
(3) Check a timer to see if it has "timed-out"
    CALL TMRCH, Parameter1, Parameter2
    Parameter1 is the software timer number (0-7)
    Parameter2 is the "running count" which will reach zero and remain there when the timing period is past
(4) Check one input line of the Digital Input Device to see whether an external switch is "on" or "off"
    CALL DIN, Parameter1, Parameter2
    Parameter1 is the line or bit number addressed
    Parameter2 is the status (0 = switch off, 1 = switch on)

The eight software timers represent nothing more than eight locations in memory that, when linked to the real-time clock service routine, will be incremented one count each time the clock interrupts the processor (i.e., on each "tick"). Subroutine TMRST sets the memory location associated with a particular timer to the (negative) length of the timing period; hence this memory location eventually will contain zero (and no longer will be incremented) when the appropriate time interval has passed. Note that, once the user program initializes or starts the timer, no additional program statements are needed to modify its contents. For all practical purposes these software timers make use of the real-time clock to emulate eight hardware interval timers. However, subroutine TMRCH does allow the user program to access the memory locations to determine whether a timing period has elapsed (a zero value for the second parameter will be returned in this case).

A BASIC program which would perform these functions is given in Table 3.2. To understand the design of the program it is best to refer to the flow chart given in Figure 3.5. A single software timer (Timer 0) is used to monitor the passing of 15 seconds of real-time (150 increments at a clock frequency of 10 Hz). Hence, after performing some initializing steps and turning on a real-time clock, Timer 0 is

Table 3.2. BASIC Program to Implement a Twenty-four Hour Digital Clock

```
0010        PRINT "THIS IS AN ACCURATE BUT RELATIVELY EXPENSIVE DIGITAL CLOCK"
0020        LET S1=0
0030        LET S2=0
0040        LET H=0
0050        LET M=0
0060        LET S=-15
0070        CALL CLON
0080        GOSUB 0500
0090        CALL DIN,1,S1
0100        IF S1=1 THEN GOSUB 1000
0110        CALL TMRCH,0,S2
0120        IF S2=0 THEN GOSUB 0500
0130        GOTO 0090

0500        REM ROUTINE TO CALCULATE & OUTPUT TIME
0510        CALL TMRST,0,150
0520        LET S=S+15
0530        IF S<60 THEN GOTO 0610
0540        LET S=0
0550        LET M=M+1
0560        IF M<60 THEN GOTO 0610
0570        LET M=0
0580        LET H=H+1
0590        IF H<24 THEN GOTO 0610
0600        LET H=0
0610        PRINT "THE TIME IS", H,M,S,"O'CLOCK"
0620        RETURN

1000        REM ROUTINE TO SET THE TIME
1010        PRINT "TYPE IN HOURS"
1020        INPUT H
1030        PRINT "TYPE IN MINUTES"
1040        INPUT M
1050        PRINT "TYPE IN SECONDS"
1060        INPUT S
1070        CALL TMRST,0,150
1080        RETURN
2000        END
```

started and the time is printed (Subroutine 500). Note that updating of Timer 0 every 0.1 second is handled automatically in the system interrupt service routine even while printing is in progress; however, in order to keep up with real time, the digital clock cannot take longer to print the time than the basic print interval (15 seconds). Even with a slow-speed terminal (30 characters/second), our print interval is sufficiently long to guarantee this.

This problem often arises in real-time operations. For example, it might be desirable to log data from a process every 5 seconds. If, however, it is not possible to output the results in the same length of time, such a short sampling interval may not be a feasible one. With a single-task program structure and one not equipped with output "spooling", the one recourse is to store the results in memory (in arrays) and output all results at the end of the experiment. In any event, our clock printer simply has to be able to keep up with real time. Initial operation of the clock is shown in the first few lines of Table 3.3.

A second problem inherent in this example is connected with our desire to be able to input starting values of time, i.e., to set the clock. The reason for the difficulty is tied to the fact that we cannot input and output information simultaneously on our single I/O device. Specifically, we must inform the system when we do wish to input new times from the terminal. (Presumably, this would be infrequently.) For this purpose an external switch, open/closed, (Switch 1) is used to modify normal program flow, i.e., cause a routine to be executed (Subroutine 1000) that requests new values of clock time via the terminal. Two examples of these changes also are given in Table 3.3. In some multitasking systems, the external switch might not have to be used; a separate task to input new values of time could be readied (and left in

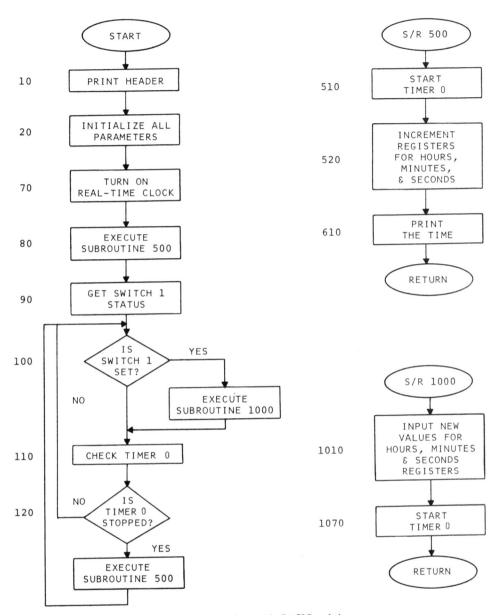

Figure 3.5. Flow chart of a sample BASIC real-time program.

the suspended state) to begin with. The input task then would not execute until someone begins typing-in values of time. An example of such a multitasking system is given in the next section.

3.4.2 A Real-Time FORTRAN Example

The digital clock problem used as an illustration in the previous section now will be used to demonstrate the properties of a simple version

of multitasking FORTRAN. One of the advantages of true multitasking software is that the user does not have to know, in advance, exactly when events will occur. For example, in the previous example, it was not possible to program the solution in BASIC in such a way that a user could "override" the timer by typing-in some new value of elapsed time unless it could be predicted in advance when the type-in would occur. The problem arises from the very structure of the BASIC program, viz., the necessity,

Table 3.3. Terminal Output from the BASIC Example Program

```
THIS IS AN ACCURATE BUT RELATIVELY EXPENSIVE DIGITAL CLOCK
THE TIME IS     0              0              0          O'CLOCK
THE TIME IS     0              0             15          O'CLOCK
THE TIME IS     0              0             30          O'CLOCK
THE TIME IS     0              0             45          O'CLOCK
THE TIME IS     0              1              0          O'CLOCK
THE TIME IS     0              1             15          O'CLOCK

TYPE IN HOURS
  ? 17
TYPE IN MINUTES
  ? 59
TYPE IN SECONDS
  ? 0
THE TIME IS    17             59             15          O'CLOCK
THE TIME IS    17             59             30          O'CLOCK
THE TIME IS    17             59             45          O'CLOCK
THE TIME IS    18              0              0          O'CLOCK

TYPE IN HOURS
  ? 23
TYPE IN MINUTES
  ? 59
TYPE IN SECONDS
  ? 0
THE TIME IS    23             59             15          O'CLOCK
THE TIME IS    23             59             30          O'CLOCK
THE TIME IS    23             59             45          O'CLOCK
THE TIME IS     0              0              0          O'CLOCK
THE TIME IS     0              0             15          O'CLOCK
```

once a program statement has been initiated, to complete execution of it before proceeding to a subsequent statement. The only way a new value of time can be typed-in is via an "INPUT" statement, and, wherever an "INPUT" is embedded in the program, the computer will wait at that point until values are typed-in. This awkward situation can only be avoided by some other procedure; for example, we used an external switch as an indicator that some value is to be typed-in and for the program to branch to the appropriate input routine.

A multitasking language permits a more natural approach, allowing the user to break the program up into segments called *tasks*. Each program segment is designed to run on cue; e.g., whenever a certain length of time has elapsed (real-time clock interrupts), a particular device interrupts, a digital input line changes state causing an interrupt, etc. In Figure 3.6, we illustrate a three-task implementation of the digital clock example that we just worked using the single-task version of BASIC. In this figure, a number of unusual flowchart symbols are used which we will define more carefully in Chapter 13. At this point, we only need to explain briefly what is depicted in this diagram.

Each user task has a priority and/or identification number. Zero represents the highest priority task; larger numbers represent lower priorities. The "bow tie" symbol is used to display the task priority and ID values (we have used the same number for both in this simple example).

The symbols with rounded sides are used to indicate calls to the operating system; in this example, we access the operating system 24-hour (time-of-day) clock, either to set it to some new starting time or to get a value for the present time. (Arguments of calls always are placed in the bottom half of the symbol.)

The trapezoidal symbols are for "task calls", i.e., calls from the user program to the multi-task scheduler. Through task calls in the user program, other tasks can be identified to the scheduler ("created"); any task can be suspended, delayed, or readied for execution; priorities can be changed, etc. At any instant, the task with the highest priority that is ready to run (i.e., has been created but is not suspended or waiting for an I/O operation to be completed) will execute. All other ready tasks must wait. Thus the tasks represent a set of program segments whose execution sequence is

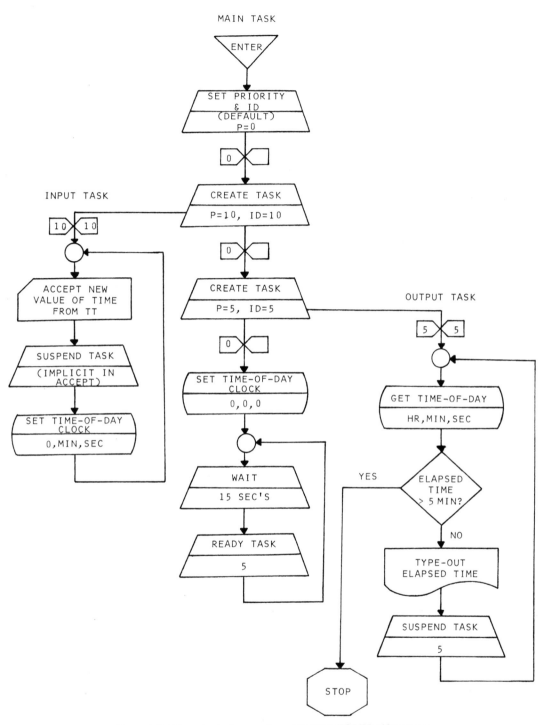

Figure 3.6. Flow chart of a sample multitasking FORTRAN program.

a function of both the user program statements and the real-time environment. In general, the task scheduler takes over the CPU and executes whenever an interrupt occurs or a task call in the user program is executed. Either of these events may result in another task becoming the executing task.

Chapter 13 is intended to cover multitasking in detail; here we merely wish to give some idea of the sorts of things that can be done with a multitasking system. Returning to our example problem shown in Figure 3.6, we let the first (or starting) task in the user program keep track of elapsed time. It will run at the highest priority level (0). A second task to output the actual time (continuously maintained by the operating system 24-hour clock) will run at the middle priority (5). And a third task to run with lowest priority (10) will input new values of time from the terminal (whenever the user wishes to type them in) and then set the system 24-hour clock to the new values. As noted above, whenever a task is not in position to run, the task scheduler suspends it. As a particular example, when the input task (10 is waiting for input from the terminal, i.e., not capable of completing the first "ACCEPT" statement with a dummy input variable, it is suspended. Whenever a terminal key is struck and no higher-priority task is running, the scheduler reactivates the input task. The time-keeping task (0) is suspended most of the time waiting for 15 seconds to pass (150 ticks of the real-time clock at 10 Hz), at which instant it always will have highest priority, begin to execute, and ready the output task (5). It then immediately initiates a new "WAIT" which causes it to be suspended. The output task (5) suspends itself after each printout of the elapsed time and, as we just noted, will only be readied for execution when the time-keeping task comes to life and makes the appropriate "READY" call.

Table 3.4 lists the FORTRAN program for this example. In this case it has been written in a nonstandard (but, hopefully, understandable) real-time FORTRAN dialect. With the exception of not showing the dummy variable printout, Table 3.3 again illustrates typical terminal output.

The concepts underlying multitask programming and operating systems are covered in detail in Chapter 13, and this example is reworked again in Chapter 14 using a multitask version of BASIC. We need only note that efficient software resources are available for most real-time computer systems, and these should be utilized where possible. Often, several days of initial work in learning the programming techniques specifically designed for real-time applications will be repaid almost immediately in the form of more efficient programs and less time required for debugging them.

3.5 SUMMARY

In this chapter we have tried to give a brief overview of real-time software, both from the system side (executive or operating system design philosophy) and from the user standpoint. On the former topic we have shown how a sequential processor, equipped with appropriate hardware, can be programmed to handle multiple "simultaneous" requests for service from the environment (process and user). On the latter topic we used two different approaches to demonstrate how a user can program a real-time computer so that the results of its execution will be keyed to process real-time and to events that cannot be predicted in advance. These are very fundamental extensions of computer programming techniques as they ordinarily are applied.

In Part V we will retrace our coverage of software and programming, mostly from the system side. There we will take a closer look at assembly language programming to help get a better notion of internal operations of the computer and, important for the case where a specialized user hardware device has to be attached to the real-time computer, to write I/O routines which can be embedded in system software. We will also discuss the software that ordinarily can be obtained from computer manufacturers and take a much more detailed look at operating systems and multitask programming.

In Part VI we devote considerable attention to user applications programming. Separate chapters are devoted to real-time versions of both BASIC and FORTRAN and a discussion of a generalized (table-driven) approach to the construction of data acquisition and control software appears, as well.

Table 3.4. A Multitasking FORTRAN Example

```
C           MAIN TASK
C           DEFINE NUMBER OF I/O CHANNELS AND NUMBER OF TASKS
            CHANTASK 2,3
            EXTERNAL INPUT, OUTPUT
            INTEGER HR,MIN,SEC
            COMMON HR,MIN,SEC
            TYPE "THIS IS AN ACCURATE BUT RELATIVELY EXPENSIVE DIGITAL CLOCK"
C           CREATE INPUT TASK WITH PRIORITY=10 AND ID NO=10
            CALL ITASK(INPUT,10,10,IER)
C           CREATE OUTPUT TASK WITH PRIORITY=5 AND ID NO=5
            CALL ITASK(OUTPUT,5,5,IER)
C           SET TIME-OF-DAY TO 0 HR, 0 MIN, AND 0 SEC
            CALL FSTIM(0,0,0)
C           SUSPEND THIS TASK FOR 15 SECONDS (I.E. WAIT OR DELAY)
       1    CALL FDELY(150)
C           READY OUTPUT TASK
            CALL ARDY(5)
            GO TO 1
            END

            TASK OUTPUT
            INTEGER HR,MIN,SEC
            COMMON HR,MIN,SEC
C           GET TIME-OF-DAY
       2    CALL FGTIM(HR,MIN,SEC)
            IF((MIN.EQ.5.AND.SEC.GT.0).OR.MIN.GT.5)GO TO 3
            TYPE "THE TIME IS", HR,MIN,SEC, "O'CLOCK"
C           SUSPEND THIS TASK
            GO TO 2
       3    STOP
            END

            TASK INPUT
            INTEGER HR,MIN,SEC
            COMMON HR,MIN,SEC
C           WAIT FOR USER TO INPUT NEW TIME AT THE TERMINAL
       4    ACCEPT "TO INITIALIZE THE CLOCK, TYPE ANY NUMBER & CR", DUMMY
            TYPE "TYPE IN HOURS"
            ACCEPT "?", HR
            TYPE "TYPE IN MINUTES"
            ACCEPT "?", MIN
            TYPE "TYPE IN SECONDS"
            ACCEPT "?", SEC
C           SET TIME-OF-DAY
            CALL FSTIM(HR,MIN,SEC)
            GO TO 4
            END
```

3.6 SUPPLEMENTARY READING

1. Brignell, J. E. and Rhodes, G. M., *Laboratory On-Line Computing*, Wiley, New York, 1975.
2. Elson, M., *Concepts of Programming Languages*, Science Research Associates, Chicago, 1973.
3. Friedman, F. L. and Koffman, E. B., *Problem Solving and Structured Programming in FORTRAN*, Addison-Wesley, Reading, Mass. 1977.
4. Harrison, T. J. (Ed.), *Minicomputers in Industrial Control*, ISA Press, Pittsburgh, Pa., 1978.
5. Pohl, I. and Shaw, A., *The Nature of Computation: An Introduction to Computer Science*, Computer Science Press, Rockville, Md., 1981.
6. Ralston, A. (Ed.), *Encyclopedia of Computer Science*, Van Nostrand Reinhold, New York, 1976.
7. Tanenbaum, A. S., *Structured Computer Organization*, Prentice-Hall, Englewood Cliffs, N.J., 1976.

3.7 EXERCISES

1. What is the potential problem or even danger involved with an unconditional "write" operation to an output peripheral device? With an unconditional "read" operation from an input device?

2. What are the advantages/disadvantages of each of the conditional transfer methods illustrated in Figures 3.1b,c?

3. There are periods of operation of any real-time computer when interrupts must be

completely disabled.   Can you think of several different reasons why this might be necessary? Is there any danger in operating with interrupts disabled? Why can't these conflicts in objectives be resolved, e.g., through redesign of hardware?

4. Are there any circumstances under which a background program should take priority over a foreground program?

5. Discuss, briefly, the advantages/disadvantages of assembly language and high-level languages such as BASIC and FORTRAN:
   a. For batch programming applications.
   b. For real-time programming applications.

6. An amusement park is considering the construction of a roller-coaster complete with on-board computers in the individual trains. Functions to be handled by the computer include the judicious use of brakes based on the train's speed (so as to prevent passengers from leaving the cars prematurely), a time function used to start the cars off at equally-spaced time intervals, the operations of dropping a cog into the chain drive used to raise the car to the top of the initial hill and then of retracting it at the top, and the release of lap restraints at the end of the ride. Note: The train must be at a full stop before dropping the cog or releasing the seat belts.

   Assume that train speed is available from on-board speedometer with voltage output and that the time for the train to reach the top

of the first hill is a constant (known) value. Describe what tasks might be required, what the function of each task might be, and how the tasks would be scheduled.

7. You have been hired as a consultant by the Local 500 Road Race Committee for the purpose of designing a computerized data acquisition system to be used in the initial time trials conducted for each potential race participant. In these time trials each driver is required to maintain an average speed of at least 100 kilometers per hour for 20 laps around a rectangular course that is exactly one kilometer long.   The required data acquisition system must measure (mechanically or electrically but without using radar) and log the averge speed of a single car at fixed 0.2-kilometer intervals around the track and must additionally compute and log the final average speed for the entire 20-lap trial.
   a. Draw a labeled diagram that indicates how you would instrument the system (sensors, measurement concepts, etc.).
   b. List and briefly describe the function of each real-time peripheral that must be included in your design for the system.
   c. Briefly describe how the system software will calculate the required speeds. Discuss the organization of tasks if you choose to use a multitasking approach.
   d. Briefly describe any inherent problems associated with your approach to the problem.

Part II
Processes, Measurements, and Signal Processing

4

Processes and Representative Applications

Thomas F. Edgar
Department of Chemical Engineering
University of Texas, Austin

4.0 OVERVIEW OF REAL-TIME COMPUTING APPLICATIONS

An understanding of the implementation of real-time computing and an appreciation of its versatility and potential are usually enhanced by a discussion of typical applications. One goal of this chapter is to illustrate different roles that a computer can assume. With this objective we shall present a limited number of representative applications rather than an exhaustive survey of the large number of actual cases where real-time computers are used.

Engineering analysis of the process interfaced with a real-time computer is required for successful design and implementation of a real-time system. The usual objective of a real-time computing application is to make decisions based on the two-way transfer of information between the computer and the process (environment). The development of rational decisions depends upon defining the relationship between the computer and the environment. Hence, a secondary goal of this chapter is to address the issue of process definition or modeling. We shall discuss some aspects of mathematical modeling, although we choose to concentrate on structural considerations rather than covering mathematical details.

Figure 4.1 depicts a typical sequence of activities performed in a real-time computing application. A measurement signal from the process environment may be passed through several devices (transmitter, signal conditioner) before reaching the digital computer. Two different modes of real-time decision-making can

then be utilized; these correspond to the passive and active computer modes as discussed in Chapter 2:

(1) Operator-initiated, where the human operator retains all power of decision-making and uses the computer to generate and assimilate information. Here the computer functions as a rapid data analyzer and presents data to the operator in some readily interpretable form. In Figure 4.1, the operator may initiate action through the operator console typewriter. This action is a controllable input to the process environment through the right-hand branch, involving some control element (e.g., valve, electric heater, etc.).

(2) Computer-initiated, where a digital computer program, which often is structured on the basis of prior analysis of the process, determines the appropriate action. This mode is often referred to as "closed-loop", since no operator intervention occurs. It is commonly used in control applications where, for example, the computer adjusts a control valve position based upon process measurements.

The circular nature of the diagram illustrates that the output from one device or block is the input to the adjacent block. Hence the environment can serve as an input to the computer (as in data acquisition), or the computer can serve as an input to the environment (as in process

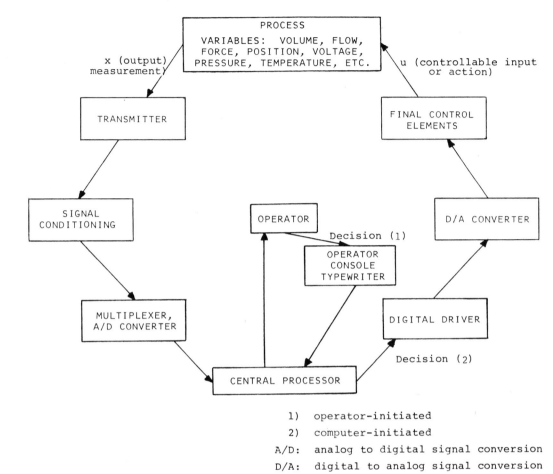

Figure 4.1. Environment/Computer Interface

control). The input–output relationships and their definition are discussed in Section 4.4.

4.1 A PROCESS EXAMPLE WITH OPERATOR-INITIATED DECISIONS

Let us consider the first mode of real-time decision-making, namely, operator-initiated. Figure 4.2 schematically depicts a digital computer connected to a chromatograph. The chromatograph is an instrument that quantitatively analyzes the composition of a multi-component gas or liquid sample; we discuss its operation later in this section.

In the chemical industry the measurement of composition is central to both the evaluation and control of unit operations involving chemical reaction and separation. In order to obtain optimum yields, meet product specifications, or

satisfy safety or environmental requirements, samples from various product streams must be taken and the results determined rather quickly. In the absence of on-line analysis, grab samples must be taken to the laboratory and then processed; this creates inevitable delays and potentially unprofitable, unsatisfactory, or unsafe operating conditions.

In the context of Figure 4.1, the chromatograph transmits electronic signals to the computer, which are then analyzed by a computer program to generate the sample composition. The human operator can monitor the computer display and take appropriate action when the process is not operating properly.

If plant and laboratory equipment operated consistently at a "steady state" corresponding to the best operating conditions, then there would be little incentive in applying real-time

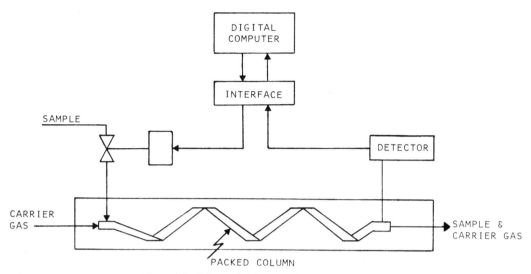

Figure 4.2. Schematic Diagram of a Chromatograph

computer analysis. However, most chemical plants do not operate this way and the plant variability often translates into a financial loss, due to "off-spec" product.

It should be recognized that the use of real-time analytical systems occasionally has its drawbacks. In essence a more sophisticated system is created where more things can go awry and where more maintenance is required. The two most significant limitations of on-line composition analyzers are accuracy and reliability, and frequent calibration checks and adjustments are required in order to insure that the instrument is reliable. To perform frequent accuracy and reliability tests manually is out of the question; therefore, the use of the computer to provide checks is a necessary but productive marriage of two pieces of equipment.

A chromatograph operates in the following way. A measured amount of the process stream sample is injected into the chromatograph column which is packed with an absorbent material. A carrier gas sweeps continuously through the column, and the sample is carried along with the carrier gas. Since the individual components are absorbed and desorbed at different rates onto the packing, the components exit the column at different times. The components at the column outlet are measured by a detector as a concentration in the carrier gas; two detection methods that are commonly used are flame ionization and thermal conductivity.

A typical detector response is presented in Figure 4.3; the succession of peaks represents individual components in the original sample, and the area under each peak is proportional to the concentration of that component. By running control samples with known concentrations, conversion factors (often called response factors) from area to concentration for each component can be determined. Chromatographs are used repetitively for a given application; therefore, significant experience is gained in selecting peaks and determining the correspondence between component concentrations and measured peaks.

While the concept of operating a chromatograph may appear simple enough, careful checks of accuracy on a routine basis must be undertaken in order to insure that meaningful data are obtained. For example, at least once a day a calibration blend of gas or liquid should be introduced into the chromatograph for testing and then compared with areas and peaks. If some deviation in a given component's peak value occurs with the calibration mixture, the response factor for that component must be adjusted. At the same time the timing of individual peaks (when they first appear) must also be compared. In order to perform the above checks manually for a large number of chromatographs (say 25), one full-time person would be required.

Of course, the above tasks can be and are per-

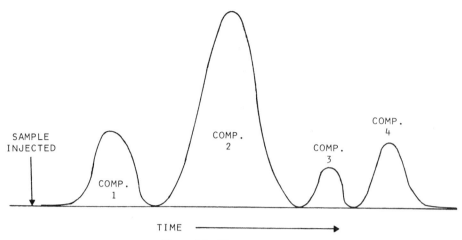

Figure 4.3. Chromatogram

formed by computers. Osborne [17] has quoted specifications for a computer-controlled gas chromatograph (GC) system for regular analysis of simultaneously operating GCs:

1. Operates the proper stream selection valve.
2. Tests for sample flow. If no flow is sensed, outputs an alarm, skips to the next stream and repeats the flow test.
3. Operates the sample injection valve to begin the analysis cycle and operates the column switching valves at appropriate times within the cycle.
4. Acquires data from the GC detector at any rate specified for each individual stream method, allowing accuracy on both conventional and high-speed analysis.
5. Processes these data at the end of each analysis cycle to separate component peak areas, identifies the components, applies appropriate response factors, and calculates the individual component concentrations.
6. Applies the following internal data consistency checks on each analysis: total area limits, sample flow, component concentration limits, and base line deviation limits. Any of these limit tests will produce an alarm. Output of further data will also be inhibited if the error is considered "fatal", as described below.
7. Transfers resulting component concentra-

tions to a data acquisition computer for use in material balance calculations, logging, etc.
8. Provides for a continuous analog output compatible with conventional control equipment for either closed-loop control or trend recording.
9. At a specified time during the day the system operates a valve to introduce a standard sample into the GC. Response factors are calculated, limits checked, and, if within the limits, used to update the appropriate method tables. A unique feature was specified for this system to provide a means of calibrating unstable components. A response factor from any component in the standard blend can be used to adjust any stream component via relative response ratios.
10. Provides dual sets of limits in standard blend methods for both response factors and retention times. A violation of the inner set results in an alarm output to analyzer personnel for the purpose of correcting minor variations before the accuracy of the analysis is affected. A violation of the outer limits is considered a fatal error and outputs an alarm plus inhibiting further output of data from the analyzer.

Note that items 6 and 9, reliability and accuracy tests, are only one part of the total package of functions performed by the computer.

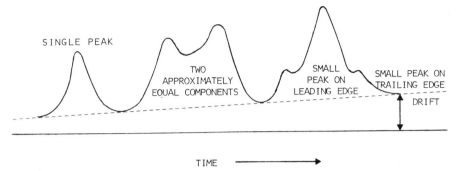

Figure 4.4. GC Peak Shapes

Item 5, which provides the major diagnostics for a GC analysis program, involves peak detection, identification, and integration of the peak area for various peak shapes, and correction for baseline variations. It should be pointed out that the peak pattern shown in Figure 4.3 is rather idealistic. In practice, peaks overlap or are "fused", which means that the identification and separation of peaks become rather sophisticated. Figure 4.4 [25] shows how the actual chromatogram can deviate from the ideal and exhibit a "drift", or steadily increasing deviation, from baseline. While the analytical procedure is somewhat arbitrary, graphical constructions, which then can be translated into mathematical formulas, are rather straightforward in their application. Specific methods of peak area calculation have been discussed by Gladney et al. [6], Ernst et al. [3], and Roberts et al. [20].

Other programmable features of a GC analysis program may include:

(1) Baseline correction for positive or negative drift, where the dotted line in Figure 4.4 becomes the baseline.
(2) Automatic increases in peak detection sensitivity for low, broad peaks late in the chromatogram.
(3) Rejection of air or solvent peaks without affecting area allocation of other peaks.
(4) Automatic rejection of "noise" spikes.
(5) Digital filtering for smoothing data; this facilitates calculation of peak slopes, which are used for peak identification.

The GC data analysis by the computer provides information in a form that allows the operator to take immediate action. For example, the

operator could rapidly correct any violation in product specification by adjusting flow rates, pressures, etc. In more complicated systems, the GC analysis could be coupled with material and energy balances in order to provide additional background data for decision-making. It should be mentioned, however, that the operator functions could be performed in a closed-loop mode, where the computer implements the decision without human intervention, as discussed below in Section 4.2.

The GC application clearly shows the need for process characterization. As discussed earlier, a gas chromatograph–computer installation would require many different algorithms, including logic for peak detection, valve opening and closing, temperature programming, periodic calibration, and many others, all of which require knowledge of system behavior. Details on real-time chemical analysis have been given by Perone and Jones [18] and Frazer et al. [5]. It is evident that process characterization of a given real-time computing system does not usually involve a single process model but rather a combination of many models and decision patterns. A further implicit process characterization is involved whenever the operator or engineer makes the decision. In this case, the decision criteria of the operator arise from a complex set of factors contained in the mind of the operator; these factors may not be readily translated into explicit mathematical form.

4.2 A PROCESS EXAMPLE WITH COMPUTER-INITIATED DECISIONS

As discussed earlier, the chromatograph plus computer provides a means of measuring the concentration distribution in a process stream. Based on the concentration measurement, this computer (or a different computer) could automatically adjust operating variables rather than requiring human intervention. As an example, suppose that we are monitoring the emission of carbon monoxide (CO) from an oil-fired heater and that the CO concentration begins to increase, signifying incomplete combustion. Under computer control, the air flow to the furnace can be increased to change the extent of combustion, thus avoiding fuel waste, air pollution, and unsafe operating conditions. This example illustrates the classical process control problem. One compares the measured concentration against the product specification; if there is an error in the concentration levels, a signal is automatically transmitted to a control device (via the computer) which compensates for the error. Compensation (such as a change in the air flow rate) is intended to bring the product stream into specification and is usually implemented by a design scheme referred to as "feedback control" or closed-loop control. There are two major types of process control strategies:

(1) Control based on transient characteristics of the process, such as feedback control.
(2) Optimization of steady state operating variables.

In both cases, the computer makes virtually all real-time decisions, illustrating the second mode of computer usage discussed in Section 4.0.

One important real-time computing application based on the latter case, steady state operation, involves the iterative adjustment of operating conditions so that some performance objective is optimized; this technique or method is often referred to as "supervisory control". Let us consider an internal combustion engine as the process. We could select as the performance objective the maximization of gas mileage (or minimization of the operation costs). On the other hand, we could choose to minimize the amount of pollutants emitted from the exhaust of the engine, which would not necessarily correspond to minimum engine

operating cost. In either case, we would adjust the engine operating conditions periodically to maintain optimum operation.

Through use of the computer, it is possible to introduce an adaptive type of control to the internal combustion engine. A fuel injection control system, which has the capability of adjusting its operation when ambient conditions change, will be available on many automobiles in the 1980s. A conventional auto combustion system does not have the capability of changing with the ambient or process conditions. However, a small on-board computer can perform changes in engine operating parameters to achieve improvements in the system performance.

Before specific design of the control logic is carried out, performance criteria for the desired system must be established. Reynolds [19] has proposed the following standards for computer-controlled combustion engine operation:

(1) Minimize carbon monoxide (CO) emission consistent with power level requested by operator. Typical performance curves for gasoline-fueled engines are presented in Figure 4.5. It should be noted that fraction of carbon monoxide decreases as the mixture is leaned. However, there is a significant reduction in brake horsepower with mixtures leaner than 13 to 1. Since brake specific fuel consumption (bsfc) increases with very lean mixtures, it appears advantageous to operate with an air–fuel ratio of approximately 17 to 1, as long as sufficient power can be developed. This ratio on a typical engine should allow approximately 80% of rated horsepower and less than 0.5% exhaust CO. If more power is required, a richer mixture will be used even though the CO emission will be above the limit. Local NO_x standards may, however, require modification of the air–fuel ratio.

(2) Provide for cold starting ease. Because of many factors, gasoline engines require a richer mixture during cold starting and warm-up periods. The injection system must provide this automatic choking capability.

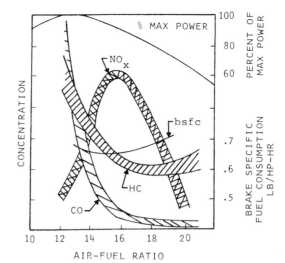

CO Carbon monoxide

NO$_x$ Nitrogen oxides

HC Unburned hydrocarbons

Figure 4.5. Typical Gasoline Engine Performance Curves

(3) Provide fuel cut-off during deceleration. During this condition the engine is used as a brake and no fuel is needed. The proposed injection controller should provide a means to turn fuel completely off during periods of maximum deceleration.

(4) Adjust to varying ambient conditions. Engines are subjected to wide temperature and barometric pressure changes which affect the operation of the carburetor. The injection controller should provide this capability automatically.

(5) Compensate for changes caused by engine and other unpredictable phenomena.

(6) Provide over-temperature and over-speed fuel cut-off.

A control system capable of meeting all of the above specifications appears to fall within the broad class of control systems known as "adaptive". Reynolds [19] proposed the following measurement variables for an adaptive engine control system: intake manifold pressure (PI), ambient pressure (PA), head temperature (T), engine speed (N), and percent carbon monoxide in the exhaust gas (CO). From these four indicators it should be possible to control the air-fuel ration (A/F) to satisfy all performance specifications listed above.

The basic operating scheme is as follows:

(1) At a specific time in each cycle a pulse generator operating at one-half crank-

shaft speed will send a signal to the computer.

(2) Upon receiving this signal the computer will read the intake manifold pressure from which the mass of air inducted in the current cycle can be computed.

(3) By using a precomputed air-fuel ratio and the mass of air computed above, the required mass of fuel to be injected is computed. A constant-pressure, fixed-cross-section injection nozzle is then opened by computer control. The length of time the injector is open is proportional to the mass of fuel desired to be injected.

(4) After fuel is injected in each cycle, the air-fuel ratio is updated for use in subsequent cycles. A small step change in the control signal (A/F in this case) is introduced. If engine speed, which is a measure of engine power, increases in response to the step change, another step change in the same direction is employed. If, however, the step produces a decreased speed, the change produces an inferior A/F, and the next step should be in the opposite direction. Repeated application of this process yields the A/F producing maximum engine speed. The interval between step changes can be set so that the controller may be tuned to give best performance.

Table 4.1. Decision Matrix For Control System Logic (Reynolds, [19])

SIGNAL CHARACTERISTIC	PROCESS CHARACTERISTIC	DESIGN DECISION	IMPLEMENTATION TECHNIQUE	FUEL MASS
DP < DP2	T > T1 T < N1	COLD STARTING	WAIT	FUNCTION OF PI, AFR
ALL DP	T > T1 N1 < N < N2	WARM UP	ADJUST (AIR/FUEL RATIO) TO MAXIMIZE ENGINE SPEED BY MEASURING RESPONSE TO SMALL STEP CHANGES IN AFR	FUNCTION OF PI, AFR
DP < DP1	T < T2 N1 < N < N2	MAXIMIZE POWER OUTPUT		
DP1 < DP < DP2	T1 < T < T2 N1 < N < N2	MODERATE POWER CONTROL CO	CO > 0.6% LEAN .4% < CO < .6% AFR OK CO < .4% RICH	FUNCTION OF PI, AFR
DP > DP2	ALL T ALL N	DECELERATION CUTOFF		
ALL DP	ALL N T > T2	OVER-TEMPERATURE CUTOFF	AFR = ∞	ZERO FUEL
ALL DP	ALL T N > N2	OVER-SPEED CUTOFF		

*Emissions data from D. A. Trayser and F. A. Creswich [26]; power and fuel consumption data from Edward F. Obert [16].

A decision matrix, Table 4.1, summarizes the logic required for this system. Based upon the interaction of the input variable (pressure drop across the throttle, DP = PA – PI) and three process variables (N, T, and CO), the decision computer selects one of four stored operating states.

More advanced electronic engine controls [2, 14] also adjust the composition of the mixture in the combustion chamber (via recirculated exhaust gas) and the timing and the energy of the spark. Measurements employed include manifold pressure, engine speed, crankshaft position, coolant temperature, air temperature, throttle position, and exhaust gas composition. Future improvements in on-board engine computers will compensate for long-term changes in components (due to wear of parts), road surface and changing tire characteristics, changes in engine friction as the engine breaks in, deterioration of the exhaust catalyst, and compliance in suspension parts. More interaction with the driver may also be provided via digital displays. For example, a miles-to-empty tank prediction could be requested; engine diagnostics and service "alarms" will also be available in automobile computers.

4.3 TYPES OF REAL-TIME COMPUTER APPLICATIONS

After having illustrated the interface between the operator, the computer, and the application environment through two examples, in this section we shall classify the different types of small-computer applications. It is often difficult to draw clear lines which distinguish one type of application from another, but the following general categories do have some distinguishing features [9]:

(1) Original equipment components
(2) Stand-along computing (batch programming)
(3) Laboratory and monitoring
(4) Process control
(5) Communications

For each category above, representative applications will be discussed.

4.3.1 Original Equipment Components

During the last decade, a large business has developed in using a central processing unit as a component in larger, more complex systems. Usually this type of application does not have an input keyboard facility and is very inexpensive, perhaps microprocessor-based; however, the user must develop the necessary interfacing. Among these applications are component testers, automatic weighing or other quality control devices, and transfer machines.

The operation of machine tools, such as lathes, mills, or hole punchers, is quite amenable to the use of small computers. Numerically-controlled hole drilling is very important in the manufacture of circuit boards. Performing the hole drilling sequentially as opposed to drilling one location repeatedly for many boards avoids problems in warping, shrinking, and stretching of boards between operations, especially for boards with high drill hole density. The automated drilling of holes can operate on absolute or incremental coordinate positioning. The coordinates to be drilled can be entered through a program, and, of course, can be changed by entering new program instructions. The use of a small computer in this application avoids extensive revisions in comparable hard-wired logic. If production demands suddenly require an increase in production of a certain circuit board, more drilling machines can be brought into service rapidly. Use of a computer also alleviates potential operator problems because all drill patterns can be stored in memory. Drilling of multiple hole sizes, where the hole positioning can be performed with respect to the specific drill to be used, can also be performed readily via computer control.

Mechanical testing of materials is another important computer application that falls in this first category. For synthetic fibers, a tester can stretch the fiber until it breaks and output analog data on the load, elongation, and stress experienced by the fiber prior to failure. Such an automated facility can provide on-line analysis of the mechanical properties of the fiber.

The testing procedure used when a computer component is not available requires separate digitizing and batch analysis of the data on a large central computer facility. In the testing of automobile tires, as much as three days is required to perform the digitizing and analysis manually.

4.3.2 Stand-alone Applications

Since this text is devoted to the subject of real-time computing, a discussion of the use of a small computer as a batch processor of computer programs is somewhat out of place. However, we should note that there are many cases where batch computing can be replaced by real-time computing to simplify procedures. For example, consider the automated digitizing and processing of the data from an experiment, as discussed in the previous section. If the "real-time" link in this system is not available, the data must be transferred to some storage device. Then, as a separate action, the data must be entered into the computer for analysis in a batch or stand-alone mode. Thus the real-time approach can streamline the overall operation and generally is preferred.

4.3.3 Laboratory and Monitoring Applications

This type of application is typical of the first mode of real-time decision-making discussed in Section 4.1. The size of the facility controlled may vary from a small piece of laboratory equipment to extremely large facilities, such as water resource or pipeline systems.

The U.S. Army Corps of Engineers has installed a supervisory system on a three-dam facility in Oregon. The system operates purely in a monitoring and data analysis mode. Temperature readings are checked to see if there are any excursions above some specified level; an alarm system is activated if any measurements exceed the tolerance level. The computer also totals generator power levels, monitors spillway-gate operating increments, and prints out malfunction data. The system monitors 150 temperature sensors and 300 alarm conditions.

Computer monitoring of a pipeline operation is a similar kind of application. Prior to the availability of inexpensive minicomputers, individual sections of a line had to be monitored by an operator examining a hardware-based display. A 2000-mile line when automated now might have 20 minicomputers used for monitoring. These machines then are connected on-line to a large central computer. The computers scan the pumping stations on a continuous basis and transmit to the central system such information as suction pressures, discharge pressures, case and holding pressures, station electrical load, gravitometer and interface detector readings, and tank gauges (where storage measurements are applicable). The minicomputers compare all readings to those last reported to the central computer. If these are beyond predefined limits, an error message concerning the discrepancies is transmitted. If control is desired in addition to monitoring, the central computer can issue commands to the smaller computers to start or stop pumping units, to sound alarms, or change setpoints.

Small computers are widely used in testing laboratories for combustion engines. The computers increase testing efficiency, which enables more tests to be performed per test stand. The results are also made available to an engineer sooner. The functions of the computer during a test include measurement of torque, speed, and fuel consumption subject to throttle and dynamometer load control. Vora [27] has described a study at General Motors Research Laboratories where 150,000 data points could be acquired in a 30-minute experiment. In one application of engine testing discussed by Jurgen [9], two 16-bit computers with shared main memory are used. Each is capable of interrupting the other and communicating via the shared memory. One CPU is assigned the measurement and control tasks for all test stands. The second processor produces test reports from the data which are stored on magnetic tape for record-keeping purposes. In addition, it can perform independent functions such as program assemblies, compilations, or executions. The setup is shown in Figure 4.6. A principal advantage of the dual CPU station is the degree of reliability afforded, since the computers can in theory back up each other in case of failure. This avoids test backlogs and

potential waste of personnel. A separation of foreground/background functions also provides some security, since interruption of one CPU by the other can occur at a preassigned priority level.

The use of minicomputers in monitoring clinical subjects in hospitals or laboratories is another important application [9]. In one case a minicomputer is used to monitor cardiac behavior and pressures in conscious animals and derives dynamic ventricular performance data. Figure 4.7 shows the measured variables, conditioned and derived signal paths, and the elements of the data acquisition and processing system used at the Cardiovascular-Renal Research Laboratory of Howard University. With a variety of implanted sensors, the investigators measure, monitor, and record the primary cardiac variables of physical dimensions, pressures, flow rate, ECG, and heart rate of test animals, acquiring the data under computer control. After a selection is made of the segment of data that is of particular interest, the data are used to calculate parameters descriptive of dynamic cardiac performance, such as cardiac output, flow, power, wall stresses, and other information. The minicomputer permits the on-line computation, printout, and display of the measured and derived quantities. It also permits the simultaneous measurement of the process variables for a rapid assessment of ventricular performance.

4.3.4 Process Control

The use of computers in controlling manufacturing processes differs from the monitoring class of real-time applications in that the overall system usually operates in the closed-loop mode; i.e., the computer analyzes the data received and then performs some action based on internal program logic (see Section 4.2). The logic, or algorithm, upon which the process control actions are based can be extremely simple or sophisticated. Process control applications occur with both continuous processes (e.g., chemical, cement, steel plant) and discrete processes (assembly line, batch processes). In general, because of the versatility required in these applications, the computers are consid-

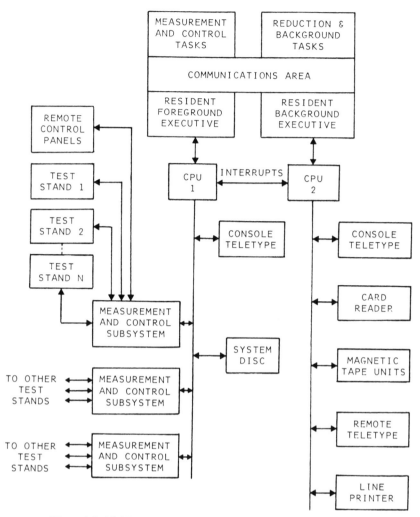

Figure 4.6. Multiprocessing System for Engine Testing (Jurgen [9])

ered to be units separate from the rest of the process, as compared with class (1) considered in Section 4.3.1, where the computer is considered to be an integrated component of the equipment (process).

The application of computer control, known as DDC (Direct Digital Control),* to a chemical plant dominated by batch processes is one where the computer can be used quite economically and effectively [11]. These batch processes are cyclic, repeatedly passing through a series of phases, such as filling a vessel, heating its contents, and discharging the product into

another vessel. Therefore, the control systems not only must provide simple feedback relationships, but also a means of varying operating conditions as a function of time; the latter is known as the operating procedure, open loop control, or sequential logic. However, open loop control is usually performed in conjunction with feedback control, in order to keep the operating variables at their designated values. Batch process control is a good example of a multitasking application, where the computer must perform parallel procedures or tasks on a single piece of equipment. Multitasking is discussed in detail in Chapter 13. Various feedback control algorithms used in computer control are discussed in Chapter 20.

*When the control computer largely replaces standard analog control devices.

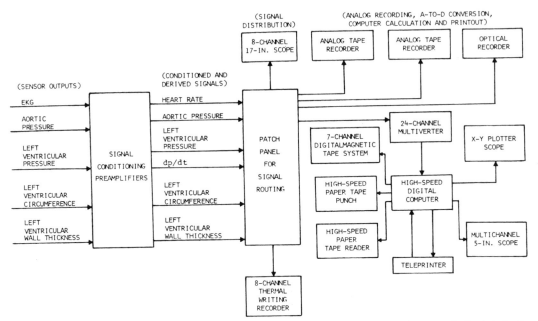

Figure 4.7. Basic Elements of a Minicomputer System for Testing Ventricular Performance in Test Animals (Jurgen [9])

Each separate batch process may have its own equipment for feed preparation, reaction, and product recovery, and each process usually has a different operating procedure. In addition, the same equipment may make several different products; hence the operating parameters and the ingredients change. Table 4.2 lists the various unit operations that can be involved in a batch process. A typical batch process is shown in Figure 4.8.

Six control states or modes can be defined as follows:

(1) Normal—process operating according to prescribed procedure.

Table 4.2. Kinds of Batch Process Systems

FEED PREPARATION	PRODUCT TRANSFER
RAW MATERIAL TANKS	PUMPS
PUMPS	HEADERS & DIVERTERS
HEAT EXCHANGER	INTERMEDIATE & FINAL STORAGE TANKS

REACTION	PURIFICATION
REACTOR	EVAPORATOR
CONDENSER	CONDENSER
HEAT EXCHANGER	DECANTER
FEED & CRUDE PRODUCT	FEED & CIRCULATION PUMPS
STORAGE TANKS	HEAT EXCHANGER
FEED & CIRCULATION PUMPS	CRUDE & INTERMEDIATE PRODUCT STORAGE TANKS

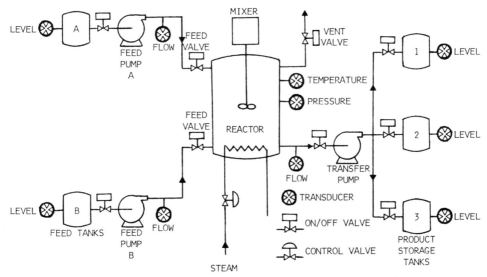

Figure 4.8. Typical Batch Processing System for Chemicals Production

(2) Hold—partial shutdown, conditions maintained near operating level.

(3) Emergency Shutdown—complete sudden shutdown.

(4) Standby Shutdown—similar to Hold, but implies total reflux or total recycle, in which operating conditions are maintained but no product is made.

(5) Restart—transitional logic to go from a Hold or Shutdown condition back to Normal.

(6) Unavailable—isolated from other systems and scheduling programs for extensive maintenance.

State #1 (Normal) can further be divided into several steps as follows:

(1) Prestart—making sure that sufficient raw materials are available to proceed with charging.

(2) Charge—charging materials A and B into the reactor.

(3) Run—maintaining the desired temperature for the proper time.

(4) Discharge—discharging reactor contents to storage, venting and cooling the reactor.

(5) Cleanup—preparing the equipment for a new task.

The operator always initiates return to Normal

from other control states. When Restart logic takes over, it usually returns to a preceding Process State to reestablish conditions that existed at the time of interruption. For instance, if the system was in Hold, and if the reactor contents cool below reaction temperature, the system must return to the heating state before the reaction can resume. The reentry point always depends on the types of actions or degree of shutdown that took place in other control states.

Complex sequential logic can be reduced to a time-ordered combination of a few basic actions. A listing of these actions with some examples follows:

(1) Operate on—off devices (pumps, valves, etc.).

(2) Activate or deactivate control loops. A particular flow loop may be used during charging of the reactor, but is not required during the reaction cycle.

(3) Open and close cascade loops. In startup of a distillation column, for example, it may be necessary to establish a specific flow rate before closing the cascade loop.

(4) Ramp setpoints to desired values. Increasing a setpoint uniformly, as opposed to changing it stepwise, may be an important factor in reestablishing stable operation or quality control.

(5) Integrate flows. Flow streams must be accurately totalized if a specified amount of material is to be added to the reactor.

(6) Change parameters. New parameters are required as plant operating conditions change.

(7) Initiate timers or delays between process steps. After a mixer is started, delay may be necessary before more materials can be added.

(8) Perform material balance calculations. To satisfy a recipe requirement, it may be necessary to calculate the desired amount of material.

(9) Compare values, measurements, and test indicators. If pressure exceeds a certain level, the control system may be called upon to open the vent valve.

(10) Initiate alarm, status, or operator messages.

(11) Release control (exit to the executive). If nothing more can be done until the next sequencing interval, control is released to the executive program. For example, suppose that the sequencing can continue only if flow is greater than a certain value. If not, control is released to the executive so that other systems can be processed.

The program for one reactor train, including the logic for all process and control states, typically requires from 500 to 2000 instructions of the types described above, depending on the complexity of the recipe. Also, each plant may have several different reactor trains, making different products. In a diversified plant, for example, feed preparation might be necessary to process or blend raw materials to form reactants A and B. As denoted in Table 4.2, crude products from a reactor might have to be processed in purification equipment, possibly consisting of an evaporator and condensor, to drive off reacted materials or solvents. The product would then be pumped from its intermediate storage location to a final destination in a tank farm. The domain of the control system could include everything from raw material to product storage.

When the computer encounters an instruction to calculate the amount of feed component, it goes to the recipe item that contains the ratio of the component weight to the total batch weight. The charge quantity is then calculated from the amount of product ordered and this ratio. Recipe items include:

(1) Operating parameters
  - temperatures, pressures, flows, etc.
  - batch totals
  - charge quantities
  - feed ratios
  - feed components
(2) Material routing parameters
  - tank valve numbers
  - valve header configurations
(3) Procedural words
  - hold for analysis after run
  - hold for manual operation
  - stripping required
  - rinse required

When the scheduling program assigns a batch to a particular processing train, it supplies necessary recipe items through a master recipe, which is stored in a bulk storage unit.

Once the various functions and interrelationships are defined, they must be coordinated and scheduled. The function of the scheduling program is to:

(1) Maintain and service queues of production requests.
(2) Assign batches to specific reaction trains, implementing the orders to minimize processing delays.
(3) Provide recipe information and special data needed for each batch.
(4) Allocate services and utilities based on plant priorities.

If the scheduling program finds a reactor train waiting for assignment, it looks for a production order requiring use of that processing system. Before it can assign a new task, it also checks availability of raw materials and storage space, spots conflicts with successive batches, and checks schedules for shared equipment. Obviously, a high level of sophistication in programming is required to perform all of the above tasks. However, the computer is suited to this task. Prior to the implementation of automation, significant manpower was required to perform these duties with considerably less efficient use of facilities.

In another example of process control, Allis Chalmers Manufacturing Company has developed a high-speed inspection station for sheet metal using a minicomputer. The sheet metal traveling past the unit is inspected for pinholes, large holes, and cracked edges. An X-ray gauge evaluates the thickness of the metal sheet. All of the above data are received by the computer and stored. When a predetermined length of sheet has passed the sensors and has been rolled, the computer issues a command to stop the system and cut the sheet. On command, the operator can obtain a printed or displayed summary of all of the measured characteristics of the completed roll of sheet metal, i.e., the number of feet for each thickness and the number of defects occurring for a particular segment. This summary can then serve as the quality control document, which is attached to the roll of sheet metal.

If the number of defects exceeds a predetermined level during processing, the visual display of this information can allow the operator to stop the roll immediately. If the defects are unacceptably close together, the operator can reverse the movement of the roll and rewind past the area of high defects; he then can issue a command to shear out the defective area and weld the sheet together again. In this case the minicomputer revises its record concerning number of defects and overall length of the roll.

4.3.5 Communications and Records Keeping

The real-time information processing and control required in the steel manufacturing industry can actually be structured on several "levels", each level performing certain tasks and using different computational facilities. As discussed by Kahne et al. [10], the steel-making plant can employ four levels of computers:

(1) Product order files; computer system for order processing, material requests, and production scheduling.

(2) File storage for work instructions, production status; computer system for daily work scheduling.

(3) On-line computer system for work instructions and data gathering.

(4) Input/output terminal and display panel for operator communications, on-line process control (positioning and sequencing).

This concept of multilevel data analysis and control (complexity and size of computing equipment decrease as we proceed from items 1 to 4 above) is often referred to as "distributed control". Distributed control has emerged as a viable production control methodology with the availability of inexpensive minicomputers and microcomputers.

A major problem in distributed control of such a system is that the overall performance of each level is affected by other levels. This means that the components of the control/ information system must be coordinated through a hierarchy of control computers which communicate with each other [28]. Malfunctions in equipment and urgent orders disturb the normal pattern of operation and require on-line coordination algorithms in order to keep the total system running smoothly. A discussion of how this would be done for another hierarchical system, the electric utility system, has been presented by Schweppe [23]; levels corresponding to national, regional, pool, and local utility operating centers exist, which can also interact with residential computers for load management.

For communication between individual computers, software packages, residing in a small machine, concentrate information which must be transmitted from one computer to another. A typical package would provide line control, code conversion, error control, and buffering of bi-directional message transmission, as discussed in detail in Chapter 9. Data are transmitted serially, and characters are buffered and read into the appropriate terminal buffer area. When the end of the message segment is received, the message can be queued for transmission to the central computer over a synchronous line. Symbols mark the beginning and end of an individual message.

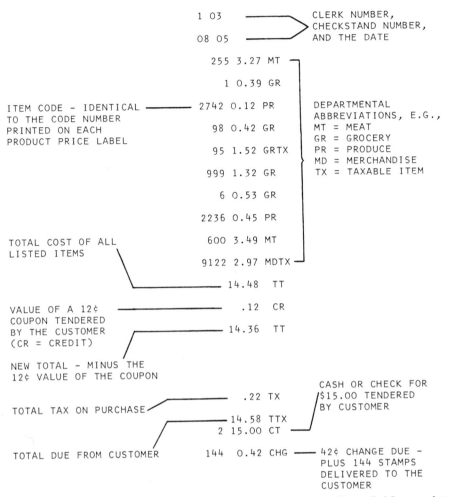

Figure 4.9. Customer's Record of Marketing Transactions at Computer-Controlled Supermarket.

The small computer also can provide an interrogation service in many applications. For example, a minicomputer programmed to operate in a question and answer mode can be used in various checkout procedures, such as in equipment testing or warehousing. Inventory control can be maintained through this means. The data so obtained can also be condensed by the computer and then sent over telephone lines to a larger central processor. The control of merchandise stock through coded labels on the packages has been implemented in many locations in the United States. Here the inventory control is performed through data collected at the checkout stand. The computer can provide an accurate record of the purchases. Recent improvement in this system has allowed the use of a portable electronic scanner to read the coded label on the commodity. The scanner automatically reads the code into the computer, which then performs the tasks mentioned earlier. A typical tape produced during processing of the contents of a grocery "basket" is shown in Figure 4.9. The net result of using a system like this is more effective management of a business, since automatic recording is provided and each store's ideal product mix (inventory) can be determined. Also, more effective cost reduction can be performed through daily and weekly sales reports by product and by department. A more detailed discussion of the use of real-time computing for data base management in inventory control has been given by Rothstein [21].

In the above example an electronic scanner was used to read coded labels. The science of picture processing and pattern recognition has now progressed to the point where, in some applications, specially coded labels are not necessarily required. For example, computerized analysis of counterfeit money by banks using such scanning devices has now become a reality. Another use of this emerging technology is in the automatic assembly and inspection of light bulbs [12]. Picture processing and pattern recognition techniques were used to process a signal provided by a television camera, to transform and extract the features of the object, to classify lead and support wires in the bulb, and to specify the coordinates of the locations of the tips of the wires. The three-dimensional location is determined by processing of two or more two-dimensional pictures. Repeatability and consistency of the system can be problematic because of dependence upon lighting, orientation of the object, and setting of the camera. However, the feasibility of such a system has been established. The extension of the general technology of three-dimensional perception also has some interesting possibilities in the robotics area [22].

4.4 THE EVOLUTION AND CLASSIFICATION OF PROCESS MODELS

Examples of real-time computing applications described above should give us some notion as to the importance of a well-defined process characterization. A model is a mathematical abstraction of a real process, and thus can often be used for computational purposes in lieu of actual process operation; i.e., we can simulate and study the behavior of the process with pen, paper, and a computer or calculator without operating the real process. The cost and time incentives offered by this approach must be weighed against the fact that the model is an abstraction, an approximate one at best, and thus does not incorporate all of the features of the real process. For example, for analytical purposes we might consider an experimental system to be at equilibrium or steady state, but in practice this condition never holds exactly,

especially in manufacturing and processing applications.

The use of mathematical models can be helpful in the following ways:

(1) Design of computer programs for real-time analysis (algorithms, logic). This is probably the most important use of a process characterization, as exemplified by the gas chromatograph and automobile engine applications discussed earlier. The algorithm is based on the calculations required and can be tested independently of the process operation.

(2) Optimization of process operations. One area where this has become a very significant activity is in supervisory process control. In a refinery, it is desirable to adjust blending operations for production of gasoline, fuel oil, and jet fuel when the physical properties of the crude oil feedstock change. A technique called linear programming, which is readily implemented on a small computer, is used extensively in the process industries to find the optimum mix of products in order to maximize profits and minimize energy consumption. Other applications of on-line optimization would include job shop scheduling and inventory control.

(3) Selection of control parameters in process control. In designing controllers to be implemented by a computer, it is usually necessary to select the controller parameters based on a model of the process [13, 15] because the selection process often is iterative, making it expensive to implement with a real process. Hence model calculations are a very cost-effective means of obtaining suitable design values.

(4) Determination of sensor locations. Sometimes it is not obvious how measurements should be selected for a given process. A trial-and-error solution using the real plant would be inordinately expensive and time-consuming. Therefore we can resort to simulation of the process to determine the best data acquisition strategy. This is especially true for the so-called distributed parameter processes (discussed below), such as monitoring

systems for air and water quality, where a potentially large number of measurement locations is required. Our intuition is often quite poor with complicated systems; hence simulation is a viable strategy for analysis.

(5) Determination of appropriate sampling frequencies. In order to estimate the best sampling rate, we must recognize that high sampling rates require greater use of the computer but yield high information content, while low sampling rates sacrifice information content for lower computational requirements. The transient behavior of the process usually determines the necessary sampling rate for retention of useful process information, and this rate can be evaluated through a process model, as discussed in Chapter 19. Filtering of process data must be approached in the same analytical way as for sampling.

The systems approach can be problematic, and there are many idiosyncrasies that must be dealt with in each disciplinary area. One benefit which derives from the systems approach is that we obtain a better understanding of cause and effect in the process, and this invariably leads to a more successful application.

4.4.1 Analysis and Development of Process Models

To be able to analyze a complicated sequence of events occurring in time, such as takes place when a computer is operating with a process, we need to introduce (or review) briefly the concept of "cause and effect". The concept of cause and effect is well known to every high school science student and, fortunately, we will be able to operate largely with such intuitive ideas. It is true that engineers and mathematicians have extended these ideas vastly; however, we largely will be satisfied to agree that, (1) for every observable effect there has to be some cause, which may or may not be measurable; (2) a given cause will yield the same effect, assuming that all other operating parameters are the same; and (3) the "effect" cannot precede the "cause" (the system is nonanticipatory).

These three cause and effect rules are intended to be applied both in analysis and design (or synthesis) of real-time systems; they are particularly important in developing process characterizations. A second problem involving cause and effect is to identify which process variables are inputs and which are outputs; this is especially important and troublesome with multivariable systems.

At the very simplest level of operation, a real-time computer is used to monitor the status of an operating process—to determine when, in what sequence, and how events occur within the process, and to measure quantitatively what is going on within the process. Both of these types of operation are characterized by the largely passive effect of the computer on the process; the computer, while taking data describing operation of the process, is not expected to affect process operations. However, we know that it will be impossible, even though desirable, to connect a computer to a process in such a way that process operations are totally undisturbed by the computer and its associated instrumentation. Nevertheless, we assume that data-logging types of operations such as those described above are essentially passive. Except for processes that function on the microscopic level, we can approach this ideal state by the proper use of instrumentation. In addition, we can characterize the process as if it were independent of the computer.

Another important issue to be addressed in building a process model is that of model complexity. A series of questions can be posed in this regard:

(1) Should the process be modeled on a macroscopic or microscopic level? Any process is driven by molecular action, and one must decide if such a microscopic description serves any purpose. In the field of fluid mechanics, the derivations of the equations of continuity and motion are based on the continuum postulate, i.e., that matter can be treated macroscopically. This obviates the need to describe activity on a molecular level. In a chemical kinetics experiment, the solution of the Schrödinger equation is not

usually needed to describe the data that will be obtained. Conversely, there are some applications where molecular considerations are important, such as in monitoring small particles in a pollution control context or in measuring molecular diffusion at extremely low pressures.

(2) Is the process too complex, even on a macroscopic level, to be completely described by the principles of chemistry and physics? For example, the existence of many chemical components and reactions does not lend itself to exact analysis. The presence of turbulence sometimes tends to confound any compact mathematical description.

(3) What is the desired accuracy of the model? How does the accuracy of the model affect its ultimate use? It has been a common practice in process control to use very simple and inexact models for determining controller settings. This practice has worked out to gain, as one industrialist has remarked, "ninety percent of the profit with ten percent of the effort".

(4) What measurements are available? What data are available for model verification? A model utilizing only measurable quantities is highly desirable, but it is difficult to attain in practice. Often unknown variables can be indirectly calculated by measuring available variables for several different sets of operating conditions.

(5) Is the complex system actually composed of small, simple subsystems? If such subsystems can be analyzed readily, then the complexity of model building is significantly reduced.

These first five items are common to any model-building exercise, both real-time and otherwise. For real-time computing applications we must answer yet a sixth question regarding model complexity:

(6) How are speed of computation requirements affected by the model complexity? This is related somewhat to item (2) above; a simple model will always reduce the computational requirements. If a decision needs to be made in real-time, the speed of computation must be faster than the speed of the process. How one estimates the "speed" of the process is discussed in Chapter 19.

4.4.2 Physical and Chemical Principles/Black Box

Once some knowledge of the necessary model complexity has been obtained, we must then decide upon the mode of model development: use of the black box approach or physical and chemical principles. The black box approach attempts to relate empirically the effect of a given input on a specific output. The acquisition of input–output data from the process of interest is a necessity in this case. The data are then correlated in some mathematical form, yielding the desired model. One of the most popular approaches used to obtain black box models is polynomial curve fitting, employing a least-squares error criterion. This method is very easy to use, and, because of the quadratic error criterion utilized, it merely requires the solution of n linear equations in n unknowns. For n unknowns (C_0, C_1, ... C_{n-1}), a polynomial equation of the form,

$$X = C_0 + C_1 U + C_2 U^2 + \cdots + C_{n-1} U^{n-1}$$

(4.1)

would result, where U is the input and X is the output. Such equations are often used to relate physical properties of a process reactant or product to the state variables of a process such as temperature and pressure. For transient systems, the unknown constants are somewhat harder to obtain, but there is sufficient computational experience in this area that no important obstacles exist [4, 24]. Note that in some cases, a qualitative understanding of how the process functions is adequate, and no explicit mathematical model such as Equation 4.1 is required. In other cases, such as discussed in Section 4.2, the process itself can serve as the "model".

By using the principles of chemistry and

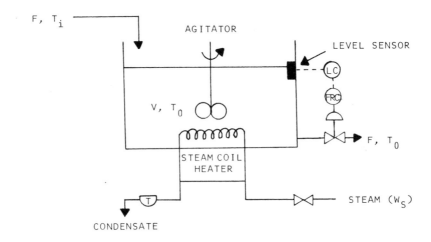

Symbols:

----	electrical or pneumatic instrument
FRC	flow recorder-controller
LC	level controller
T	steam trap

Figure 4.10. Agitated Heating Tank

physics, one can in theory construct a physical model prior to collecting process data. However, it is rarely the case that the original model does not need to be modified once actual process data have been acquired. The chief advantage of these models is that one can readily interpret the results in terms of the process properties, whereas a black box model contains only numerical values for its coefficients. In addition, a black box model is usually accurate only over limited ranges of operating conditions, while a physical model is applicable over wide operating ranges.

To illustrate the physical model approach, consider a heating system where the fluid to be heated is passed through a stirred tank with an internal heating coil containing steam (Figure 4.10). Suppose we have a flow valve that automatically controls the level in the tank at some specified value. For example, if the feed rate to the tank increases for some reason, the outlet valve opens slightly to compensate for the change. We can write a steady state or static material balance for this problem, but it has a trivial answer, i.e.,

$$\text{flow rate in} = \text{flow rate out} = F$$

We can also write steady state energy balance for the fluid:

$$\text{energy in} = \text{energy out}$$

Using constant heat capacities (C_p) to express energy content of the fluid with respect to some reference temperature $\overline{T}$, we can approximate the energy balance as

$$FC_p(T_i - \overline{T}) + W_s\lambda = FC_p(T_0 - \overline{T}) \quad (4.2)$$

or:

$$FC_pT_i + W_s\lambda = FC_pT_0 \quad (4.3)$$

where W_s is the steam flow rate and λ is the heat of vaporization of the steam. The above equation is in effect a physical model relating the major variables (F, T_i, W_s, T_0) as a function of physical properties of the fluid (λ, C_p). It is clear that the determination of this relationship for well-known fluids (e.g., water) would be rather easy; hence this approach is certainly preferable to fitting a large amount of data to a general equation which may be applicable only for a narrow range of conditions and not subject to extrapolation.

There is an intermediate approach in process modeling which lies between black box and physical modeling methods. This technique, often called phenomenological modeling, utilizes general equations of the form of Equation 4.3, but leaves some of the parameters unspecified or undetermined. For example, given values of F, T_i, W_s, and T_0 for several sets of operating conditions, we could calculate values for the "unknown" parameters λ and C_p. This method avoids an independent determination of these parameters as well as compensating for errors in the physical model (e.g., incomplete mixing, two-phase flow through the steam trap, etc.). This approach is usually quite successful, for it combines the fundamental theory, which is so necessary for interpretation of the process, with empiricism, which usually is required to compensate for small modeling errors.

The in-depth development of process models is necessarily touched upon lightly in this chapter. This discussion and that in the sequel is not intended to provide instruction on how models are developed, since the identification of both black box and physical/chemical models is a rather lengthy subject. The interested reader should refer to the textbooks by Eykhoff [4], Himmelblau and Bischoff [8], Luyben [13], Seinfeld and Lapidus [24], Martens and Allen [15], and Bryson and Ho [1] for explicit treatment of the development of process models and the estimation of unknown parameters.

4.5 MATHEMATICAL MODEL STRUCTURES

There are a number of different model dichotomies which must be considered in selecting the structure of a mathematical model. Below we discuss five structural categories and give some examples of their usage.

4.5.1 Algebraic/Differential

The first type of model dichotomy is algebraic/differential, or static/dynamic. Static systems remain at, or very close to, a steady state, such as the stirred tank example. The state variables (force, concentration, temperature, etc.) are in equilibrium with each other and do not change with respect to time. A system that exhibits

both static and dynamic behavior can be expressed as

$$dx/dt = f(x, u) \qquad (4.4)$$

where x is the output and u is the input. The time-varying behavior of x(t) can be found by integrating Equation 4.4 for a given u(t). In the previous stirred tank example, we could allow for an accumulation of energy in the tank itself. If M is the mass of fluid in the tank, which is held constant as a consequence of the level controller, then the unsteady state energy balance can be written as

energy accumulation = rate energy in
$$- \text{rate energy out}$$

$$MC_p \frac{dT_0}{dt} = FC_p(T_i - \overline{T}) + W_s\lambda$$
$$- FC_p(T_0 - \overline{T}) \qquad (4.5)$$

Note that the temperature of the fluid in the tank is assumed to be homogeneous and at T_0 the temperature of the effluent stream. This is a first-order nonhomogeneous ordinary differential equation, and thus can be solved analytically, since the variable T_0 appears linearly in the equation. This type of system is commonly referred to as a "first-order system" or "first-order filter". More details on the use and solution of linear ordinary differential equations, particularly for the important task of specifying an appropriate sampling time for the process, are given in Chapter 19.

4.5.2 Continuous/Discontinuous

The second type of model dichotomy is continuous vs. discontinuous. The model given above is a continuous time representation. However, the use of real-time computers in conjunction with a process requires a basic change in the analytical approach for the overall system (computer plus process). A process that is continuous in time must be analyzed in discontinuous or discrete time terms once the computer is connected. A digital computer does not deal with pure analog signals, but must sample the process signals periodically and then convert the signals into a digital (discrete) representation. It certainly can be argued that as the number of digits stored in the computer increases and the sampling time becomes infinitesimal, the analog analysis is virtually equivalent to the digital one. However, equipment speed, storage, and accuracy limitations dictate that the discrete time process and the continuous time process must differ somewhat. Thus, there is some motivation for understanding how a discrete model is related to a continuous process characterization. A discussion of the mathematics of sampled-data systems is given in Chapter 19.

The key to the conversion from continuous to discontinuous is the analog to digital converter. The A/D converter is analogous to a car's speedometer, in that a continuous signal is converted to another signal which assumes a discrete or integer value; in this case the driver interprets the signal as an integer-valued miles per hour (because the speedometer is not calibrated to decimal fractions). The electronic operation of the A/D converter is discussed in detail in Chapter 10; other aspects of discrete-valued signals are discussed in Chapter 2.

An example of a discrete equation, one which often is used for smoothing or filtering process data, has the form

$$x_{k+1} = \alpha x_k + (1 - \alpha)u_k \qquad (4.6)$$

where x_k is the filtered variable at discrete points in time, k and k + 1, u is the raw data point received at time step k, and α is a weighting parameter which can be set between zero and one. The digital filter takes a weighted average of the current measurement (u_k) and the most recent filtered measurement (x_k) to obtain the new "filtered" measurement, x_{k+1}. This new filtered value is then used in the generation of the filtered value, x_{k+2}, for a new raw data point, u_{k+1}. Filtering algorithms are discussed in more detail in Chapter 19.

4.5.3 Deterministic/Stochastic

All systems discussed until now have been assumed to be deterministic in nature; however, many important real-time applications require

the interpretation of a stochastic process, where some of the variables have a random nature, i.e., do not appear to follow "Rule 2" of the cause and effect relationships discussed in Section 4.4.1. With deterministic models, each parameter is a fixed number for a specified set of operating conditions. We assume that there is no inherent randomness in the mathematical equation, although the error between process and model may assume a probabilistic character. Processes such as those involving radioactive particle emission are distinctly stochastic, as is the interpretation of fluid turbulence. The latter phenomenon is especially important in the monitoring and analysis of air and water pollution phenomena. The modeling of such processes requires a deterministic component and a probabilistic component; the probabilistic component often is specified as some mean value, $\bar{x}$, with standard deviation σ, assuming a Gaussian (normal) probability density function. Often white noise is added directly to the process model (to indicate our ignorance of the precise mechanism of the process noise); we also can add a noise term directly to the model for our measuring device. Such a term can compensate for unexplained variations in the accuracy of the measurements.

Another application of statistical models is in the utilization of statistical control charts [7] for monitoring process data; these charts determine if a newly-recorded data point is statistically consistent with previous data. Since all real processes exhibit some drift and noise, we need to be able to determine whether a new data point indicates a "change" in the process. A control chart is a type of black box model, which does not rely on the physical parameters of the process. A real-time computer can easily perform the necessary statistical tests of significance, update means and variances, and output a signal to the operator when the process appears to experience an abnormality.

4.5.4 Ordinary Differential/Partial Differential

Another dichotomy to be discussed in the analysis of mathematical model equation structure is the use of ordinary vs. partial differential equations (o.d.e. vs p.d.e.). This distinction lies merely in the number of independent variables. O.d.e.'s contain a single independent variable (usually time in real-time computing applications), while p.d.e.'s contain more than one independent variable (usually time and geometrical dimensions in real-time applications). In characterizing a process such as the heated stirred tank, we may have some conceptual difficulty in eliminating spatial variation in temperature from consideration, since there will at least be turbulent fluctuations in temperature from point to point; however, in order to simplify the analysis, we choose to consider such a system as having lumped parameters, so that the variables represent some aggregate or average behavior for the complete tank.

One of the difficult problems facing us when a real-time computer is to be applied to a distributed process is the selection of the number and location of the measurement points. If a process variable is changing in the spatial direction, we must track it at several points. Consider the problem of a city natural gas distribution network, which is subject to upsets from both the demand and the supply side. We might wish to monitor the gas pressure and flow rate as a function of position and time; it is difficult to decide where such monitors should be placed so that a potentially unsafe situation can be prevented or certain parts of the town not be cut off from the gas supply due to large demands in other parts of town. The problem of monitoring pollutant levels in a river basin is very similar; here we must consider variations in three spatial dimensions as well as time.

4.5.5 Sequential/Parallel

The final model structure that is of interest in real-time computing is sequential vs. parallel. This dichotomy is pertinent primarily because of the electronic circuitry employed in digital computer systems. However, a secondary consideration is that methods of analyzing both the static and dynamic behavior of such systems are facilitated by the sequential or parallel nature of the system. Block diagram analysis, which is illustrated in Chapter 19, is a technique that relies quite heavily on converting the system description into a group of parallel or series "block" operations. This technique is also applicable to the "process" side of real-time computing; hence it is an extremely valuable tool for visualizing complex interrelationships in space or time.

Having given a framework for classifying the different kinds of mathematical equations used for process modeling, the general discussion should be concluded with a caveat regarding model verification. One should be very careful that a model developed from physical principles conform with reality; it is not uncommon that when physical parameters are estimated from acquired experimental data (and indirectly computed based on the postulated model), they do not agree with direct measurements of the parameters. For example, in petroleum reservoir engineering, we can estimate parameters such as porosity and permeability of the reservoir porous media either by running a pressure test on a production well or by obtaining more expensive core samples and running laboratory tests. The disparity of results between the two methods continues to be a persistent problem.

4.6 SUMMARY

In this chapter a number of real-time applications, spanning several disciplines, have been presented. These applications, which illustrate the versatility of real-time computers, can be classified as operator-dependent or operator-independent. In addition, several classifications dealing with the nature of the application, i.e., original equipment component, monitoring, stand-alone, process control, and communications, have been presented.

The need to develop process characterizations for application of real-time computing to data acquisition and process control has been defined in this chapter. Modeling of actual processes must be undertaken with some care, and this chapter has defined an overall modeling structure so that process characterization can be performed in an organized fashion. The all-important first step in such a procedure is to define the system boundaries and the system inputs and outputs. Once this is done, the philosophical question of model complexity must be posed. Thereafter, the model classification must be determined. With this classification, one can utilize standard mathematical techniques, as described in the references, to arrive at the model.

However, it is misleading to think that modeling is a cookbook procedure that requires no judgment. In fact, the combination of intuition and judgment with mathematical tools is the most successful approach for developing process characterizations. It has been said that, with enough adjustable parameters, a person can curve-fit anything. However, by using good judgment and intuition, one can obtain a process model that is meaningful, offers critical insight into the operation of the system, and exhibits parsimony in the mathematical description.

4.7 REFERENCES

1. Bryson, A. E. and Ho, Y. C., *Applied Optimal Control*, Blaisdell, Waltham, Mass., 1969.
2. Coleman, D., "Sensors for Automotive Engine Control", *Automotive Engineering*, *85* (5), 36, 1977.
3. Ernst, R., Freeman, R., Gestblom, B., and Lusebrink, R., "Detection of Very Small Nuclear Magnetic Resonance Spin Coupling Constants by Resolution Enhancement", *Molec. Phys.*, *13*, 283, 1967.
4. Eykhoff, P., *System Identifiction—Parameters and State Estimation*, Wiley, New York, 1974.
5. Frazer, J. W., Rigdon, L. P., Brand, H. R., and Pomernacki, C. L., "Characterizing Chemical Systems with On-Line Computers and Graphics", *Anal. Chem.*, *51*, 1739 1979.
6. Gladney, H. M., Dowden, B. F., and Swalen, J. D., "Computer Assisted Gas-Liquid Chromatography", *Anal. Chem.*, *41*, 883, 1969.
7. Himmelblau, D. M., *Process Analysis by Statistical Methods*, Wiley, New York, 1970.
8. Himmelblau, D. M. and Bischoff, K. B., *Process Analysis and Simulation*, Wiley, New York, 1968.
9. Jurgen, R. K., "Minicomputer Applications in the Seventies", *IEEE Spectrum*, p. 37, Aug. 1970.
10. Kahne, S., Lefkowitz, I., and Rose, C., "Automatic Control by Distributed Intelligence", *Sci. Am.*, *240* (6), 78, 1979.
11. Kendall, D. C. and Howard, J. R., "How to Approach Direct Digital Control for Batch Processes", *Inst. Tech.*, p. 47, Oct. 1967.
12. Lin, W. C. and Chan, C., "Feasibility Study of Automatic Assembly and Inspection of Light Bulb Filaments", *Proc. IEEE*, *63*, 1437, 1975.
13. Luyben, W. L., *Process Modeling, Simulation, and Control for Chemical Engineers*, McGraw-Hill, New York, 1973.
14. MacDonald, H. C., "On Board Computers May Provide Dash Display Inputs", *Automotive Engineering*, *87* (7), 50, 1979.
15. Martens, H. R. and Allen, D. R., *Introduction to Systems Theory*, Merrill, Columbus, Ohio, 1969.
16. Obert, E. F., *Internal Combustion Engines*, International Textbook, New York, 1950.
17. Osborne, J. E., "Dedicated Computer for Pilot Plant Analyzers", AIChE Meeting, Tulsa, Okla., April 1973.
18. Perone, S. P. and Jones, D. O., *Digital Computers in Scientific Instrumentation*, McGraw-Hill, New York, 1973.
19. Reynolds, E., "Design Optimization by Adaptive Control", Master's thesis, Department of Mechanical Engineering, The University of Texas at Austin, Austin, Tex., 1972.
20. Roberts, S. M., Williamson, D. H., and Walker, L. R., "Practical Least Squares Approximation of Chromatograms", *Anal. Chem.*, *42*, 886, 1970.
21. Rothstein, M. F., *Guide to the Design of Real-Time Systems*, Wiley-Interscience, New York, 1970.
22. Rosenblatt, A., "Robots Handling More Jobs on Industry Assembly Lines", *Electronics*, *46*, 93, 1973.
23. Schweppe, F. C., "Power Systems 2000: Hierarchical Control Strategies", *IEEE Spectrum*, p. 42, July 1978.
24. Seinfeld, J. H. and Lapidus, L., *Process Modeling; Estimation and Identification*, Prentice-Hall, Englewood Cliffs, N.J., 1974.
25. Shah, M. J., "Automation of Remote Instrumentation with an On-Line Process Control Computer", 64th Annual AIChE Meeting, San Francisco, Cal., Nov. 1971.
26. Trayser, D. A. and Creswick, F. A., *Effect of Induction System Design on Automotive Engine Emissions*, ASME, New York, 1969.
27. Vora, L. S., "Computerized Engine Mapping", *Automotive Engineering*, *85* (4) 54, 1977.
28. Wilhelm, R. G., "A Methodology for Auto/Manual Logic in a Computer-Controlled Process", *IEEE Trans. Auto. Cont.*, *AC-24*, 27, 1979.

4.8 EXERCISES

1. In Section 4.2, an advanced electronic engine control system was described. The measurements employed include manifold pressure, engine speed, crankshaft position, coolant temperature, air temperature, throttle position, and exhaust gas composition.
   a. Discuss how you would use the information from each measurement to improve

the engine performance. (You need to define some criterion or criteria for engine performance.)

b. What kinds of mathematical models would be needed to relate measurement variables to performance?

c. Classify each model discussed in (b) using the scheme presented in Section 4.5.

2. The sequential logic of batch process control described in Section 4.3.4 entails 11 steps. Describe what measurements would be needed for each step, and give examples of FORTRAN or BASIC commands that might be employed.

3. Consider a liquid storage tank with a constant displacement effluent pump and an entering stream of w_i kg/hr. The pump speed can be adjusted to yield an effluent rate of w_0 kg/hr.

a. Write a differential equation for the change in the level (h) as a function of time when $w_i \neq w_0$.

b. For an initial level (h) of 5 meters, $w_i = 6$ kg/hr and $w_0 = 5$ kg/hr, plot the change in level as a function of time. Assume that the tank is cylindrical with a circular (end) cross section of 3 square meters, and the liquid is water. If the height of the tank is 10 meters, at what time will the tank overflow?

c. How could you control the tank level so that it stays close to 5 meters?

5

Measurements, Transmission, and Signal Processing

Joseph D. Wright
Xerox Research Centre
Mississauga, Ontario

5.0 INTRODUCTION

The successful operation of any system comprising processes and real-time computers, whether the objective is data acquisition or process control, depends critically upon the measurement and transmission of process data. This chapter is intended to give an overview of the entire measurement system from the process to the computer and, for control applications, back to the process once again.

It is important to consider each aspect of a measurement system in its proper perspective. That is, the output of the overall system, say a data acquisition system, can be no better than the quality of the measurements. This is not simply a matter of choosing good transducers. The means by which the signals from transducers are carried to the computer can also greatly affect system performance. Finally, signals arriving at the computer interface must be properly conditioned in order to give a satisfactory reading to the analog to digital converter. The problem does not even stop here. Signals measured by the computer are inherently sampled. The sampling process alters the signal and cannot help but lose some useful information. Proper design of any data reconstruction algorithm is also important in a sampled data system. Figure 5.1 shows schematically the components of a measured system.

The components of a measured system can be broken down into a sequence of seven subsystems between the process and the computer:

Transduction—the conversion of a physical property or measurand in the process into some alternative form, usually electrical for subsequent transmission to the computer system.

Transmission—the means by which the output of the transducer or transmitter is brought to the computer system. This may be done by single wires, pairs of wires, or other techniques such as telemetry.

Termination—The wires bringing electrical signals to the system must be terminated in some fashion to complete the required electrical circuits.

Signal Conditioning—In general this term applies to any sort of signal modification except amplification. Various modifications such as filtering, attenuation, level shifting, linear or nonlinear compensation may be applied to the signals.

Multiplexing—Large numbers of signals of different types must often be processed by the computer. Various techniques for switching these multiple inputs so that they may share a fewer number of system input channels must be applied. Multiplexers may be high- or low-speed and may also switch high (voltage)-level or low-level signals either before or after amplification.

Amplification—Many transducer signals are of too low a level to make accurate use of analog to digital conversion units on computer systems. They must be amplified to make optimum use of the input facilities. Various other functions provided by amplifiers in-

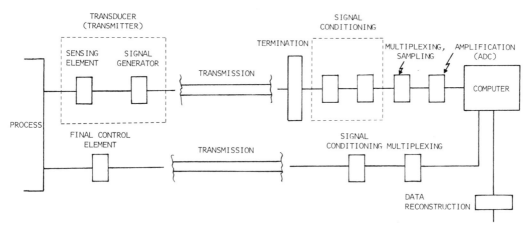

Figure 5.1. Overall Process-Computer System

clude buffering, for impedance matching, and sampling, for critical sample-and-hold timing conditions.

Analog to Digital Conversion—The analog to digital converters (ADCs) provide a digital representation of the analog signal for use in the computer systems.

Each of the above subsystems with the exception of analog to digital converters is discussed in more detail in separate sections. Analog to digital converters along with digital to analog converters are discussed in Chapter 10. Finally the loop is closed from the computer back to the process with a discussion of output signals and final control elements so that the analysis and computation done as a result of the process measurements may be brought back to affect the process operation.

Discussions of sampled data analysis, data reconstruction and sampled data control are presented in Chapters 19 and 20.

There are a number of useful reference books for this chapter. In particular, the reader should be aware of the three volumes by Liptak [8-10], entitled *Instrument Engineers Handbook*, the volume by Harrison [5] *Handbook of Industrial Control Computers*, the *ISA Transducer Compendium* by Minner [12] and Harvey [6], *Process Instruments and Controls Handbook* by Considine [3], the *Handbook of Applied Instrumentation* by Considine and Ross [4], *Instrumentation in Process Control* by Wightman

[14], and many more specific references for particular instrumentation applications.

5.1 TRANSDUCERS AND TRANSMITTERS

Although a very wide variety of process variables may be measured, the great majority of process data acquisition systems are concerned with measurements of temperature, pressure, density, flow, pH, displacement, and concentration. Mechanical systems often require transducers measuring displacement, velocity, and acceleration; and electrical systems may require measurements of voltage, current, and power. An excellent and thorough discussion of the transducers required for these applications and many more is provided by Liptak [8-10]. Standard manufacturers are also listed. In this section we restrict ourselves to a general description of transducers and their properties.

Almost all transducer outputs are analog signals, although some have outputs of resistance, capacitance, or inductance from which useful voltage or current signals must be obtained. Basically, a continuous input from the process results in a continuous output in the form of a voltage or current from the transducer. In many cases this output is linearly proportional to the value of the process variable although a notable exception occurs with some types of flow measurements in which a square-root relationship results. It is important to realize, however, that some types of measurements are inherently digital in nature, particularly with the advent of digital voltmeters and other types of counting instrumentation. In these, the transducer output is in discrete form, either as a series of pulses or as a binary or coded binary number.

For most types of process measurements a range of transducer outputs is possible. The continuous ones include signal levels from dc millivolts (thermocouples), 1–5 milliamps (ma), 4–20 ma or 10–50 ma (the standard industrial transmitter ranges), 1–5 volts dc, 0–10 volts dc, and a series of various types of ac output with frequency varying as a function of the input. The problem that arises in a general data acquisition system is that no great degree of standardization is obtained, and many different signal types must be handled. We will restrict most of the discussion to transducers that provide an electrical output, since we must provide an electrical signal to the computer. Pneumatic instrumentation in many process installations can be interfaced through the use of pneumatic to electrical converters.

Most electrical output transducers consist of two parts: the sensing element and the signal generator. When the pair operate together we normally refer to the unit as a transmitter. Some of the most basic types of transducers are used to measure changes in physical properties directly. Typically this property would be related to an electrical property such as resistance or inductance. The obvious examples are thermocouples, resistance thermometer bulbs, and thermistors. Other examples include piezoelectric crystals which are sensitive to distortion and photocells which are sensitive to light intensity; capacitance devices which generate changing dielectric properties; and so on.

Certain fundamental transducers reflect changes in chemical properties. The best examples of this are pH electrodes, but electrical conductivity cells in which the conductivity is proportional to ion concentration also qualify.

More complex transmitters are based upon principles of force or motion balancing devices which generate electrical signals proportional to these properties. In a force balancing transmitter, for example, a ferrite disk is displaced by a pressure measurement. This displacement

moves it closer to a transformer coil, causing an imbalance in an amplifier circuit. The imbalance in turn is used to drive a force motor which counteracts the pressure on the ferrite disk until the two are in balance.

Position balancing transmitters are used in a similar fashion except that the motion caused by the process variable change (pressure, perhaps) is directed against a calibrated spring.

Industrial transmitters for pressure, temperature, flow and other common measurements usually provide a standard output signal which is either 4–20 ma or 10–50 ma and can be taken directly to the computer room. Laboratory transducers more often provide a voltage output which is in either the range 1–100 mV or 0–10 volts. The former are provided by standard chromatographs or other composition measuring devices, whereas the latter are often provided by laboratory type pressure transducer/transmitters. If instrumentation amplifiers, which are discussed in Section 5.6, are used, we most often provide an output signal that is directly compatible with the analog to digital converters.

It is beyond the scope of this chapter to attempt to discuss in detail the many different kinds of transducers and transmitters available. However, since temperature and pressure measurements form the basis for so many systems, it is worthwhile devoting some discussion to these particular measurements as examples.

5.1.1 Temperature Measurement

Although there are many different temperature measuring devices available, we will restrict discussion to the basic transducers which can provide electrical signals, i.e., thermocouples, resistance bulb thermometers, and thermistors.

Thermocouples. The basic principle of operation of thermocouples is that when two dissimilar metals are connected as shown in Figure 5.2, where the two junctions are maintained at different temperatures, an electric current will flow in the circuit. If the temperature of the reference junction is known, the emf which causes the current to flow is a measure of the temperature difference between the measure-

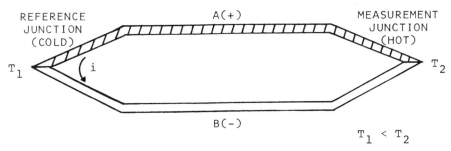

Figure 5.2. Thermocouple Seebeck Effect or Thermoelectric emf

ment junction and the reference junction. Despite the fact that an extremely large number of combinations of metals could be used, only a few are in standard use. These are discussed in any book on thermocouples, but a detailed discussion may be found in Baker et al. [1,2].

Because thermocouples produce very low-level signals and also because *any junction* of dissimilar metals produces a thermoelectric effect, great care must be taken in using thermocouples for measurement systems. Several fundamental laws are very useful.

(1) Law of the Homogeneous Circuit: With homogeneous conductors the thermoelectric effect for a given pair of conductors is a function only of the temperatures at the junctions and not of the wires themselves. Hence, we need only be concerned with junctions (connections).

(2) Law of Intermediate Temperatures: If two pairs of thermocouples are used as shown in Figure 5.3 where the intermediate temperature, T_2, is common to the two pairs, then this is equivalent to one pair of thermocouples measuring the emf between T_1 and T_3.

(3) Law of Intermediate Metals: If a thermoelectric circuit containing wires A and B

is broken and a third wire C is added, the thermoelectric emf is not changed if the two new junctions are held at a constant temperature even though this temperature is different from the reference temperature and the measurement temperature. This is shown in Figure 5.4.

This final law is probably the most significant, since inadvertently it affects every thermocouple system. Typically, we would wish to measure a process temperature as shown in Figure 5.5 relative to some reference temperature. Most often in a laboratory environment the reference temperature is an ice bath (or nowadays an electronic equivalent) somewhere near the process. We might then wish to carry the emf some distance to a digital voltmeter or to a computer system using standard copper wire as opposed to thermocouple wire. Provided the connections shown at point C are maintained at a constant and equal temperature, no problem will result. However, if these two connections are at different temperatures, a thermoelectric signal will be generated. Additional considerations must be taken into account for the copper wire such as using twisted shielded conductors. This is discussed in Section 5.2.3. A practical limit on length of cable is about 200 feet, although much longer

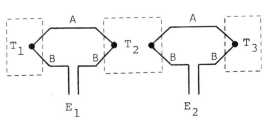

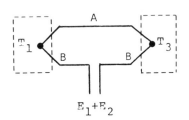

Figure 5.3. Law of Intermediate Temperatures

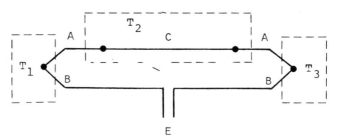

Figure 5.4. Law of Intermediate Metals

runs can be made provided one uses extremely good wiring practice.

For computer control applications in large installations it is clearly impossible to maintain ice baths for the reference temperature. In these cases all thermocouples are brought to a common reference block (junction) usually near the computer room. The block is maintained at a constant temperature and is thermally connected to, but electrically isolated from, the binding posts for the thermocouples. The block temperature is measured separately, usually by a resistance thermometer bulb. This case is illustrated in Figure 5.6. Note that if one of the thermocouple materials is copper, then only one reference junction is required. It is also possible to purchase compensated reference temperature junctions from a number of suppliers of thermocouple transmitters. More will be said about this in Section 5.6.2 after some of the current trends in multiplexers and instrumentation amplifiers are discussed.

Resistance Thermometers. The principle of temperature measurement by resistance thermometers (RTDs) is based upon the variation of resistance with temperature in certain metals, commonly platinum, nickel, tungsten, and copper. Essentially a length of the active wire is encased in an enclosure (probe or disk) with two, three, or four connecting leads for hookup to the readout device which is commonly a wheatstone bridge or equivalent. A typical system is shown in Figure 5.7. If an RTD were used to monitor the temperature reference block, then both the power supply voltage and the bridge imbalance voltage would have to be transmitted to the computer system.

Thermistors. Certain mixtures of semiconductor materials have very high temperature coefficients for resistance as a function of absolute temperature. The temperature coefficients are usually negative, and the sensitivity is very high, making them extremely useful for narrow span temperature measurements. Differential temperatures as small as $10^{-3}\,°F$ can be measured. The wheatstone bridge circuit shown in

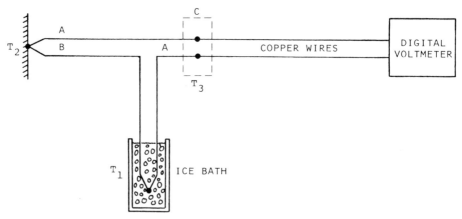

Figure 5.5. A Typical Thermocouple System

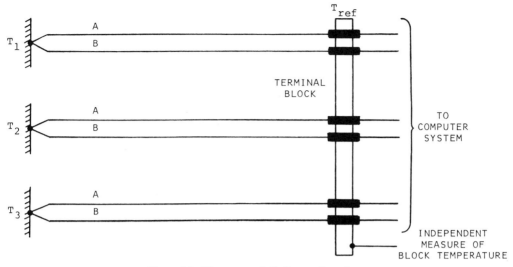

Figure 5.6. Thermocouple Reference Junction

Figure 5.7 can be used to measure thermistor output, but a simpler circuit using a microammeter may also be used. Care must be taken to pass only a very low current through the thermistor or it will heat up, changing its own resistance and hence any inferred temperature. One problem with thermistors is that, although the electrical dependence on temperature for a 'particular device does not change, the intercept can vary greatly, causing a need for recalibration. Commercial RTDs tend to be better in this regard than individual units incorporated in simple laboratory systems.

Semiconductor Temperature Sensors. Recent developments in semiconductor technology allow temperature measurements using integrated circuit devices. An example of this technology is provided by the LM/135 series of precision temperature sensors manufactured by National Semiconductor. These devices are very inexpensive and operate over the range $-55°C$ to $150°C$ with an error claimed at less than $1°C$ over a span of $100°C$. The units must be waterproofed to operate in liquid environments and also require operational amplifiers to complete the circuits. Nevertheless they provide interesting alternatives to thermocouples for some applications.

5.1.2 Pressure Measurements

Many different types of pressure measuring devices exist, but discussion in this section will be confined to electronic pressure sensors. As with thermocouples and temperature measure-

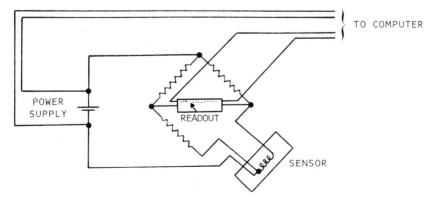

Figure 5.7. Resistance Thermometer-Wheatstone Bridge Circuit

ments, a detailed presentation is available in the references on measurements, notably Liptak [8–10]. Most pressure transducers combine a bellows or diaphram of some sort with a sensing element.

Strain Gauge Transducers. Strain gauges are metallic conductors that change their resistance when subjected to mechanical strain. Strain gauges may be either bonded to a diaphragm directly or mounted on a mechanical frame which moves when subjected to a force. The most common sensing element is a wheatstone bridge circuit as shown above, but with each arm of the bridge containing a strain gauge element. Strain gauge transducers can be used to measure torque, weight, horsepower, velocity, or acceleration in addition to pressure, a versatility that makes them a very common sensing element.

Electronic Pressure Transmitters. Many commercial pressure transmitters make use of variable reluctance transformers, differential transformers, or variable capacitance units instead of strain gauges. The variable reluctance transformer units have a ferrite core which is moved relative to a transformer carrying an oscillating signal. The motion changes the inductance of the system which is sensed by a feedback system running a force drive motor. The motor moves until its force balances the applied force; the motor current then gives an indication of the pressure. Differential transformer systems operate in a similar fashion. Capacitance type pressure transmitters use a high frequency oscillation to energize the sensing elements which consist of one or two capacitor plates. Relative

stiffness of the plates determines the pressure range of the unit. An applied pressure will change the gap in the capacitive plates, and the change in capacitance in the oscillating circuit may be detected by a bridge current.

Solid State Pressure Transducers. There are two basic types of solid state transducers for pressure measurements. One depends on the piezoelectric effect which is basically very similar to the effect in strain gauges except that the material is a ceramic instead of a conductor. Very high sensitivities can be generated by these materials. The other type of transducer is a pressure-sensitive transducer known as a Pitran,* which is basically a transistor having its emitter base junction mechanically coupled to a diaphragm to which pressure is applied. The pressure causes changes in the electrical characteristics of the transistor. A very simple circuit is illustrated in Figure 5.8. Typical operating regions include V_{CE} 1–3 volts with I_{CE} setting the pressure range.

More details are available from manufacturer's data sheets, and more complex circuits with improved characteristics can be designed.

5.1.3 Transducer and Measurement Errors and Signals

The differences between perfect measurements and the measurements actually made by the system are generally referred to as system errors and are expressed in many ways, such as a percent of full scale (% F. S.) or as referred to the

---

*Pitran is a registered trademark of Stow Laboratories, Inc.

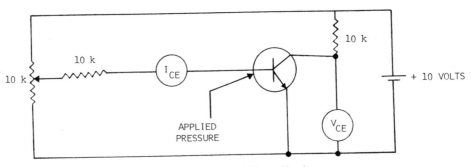

Figure 5.8. A Typical Pitran Circuit

input (R. T. I.). There are many sources of error including those caused by tolerance of the components in the system. For low-level voltage systems in particular, two important sources of error include common mode voltage signals and thermoelectric signals. These are discussed in more detail in other sections.

It is useful to define a series of standard terms for use in considering transducer performance:

(1) *Error*—the difference between the value of a variable indicated by the instrument and the true value at the input.

(2) *Accuracy*—a qualitative term used to indicate the performance of a transducer in probabilistic terms, i.e., the relationship of the output to the true input within certain probability limits. It is a function of nonlinearities, hysteresis, temperature variation, and drift.

(3) *Repeatability*—the closeness of agreement within a group of measurements at the same input conditions.

(4) *Drift*—the change in output that may occur despite constant input conditions. Usually changes due to temperature effects are considered separately.

(5) *Resolution*—the smallest change in input that will result in a significant change in transducer output. Precision is related to resolution and usually is stated in terms of the number of significant figures or bits in the transducer output. Sensitivity, which relates to the ratio of the change in output to the change in input, is not necessarily the same as resolution.

(6) *Threshold*—the minimum change in input required to change the output from a zero indication. For digital systems this is the input required for 1 bit change in output.

(7) *Backlash*—the maximum change of any one link in a physical or mechanical system that can be made without moving the other elements.

(8) *Hysteresis*—the algebraic difference at various points of measurement when the points are approached from above and below their values.

(9) *Zero Stability*—a measure of how well an instrument will return to zero with no input after an input has been applied.

(10) *Stiction*—the static friction which must be overcome to move an instrument from rest. It is usually expressed as a percentage of full scale.

(11) *Coulomb Friction*—dynamic friction, independent of velocity.

(12) *Viscous Friction*—a velocity-dependent friction term.

(13) *Live Zero*—applicable to systems such as standard industrial transducers where zero is represented by, say, 4 ma.

(14) *Suppressed Zero*—a means of adjusting the working range of an instrument to a particular region of interest to increase sensitivity.

(15) *Range*—the maximum range of input variable over which the transducer can operate.

(16) *Span*—the working range of an input variable over which an instrument is operating. Note that the span is included within the bounds of the range.

Detailed analysis of transducer performance in the above list of definitions may be found in many standard books on instrumentation [Ref. 8, for example].

An interesting example on interpreting performance specifications has been presented by Wightman [14] where a digital voltmeter specification has been set out as:

Accuracy	± 0.01% ± 1 digit
Resolution	10 μV
Scale	0–9999
Input Ranges	100 mV, 1V, 10V, 1000 V
	(by switched attenuators)
Zero Stability	0.01% of reading per °C

Casual interpretation might suggest that a reading could be made on the 100 mV scale for any voltage up to the maximum with an error of 0.01%. At full scale this is true, since 0.01% of 100 mV is 10 μV, and this is within the resolution of the instrument. However, at the low end of the scale, say, for example, 1 mV, the

MILLS COLLEGE
LIBRARY

10 μV resolution becomes significant. The error is $\pm$ 1 bit (10 μV) which therefore gives the reading 1 mV $\pm$ 1%. (Strictly speaking there is a further 10 μV uncertainty inherent in the instrument design which would double this error, but this is a rather subtle point and is not critical to the understanding of the discussion above.) If in addition the room temperature changes (from instrument design specification levels), we must add a further 0.01% of reading per °C. At 50 mV this adds 0.01% $\times$ 50 mV or 5 μV/°C. Hence a 2°C change in temperature adds 10 μV error.

Finally, no improvement is obtained on the higher ranges because the 10 μV resolution is also amplified to 100 μV (10 μV $\times$ 10) on the 1 volt scale, etc., and the other errors follow accordingly. Similar analyses may be done on transducer specifications and care must be taken if high accuracy is a requirement.

Another source of error in transducer use which often is not well understood by nonelectrical engineers is due to loading of transducers by the measurement system, a voltmeter or amplifier, for example. The problem here is one of impedance matching. Figure 5.9a shows a current source with impedance R_1 generating current i and feeding load resistance R_2. Since some of the current passes through R_1, the current reaching R_2 is reduced by $R_1/(R_1+R_2)$. Thus for errors to be less than 0.1%, R_1/R_2 must be greater than 1000. Similarly, Figure 5.9b shows a voltage source feeding a load R_2. The voltage drop across the load is reduced by $R_2/(R_1+R_2)$. In this case R_2/R_1 must be greater than 1000 for errors to be less than 0.1%. Thus, in summary, the load impedance must be at least 1000 times the source im-

pedance in voltage transducers, whereas the source impedance must be greater than 1000 times the load impedance in current-based systems.

The types of signals that analog transducers produce have already been mentioned briefly. They include:

- Direct voltage, several volts (potentiometric devices)
- Direct current, 1-5 ma, 4-20 ma, 10-50 ma (process control signals)
- Low-level dc, millivolts (thermocouples, strain gauges, chromatograph ion detector outputs)
- Low-level ac, millivolts (magnetic flowmeters)
- Demodulated ac, volts (displacement devices)
- Frequency (density meters)

Notice that two distinct levels of signals arise: high-level and low-level. In general, signals that are (or can produce) voltages in the range greater than 1 volt are considered to be high-level, whereas those in the range of about 0-500 millivolts are low-level. Low-level signals present much more difficulty in the area of signal transmission than do high-level ones. Various aspects of this problem are discussed in Section 5.2.

5.1.4 Digital Transducers

Digital transducers generally produce output in the high-level voltage range. The most common types of transducers are those that measure either frequency or period. Frequency is mea-

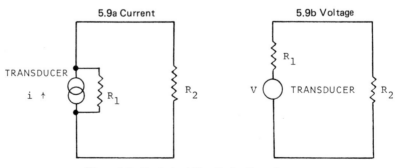

Figure 5.9. Equivalent Circuits for Transducers

sured by counting crossings of an axis over a fixed time interval, whereas period is measured by counting pulses produced by a stable clock generator between two successive process input pulses. Rotating flowmeters, tachometers, geiger counters, or neutron detectors are examples of digital output devices.

Other types of digital output transducers include those in which mechanical switches are closed or opened depending on the position of a rotating shaft. Shaft position encoders and liquid level sensors using magnetic float pickups (which open different switches as a function of position) are examples. It should be noted here that straight binary or binary coded decimal (BCD) would be used for the period and frequency measurements, but Gray code, in which only a single bit changes for each increment, is normally used for position.

Contact sense inputs to a computer system also represent a digital input device. Here the presence or absence of a voltage on a single line is read as a one or a zero by the computer system. Usually contact sense inputs are grouped into the same size register as that of the computer word and are all read simultaneously into an input register.

One example of a new digital transducer is provided by the photoelectric liquid level sensor developed by Hi-G Electronics. The device basically consists of a light emitting diode which directs a beam of light down a shaft terminated by a 45° cone. Under normal conditions light is reflected back up the shaft to an infrared light-activated transistor. When liquid touches the cone, however, the refractive index changes, and light is no longer reflected to the photo transistor. The result is an electronic on/off detector for liquid level.

Another example which has broad usage for thickness or positon measurements is the linear photodiode array. EG&G Reticon manufactures one range of devices which contain from 64 to 2048 individual diode elements (up to 1024 elements per inch). The devices consist of a row of photodiodes which can respond to an incident light beam. Simple scanning and counting circuits are used to determine the presence or absence of light on a particular diode, which then can be related to the position or size of an object in front of the light beam. Similar devices are available for rectangular arrays, with up to 10,000 individual points being detectable.

5.2 SIGNAL TRANSMISSION

No matter what type of transducer has been chosen, the way in which it is connected to the computer critically affects the performance of the data acquisition or control loop. The following discussion assumes that the signal source is reasonably close (say less than 1 mile) to the computer to allow for economic transmission over a pair of wires. For greater distances various types of telemetry systems should be considered.

The measurement as it reaches the computer will have two components: the signal and noise. A high ratio of signal to noise is required to maximize the performance of the system. Noise may occur because of high frequency variations in the process, turbulence (flow, for example) around the sensor, instrument noise, or stray pickup due to electromagnetic interference along the transmission lines. Good instrument practice is required to minimize these, particularly the latter type. Proper shielding, screening, grounding, and routing of measurement wires is essential. Low-level signals are particularly difficult to transmit without noise effects.

It is useful to define the various types of noise which can appear during signal transmission. Interference is any noise emanating from sources outside the process, but induced into it by various electromagnetic mechanisms. There are several common sources of noise in transmission systems. Inductive coupling results from the inductive action of 60 Hz power lines or 120 Hz fluorescent lighting. Switching inductive loads on and off can generate very large line transients which give rise to noise by a transformer type of action. Common impedance coupling is caused by placing more than one ground on a signal circuit. We will discuss this item in more detail in Section 5.2.3. Capacitive coupling occurs because of a distributed capacity between any conducting surfaces separated by a dielectric. It occurs most frequently between long transmission lines

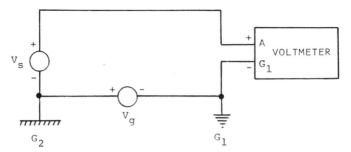

Figure 5.10. Single-ended Measurement System

and ground, between transducer cases and ground, or between computer chassis and ground. High frequency signals are most disturbed by this type of interference.

5.2.1 Thermally-Induced Noise

We have already described in detail the phenomenon giving rise to thermoelectric emf's when two dissimilar conductors are connected. In transmission wires connected to transducers this type of noise may be generated at each junction in the circuit. Care must be taken to keep pairs of junctions (as in a connector or on a terminal strip), which can make a thermoelectric circuit, at the same temperature. It is relatively easy to have terminal strips near hot equipment where a substantial temperature gradient can exist between two terminal posts even a short distance apart. If the pair of wires is carrying low-level signals, serious thermal noise will be generated, possibly masking the useful transducer signal completely.

5.2.2 Common Mode Noise

A single-ended measurement system as shown in Figure 5.10 is one where the transducer sig-

nal, V_s, is connected to one ground, G_2, while the measuring device is grounded at G_1. Two separate grounds are seldom, if ever, at the same potential, and this causes current to flow through the conductor. The overly simple situation depicted in Figure 5.10 shows that if V_g is zero, the voltmeter will read correctly the signal voltage, V_s. However, if it is not, the voltmeter will read $V_s + V_g$.

For a low-level measurement system a better alternative is to use a differential measuring device such as shown in Figure 5.11. Here it is assumed that the voltmeter ground is separate from the differential input terminals, A and B. The voltage difference between the two grounds is shown as V_{cm} and is called the common mode voltage. If G_1 is perfectly isolated from the differential inputs, A and B, then the voltmeter reads exactly the correct signal, V_s. However, this is never quite the case, and so a current due to the common mode voltage flows through the two signal lines. A simplified equivalent circuit for this system is shown in Figure 5.12 where R_1 and R_2 represent the line resistances, R_{g1} and R_{g2} represent the resistances to ground from the two conductors, and R_s represents the source resistance. (In fact all

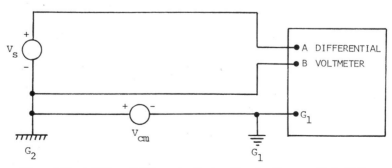

Figure 5.11. A Differential Measuring System Showing Common Mode Voltage

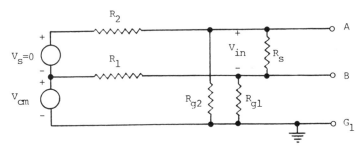

Figure 5.12. Equivalent Circuit for Figure 5.11

these are impedances which affect the frequency response, but not the steady state analysis shown here.) In deriving the effect of the common mode voltage on the measured signal, we have taken the input signal voltage, V_s, to be zero.

Generally R_s is very large and R_{g1} and R_{g2} are even larger, so that a simplified expression for V_{in} due to V_{cm} (since V_s is assumed zero) can be derived as:

$$V_{in} = V_{cm} \frac{(R_1 R_{g2} - R_2 R_{g1})}{(R_2 + R_{g2})(R_1 + R_{g1})}$$

and since both R_{g1} and R_{g2} are generally orders of magnitude greater than R_1 and R_2, a simplified rule of thumb expression for differential systems is:

$$V_{in} \approx V_{cm} \frac{(R_1 - R_2)}{R_g}$$

where R_g is the average of R_{g1} and R_{g2}.

A design parameter or measure of system performance is called the common mode rejection ratio, CMRR. This is defined as the ratio of the common mode voltage to the voltage it induces at the input, V_{in}. For the simple analysis just shown, this is defined as:

$$CMRR = \left| \frac{(R_2 + R_{g2})(R_1 + R_{g1})}{R_{g1} R_2 - R_{g2} R_1} \right|$$

$$CMRR \approx \left| \frac{R_g}{R_1 - R_2} \right|$$

and usually expressed in decibels so that

$$CMRR \, (db) = 20 \log_{10} (CMRR)$$

Typical values for good systems range from 100 db to 160 db with line imbalances, i.e., $R_1 - R_2$, of up to 1000 Ω.

It must be emphasized again that this is a steady state analysis. Frequency response considerations can be very important if high speed transient signals are being monitored. A simplified analysis can be performed on a circuit similar to that shown in Figure 5.12 with R_{g1} and R_{g2} being replaced by capacitors C_1 and C_2. The AC common mode rejection ratio (approximate) result is:

$$AC \, CMRR \approx \left| \frac{1}{(R_1 + R_2)\omega C} \right|$$

where ω is the noise frequency.

Typical values for high quality systems are of the order of 100–120 db at 60 Hz with line imbalances of up to 1000 Ω.

One of the most severe noise problems occurring as a result of common mode voltage is due to common mode conversion. This is the process by which the common mode input noise is converted into normal mode noise (V_{in} in our analysis) which is indistinguishable from the desired signal. Thus the presence of V_{in} cannot be detected from measurements in the system.

5.2.3 Grounding and Shielding Practice

Much of the electromagnetically generated noise and common mode voltage noise can be reduced by following good practice in grounding and shielding. If the noise is not controlled properly, the entire analog "front-end" of a computer system may fail to operate. That this

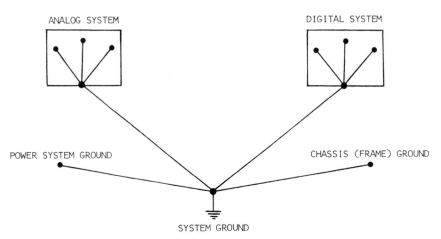

Figure 5.13. Single Point Grounding System

is so important can be seen through a comparison of a traditional measurement system with a computer system. The older equipment will attenuate all but low frequencies and hence filter out much of the high frequency noise. Computer systems on the other hand can sample signals easily at the rate of 50 kHz. Detecting noise induced from 60 Hz line voltage sources is very easy, and hence this type of noise must be prevented from corrupting the signal.

Of prime importance to system operation is the proper establishment and maintenance of the grounding systems. There are generally four systems involved: chassis grounds, power system grounds, analog system grounds, and digital system grounds. The best general philosophy is to establish separate grounding systems for each of these which come together at a common single point system ground. This is illustrated in Figure 5.13. Harrison [5] has an excellent discussion on the problems of grounding in computer systems. The most prevalent problem and also the most difficult to troubleshoot is caused by ground loops. These are circuits created when different grounds are connected and through which currents flow due to differences in ground potentials. A major difference in philosophy between analog and digital designers often causes problems. Whereas digital design requires multiple ground connections to reduce transients, analog practice requires minimum interconnection of grounds. Shields on cables are a common source of

ground loops in analog systems. Multipoint grounding frequently causes problems. Figure 5.14 illustrates the effect of common mode voltage causing a ground loop in a single-ended system. Figure 5.15 illustrates how this problem can be eliminated by using a shielded cable in conjunction with a differential amplifier. Note that the shield is only connected to *one* ground, and this is preferably the source or signal ground and *not* the system ground. More specifically the shield should be connected to the source of the common mode noise. Reversing the shield ground connections in Figure 5.15 would reduce the effectiveness of the shielding action.

There are two basic types of shielded cable available. The first is coaxial cable which has a braided wire mesh shield covering a single internal connector. The disadvantage of coaxial cable is that the shield is required as a conductor and hence cannot be used as a guard shield as shown in Figure 5.15. This will cause some loss in common mode voltage rejection but the shield is reasonably effective in reducing electrostatic induction by about 85%. If two coaxial cables are used with one center wire for each of the conductors, and if the shields are connected together only at one point, the CMRR is not affected.

The more commonly used cable for data acquisition systems is twisted shielded pair cable. Individually insulated conducting wires are twisted together in pairs along with a conducting bare wire (low resistance drain) also twisted

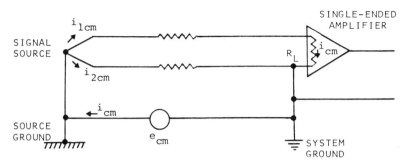

Figure 5.14. Common Mode Voltage and Current Flow

around the pair but inside a foil wrapper which covers the wire and is in contact with the bare wire. The shielding effectiveness is almost 100% for electrostatic induction in this type of cable. Standard cables come with 1, 2, 3, 6, 11, 15, 19, 27, and 50 pairs of conductors in a single insulated cover. Typically wire in the cables is 22 AWG (7 $\times$ 30 stranded) or 18 (16 $\times$ 30 stranded). The stranded wire is better for flexibility in use, but has a minor disadvantage for certain types of connectors.

The most important reason for using twisted shielded pairs of wires is to eliminate electromagnetically induced noise. If a changing magnetic flux induces current in one loop of the twisted pair, an equal and opposite-polarity current is induced in the adjacent loop, thus cancelling the effect. The usual number of twists per foot for this type of wire is 6.

In concluding this section on grounding and shielding it may be useful to summarize some of the important ideas. Some general points on low-level signal wiring are listed below:

- Do not use a shield as a signal conductor.

- Do not splice a low-level pair of signal wires.
- Maintain a minimum length of untwisted wire at the signal termination end.
- Connectors in low-level cables should carry the shield in pins adjacent to those of the signal pair.
- Keep low-level circuits at least 1 foot away from noisy circuits and power cables.
- Cross low-level cables and power cables at right angles.
- Use individually shielded twisted pairs, one set for each transducer. Multiple sets may be included in the same cable.
- Connect unused pairs of cables together and ground them at the opposite end of the system from that of their respective shields.
- Do not put power cables and signal-carrying cables in the same cable channel or conduit.

The usual ways of reducing common mode voltage are summarized:

Single Point Grounding—Insures that the signal

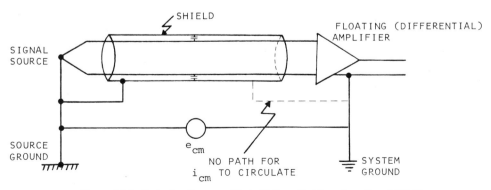

Figure 5.15. Reducing Common Mode Voltage Problems by a Guard Shield

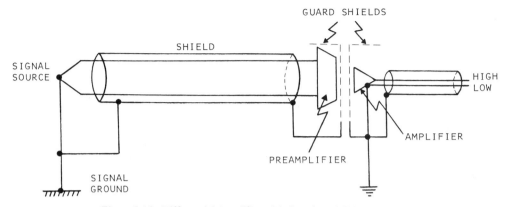

Figure 5.16. Differential Amplifier with Guards and Shielded Cable

is grounded at the transducer or at the amplifier but not at both.

Total Isolation—Perfect isolation of transducer (sensor) or computer input amplifier or both can effectively reduce common mode voltage. However, it is virtually impossible to prevent process sensors from contacting ground. Even water can provide enough of a leakage path to ground to cause severe common mode problems (with thermocouples contacting pipes, for example).

Shielding and Guarding—Most differential amplifiers now have an internal "guard" shield which surrounds the entire input section. The principle behind this is that the guard shield or guard is maintained at the maximum common mode voltage which would be seen at the inputs. This is accomplished as shown in Figure 5.16 by connecting the signal cable shield to the guard input, insuring at the same time that it is insulated from system (chassis) ground. The signal source ground is connected to the shield as closely as possible to the source. Similarly, the signal cable pair is grounded only at the one point. The amplifier chassis, the low side of the now single-ended output, and any internal shielding are connected to the system ground. More will be said about special amplifiers in Section 5.8.

5.3 TERMINATION

The termination facility in a data acquisition system provides a convenient place and means for terminating wires coming from the process

to the computer. Different considerations apply in a small research laboratory environment compared with a large industrial computer system where thousands of wires may require termination. Nevertheless some general observations as to ease of access, contamination, physical vibrations, and so on apply to both. Harrison [5] has a comprehensive discussion on termination procedures, and we will restrict comments here to a few key points. Some of the major factors which must be taken into account are noise, contact and leakage resistance, thermal potentials, common mode effects, and grounding.

The termination area should be free of interference from large motors, fluorescent lighting fixtures (particularly for low-level systems), and other noise-producing equipment. Contact resistance does not greatly affect voltage circuits, but can have a significant effect on signals from current transmitters. Brass studs are the most common form of terminal post. Even such apparently minute details as the material of construction of the washers separating the stud from the terminal panel are of importance in low-level signals. For thermocouple termination this is particularly important because on the one hand we require electrical insulation from the reference block (usually made of aluminum plate), but on the other we require good thermal connections. The result is that the nonelectrically conducting washers provide a capacitance type of action which reduces the AC CMRR of the system.

Grounding of a termination panel presents difficult problems. Normally it is grounded to its enclosure to reduce leakage and cross-talk between the input signals, but then the grounding can cause common mode voltage problems.

In small applications great care must be taken when AC power control circuits (Section 5.7) and low-level measurement signals come to the same termination area. It is very easy to inadvertently cross a 110-volt AC line with a low-level or logic line with disastrous consequences. Optical isolators should be provided wherever possible to prevent damage, particularly to the logic inputs of the computer system. Accidental shorts in process equipment can very easily occur, and severe damage to the computer equipment may result.

Industrial systems, particularly in hazardous (explosion-proof electrical) environments, frequently require the use of zener barriers in the termination circuit. These are devices that pass a low voltage current (e.g., 4–20 ma) or voltage with no problem, but which prevent high power surges. Typically, a zener barrier will short to ground above 27 volts or, alternatively, above 60 ma. It should be realized that many individual types can be purchased, depending upon the exact application. For sustained high power surges the zener barrier actually operates as a fuse and self-destructs. These devices prevent high power signals from passing through a circuit and can be used either to protect the computer analog subsystem or the field instrumentation.

Flexibility of termination facilities is important, particularly in small, research-oriented applications where many different measurement variables may be connected to the computer system over a period of time. One approach that has been very successful is to insert a patch panel terminal block between cables coming from the process areas and the analog front end of the computer system. Individual users of the system can jumper the cable lines to the computer input lines on their indi-

vidual panels to suit their own applications in much the same way as patch panels on analog computers have been used for years. Patch panels are commercially available having from one or two hundred connections to several thousand to suit large applications. Care must be taken not to jumper current-carrying lines into voltage inputs or high-level voltage signals into low-level inputs unless extra buffering and protection have been added.

Larger systems are more likely to be designed with conventional binding posts for connection of process inputs to the computer systems. Connections are hardwired to the system, and some over-design in input multiplexing capabilities (Section 5.5) may be required to reduce the frequency of change in connections to the system.

5.4 SIGNAL CONDITIONING

Signal lines from process transducers or other instrumentation are terminated both physically and electrically at some panel near the computer system. A variety of terminology is used to describe the conditioning functions applied to the signals at this point. Typical signal conditioning functions include signal conversion and standardization (usually from milliamps to volts, but also including AC to DC, impedance to voltage, conversion of voltages to a standard range, and so on), impedance matching, filtering, and multiplexing. The last topic, multiplexing, is discussed in Section 5.5. Signal conversion is primarily aimed at converting a current signal to a voltage signal for input to the analog to digital converter (ADC). Figure 5.17 shows a typical circuit where the shunt resistor R is chosen so as to provide the correct voltage range for the ADC amplifier input. The low resistances required here generally do not affect the high impedance source current ampli-

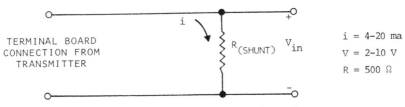

$$i = 4\text{--}20 \text{ ma}$$
$$V = 2\text{--}10 \text{ V}$$
$$R = 500 \ \Omega$$

Figure 5.17. Current to Voltage Conversion

fiers, but they must be precision resistors to provide the required accuracy. (Note that the accuracy of current transmitters is not better than about 1%.) Other conversions such as AC to DC are provided by rectifier circuits; impedance inputs require bridge circuits.

Various approaches are used to standardize DC input voltages. Resistor divider networks using standard precision resistors may be built to reduce voltage to a standard range. In using the resistor networks some attention must be paid to the relative values of the source (transmitter) and input (ADC amplifier) impedances because the circuits are nonlinear. Balancing techniques using differential attenuators should be used when high common mode rejection properties are desired.

Impedance matching (see comments in Section 5.1) between signal source plus transmission line and the input to the computer system is essential to maintain satisfactory system operation. Incorrect matching may result in loading of the transmitters or lead to excessive voltage drops in the lines. Recall that for voltage systems the load impedance should be at least 1000 times higher than the source impedance for 0.1% accuracy, whereas for current-based systems the source impedance must be at least 1000 times higher than the load impedance. Amplifiers are commonly inserted into the lines to provide proper impedance matching even though the voltage levels may be correct for the ADC.

Filtering using analog filters is often necessary, particularly to remove high frequency noise components and noise at the 60 Hz frequency level due to pickup from power lines. The simplest filter is the balanced resistance-capacitance filter shown in Figure 5.18. It is particularly effective in reducing high frequency noise exponentially with increasing frequency. The amplitude ratio for a filter of this type is approximately

$$A \text{ (db)} = 20 \log \left| \frac{V_{in}}{V_s} \right| = -10 \log (1 + \omega R_f C_f)^2$$

which can provide filter action for common mode noise as high as 40 decibels. Practical considerations require that high quality electrolytic capacitors be used. In this case the user must be careful to maintain the correct polarity in his circuit.

Design techniques for RC filters depend upon specification of the desired degree of attenuation. Typically the resistor might be of the order of 1000 Ω and C calculated to give the desired degree of attenuation. Harrison [5] gives a series of nomographs for design depending on the specific form of the RC filter. Filtering action, however, introduces time lags into the system with a time constant of the order of RC. The RC time constant is often of the order of a second or more, which may be excessive for some applications.

An inductance-capacitance (LC) filter as shown in Figure 5.19 may be used to provide equally good filtering action with a much reduced time constant equal to LC. The cost of this filter is considerably more than that of the RC filter, and the inductors are large and heavy. Sometimes higher noise rejections (60 db) are required. These are difficult to obtain with the simple passive filter circuits shown above, and

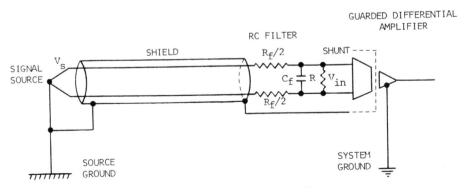

Figure 5.18. Signal Conversion and RC Filtering

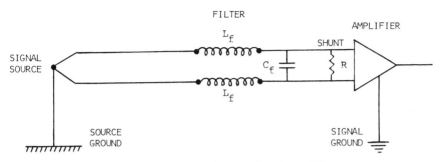

Figure 5.19. Simple Inductance-Capacitance Filter

active circuit elements* (transistor circuits) must be included at higher expense to provide this level of noise rejection.

Frequently, numerical filtering is applied to the data once they are in the computer as an alternative to the analog filters discussed above. Numerical filtering is simply the numerical equivalent of the RC or LC filtering action and is implemented by passing a measured input through a transfer function (or its discrete mathematical equivalent) to calculate the output. Details of this filtering approach may be found in Chapter 19. It should be pointed out here, however, that where specific electrical noise is present (e.g., 60 Hz noise), electrical filters are more useful than digital filters. The primary reason is that, for digital filters to work, one must sample the signal at twice the minimum frequency of desired signals. This could put a very heavy load on the analog sub-system–computer communications which the analog filters would minimize.

5.5 MULTIPLEXING

Multiplexing in a data acquisition system implies the sharing of a series of inputs with one re-source, ultimately the CPU. However, before the data reach the computer, several levels of multiplexing may have occurred. Usually a set of inputs is multiplexed to one analog to digital converter. Low-level signals are often multi-plexed to one differential amplifier before being transmitted as high-level single-ended

signals to the ADC. Signal conditioning and filtering (see Section 5.4) may occur before and/or after the multiplexer.

Multiplexing may be accomplished either mechanically or electrically, and the switching may be done randomly or sequentially. Electro-mechanical devices use metallic contacts which are switched by means of solenoids. The low resistance (less than 1 ohm) for closed switches and virtually infinite resistance for open ones makes mechanical relay types highly accurate and provides very little leakage between input lines.

A very early type of electromechanical multiplexer was a so-called crossbar switch, which consisted of a three-dimensional array of contacts. The x and y coordinates selected the set of input lines (z-axis) to be input to the system. The crossbar switches operate at a maximum rate of about 100 points/second. Motor-driven rotary switches were also an early type of multiplexer. These operate even slower than the crossbar switches at rates of about 5 to 10 points/second.

5.5.1 Reed Relays

Multiplexers in low-level systems today are frequently constructed of reed relays which are electromechanical magnetically operated switches. The reed switch is enclosed in a sealed glass capsule with the reeds being con-structed of a nickel–iron alloy so that they have a high magnetic permeability. The actual contact portion of the reeds is coated with another metal to provide better contact. A coil surrounds the glass capsule which activates the contact when a current is passed through it. A sketch of a reed relay is shown in Figure 5.20.

---

*For those persons not familiar with active and passive circuits a good set of analogies is given in T. W. Weber, *An Introduction to Process Dynamics and Control*, Wiley-Interscience, 1973.

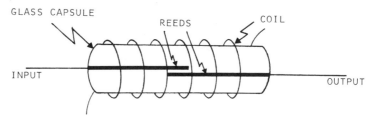

GLASS CAPSULE

REEDS

COIL

INPUT

OUTPUT

Figure 5.20. A Reed Relay

Contact resistance of the closed relays can be as low as 50 mΩ. Closing speed of the relays is quite high although they are seldom used at sampling rates higher than 250 points/second. A serious problem from back emf exists in the drive coils, and this should always be reduced or eliminated with diode suppressors. Another problem is relay bounce. Any electromechanical switch will bounce several times before settling, and with high-speed systems the user must be certain that a stable condition exists in the relay at the time its output is used.

The operating life of reed relays is very high, typically of the order of 10^6 to 10^9 cycles. Loading of the relay (resistive loading) has a strong effect on life. Care must be taken to assure that the correct ratings of the relay are specified. This is most important when high common mode voltages exist.

5.5.2 Mercury-Wetted Contact Relays

The problem of relay bounce can be almost eliminated through the use of mercury-wetted contacts.   Structurally   the   mercury-wetted relays are similar to reed relays except that they must be operated in a vertical or near vertical position so that the small pool of mercury remains at one end. The contact point is covered by a thin film of mercury which forms a mercury filament as contact is broken or made. In the breaking action the filament beads, thus breaking the contact, whereas in the making action the filament maintains a contact even though some bounce is present. The electrical and mechanical properties of the mercury-wetted relays are superior to the dry reed relays.

5.5.3 Semiconductor Switches

Electrical or electronic multiplexers operate in principle in exactly the same way as do mechanical multiplexers. They have higher leakage than mechanical switches, but on the other hand can be driven at very high speeds (of the order of microseconds per input).   One of the main differences between semiconductor switches and mechanical relays is that in many cases there is a minimum voltage level required to overcome biases in the switch, and this can create offset errors in the output.   Certain types of semiconductor switches do not suffer this defect and can be used for high- or low-level switching. New technology in CMOS or VMOS integrated circuits has enabled the development of switches for very low voltage or, at the other end of the spectrum, for very high power with no adverse effects.

Bipolar Transistors. A common type of semiconductor switch can be made from bipolar transistors. A basic circuit is shown in Figure 5.21 where a pulse to the transformer allows the transistor to conduct current momentarily from the source to the output. The switches may be designed in either a normal or inverted mode with the inverted form having better switching   characteristics   but   lower   gains. Detailed analysis of bipolar transistor switches is given in Harrison [5] and in many basic books in electrical circuits. Even the best circuits have offset voltages in the range of 0.5–2 mV which would preclude their use for thermocouple multiplexing, for example. An extension to the circuit shown in Figure 5.21 called a Bright transistor switch reduces somewhat the offset voltage and more importantly makes the switch itself bipolar, i.e., capable of switching both positive and negative voltages. However, for low-level multiplexing one must use other technologies.

Field Effect Transistors. FETs and Metal Oxide semiconductor FETs (MOSFETs) overcome the

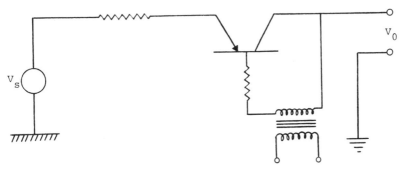

Figure 5.21. An Inverted Bipolar Transistor Switch

problem of bias voltages in bipolar transistor switches. Basically the switch can be viewed as a channel which is opened when the device is conducting and "pinched off" when it is not. For analog multiplexing the MOSFET switches are preferred to FET switches. General problems occur in leakage currents and contributions to common mode errors due to differences between multiplexer channels; however, CMOS technology has eliminated most of these problems. Current multiplexers can be used with confidence even for very low-level signals.

Photocoupled Devices. One of the most difficult problems to overcome in solid state multiplexer circuits is the isolation of the device voltages from the signal voltages. Photoactivated diodes or transistors are used instead of the normal transistors, thus eliminating any effect from drive voltages. Unfortunately these switches cannot be used for low-level signals where the most severe problems exist. They can, however, be used very successfully for electrical isolation of digital signals between the process and the computer system. A typical photocoupled device is illustrated schematically in Figure 5.22. In this case a light emitting diode is used in conjunction with a phototransistor to obtain almost total electrical isolation between the input and output circuits and fast switching times (on the order of one microsecond). Note that it is important to use separate power supplies for the light emitting and the light detecting stages of the isolator if full separation is to be maintained.

5.5.4 Multiplexer Configurations

There are many possible approaches for multiplexer systems. These include single throw switches (with common ground or return lines), differential double pole single throw systems, and many others. Apart from these there are two basic important types of electromechanical multiplexing systems, particularly in the area of low-level signal multiplexing. The first is a so-called flying capacitor coupled multiplexing system. Figure 5.23 shows a sketch of a set of signal wire pairs coming in, each with some analog filtering and the capacitor samplers. The inputs are normally connected to the capacitors which become charged to the value of the input voltage. When a channel is selected, the capacitor is disconnected from the input side (measurement side) and connected to the amplifier side. This circuit provides some input attenuation (small) and phase lag, but it provides high isolation or noise rejection from the process. Usually the amplifier is shorted out during the transistion period to avoid stray noise pickup due to the high-impedance open circuit.

The second multiplexing scheme allows for carryover of the shield through the multiplexer to the amplifier guard. This system is used for slightly higher frequency requirements (1–10 Hz) than that which the flying capacitor system

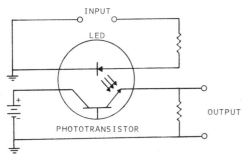

Figure 5.22. Optical Isolator Circuit

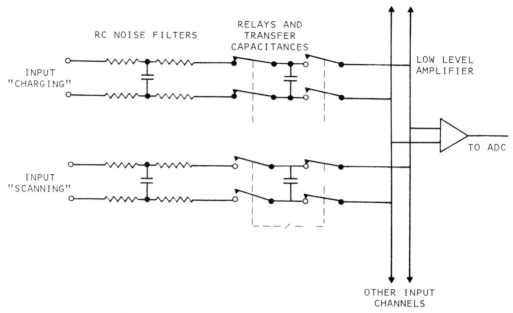

Figure 5.23. Flying Capacitor Sampler and Multiplexer Circuit

can handle (0.5–1 Hz). This is shown in Figure 5.24. The entire ADC subsystem is enclosed within the guard. Digitized output and input control signals are transmitted through the guard by pulse transformers.

5.5.5 Scanivalves*

Although most of the discussion on multiplexing has been on the assumption that electrical signals are being switched, one notable exception must be considered in view of its wide-

spread use and the cost of electronic pressure transmitters. This is the Scanivalve which is used for multiplexing pressures into one pressure transducer. Figure 5.25 illustrates its general layout. The valve itself can be switched manually, or with solenoid drive valves or by stepper motors which can be computer-driven. Outputs from the single pressure transducer are interfaced to the computer system in the normal way.

5.6 AMPLIFIERS

The subject of amplifiers is very broad and has been well covered in many books. However, a

---

*Scanivalve is a registered trademark of the Scanivalve Corporation.

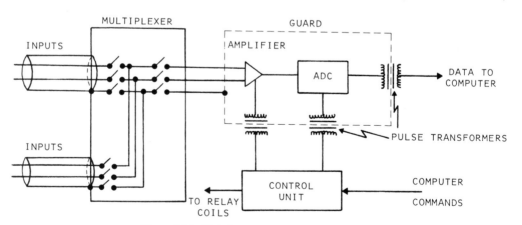

Figure 5.24. High Common Mode Rejection System

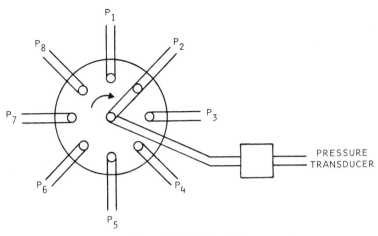

Figure 5.25. A Scanivalve

few basic design features are discussed here. The basic building block for most amplifier systems is the high gain operational amplifier. A detailed analysis of high gain amplifiers and applications is presented by Tobey et al. [13]. The best source of current information is from the catalogues and applications manuals by manufacturers of packaged amplifiers. New products are constantly announced in electronic and instrumentation trade journals. In addition, almost all manufacturers of operational amplifiers provide applications data sheets on request. The critical factors in amplifier selection are gain and sensitivity, bandwidth and settling time, temperature coefficient and drift, and common mode rejection.

Gain normally refers to the closed-loop gain, which in process applications can vary from 1 to 10,000 (the latter is not usually attainable with a single operational amplifier). Gain variation and adjustment are also important parameters.

The bandwidth and settling time of the amplifier are tied up with the sampling rate of the computer system. The basic reason is that as inputs are switched into the amplifier by the multiplexer circuits, we must be able to guarantee that transient effects are settled to within the quantizing level of the ADC before the conversion is initiated. Thus high-resolution ADC systems (14 bits) put much tighter constraints on the amplifiers than low-resolution systems. The "time constant" of the amplifier is also critical if high-speed transients are being monitored. For example, if high-speed heat transfer dynamics were being monitored with conventional temperature transmitters which have time constants of the order of 1 or more seconds, no useful results would be obtained because of dynamic errors. This is a case where special amplifier circuits would have to be used to permit collection of accurate data.

Temperature variation of amplifier gain has an important effect on the accuracy of data acquisition systems. Drift of the amplifier due to zero offset as a function of temperature is also very critical.

Finally, the common mode rejection characteristics of the amplifier are important. As has already been pointed out, most data acquisition systems exhibit substantial common mode voltage effects. Thus the amplifier must have a high CMRR, or else techniques such as flying capacitor sampling must be used.

5.6.1 Specifying Amplifiers in Analog Front End Systems

By and large the normal user of computer systems for data acquisition will not require a sophisticated knowledge of amplifier characteristics. Amplifiers will already have been selected by the manufacturer. However, an option is always available as to whether to purchase a single-ended or a differential input system. In general, for process and pilot plant work an amplifier with differential inputs is required. Some transmitters as well as chromatograph outputs have differential or isolated outputs already, in which case a single-ended system would be satisfactory.

A second feature which may be specified is whether the system amplifier has a programmable gain or not. For high-level systems this is not too important as the shunt resistors for current transmitters may be used to vary the input voltage to the system. Low-level input systems are more likely to have programmable gain amplifiers. These are extremely useful as the range of low-level signals coming from various process sensors can be very broad. Some systems also have autoranging amplifiers which set the gain automatically according to the level of the input signal. One point to note is that amplifiers with programmable gains often require several milliseconds of settling time following a gain change. This can greatly reduce throughput speed in multiplexed systems with changing input voltages.

5.6.2 Instrumentation Amplifiers

Instrumentation amplifiers are packaged units using operational amplifiers. They have differential inputs, high common mode rejection characteristics, high impedance, a wide selection of available gains, and high accuracy. They are an ideal amplifier for providing high gains in the presence of large common mode voltages for signals from strain gauges, thermocouples, and other transducers. Most of the necessary circuitry is inside the block, but· the user will have to provide gain setting resistors, zero offset adjustment resistors (if desired), and a power supply. Additional features often include CMR trim inputs, guard shield connection (see Figure 5.18), and external balance adjustments for the differential amplifiers. A typical simplified system is shown in Figure 5.26. The power supply connections are not shown. The expression for output can be modified by adding an additional resistor, R, in the line from E_0 to output sense so that the overall gain is changed by the ratio $(R + R_0)/R_0$.

A small number of instrumentation amplifiers can be much cheaper than the equivalent number of process transmitters (for thermocouples, for example) and interfacing low-level

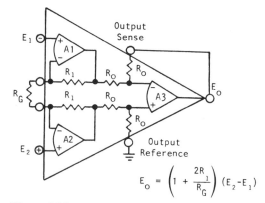

$$E_O = \left(1 + \frac{2R_1}{R_G}\right)(E_2 - E_1)$$

Figure 5.26. A Typical Instrumentation Amplifier

signals from laboratory instruments to high-level input systems is very easy using these amplifiers.

New trends in integrated circuit design have led to many improvements over the more traditional mechanical forms of multiplexers and switching circuits. The reader is urged to consult with experts in these fields concerning their particular applications, since a comprehensive survey of new switches is beyond the scope of this introductory material. Nevertheless, one example will illustrate some of the potential advances. Analog Devices now manufactures a four-channel, isolated thermocouple/mV conditioner for use with thermocouples in temperature measurement and control systems. The signal conditioner is shown schematically in Figure 5.27. It is used in conjunction with a universal cold junction compensator. In the example, thermocouple types J, K, and T are used with an external temperature measurement of the terminal strip to which the thermocouple leads are connected. The output is 0–5 V as shown, but could as well be 4–20 ma using slight variations in the circuit. Compensation for the different thermocouple types is accommodated by digital inputs to select amplifier gains. Nonlinear compensation would be carried out in the computer system used to monitor the temperatures.

5.7 FINAL CONTROL ELEMENTS

Most real-time computer applications fall into two categories: data acquisition, and process control or general servomechanism control,

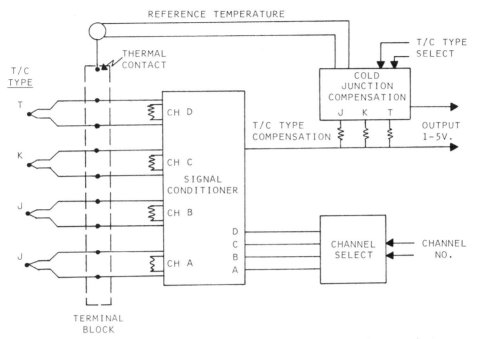

Figure 5.27. A Thermocouple Signal Conditioner with Cold Junction Compensation

i.e., feedback systems in which some form of signal based on measurements input to the computer is brought back to another process element. These elements which accept signals from the computer are called final control elements. Typically they are devices to regulate the flow of fluids (materials) or energy into the process.

5.7.1 Control Valves

By far and away the most common final control element in process applications is the control valve. A simple sketch is shown in Figure 5.28. The main function of the control valve is to accept a signal within a certain range from the computer and translate this into a valve posi-

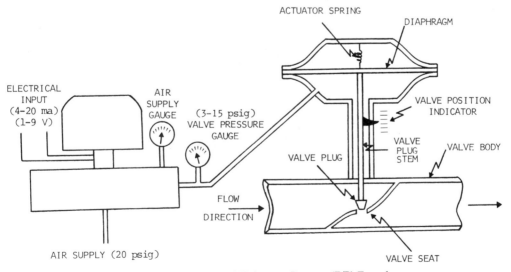

Figure 5.28. Control Valve and Voltage to Pressure (E/P) Transducer

tion or status ranging from fully closed to fully opened. A comprehensive discussion on control valves would fill an entire textbook, and that is not the purpose here. Briefly an electrical analog signal from the computer activates an electropneumatic transducer. In most instances the full range of the input signal (1-9 volts or 4-20 ma DC) is converted to an equivalent air signal between 3 and 15 psig. This air signal operates on a diaphragm to open or close a valve. A spring counterbalances the diaphragm air pressure to force return toward the normal position as air pressure is released. Typical properties that must be specified for control valves are: valve coefficient, size, material of construction, air to open or air to close, valve characteristics (linear, equal percentage, or quick opening), and others. Comprehensive details may be found in the *ISA Handbook of Control Valves* [7].

The valve coefficient or C_V is a measure of the flow rate of fluid that will pass through the valve and is, by definition, the flow rate of 60°F water in gallons per minute that will flow through the valve with a pressure drop of 1 psi.

Valve characteristics relate the flow rate through the valve as a function of the stem position. Figure 5.29 shows some typical valve characteristic curves. Most research laboratory control valves are either linear or equal percentage, the latter being used most often in conjunction with flow controllers because of the inherent linearization of the pair operating together.

5.7.2 Electrical Power Control

Electrical energy is controlled in two basic ways. The first is by on-off control using relay switches activated by the computer. The other techniques all fall into a classification of proportional control. Examples of these are silicon controlled rectifiers (SCRs), saturable core reactors, and power amplifiers. The most useful of these is probably the SCR owing to the ease with which they can be controlled using solid state electronic circuits. The principle of operation of an SCR is that a portion of a 60 Hz power signal is activated with the power being dissipated during a fraction of each cycle. Figure 5.30 illustrates one cycle of an AC signal in which two SCRs are controlling the conduction state—hence current flow to some form of resistance load. The cross-hatched area is the portion of the cycle in which current is flowing. Details of this and the other devices may be found in many basic electrical engineering textbooks.

Power loads can be highly inductive or purely resistive. SCRs must be chosen to match the type of load being controlled.

An extension of the SCR principle for power control is the TRIAC which differs from an SCR in that it can conduct current in either direction. A typical circuit is shown in Figure 5.31 where the most useful feature, apart from good power control, is that the transistor switch is photoelectrically activated. The advantage of this is that the power circuit can be completely isolated from the computer system and run by a low voltage output to a lamp which activates the phototransistor switch for the TRIAC.

5.7.3 Other Final Control Elements

There are many other ways in which process input variables may be adjusted by signals from the computer system. Often variable-speed pumps and compressors activated by variable-speed drives or incremental speed changers may be used to adjust the flow of material. Stroke length in reciprocating pumps can be adjusted on signal from the computer. Slots or openings feeding conveyor belts as well as the speed of the belts may be adjusted.

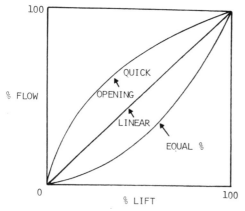

Figure 5.29. Typical Control Valve Characteristic Curves

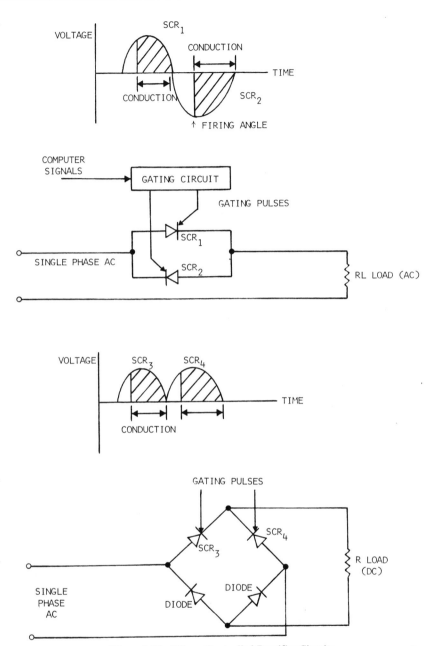

Figure 5.30. Silicon Controlled Rectifier Circuits

Variable-speed DC motors may be used to run the pumps with speed adjustment obtained by means of shunts and variously excited armatures. Probably the most useful variable-speed motor in process control is the shunt motor in which the field magnetic flux, the armature voltage, or the armature resistance is adjusted by signals coming from the computer.

One easy way to adjust rheostats or mechanical devices is by use of stepper motors. These accept a logic level pulse or series of pulses from the computer at specified maximum rates and move one increment of rotation for each pulse. These are most useful in scanning types of laboratory instrumentation such as scanning UV spectrophotometers where the computer

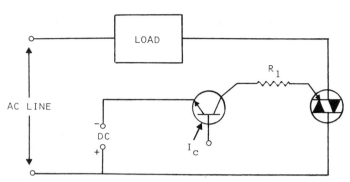

Figure 5.31. Triac Power Control

provides pulses that cause a stepper motor to move the unit through a series of wavelengths. Stepper motors may also be used for providing fail-safe operation of process control valves. In the event of a power failure the valves remain at the last position transmitted, since the stepper motor locks unless activated by a pulse. Stepper motors driving rheostats provide an alternative means of power control to Triacs, and for computers with only logic level outputs provide a convenient means of manipulating process variables.

5.8 NEW TECHNIQUES IN INTERFACING

The relatively comprehensive overview of a data acquisition system that has been presented so far represents a rather traditional approach to computer/process interfacing. Solid state electronics, digital systems, and microprocessor technology are advancing at an incredible pace and are providing new alternatives. The multitude of electronic trade journals in digital electronics provides a good source of current product information.

5.8.1 New Devices

We have mentioned already that moderate-quality instrumentation amplifiers now can be obtained at reasonable cost and provided for each measuring thermocouple. This technique reduces costs and greatly reduces problems due to transmission line noise when compared to the two more traditional approaches we discussed as alternatives, viz., a low-level analog input system with expensive cabling and multiplexing techniques required, or an expensive electronic transmitter for each thermocouple.

An alternative to transmitting analog signals over multiple pairs of wires is to place remote analog to digital converters and multiplexers near the process and transmit digital control signals to the unit and receive digital signals back from the ADC. Digital signals can be transmitted virtually noise-free by contrast with analog signals and can also be optically isolated from the process or computer.

Because not every computer has either digital or analog input/output capabilities, but virtually all computers are equipped with console CRT or typer (current loop or EIA) ports, some manufacturers have developed asynchronous serial data exchange modules which utilize standard terminal software and hence communicate using the ASCII code. (The use of ASCII code to represent alphanumeric information is discussed in Chapter 6.) A few key ASCII symbols from the computer are used to activate the remote unit which has an ADC, a multiplexer with several channels, and a limited number of DAC channels as well. These units are especially well suited for small laboratory environments where the user is relieved of many of the problems of interfacing and also of developing software, since the units can be run using standard BASIC or FORTRAN read and write commands. The maximum distances the units can be run without special equipment is dependent on transmission speed but can reach 10,000 feet at less than 400 baud over twisted shielded pairs. The sampling rate can be as high as about 50 samples per second at 4800 baud transfer rates. (For a discussion of baud rates see Chapter 9.)

Finally, the development of microprocessor and microcomputer system technology has spawned a whole new generation of process data acquisition and control units. Complete analog and digital subsystems for data acquisition and control can now be purchased and placed directly on (or near) the process as a miniaturized version of the larger systems we have been discussing. Communications from these microprocessors to the minicomputer (or other) systems is carried out using standard intercomputer communications technology.

One of the most recent trends is that manufacturers of "front-end" systems, integrated circuits, operational and instrumentation amplifiers, analog to digital converters, etc., are now packaging microcomputer-based data acquisition systems. These typically cost from $5,000 to $20,000 and use multitasking BASIC as the system programming language. Subroutine calls to initiate analog input/output are available. The analog subsystems can be tailored in a very flexible way to meet user requirements from compensated thermocouple inputs to high-level voltage or digital inputs or to industrial 4-20 ma systems. Readers should contact a number of manufacturers of such systems for current data on their products. It is noted here that technology is developing so rapidly in this area that only current information has relevance for potential users contemplating a new system.

5.8.2 Standardization of Process Computer Interfaces

One of the most serious problems which has arisen with the advent of so many new and different computer data acquisition systems is that the standardization in interfacing of measurement equipment which used to exist has disappeared. Most process-oriented instrumentation used to operate in the range of 3-15 psi and then 4-20 or 10-50 ma. The result was that process instrumentation was compatible both with other instruments from the same manufacturer, and also with those of other manufacturers. Now we have high-level instruments, low-level instruments, an unbelievable proliferation of cable connectors, and it is a nontrivial task to interface process equipment to computer systems. Two systems are emerging as candidates for a new standardization. These are the IEEE 488 bus originally developed as the Hewlett-Packard Interface Bus (HP-IB) and the CAMAC system.

IEEE 488 Bus. Briefly the IEEE 488 Bus is a 16-wire interface capable of transmitting byte-serial data at rates as high as 1 megabyte/second. Up to 15 devices can be interfaced in a network. The 16 lines are grouped into sets of 8 lines for data transfers (byte serial), 3 lines for data transfer control, and 5 lines for bus control. Many new instruments are being manufactured with IEEE 488 connections. Details of the bus are provided in IEEE Standard 488 and ANSI Standard MCl.1. This interface is useful for short-distance connections between laboratory instruments and computers, but is not useful for large-scale process control systems in chemical or manufacturing plants.

The CAMAC Interface. The other system, the CAMAC interface crate, uses a data highway comprised of 24 bits in either parallel or serial form. The parallel system connections between crates are over a 66-pair cable which can transmit data at rates of up to 24 megabaud for distances up to 300 feet. The serial connections are over twisted pairs or coaxial cable and operate up to 5 megabaud with no practical limit on distance. The CAMAC system puts all interfaces in a crate which can accommodate up to 25 devices. Crates are connected to the computer system I/O bus by a variety of techniques depending on the installation. The crates can contain devices performing signal conditioning, analog to digital conversion, providing control outputs, and so on. Details of the CAMAC system are available in IEEE Standards 583, 595, 596, and also IEC Standard 516.

A brief discussion comparing HP-IB with CAMAC systems has been presented by Merritt [11]. In the future it is likely that many manufacturers of instruments and sensors will be providing interface connections which are compatible with one or the other of these systems. Furthermore, it is likely that the ANSI Standard FORTRAN extensions for Process Control will address the problem of standardization of software for these devices as they become more popular.

5.9 ANALOG TO DIGITAL AND DIGITAL TO ANALOG CONVERTERS AND DIGITAL I/O

In the preceding discussion very little has been said about the analog to digital converters

(ADCs) or the digital to analog converters (DACs) for the data acquisition and control system. These devices are discussed in detail in Chapter 10. However, for completeness here it is sufficient to define the resolution of ADC inputs for typical systems. Most process control ADC systems should contain 12-bit converters to maintain reasonable accuracy of the converted input. High-precision analytical instrumentation is capable of more accuracy, and up to 15 bits may be required but at greatly increased cost. The speed of most ADCs today is in the order of 50,000 samples per second which is more than adequate for most applications.

Similarly the DACs should be 12-bit resolution to be consistent with the ADCs although control valves require only about 10 bits of precision for satisfactory operation. For certain applications sufficient accuracy for control valves can be obtained using 8-bit DACs. The advantage of these is that they can be operated with 8-bit (byte) oriented computer output hardware using either serial or parallel transmission techniques.

5.10 DIGITAL INPUT AND OUTPUT

As with ADCs and DACs, this topic will be discussed in Chapter 10 where the generalized computer/process interface is described in detail. However, we should point out here that digital (binary) signals from and to the process may take two forms: (1) single pole/single throw relay states; (2) general logic level signals. In the first case a process signal would open or close a particular relay; the computer would detect the process state by measuring whether or not current flows through the relay's contacts (closed or open). Similarly, a digital signal from the computer would close or open a particular output relay (which might switch AC power off or on to some part of the process). In the second case signals from and to the process might consist simply of transistor/transistor logic (TTL) compatible signals, 0 and +5 VDC, which would go directly from the process into the Digital Input Device or from the Digital Output Device directly to the process.

With relay-based systems, a certain degree of isolation between process and computer is obtained; this is not the case if logic level signals are routed back and forth directly. There is always a possibility that a high voltage line inadvertently may be attached to lines connected to the Digital I/O Devices and cause extensive damage to the computer system. Even with relatively safe relay I/O this possibility exists, for example, if a relay malfunctions. Particularly if 110 or 220 VAC circuits are being switched in and out, additional protection should be provided. Optical couplers of the type described in Section 5.5.3 should be used to provide as nearly complete isolation as possible for logic level signal lines between the computer and the process (or process relays).

5.11 SUMMARY

In this chapter we have attempted to cover a very large and important area, one that requires attention and specialized knowledge in many applications. We have attempted to illustrate several important classes of measurement, e.g., temperature and pressure, and have dealt with the many problems of signal transmission at length. That particular area often is ignored or neglected, invariably with negative consequences for the real-time user. (He may find, for example, that he is measuring a common mode voltage instead of the process temperature he thought he was measuring.) Careful attention to system design and to wiring practices certainly is warranted in the signal transmission area.

We also have discussed those techniques of multiplexing and signal conditioning that ordinarily are not accommodated by the generalized process/computer interface described in Chapter 10. There we will deal exclusively with electronic switching devices.

Finally, we have mentioned several new techniques for process interfacing. It is likely that bus systems of the IEEE 488 form, used to couple the real-time computer and process through "smart" analog front ends (i.e., microprocessor-controlled ADC/DAC units), will quickly replace much of the present technology in the signal transmission area. Also, "smart" instruments and final control elements will eliminate much of the imprecision and un-reliability that presently exist in the measurement and process manipulation areas. Nevertheless, despite the availability of this new technology, it still will be necessary to pay close attention to the problems encountered in connecting the process to the computer.

5.12 REFERENCES

1. Baker H. D., Ryder, E. A., and Baker, N. H., *Temperature Measurement in Engineering*, Vol. 1, Omega Press, Stamford, Conn. 1975 (a).
2. Ibid., Vol. 2, Omega Press, Stamford, Conn., 1975 (b).
3. Considine, D. M., *Process Instruments and Controls Handbook*, McGraw-Hill, New York, 1967.
4. Considine, D. M. and Ross, S. D., *Handbook of Applied Instrumentation*, McGraw-Hill, New York, 1964.
5. Harrison, T. J., *Handbook of Industrial Control Computers*, Wiley-Interscience, New York, 1972.
6. Harvey, G. F., Ed., *ISA Transducer Compendium*, 2nd Ed., Part 1, IFI/Plenum, New York, 1969.
   Ibid., Part 2, ISA Press, Pittsburg, 1970.
   Ibid., Part 3, ISA Press, Pittsburg, 1972.
7. Hutchinson, J. W., *ISA Handbook of Control Valves*, 2nd Ed., ISA Press, Pittsburgh, Pa., 1976.
8. Liptak, B., *Instrument Engineers Handbook*, Vol. 1, *Process Measurement*, Chilton, Philadelphia, 1969.
9. Ibid., Vol. II, *Process Control*, Chilton, Philadelphia, 1970.
10. Ibid., Supplement I, Chilton, Philadelphia, 1972.
11. Merritt, R., "Universal Process Interfaces", *Inst. Tech.*, *23* (8), 29, 1976.
12. Minner, E. J., *ISA Transducer Compendium*, Plenum, New York, 1963.
13. Tobey, G. E., Graeme, J. G., and Huelsman, L. P., *Operational Amplifiers—Design and Applications*, McGraw-Hill, New York, 1971.
14. Wightmen, E. J., *Instrumentation in Process Control*, CRC Press, Cleveland, OH, 1972.

5.13 EXERCISES

1. Explain the difference between a transducer and a transmitter.
2. Based on your own experience, list a number of *physical* techniques that might be used to measure:
   a. Liquid or gas flow.
   b. Liquid level.
   c. Position (both small and large mechanical displacements).
   d. Linear velocity (speed) and acceleration.
   e. Rotational (shaft) speed and acceleration.

   Which of these techniques would be most convenient for use with a data acquisition computer, i.e., would yield a measured variable as an analog voltage or in digital representation directly?
3. An experimental process has been built and its output variables instrumented. You run a set of calibration tests on one of the temperature transmitters and obtain data which are shown in the figure. With respect to this transmitter, comment briefly on:
   a. Its linearity.
   b. Its reproducibility.
   c. Its precision.
   d. Its accuracy.
   e. If, by some procedure, the data scatter could be eliminated, could you utilize such results to obtain a useful calibration? Explain.

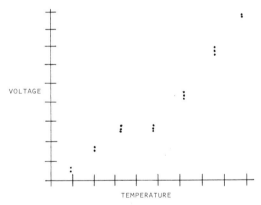

Figure for Exercise 3

4. A thermocouple, composed of the two dissimilar metals, Chromel/Alumel, is connected to an instrumentation amplifier by a pair of copper wires. Briefly list the points of good instrumentation practice that should be observed in making the connections. You may assume that the environment is not electrically-noisy and that the length of the copper leads is short.
5. A low-level signal, such as obtained from a thermocouple, must be transmitted over a distance of approximately 10 meters to an instrumentation amplifier and thence, over a distance of 100 meters to an ADC. What

can you say about the use of shielded cable in this case; i.e., is it necessary at all since the instrumentation amplifier has been used to amplify the thermocouple output?

6. A thermocouple can work quite well with its exposed "bead" making contact with an electrically-conducting process element such as a metal vessel wall. There is the possibility of problems associated with this usage, related to grounding. What might they be, and how could they be minimized?

7. What are the advantages and disadvantages of mechanical (relay-based) multiplexers discussed in this chapter and those based on electronic switches discussed in detail in Chapter 10?

8. An older process plant, using 3–15 psig instrumentation exclusively, is to be retrofitted with a data acquisition computer. How would you evaluate the alternative ways of bringing process instrument output information into the computer? Can you think of any reason(s) that might justify a conversion of the instrumentation from pneumatic to electronic for all or part of such a plant?

9. Photocoupled devices (a light emitting diode plus phototransistor) may be used to isolate the computer system from the process by placing one device in each digital input or output line. In order to guarantee safety of the computer, a separate power supply should be used for each side of the protective devices. Show by means of a schematic drawing, how this would be done. Why is it necessary? Can a similar approach be employed to protect analog input and output circuits, e.g., using operational amplifiers? How?

10. A simple scheme has been developed for controlling a small fixed-bed chemical reactor in a laboratory. Four internal temperatures of the bed are measured using thermocouples. The thermocouple outputs (-10 to $+10$ millivolts) are linearly related to the measured temperature over the range 20 to 500°C. A resolution of the temperature measurements of ± 0.1°C is needed.

It has been determined that a single heating element at the entrance to the reactor is sufficient for control purposes. An SCR is available that furnishes power through the element, linearly, in the range 0 to 10 watts as the SCR input varies from -10 to $+10$ volts. It is desired to have a resolution of 0.005 watt to the heater.

Assume that the range of conversion equipment (both ADC and DAC) for the computer is -10 to $+10$ volts and that the control method used is unimportant.

a. Determine the specific instrumentation hardware required for successful data acquisition and manipulation of the reactor heater.

b. The resolution of an ADC or DAC is 1 part in 2^N where N is the number of bits in the ADC (DAC). How many bits would be required in the ADC and the DAC to obtain the desired resolution on input and output?

Part III
Introduction to Digital Arithmetic and Hardware

6

Representation of Information in a Digital Computer

D. Grant Fisher

Department of Chemical Engineering
University of Alberta

6.0 INTRODUCTION

Most people have learned to handle information in a wide variety of forms: integer numbers, real numbers, Roman numerals, symbols, lists of instructions, etc. However, almost all digital computers in use today function *internally* with a binary, fixed-word-length representation of all data and instructions. Therefore if a digital computer is to be used to assist with an application that involves information in any form other than binary, then obviously some means, or convention, must be found so that all the pertinent information can be represented by fixed-length binary words. This is shown schematically in Figure 6.1 which implies that all information passing between the user and the digital computer must be converted from "external" to binary format.

In most large computer systems the conversion of all input/output data is done by the computer system so that, even though the internal computer operations are done in binary form, it appears to the user that he is working with a computer system that uses information in the same format that he does. Also the programming of engineering or business applications on a large computer is usually done in a high-level language such as FORTRAN or COBOL. High-level languages have a vocabulary and syntax that is familiar and convenient for the user, and he does not have to be concerned with how the computer actually performs an operation such as "COST = A*B/INDEX". Similarly a large operating system (i.e., executive program) is normally available to "automati-

cally" handle scheduling of input/output, operations, error conditions, etc. In other words, the internal operations of the computer are "invisible" to the user, and he need be concerned only with application-oriented, functional operations such as: read data, do calculations, print results, etc.

However, with most smaller computer installations—particularly real-time, sensor-based minicomputer and microcomputer applications—the user cannot assume such a high level of hardware/software support and must become familiar with the internal operations and data representation of the computer. For example:

(1) The computer vendor might not supply all the hardware and/or software necessary to handle the conversion of process input/output data from the format produced by the interface instrumentation to the binary form used by the computer, or vice versa. Therefore the user must either do it manually; write appropriate software to do the conversions; or select, buy, or build suitable hardware.

(2) High-level languages, such as FORTRAN, are usually not powerful enough, or efficient enough, for programming all real-time applications, so some coding must be done in assembler language, or machine code, both of which require an understanding of binary representations and basic computer operations.

(3) Interfacing peripheral devices and/or

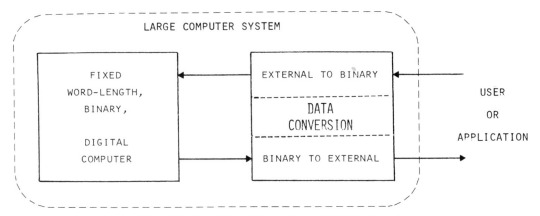

Figure 6.1. Separation of Computer and Interface Operations

instruments not supplied by the computer vendor usually requires a detailed knowledge of internal (binary) data representations and computer operations, in order to do the necessary programming and/or hardware construction.

(4) Debugging a real-time computer application frequently requires stopping the computer at specific points during program execution to examine the contents of the computer registers and/or memory; all this information is in binary format.

(5) Because real-time computer applications differ so much from one installation to another, it is difficult to develop a standard operating system. Hence the user must frequently write it himself or at least learn to work with "internal" functions such as "overflow indicators", "interrupt priorities", "device status words", "masks", etc.

The most common examples of converting information to binary format are listed in Table 6.1. Note that *every* piece of information to be processed by the computer must be converted from the external form, such as listed in the left column of the table, into a unique representation consisting of ones or zeros. (These binary representations are then stored or manipulated in the form of "words" which contain a specified number, N, of bits where N is determined by the particular computer hardware.) The inverse problem is perhaps more dramatic:

$$0110 \ldots 110 = ???$$

Note that there are no plus or minus signs, decimal points, alphabetic characters, or specific indication of what type of information the string of binary ones and zeros represents. All of this must be determined by the codes and conventions illustrated in this chapter and the related discussion of computer hardware, programming, etc., in other chapters.

The following sections deal specifically with various means of representing information in an appropriate binary form and with techniques for converting from one representation to another.

6.1 NUMBER SYSTEMS

We are all so familiar with the decimal number system that we can usually process numerical information without any conscious thought about the definitions and principles that are involved. However the construction of digital computers and other devices requires, or makes it more convenient to work in other number systems such as the binary number system. Rather than treat every number system independently, it is much more desirable to identify the general principles that apply to all number systems, since we can then generalize our experience with the decimal number system and apply it to other systems.

6.1.1 General Concepts

A number system is simply a means of representing numeric quantities using a set of sym-

Table 6.1. External Vs. Binary Representation of Information

Type of Information	Example	Binary Form*
NUMBERS		
Integers	1,32767	
Negative integers	-1,-32768	
Floating point	-1.,-32.768	
Scientific	-0.3267E+2	
Complex	32 + 23i	
Double Precision	12345.6789	
LOGICAL VALUES	TRUE., FALSE.	
ALPHANUMERIC		
Alphabetic	A...Z, a...z	
Numeric	0, 1, 2...9	
Special	$, @	
INSTRUCTIONS		
Arithmetic	+, -, *, /	
Logical	AND, OR, EOR	
Operations	input, output	
MISCELLANEOUS		
Status of component	on, off, busy	
Error	yes, no	

Binary Form diagram:

```
 1  2  3  4            N
+--+--+--+--+--  --+--+
| 0| 1| 1| 0|  ...  | 0| 1|
+--+--+--+--+--  --+--+
N = no. of bits
  = word length
```

* In order to be processed by a fixed word-length, binary, digital computer each piece of information must be represented by a "string" of n bits. Various terms are used to refer to different length bit strings: bit (n=1); byte (usually n=8); word (n=N, a value defined by hardware); multiword (n/N=integer); string (n=variable), etc.

bols. The *base* or *radix* of a number system is the number of symbols it contains, usually including zero. For example the decimal number system has a base or radix of ten and makes use of the ten symbols: 0, 1, 2, 3, 4, 5, 6, 7, 8, 9. We can only speculate as to why the base ten was chosen, but it seems reasonable to associate it with the fact than man has ten fingers that can be used to indicate, or symbolize, all ten values in the decimal system. In electronic or mechanical systems, however, it is usually much more convenient to use a binary system because the two symbols, or values, 0 and 1 can be represented by any device or convention that has two distinct states. For example: a switch is open or closed; a light is off or on; a hole in paper tape or in a computer card is absent or present; voltage is absent or present (or negative vs. positive); material is demagnetized or magnetized; etc. The number systems and symbols that are normally used in computer applications are summarized in Table 6.2.

When the number system or radix is not

Table 6.2. Typical Number Systems Used in Computer Applications

Number System	Base or Radix	Symbols
Decimal	10	0,1,2,3,4,5,6,7,8,9
Binary	2	0,1
Octal	8	0,1,2,3,4,5,6,7
Hexadecimal	16	0,1,2,3,4,5,6,7, 8,9,A,B,C,D,E,F

obvious from the context, then it is common to use a subscript to indicate the number system. For example, we can show that:

$$256_{10} = 100000000_2 = 400_8 = 100_{16}$$

whereas without the subscripts the equalities do not hold.

Table 6.3 compares the representations of the decimal numbers 0 through 16 as expressed in the decimal binary, octal, and hexadecimal number systems.

It should be obvious that we could define an infinite number of different systems by using different base values, and that the symbols are independent of the base. For example, Roman numerals are different symbols that were once commonly used with the decimal number system, and we could have created new symbols such as a $\underline{0}, \underline{1}, \underline{2}, \underline{3}, \underline{4}, \underline{5}$ to represent the values ten through fifteen in the hexadecimal number system. These "new" hexadecimal symbols would be equally, or perhaps more convenient for humans to work with, but could not be handled as conveniently by existing devices such as typewriters. Therefore "flexible" man defers to "inflexible" machines!

Once we have learned to work with one number system, it is usually relatively easy to learn to work with other number systems or to translate information from one system to another. All that is required is a suitable "dictionary", or list of rules containing the definitions, structures, and allowable operations plus a lot of careful, methodical work. In fact

the problem is so well structured, and the procedures are so methodical, that computers are normally built, or programmed, to do the necessary conversions whenever required (cf. Figure 6.1). A little thought will confirm that all the common information-processing operations, such as arithmetic, counting, etc., are independent of the symbols, or representation used. Thus the same result is obtained by adding decimal (integer) numbers directly as is obtained by: converting each decimal number to its binary equivalent; adding the binary numbers; converting the result from binary to its equivalent decimal representation. We would probably prefer to work with decimal numbers directly, but, as indicated by Figure 6.1, a digital computer follows the latter procedure. (An operation that is not invariant under a change of number systems is, for example, counting how many times the symbol 1 occurs in a given set of numeric values.)

Once the definitions of base or radix, the numeric symbols, and the value zero are understood, it is important to grasp the concept of relative position. For example, the symbol 7 implies something different in each of the numbers 567, 576, and 765; in 567 it represents 7 units, in 576 it represents 70 units, and in 765 it represents 700 units. In other words, the value represented by a particular digit (symbol) depends on its relative position in the number.

The most general way of defining a number system is to use a table to associate a weight or position coefficient with the relative position of each symbol in a number. The value repre-

Table 6.3. Comparison of Different Number Systems

Decimal	Binary	Octal	Hexadecimal
0	0	0	0
1	1	1	1
2	10	2	2
3	11	3	3
4	100	4	4
5	101	5	5
6	110	6	6
.7	111	7	7
8	1000	10	8
9	1001	11	9
10	1010	12	A
11	1011	13	B
12	1100	14	C
13	1101	15	D
14	1110	16	E
15	1111	17	F
16	10000	20	10

sented by that symbol is then found by multiplying the defined value of the symbol by its weight or position coefficient. For example:

$$7707 = 7 * \underline{1000} + 7 * \underline{100} + 0 * \underline{10} + 7 * \underline{1}$$

or

$$7707 = \text{seven } \underline{\text{thousand}}, \text{ seven } \underline{\text{hundred}},$$
$$\text{and seven } \underline{\text{units}}$$

or

$$7707 = 7 * \underline{10^3} + 7 * \underline{10^2} + 0 * \underline{10^1}$$
$$+ 7 * \underline{10^0}$$

where the underlined values are the position coefficients or weights. Table 6.4 compares the position coefficients used in different number systems. By using this table the reader should confirm that:

$$1101_2 = 1 \times 2^3 + 1 \times 2^2 + 0 \times 2^1$$
$$+ 1 \times 2^0$$
$$= 13_{10}$$

$$1101_8 = 1 \times 8^3 + 1 \times 8^2 + 0 \times 8^1$$
$$+ 1 \times 8^0$$
$$= 577_{10}$$

$$1101_{16} = 1 \times 16^3 + 1 \times 16^2 + 0 \times 16^1$$
$$+ 1 \times 16^0$$
$$= 4353_{10}$$

Position Coefficients—For most modern number systems, the position coefficient of a particular numeral is simply the radix raised to a power equal to its position (counted, starting with zero, from the right). Thus, as long as this convention is followed, it is not necessary to actually define position coefficients by a table.

Most/Least Significant Digits—The leftmost, or

Table 6.4. Position Coefficients for Different Number Systems

Position*	Number System				
	Decimal	Binary	Octal	Hexadecimal	4221 BCD**
0	$1 = 10^0$	$1 = 2^0$	$1 = 8^0$	$1 = 16^0$	1
1	$10 = 10^1$	$2 = 2^1$	$8 = 8^1$	$16 = 16^1$	2
2	$100 = 10^2$	$4 = 2^2$	$64 = 8^2$	$256 = 16^2$	2
3	$1000 = 10^3$	$8 = 2^3$	$512 = 8^3$	$4096 = 16^3$	4
4	$10000 = 10^4$	$16 = 2^4$	$4096 = 8^4$	...	
...	...	...	...	...	...
N	10^N	2^N	8^N	16^N	?

* The position of the symbol (numeral) in the number is counted, starting with zero, from right to left and expressed in decimal notation. The letter N is used to represent any finite decimal integer.

** "4221 BCD" is a (seldom used) Binary Coded Decimal representation included to point out that position coefficients don't always have to be integer powers of a radix.

"first", nonzero digit in a number is referred to as the most significant digit (because it is associated with the largest position coefficient) and is commonly abbreviated MSD. Similarly the rightmost, or "last" digit in an integer number is referred to as the least significant digit (LSD) and may have the value zero. Although the word "digit" is associated in our minds with the decimal number system, the terms MSD and LSD are applied to all number systems. Thus in the following examples the numerals with a double underline are MSDs and those with a single underline, LSDs.

$$\underline{\underline{256}}_{10} = 00\underline{\underline{1}}00000000\underline{0}_2 = \underline{\underline{4}}00_8 = \underline{\underline{1}}00_{16}$$

Note that the term MSD implies nothing about the value represented by that digit (e.g., the MSD in the binary and hexadecimal examples given above is 1 in both cases), but the first represents $1 * 2^8$ and the latter $1 * 16^2$. Also, note that the term *number of significant digits* (NSD) refers to something quite different. The number of significant digits in a set of integers in usually the largest number obtained by counting the digits from the MSD to the LSD

inclusive, for each number in the set. However, "trailing" zeros may or may not be significant; hence it is best to define the number of significant digits by a separate statement. A very common "error" in computer applications is to assume that all the digits printed out are significant. Fixed-word-length, binary computers (normally) use the same number of binary digits in all internal operations, but the NSD in the result depends on the NSD in the input data and often on the calculation procedure used. The user of computers should always give careful consideration to the NSD in the input data, in each value as it is processed internally, and in the final result.

Count and Carry—Individual items can be "labeled" or "numbered" using any number system. If only a single digit is to be used, then the maximum value that can be represented is equal to the base or radix of the number system being used. Most people "labeling" or "numbering" items in the decimal system will use only the digits from 1 to 9. If they used all the values from 0 to 9, they could label or number ten items instead of nine. The loss in efficiency by "neglecting" zero when numbering in the decimal system is not too serious, but in the binary number system where the only states (symbols) are 0 and 1, it is obvious that zero must be used. IN COMPUTER APPLICATIONS, LABELING AND NUMBERING USUALLY START WITH 0 RATHER THAN 1. Thus sixteen items may be labeled or numbered from 0 through 15, rather than from 1 through 16, (and in some cases may even be numbered "backwards", i.e., from right to left or from bottom to top). The user should *always* check to make sure that the same convention has been used consistently throughout all phases of his computer program or application!

When the number of items to be counted, or numbered, exceeds the base or radix of the number system being used, then it is necessary to "carry" one into the next more significant position. For example:

$$9 + 1 = 10_{10} \qquad F + 1 = 10_{16} \qquad 1 + 1 = 10_2$$

This should be obvious from the definition of position coefficients, discussed above, and can be seen in Table 6.3, which compares values in different number systems.

Radix Point—The symbol that separates the "whole part" from the "fractional part" of a given number is referred to as a radix point. In the decimal number system it is commonly called a decimal point and is represented by a period. In the binary number system it is usually called a binary point and represented either by a period or by the caret symbol ($\wedge$).

If a number is written or displayed without a radix point (and one is not implied by convention or context), then it is assumed that it is located to the right of the least significant digit.

In the decimal system the base is 10 and the numbers to the left of the decimal point are weighted by factors of 10^0, 10^1, 10^2, etc. The numbers to the right of the decimal point represent 10ths, 100ths, 1000ths, etc., or, mathematically speaking, weightings of 10^{-1}, 10^{-2}, 10^{-3}, etc. For example:

$$543.21 = 5 \times 10^2 + 4 \times 10^1 + 3 \times 10^0$$
$$+ 2 \times 10^{-1} + 1 \times 10^{-2}$$

The same approach can be used with any number system whether it is decimal, binary, octal, or hexadecimal. In the binary number system, the first binary bit to the left of the binary point has a weight of 2^0, the next bit has a weight of 2^1, and so on. The first bit to the right of the binary point has a weight of 2^{-1}, and the next bits are weighted 2^{-2}, 2^{-3}, etc. As an example, the binary number 1001.011 would be equivalent to the decimal number 9.375, since,

$$1001.011_2 = 1 \times 2^3 + 0 \times 2^2 + 0 \times 2^1$$
$$+ 1 \times 2^0 + 0 \times 2^{-1}$$
$$+ 1 \times 2^{-2} + 1 \times 2^{-3}$$

$$= 8 + 0 + 0 + 1 + 0 + \tfrac{1}{4} + \tfrac{1}{8}$$

$$= 9.375$$

When information is stored within a binary computer system, there is no way to indicate physically the presence or absence of a binary point. As a result, it is necessary to define standard methods of representing numbers that contain whole and fractional parts. The method of representing floating point numbers in the binary number system discussed in a later section is a particular example of this. In more advanced applications, such as direct digital control, it is sometimes advantageous to use different standard forms in different parts of the program. This requires greater care and understanding on the part of the programmer, but can result in more efficient execution and more efficient utilization of the computer system.

6.1.2 Binary Number System

The important features of the binary number system follow directly from the preceding general discussion, but will be summarized in this section for convenience and as a review.

Radix and Symbols—The binary system has a radix or base of 2 and uses only the symbols 0 and 1.

Position Coefficient—The position coefficients follow the convention used in most modern number systems and can be used directly as a means of converting binary values to their decimal equivalent; for example:

$$11101_2 = 1 \times 2^4 + 1 \times 2^3 + 1 \times 2^2$$
$$+ 0 \times 2^1 + 1 \times 2^0$$
$$= 29_{10}$$

Note that the power to which the base 2 is raised is equal to the position of the binary digit when counted, starting with zero, from the least significant digit (LSD) and proceeding leftward to the MSD.

Count and Carry—Counting normally starts with the value one and proceeds to some maximum value; for example, we count the number of fingers on one hand and get a total of five. We could label or number our fingers as one through five or as zero through four. Both approaches are "correct" as long as they are clearly defined and consistently used. As mentioned previously, both approaches are used in computer applications and can be a source of confusion. In either case the binary equivalents of the values zero through five are:

$$0, 1, 10, 11, 100, 101$$

The concept of "carrying one" to the next more significant position when the value in a given position equals the radix or base value should be obvious from the above sequence or by analogy to the decimal system. Thus:

$$1 + 1 = 10_2 \qquad 11 + 1 = 100_2$$
$$111 + 1 = 1000_2, \text{etc.}$$

There is a maximum number of bits (i.e., positions in a binary number) that can be handled, or stored, in a single step in a fixed-word-length binary computer. If this maximum number is four, then the largest number that can be represented is 1111_2. Adding one to this value would (normally) produce 0000_2 with an "overflow" or "carry" indicator turned on to indicate that the capacity of the computer has been exceeded and possibly that significant information has been "lost" or "destroyed".

Radix Point—Binary digits to the right of a binary point have a weighting of 2^{-n} where n is the position counted, starting with 1, from the binary point and proceeding to the right. Thus:

$$10_\wedge 1 = 1 \times 2^1 + 0 \times 2^0 + 1 \times 2^{-1}$$
$$= 2.5_{10}$$

$$1_\wedge 01 = 1 \times 2^0 + 0 \times 2^{-1} + 1 \times 2^{-2}$$
$$= 1.25_{10}$$

$$_\wedge 101 = 1 \times 2^{-1} + 0 \times 2^{-2} + 1 \times 2^{-3}$$
$$= 0.625_{10}$$

$$101 = 1 \times 2^2 + 0 \times 2^1 + 1 \times 2^0$$
$$= 5$$

It must be remembered that *only* two states are transmitted or stored by computer components and that we represent these by the binary symbols 0 and 1. Therefore binary points, plus or minus signs, etc., must be handled internally by a standard convention or procedure rather than indicated by the actual physical presence of $\wedge$, +, – signs. Thus, inside the computer the value is simply 101. If no other convention is indicated, binary numbers are normally interpreted as positive integers, i.e., with the binary point assumed to be to the right of the LSD as assumed in the last of the above examples.

Changing from one number system to another and/or the adoption of specific conventions can, in some cases, introduce unavoidable, but in most cases small, errors. For example, if we assume a maximum of four binary digits and an implied binary point left of the MSD then "round-off" or "approximation" errors can be introduced when decimal values are converted to their closest binary equivalent. For example,

$$0.5_{10} \longrightarrow {}_\wedge 1000_2 = \tfrac{1}{2} = 0.5$$

$$1.0_{10} \longrightarrow {}_\wedge 1111_2 = \tfrac{15}{16} = 0.9375$$

$$0.2_{10} \longrightarrow {}_\wedge 0011_2 = \tfrac{3}{16} = 0.1875$$

Using more binary digits (a longer-word-length computer) would reduce the error in both of the last two examples, and moving the binary point one position to the right would have the following result:

$$0.5_{10} \rightarrow 0_\wedge 100 = \tfrac{1}{2} = 0.5$$

$$1.0_{10} \rightarrow 1_\wedge 000 = 1 = 1.0$$

$$0.2_{10} \rightarrow 0_|0_\wedge 010 = \tfrac{1}{4} = 0.25$$

Most minicomputers use 16 bits, and most data processing computers use at least 32 bits, so the round-off errors due to conversion are usually negligible.

6.1.3 Octal Number System

The octal number system is not normally used for information *processing* by computers or by humans. It is normally used as simply a more convenient and efficient means of *representing* binary data. For example a 12-bit binary number can be represented by only four octal numbers.

To convert the binary number 11110101100 into octal it is first separated into 3-bit groups by starting with the LSD end of the number and supplying leading zeros if necessary:

$$011 \ 110 \ 101 \ 100$$

The binary groups are then replaced by their octal equivalents:

$$011_2 = 3_8$$

$$110_2 = 6_8$$

$$101_2 = 5_8$$

$$100_2 = 4_8$$

and the binary number is converted to its octal equivalent:

$$3 \ 6 \ 5 \ 4$$

Conversely, an octal number can be expanded to a binary number using the same table of equivalents.

$$4563_8 = 100 \ 101 \ 110 \ 011_2$$

The conversion between the octal and binary number systems is always exact, and the procedure is so simple that most people, after a little practice, can work faster and with fewer mistakes using octal rather than binary representations. For example it is easier to set 12 computer console switches to 3654_8 than to remember 011110101100! It also takes only one third the space to print out a number in octal rather than binary.

6.1.4 Hexademical Number System

Most minicomputers use a 16-bit word-length to represent information internally, but most people find binary numbers like 0111011101011011 cumbersome to handle. Therefore, for the same reason and in the same way that octal numbers are used as a "shorthand" way of writing 12-bit numbers, people frequently use hexadecimal numbers to represent 16-bit binary numbers.

One hexadecimal digit can represent all possible combinations of four bits, and four hexadecimal digits can represent all possible combinations of 16 bits by taking the bits four at a time. For example, the 16 binary bits 0111111101011011 can be broken up into the four subpatterns of 0111 1111 0101 1011 which can be represented by the hexadecimal digits 7, F, 5, and B. So 0111111101011011 can be represented by 7F5B in hexadecimal, and hexadecimal becomes a shorthand notation for binary numbers or bit patterns. It is possible to count, and do operations such as arithmetic, in the hexadecimal number system in a manner analagous to the familiar decimal number system. However, except for very simple operations, most people prefer to convert back and forth to the decimal (or binary) system before doing arithmetic operations.

As we shall see later, "strings" of four bits are used for binary coded decimal data (BCD), and eight-bit "strings" (commonly called "bytes") are usually used for encoding alphanumeric data. These can be conveniently represented by one or two hexadecimal digits respectively.

6.2 CONVERSION BETWEEN NUMBER SYSTEMS

Any given numeric value can be converted from one number system to another by the following means:

(1) Hand calculations using the basic principles defined in the preceding sections.
(2) Tables of equivalent numbers (e.g., Table 6.3).
(3) Specially designed hardware (e.g., a hand calculator).
(4) A computer program.

6.2.1 Any Number System to Decimal

Binary, octal, or hexadecimal numbers can be converted to an equivalent decimal value by using the position coefficients given by a table such as Table 6.4, or simply by remembering that the position of each digit in the number specifies the power of the radix that is associated with that digit. The presence of a radix point indicates that the position of each digit should be taken relative to that point. For example:

$$110.1_2 = 1 \times 2^2 + 1 \times 2^1 + 0 \times 2^0$$
$$+ 1 \times 2^{-1}$$
$$= 6.5$$

$$110.1_8 = 1 \times 8^2 + 1 \times 8^1 + 0 \times 8^0$$
$$+ 1 \times 8^{-1}$$
$$= 72.125$$

$$110.1_{16} = 1 \times 16^2 + 1 \times 16^1 + 0 \times 16^0$$
$$+ 1 \times 16^{-1}$$
$$= 272.0625$$

The reader can practice number conversions by using the procedures developed in Section 6.2.2 to convert any arbitrary decimal number to the desired number system and then using the above procedure to convert it back to the original decimal value. (It is suggested that the reader start with integer numbers because, as shown previously, the conversion of numbers with fractional parts is more difficult and sometimes results in round-off errors.)

6.2.2 Decimal to Other Number Systems

From the general conventions that apply to all modern number systems it is known that the value of a number containing $n + 1$ digits $m_n \ldots m_1 m_0$ is equal to the sum of all terms $\{m_i \times R^i, \ i = 0, \ 1 \ldots n, \ 0 \leqslant m_i < R\}$ where $R = $ radix. For example:

$$123_{10} = 1 \times 10^2 + 2 \times 10^1 + 3 \times 10^0$$
$$= 123_{10}$$

$$123_8 = 1 \times 8^2 + 2 \times 8^1 + 3 \times 8^0$$
$$= 83_{10}$$

This feature leads directly to a procedure for converting a binary number to its equivalent decimal value, as discussed previously. Thus:

$$1101_2 = 1 \times 2^3 + 1 \times 2^2 + 0 \times 2^1 + 1 \times 2^0$$

$$= 1 \times 8 + 1 \times 4 + 0 \times 2 + 1 \times 1$$

$$= 8 + 4 + 0 + 1 = 13_{10}$$

To convert a decimal integer to another number system this procedure is essentially just reversed. In general terms the procedure is:

Step 1: Compute all the values $\{R^i, i = 0,1 \ldots n\}$ where R is the radix of the new number system and n is an integer such that the number, C, to be converted is in the range $R^n \leqslant C < R^{n+1}$.

These powers of the radix, R, will be used to expand the decimal number, C, in the form:

$$C = m_n \times R^n + m_{n-1} \times R^{n-1}$$
$$+ \cdots + m_0 \times R^0$$

The coefficients $\{m_i, i = 0, 1 \ldots n\}$ are determined in Step 2.

Step 2a: Calculate m_n which is equal to the maximum number of times that R^n will divide evenly into the number to be converted (or alternatively the maximum number of times that R^n can be subtracted from C and still have a remainder $\geqslant 0$).

Example 6.1. Convert 13_{10} to binary.

Step 1:   The radix of the binary system is 2 and the required powers of 2 are:

$$2^0 = 1, \quad 2^1 = 2, \quad 2^2 = 4, \quad 2^3 = 8 \quad (2^4 = 16 > 13).$$

The value of n = 3.

Step 2a: Begin the subtractions starting with R^n

$$13 - 8 = 5 \qquad m_3 = 1$$

Step 2b: Continue the subtractions for all lower powers

$$5 - 4 = 1 \qquad m_2 = 1$$
$$1 - 2 < 0 \qquad m_1 = 0$$
$$1 - 1 = 0 \qquad m_0 = 1$$

Step 3:   The desired result is $m_3 m_2 m_1 m_0 = 1101_2$
Note that all values of $\{m_i, \ i = 0, 1 \ldots n\}$ are in the required range $0 \le m_i < R$.

Step 2b: Calculate $\{m_i, \ i = n - 1, \ \ldots 0\}$ by division (or repeated subtraction) of R^i from the remainder left after the preceding operations. If a value R^i cannot be divided or subtracted from the current value of the remainder, set $m_i = 0$ and proceed to the next lower value of R^i.

Step 3: The desired value in the new number system is the n + 1 digit value: $m_n \ldots m_1 m_0$

Examples 6.1–6.4 illustrate the method. Subtraction rather than division is emphasized in the examples because this method is frequently called the *subtraction of powers method.*

6.2.3 Decimal to Binary by "Successive Division"

A decimal number may be converted to binary by repeated integer division by 2 as described below.

Divide the decimal number by 2. If there is a remainder, put a 1 in the *LSD* of the partially formed binary number; if there is no remainder, put a 0 in the LSD of the binary number. Divide the quotient from the first division by 2, and repeat the process. If there is a remainder, record a 1; if there is no remainder, record a 0. Continue until the quotient has been reduced to 0.

The result of Example 6.5 is $49_{10} = 110001_2$.

Example 6.2. Convert 43_{10} to binary.

Step 1:   The required powers of 2 are:

$$2^0 = 1, \ 2^1 = 2, \ 2^2 = 4, \ 2^3 = 8, \ 2^4 = 16, \ 2^5 = 32.$$

Step 2:

$$43 - 32 = 11 \qquad m_5 = 1$$
$$11 - 16 < 0 \qquad m_4 = 0$$
$$11 - 8 \ = 3 \qquad m_3 = 1$$
$$3 - 4 \ < 0 \qquad m_2 = 0$$
$$3 - 2 \ = 1 \qquad m_1 = 1$$
$$1 - 1 \ = 0 \qquad m_0 = 1$$

Step 3:   By definition $43_{10} = 101011_2$

Example 6.3. Convert 13_{10} to octal.

Step 1: The required powers of the radix, 8, are:

$$8^0 = 1, \ 8^1 = 8$$

Step 2: The values represented by the above powers of 8 are then subtracted from 13.

$$13 - 8 = 5 \qquad m_1 = 1$$

$$\left.\begin{array}{l} 5 - 1 = 4 \\ 4 - 1 = 3 \\ 3 - 1 = 2 \\ 2 - 1 = 1 \\ 1 - 1 = 0 \end{array}\right\} \qquad \begin{array}{l} \text{since 1 is subtracted five times this} \\ \text{indicates that } m_0 = 5 \text{ (division of} \\ \text{the remainder 5 by } 8^0 \ (= 1) \text{ would give the} \\ \text{same result more quickly)} \end{array}$$

Step 3: The desired octal representation of 13_{10} is 15_8

Note that the LSD is calculated *first* and the MSD *last*. Do not write the binary number down backwards!

6.2.4 Decimal to Any System by "Successive Division"

The "successive division" method is a little more difficult to relate to the basic properties of number systems than the subtraction of powers method. However, the following development leads to a generalized procedure and also illustrates the basic properties of number systems.

A decimal integer C, to be converted to a number system with radix, R, can be written as:

$$C = m_n \times R^n + \cdots + m_2 \times R^2 + m_1 \times R^1 + m_0 \times R^0$$

Noting that $R^0 = 1$ and factoring R^1 out of the first n terms gives:

$$C = [m_n \times R^{n-1} + \cdots + m_2 \times R^1 + m_1]R + m_0$$

This shows that C can be expressed as a multiple of R plus a remainder m_0. If C is divided by R, the quotient, Q, is equal to the value in brackets, and the remainder is equal to m_0.

Noting that the quotient, Q, from the preceding division is equal to the term in brackets and factoring out a common factor, R, from the first n − 1 terms gives:

$$Q = [m_n R^{n-2} + \cdots + m_2]R + m_1$$

Thus division of Q by R will give a new quotient plus a remainder equal to m_1.

Example 6.4. Convert 43_{10} to octal.

Step 1: The required powers of the radix are:

$$8^0 = 1, \ 8^1 = 8$$

Step 2: Subtraction of the above values gives:

$$\left.\begin{array}{l} 43 - 8 = 35 \\ 35 - 8 = 27 \\ 27 - 8 = 19 \\ 19 - 8 = 11 \\ 11 - 8 = 3 \end{array}\right\} \qquad \text{or } 43 \div 8^1 = 5 \qquad m_1 = 5$$

$$\left.\begin{array}{l} 3 - 1 = 2 \\ 2 - 1 = 1 \\ 1 - 1 = 0 \end{array}\right\} \qquad \text{or } 3 \div 8^0 = 3 \qquad m_0 = 3$$

Step 3: The desired octal representation of 43_{10} is 53_8

Example 6.5. Convert 49_{10} to binary.

Division by 2		Quotient		Remainder		
$49 \div 2$	$=$	24	$\longrightarrow$	1	$=$	m_0
$24 \div 2$	$=$	12	$\longrightarrow$	0	$=$	m_1
$12 \div 2$	$=$	6	$\longrightarrow$	0	$=$	m_2
$6 \div 2$	$=$	3	$\longrightarrow$	0	$=$	m_3
$3 \div 2$	$=$	1	$\longrightarrow$	1	$=$	m_4
$1 \div 2$	$=$	0	$\longrightarrow$	1	$=$	m_5

This procedure can be repeated for $\{m_i, i = 2, 3 \ldots n\}$, at which point the quotient finally becomes zero. This is illustrated in Examples 6.5 and 6.6.

Thus $1101_{10} = 2115_8$ which can be proven by the expansion:

$$2115_8 = 2 \times 8^3 + 1 \times 8^2 + 1 \times 8^1 + 5 \times 8^0$$
$$= 1024 + 64 + 8 + 5 = 1101_{10}$$

6.2.5 Conversion of Fractional Values

If the number to be converted contains a radix point, then the conversion is done in three separate steps:

(1) The "whole" part of the number, i.e., that part of the number to the *left* of the radix point, is converted using the procedures outlined in the preceding sections.

(2) The "fractional" part of the number, i.e., that part of the number to the *right* of the radix point, is converted using procedures analogous to those discussed above. Note that the conversion may not be exact as illustrated previously.

(3) The desired value is simply the two parts calculated in Steps 1 and 2 separated by a radix point.

The reader should consult the references for detailed examples or make up his own and check them by converting both ways—e.g., from decimal to binary and then binary back to decimal.

6.3 ADDITION AND SUBTRACTION IN DIFFERENT NUMBER SYSTEMS

Addition and subtraction in binary and other nondecimal number systems is done in a manner analogous to the familiar operations with decimal numbers. The LSD are aligned and each column added separately. If the sum in any one column is $\geqslant n \times R$, then n is carried over to the next column. If the total is less than the numerical base, R, then no carry is generated. For example:

Example 6.6. Convert 1101_{10} to octal.

Division by R		Quotient		Remainder		
$1101 \div 8$	$=$	137	$\longrightarrow$	5	$=$	m_0
$137 \div 8$	$=$	17	$\longrightarrow$	1	$=$	m_1
$17 \div 8$	$=$	2	$\longrightarrow$	1	$=$	m_2
$2 \div 8$	$=$	0	$\longrightarrow$	2	$=$	m_3

$$\begin{array}{rr} 1\,1\,0\,1\,0\,1_2 & 6\,5_8 \\ +1\,1\,1\,1\,0\,0_2 & +7\,4_8 \\ \hline 1\,1\,1\,0\,0\,0\,1_2 & 1\,6\,1_8 \end{array}$$

It should be noted that if the first addition were done in a 6-bit arithmetic unit (computer), then the last carry would "overflow" and hence the MSD of the sum would be lost. (Most computers test for overflow of this nature and generate an overflow error signal, i.e., set an overflow bit or generate an interrupt.)

Subtraction is similar except that when a "borrow" is required, care must be taken to use the base value. For example:

$$\begin{array}{rr} 1\,1\,1\,1\,0\,0_2 & 7\,4_8 \\ -1\,1\,0\,1\,0\,1_2 & -6\,5_8 \\ \hline 0\,0\,0\,1\,1\,1_2 & 0\,7_8 \end{array}$$

Other arithmetic operations such as multiplication or division are normally not performed manually in nondecimal number systems although it is sometimes convenient to be able to do so. Basically it requires that the person memorize the necessary multiplication tables in each number system. It is suggested that the reader leave the question of multiplication, etc., in nondecimal number systems until he feels that the extra effort would be justified by his own requirements.

6.4 BINARY REPRESENTATIONS

6.4.1 Negative Integer Numbers*

Negative numbers and subtraction are usually handled in the binary system in either of two ways: direct binary subtraction (as shown above) or more commonly using two's complement arithmetic.

To see how negative numbers can be handled in the computer, consider a mechanical register, such as a car mileage indicator, being rotated backwards. A 5-digit register approaching and passing through zero would read the following:

---

*This section is based on material in Ref. 1, and the reader is referred there, or to an equivalent manual for additional details.

00005
00004
00003
00002
00001
00000
99999
99998
etc.

It should be clear that the number 99998 corresponds to -2. Further, if we add:

$$\begin{array}{r} 00005 \\ 99998 \\ \hline 00003 \end{array}$$

and ignore the carry to the left, we have effectively performed the operation of subtracting.

$$5 - 2 = 3$$

The number 99998 in this example is described as the ten's complement of 2. Thus in the decimal number system, *subtraction* may be performed by *adding* the ten's complement of the number to be subtracted.

If a system of complements were to be used for representing negative numbers, the minus sign could be omitted in negative numbers. Thus all numbers could be represented with five digits; e.g., 2 represented as 00002, and -2 represented as 99998. Using such a system requires that a convention be established as to what is and is not a negative number. For example, if the mileage indicator is set to 48732, is it a negative 51268 or a positive 48732? With an ability to represent a total of 100,000 different numbers (0 to 99999), it would seem reasonable to use half for positive numbers and half for negative numbers. Thus, in this situation, 0 to 49999 would be regarded as positive, and 50000 to 99999 would be regarded as negative.

In this same manner, the two's complements of binary numbers are used to represent negative numbers, and to carry out binary subtractions, in most computers. For example in a 12-bit computer the numbers from 0000_8 to 3777_8 are regarded as positive, and the numbers from

4000_8 to 7777_8 are regarded as negative. If these numbers were written in binary notation, it would be noticed immediately that all the positive numbers have a zero in the first (i.e., leftmost or MSD) position, while the negative numbers have a 1 in the first position. (Thus a MSD = 1 indicates that the value is negative, but it is not correct to say that the 1 in the MSD represents a minus sign.)

The two's complement of a number is defined as that number which when added to the original number will result in a sum of zero (when the carry beyond the MSD of the values being added is ignored). The binary number 110110110110 has a two's complement equal to 001001001010 as shown in the following addition.

$$
\begin{array}{r}
110\ 110\ 110\ 110 \\
001\ 001\ 001\ 010 \\
\hline
1\ 000\ 000\ 000\ 000
\end{array}
$$

The easiest method of finding a two's complement is to first obtain the one's complement, which is formed by setting each bit to the opposite value.

101 000 110 111	Number
010 111 001 000	One's complement of the number

The two's complement of the number is then obtained by adding 1 to the one's complement.

110 001 110 010	Number
001 110 001 101	One's complement of the number
+1	Add 1
001 110 001 110	Two's complement of the number

Subtraction is then performed by using the two's complement method. That is, to subtract A from B, A must be expressed as its two's complement and then the value of B is *added* to it. Example:

010 010 010 111	A
101 101 101 001	Two's complement of A

(carry is	011 001 100 010	B
ignored)	1 000 111 001 011	B - A

When negative (two's complement) numbers are involved, or when subtraction operations are carried out by addition of the two's complement of a number, the detection of overflow conditions is not as straightforward as we indicated earlier. For example, if we consider four-bit integers again but now assume that negative numbers are permissible, the range of integers which can be handled is:

$$\text{Max} = 7_{10} = 0111_2$$

$$\text{Min} = -8_{10} = 1000_2$$

Some typical numerical operations might be as shown in Example 6.7. We see from this example that determining whether a result is correct is much more complicated than when only addition of positive numbers is involved. In case (a) we obtain an erroneous result with no overflow; in (b), the correct result with no overflow. In case (c) we come to an erroneous result with overflow; however, in (d) we obtain a correct result with an overflow. Clearly the rule for detection of an overrange situation will have to depend on more than a check of the carry or overflow bit. After investigating a number of possibilities we would determine that the following rule is appropriate:

If the most significant bit of each addend is the same (both zeros or both ones) and the most significant bit of the sum is different from them, then overrange has occurred.

Hence overrange does not depend on the carry or overflow bit at all. This situation may appear strange at first glance, and the student is encouraged to test a few cases on his own. We conclude that, if integer addition is to be performed correctly, the computer must be able to test the most significant bits of numbers added (or subtracted) and of results, and to make "logical" decisions such as the one embodied in the overrange rule. Logical variables and operations of this sort will be discussed in Section 6.4.3.

Example 6.7. Detection of Overrange Condition

(a) $4_{10} + 5_{10}$:

$$4_{10} = 0100_2$$
$$5_{10} = 0101_2$$
$$9_{10} \quad 0 \quad 1001_2$$

No overflow ———⤴   ⤵——— Result = -7_{10}

(b) $4_{10} + (-5_{10})$:

$$4_{10} = 0100_2$$
$$-5_{10} = 1011_2$$
$$-1_{10} \quad 0 \quad 1111_2$$

No overflow ———⤴   ⤵———Result = -1_{10}

(c) $-4_{10} + (-5_{10})$:

$$-4_{10} = 1100_2$$
$$-5_{10} = 1011_2$$
$$-9_{10} \quad 1 \quad 0111_2$$

Overflow ———⤴   ⤵——— Result = 7_{10}

(d) $-4_{10} + (-4_{10})$:

$$-4_{10} = 1100_2$$
$$-4_{10} = 1100_2$$
$$-8_{10} \quad 1 \quad 1000_2$$

Overflow ———⤴   ⤵———Result = -8_{10}

6.4.2 Floating Point Numbers

In most of the preceding discussion in this chapter, we have assumed that the numbers in "fixed-word-length binary form" are integers, i.e., whole numbers without fractions. Most practical problems, particularly in engineering applications, involve fractional or "real" numbers. Integers such as 27 or 129 might be used to express the number of people in a room but so-called real numbers such as 4.038 or 251.86 would be needed to describe accurately the length of a stick or the volume of a bottle.

Any real decimal number can be represented as the product of two other numbers, one of which is a decimal fraction and the other a power of ten. For example, 32.767 can be represented as .32767 times ten to the second power. If we use the letter E to mean "times ten to the power", then 32.767 can be written as .32767E2. This method of representing real numbers is called exponential or scientific notation. The table below shows how several real numbers appear in scientific notation.

Common Notation	Scientific Notation
31,415,926.	.31415926E8
.000314159	.314159E-3
3.1415	.31415E1
-314.0	-.314E3
-.0314	-.314E-1

The above examples of scientific notation indicate that the power of ten or the exponent (the number to the right of E) is simply the number of places that the decimal point must be shifted to convert to common notation. A negative exponent such as E-3 indicates that the decimal point should be shifted three places to the left; a positive exponent such as E8 indicates that the decimal point sould be shifted eight places to the right.

Floating point notation in most computer systems consists of using two or more binary words to represent the fractional part and the exponent part of numbers written in scientific notation. The binary number representing the fractional part of the original number is called

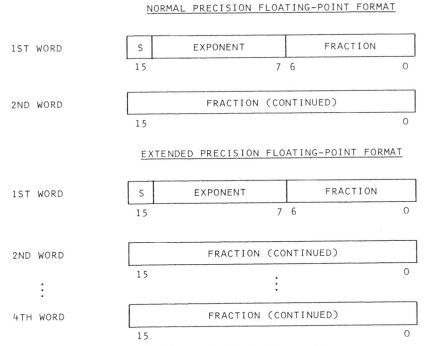

Figure 6.2. Floating Point Number Representations

the *mantissa* or *fraction*; the binary number representing the exponent is called the *characteristic* or *exponent*. Each of these components uses a predefined number of bits. The number of bits making up the characteristic will determine the maximum and minimum magnitude of the floating point number that can be represented. The number of bits making up the mantissa will determine the number of significant digits that can be represented.

Many computer systems offer expanded capability commonly called "Double" or "Extended" precision. This allows a user to specify a larger (more bits) mantissa which increases the number of significant digits that can be represented. This is particularly important in applications where repeated arithmetic manipulation may cause significant round-off errors.

As an example, Figure 6.2 gives the normal and extended floating point formats for the DEC PDP-11 line of computers.*

*This discussion is adapted from Ref. 1. The reader should consult the operating manuals for the computer system he is using because the conventions for floating point representations are system-dependent.

Note that in DEC 16-bit computers the leftmost bit (MSB) is bit 15, while the rightmost bit (LSB) is bit 0. This is the reverse of the bit numbering system used with some other small computers.

For positive numbers the sign (S) bit is 0, and for negative numbers the sign bit is 1 (note that the mantissa does not use the two's complement form for negative numbers). A different convention is used for expressing positive and negative exponents in the 8-bit field. The exponents are expressed in "excess 128_{10} notation" which means essentially that 128_{10} (200_8) is added to the binary exponent. Thus exponents from -128_{10} to $+127_{10}$ in standard notation become 0 to 255_{10} in excess 128_{10} notation (cf. example below).

The mantissa or fractional part of floating point numbers is usually assumed to be in "normalized" form; i.e., the decimal number is converted to binary, and then the exponent is adjusted so that the MSB of the fractional part (immediately to the right of the implied binary point) is nonzero. This means that as many significant figures as possible are preserved. DEC computers use one additional convention which effectively increases the number of

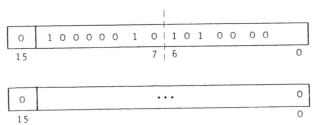

Figure 6.3. Floating Point Representation of +3.25

significant bits by one: since the MSB in a normalized binary number is always one, DEC omits it from the binary representation (but of course allows for it in all calculations and conversions). This is best illustrated by an example.

Consider the decimal number +3.25:

- In binary form (with implied binary point): $+11_\wedge01$.
- In normalized binary form: $+_\wedge1101$ with binary exponent 2.
- In DEC normal precision, memory image form, with exponent 130_{10}, we have the representation given in Figure 6.3.

Reconversion of the memory-image form to decimal proceeds as follows:

- Sign = positive
- Fractional part (allowing for the fact that the MSB is omitted but is always 1) = $_\wedge1101_2 = \frac{1}{2} + \frac{1}{4} + \frac{1}{16}$
- Exponent: $1\ 000\ 0010 = 2^7 + 2^1 = 130_{10}$
               (excess 128)
               $= 2_{10}$
- Thus: $+(\frac{1}{2} + \frac{1}{4} + \frac{1}{16}) \times 2^2 = +3.25$

The memory image representation of zero in normal precision floating point form is simply two words with every bit set to zero. A few examples done manually by the reader will clarify the steps involved in floating point conversions (and also emphasize the advantages of having the computer do all the conversions).

A common problem in both hand calculations and programming where real numbers are involved is how to keep track of the decimal point in a series of arithmetic operations. Although it is possible to note the position of the decimal (or binary) point and to remember

the position of the point in simple calculations, complex problems involving real numbers of widely differing sizes make it necessary to provide some automatic scheme to insure the proper positioning of the decimal point after each operation.

Floating-point addition and subtraction require the computer to perform the following operations:

(1) Compare characteristics (exponents).
(2) Shift mantissa of the smaller number so that both numbers have equal characteristics.
(3) Add or subtract the mantissa.
(4) Normalize the results, i.e., shift the mantissa so that the MSD is to the right of the radix point, and adjust the characteristic accordingly.

Floating point multiplication and division are much simpler:

(1) Multiply or divide mantissas.
(2) Add or subtract characteristics.
(3) Normalize the results.

These operations may be implemented with software (subroutines), with firmware (microprograms), or with hardware, and on most large computing systems are "invisible" to the user. However, with smaller computer applications, such as with a microprocessor, the user might have to program floating point operations himself.

Perhaps this is the place to emphasize that this chapter deals only with the "data" or "information" bits stored in the computer. The bits used to represent one word of data in binary form (typically 16 bits per word) may be augmented at various points within the computer system for specific purposes. For example:

(1) It is common to add an extra *parity* bit to every word (or in some cases to every byte [8 bits]). The parity bit would be set to zero or one in order to make the total number of one bits an odd number (odd parity) or an even number (even parity). The parity bit makes it possible to detect an odd number of errors in a "string" of bits—e.g., if one bit in a word is transmitted as a 1 instead of a 0, this will generate a "parity error".

(2) Several of the most recent computer systems, with high-speed semiconductor memory, add several bits (e.g., 5) to each word which permits the *detection* of most errors in a bit string and provides a basis for the automatic *correction* of all one-bit and most two-bit errors.

(3) Some computer systems attach an extra bit to each word of memory to permit "flagging" individual words in memory as "read-only" areas.

(4) Hardware memory management systems typically expand the "effective address" contained in any instruction from 16 bits to up to 20 or more address bits, so that additional memory can be addressed.

However, the implementation of these internal hardware features is invisible to most users, and hence it is not necessary to be familiar with the details of how these extra features are implemented unless one works at the detailed hardware or systems level.

6.4.3 Logical Variables and Operations

Logical variables can have only two possible values, TRUE or FALSE. In binary systems these are conveniently represented by 1 or 0 respectively, and hence the conversion problem is trivial. Note that many FORTRAN compilers use a whole word to represent a single logical value, e.g., FALSE = 0 and TRUE $\neq$ 0, whereas it often is the case with real-time systems that each bit in the word will represent a single logical state, e.g., FALSE = 0, TRUE = 1.

The logical operations most commonly used in computer applications are directly related to concepts involved in simple electrical switching circuits and to the formal mathematical definitions and relationships defined in "Boolean Algebra". Thus the description of the actual logical operations can be just as well included with the discussion of:

(1) Computer programming, since they are usually included as part of the assembly language instruction set.

(2) Computer hardware because circuits that perform these logical operations are the basic building blocks of digital computers.

(3) Formal mathematical techniques used by computer-oriented people.

The most commonly used logical operations are defined in the following sections to emphasize that a string of binary bits can represent *logical* information as well as *numerical* information. Also, as shown in one of the examples, logical operations can be used to convert information from one binary representation to another (i.e., alphanumeric code to a numeric value).

Logical AND Operation—The logical AND operation is analogous to a conductor with two switches A and B. In order that current may flow through it to C, both switches must be closed. We show this situation in Figure 6.4.

Logically this is written as follows (note that the symbols for logical operations are not the same in all references, e.g., AND is represented as • or $\wedge$, but not + which usually indicates an arithmetic operation or a logical OR operation):

$$A \bullet B = C$$

which means that C is true if A *and* B are true. In the digital computer a true condition is signified by the presence of a 1 bit, whereas a false

Figure 6.4. AND Operation Implemented with Two Switches

condition is signified by a 0 bit. The following table summarizes all possible outcomes of "ANDing" A and B:

A	B	C
0	0	0
1	0	0
0	1	0
1	1	1

An example of the AND operation using 8-bit bytes is:

$$\begin{array}{rl} \text{A} & 1\,0\,1\,1\,0\,1\,0\,1 \\ \text{AND} \quad \text{B} & 0\,0\,0\,0\,0\,1\,1\,1 \\ \hline \text{C} & 0\,0\,0\,0\,0\,1\,0\,1 \end{array}$$

Logical Inclusive OR Operation—A second logical operation is the inclusive OR. This an analogous to a circuit with two switches, A and B, in parallel as shown in Figure 6.5. Current may pass to C if either A or B is closed, or if both are closed.

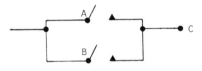

Figure 6.5. OR Operation Implemented with Two Switches

Logically this is written

$$\text{A} + \text{B} = \text{C}$$

which says that C is true if either A or B, or both, are true. Using a 1 bit to signify true and a 0 bit for false, we can construct the following table which shows all possible outcomes of "inclusively ORing" A with B:

A	B	C
0	0	0
1	0	1
0	1	1
1	1	1

An example of the inclusive OR operation using 8-bit bytes is:

$$\begin{array}{rl} \text{A} & 1\,0\,0\,0\,1\,1\,1\,0 \\ \text{OR} \quad \text{B} & 1\,0\,1\,0\,0\,1\,1\,1 \\ \hline \text{C} & 1\,0\,1\,0\,1\,1\,1\,1 \end{array}$$

Logical Exclusive OR Operation—The logical exclusive OR is analogous to a circuit with two switches A and B in parallel, but mechanically linked so that only one may be closed at a given instant (Figure 6.6). Logically this is written:

$$\text{A} \oplus \text{B} = \text{C}$$

The following table summarizes the results of all possible outcomes of performing an exclusive OR operation on A and B:

A	B	C
0	0	0
1	0	1
0	1	1
1	1	0

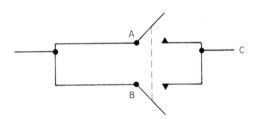

Figure 6.6. Exclusive OR Operation Implemented with Two Switches

An example of the exclusive OR operation follows:

$$\begin{array}{rl} \text{A} & 1\,0\,0\,1\,0\,1\,0\,1 \\ \text{EOR} \quad \text{B} & 1\,1\,0\,0\,0\,0\,1\,1 \\ \hline \text{C} & 0\,1\,0\,1\,0\,1\,1\,0 \end{array}$$

Other Logical Operations—Other logical operations can be defined, but the AND, OR, and EOR are the ones most commonly used in programming applications. For example, the logical NOT operator complements a binary bit or string of bits, i.e., forms the one's complement we defined earlier. For this reason the one's complement often is called the *logical* complement. The NEGATED or INVERTED logical operations, such as NAND (not And) and NOR (not Or), are sometimes used in describing, or implementing, computer applications. The reader may easily construct the appropriate truth tables by noting that C will simply be the negated value or complement of the values given in the AND and OR tables. Examples and further discussion are included in Chapter 7.

Applications—Most people are less familiar with logical operations than with standard arithmetic operations, and hence fail to use them as often and as effectively as they should. The following are examples of typical programming methods of manipulating bits within a given binary word.

(1) Specified bits in a given binary number can be "zeroed", without changing the other bits, by ANDing the word with a "mask" which contains 0's in the specified bit locations and 1's everywhere else.

(2) Specified bits in a given binary number can be set to 1 without changing the other bits by ORing the number with a mask containing 1's in the specified bit locations and 0's everywhere else.

(3) Selected bits in a given binary number can be complemented by EXCLUSIVELY ORing the number with a mask which has 1's in the selected locations and 0's everywhere else.

6.4.4 Alphanumeric Information

The two most common alphanumeric codes in use today are ASCII (American Standard Code for Information Interchange) and EBCDIC (Extended Binary Coded Decimal Interchange Code). These are 7- and 8-bit codes respectively, and the corresponding tables can be found in Table 6.5, and in most reference books. EBCDIC is used by most IBM computers, whereas ASCII is used by most other computer systems.

Any computer system consists of a large number of components such as the central processing unit, peripheral units such as a line-printer or a card reader, or sensor-oriented peripherals such as an analog to digital converter. Let's consider the familiar task of entering one instruction from a computer program into the computer using punched cards. A keypunch can be used to convert the instructions we write down in our computer program into holes on a set of punched cards. These punched cards are read into the computer system using a card reader which converts the information represented by the punched holes into a Hollerith card code. The Hollerith code can then be converted into ASCII or EBCDIC code.

These code conversion steps can be done by special electronic circuits added to the individual peripheral units instead of using computer subroutines. The decision of whether to use hardware or software is made by the system designer based on the tradeoff between the cost of the special electronic circuit vs. the implied cost in execution time and storage space required by the code conversion subroutines.

On some computer systems there are several different codes that are used by different devices, and a set of subroutines is required to convert from one to the other. In some real-time computing applications, fast responses may be so important that it is necessary to use hardware in order to avoid the delay caused by executing code conversion subroutines every time data is inputted or outputted from the system. (See Louden [2], pp. 326–327, Section 13.7.)

As a particular example of the difficulties that can arise due to the use of software code conversion subroutines, consider what happens when the "A" format is used in a FORTRAN program to be executed on a digital computer. If alphanumeric characters are stored internally in EBCDIC, then each character requires 8 bits. If the computer uses 16-bit words, then the maximum number of characters that can be read into a variable is as follows:

Integer (1 word)	2
Real (2 words)	4
Extended Precision (4 words)	8

If a variable has space left over, it is filled with

Table 6.5. ASCII and EBCDIC for Alphanumeric Symbols

EBCDIC (8-bit)		USASCII (7-bit)			EBCDIC (8-bit)		USASCII (7-bit)		
Hex	Name	Octal	Name	Teletype	Hex	Name	Octal	Name	Teletype
00	NUL	000	NUL	CT-SFT-P	86	f	146	f	NONE
01	SOH	001	SOH	CTRL-A	87	g	147	g	NONE
02	STX	002	STX	CTRL-B	88	h	150	h	NONE
03	ETX	003	ETX	CTRL-C	89	i	151	i	NONE
05	HT	011	HT	CTRL-I	8B	{	173	{	NONE
07	DEL	177	DEL	RUBOUT	91	j	152	j	NONE
0B	VT	013	VT	CTRL-K	92	k	153	k	NONE
0C	FF	014	FF	CTRL-L	93	l	154	l	NONE
0D	CR	015	CR	RETURN	94	m	155	m	NONE
0E	SO	016	SO	CTRL-N	95	n	156	n	NONE
0F	SI	017	SI	CTRL-O	96	o	157	o	NONE
10	DLE	020	DLE	CTRL-P	97	p	160	p	NONE
11	DC1	021	DC1	CTRL-Q	98	q	161	q	NONE
12	DC2	022	DC2	CTRL-R	99	r	162	r	NONE
13	DC3	023	DC3	CTRL-S	9A	NONE	140	Critic	NONE
15	LF	012	LF	LINE FEED	9B	}	175	}	ALT MODE
16	BS	010	BS	CTRL-H	A2	s	163	s	NONE
18	CAN	030	CAN	CTRL-X	A3	t	164	t	NONE
19	EM	031	EM	CTRL-Y	A4	u	165	u	NONE
1C	IFS	034	FS	CT-SFT-L	A5	v	166	v	NONE
1D	IGS	035	GS	CT-SFT-M	A6	w	167	w	NONE
1E	IRS	036	RS	CT-SFT-N	A7	x	170	x	NONE
1F	IUS	037	US	CT-SFT-O	A8	y	171	y	NONE
25	LF	012	LF	LINE FEED	A9	z	172	z	NONE
26	ETB	027	ETB	CTRL-W	AA	NONE	136	^	SHIFT-N
27	ESC	033	ESC	CT-SFT-K	AD	[	133	[	SHIFT-K
2D	ENQ	005	ENQ	CTRL-E	BA	NONE	134	\	SHIFT-L
2E	ACK	006	ACK	CTRL-F	BD	]	135	]	SHIFT-M
2F	BEL	007	BEL	CTRL-G	C1	A	101	A	A
32	SYN	026	SYN	CTRL-V	C2	B	102	B	B
37	EOT	004	EOT	CTRL-D	C3	C	103	C	C
3C	DC4	024	DC4	CTRL-T	C4	D	104	D	D
3D	NAK	025	NAK	CTRL-U	C5	E	105	E	E
3F	SUB	032	SUB	CTRL-Z	C6	F	106	F	F
40	Space	040	Space	Space	C7	G	107	G	G
4B	.	056	.	.	C8	H	110	H	H
4C	<	074	<	<	C9	I	111	I	I
4D	(	050	(	(	D1	J	112	J	J
4E	+	053	+	+	D2	K	113	K	K
4F		174		NONE	D3	L	114	L	L
50	&	046	&	&	D4	M	115	M	M
5A	!	041	!	!	D5	N	116	N	N
5B	$	044	$	$	D6	O	117	O	O
5C	*	052	*	*	D7	P	120	P	P
5D	)	051	)	)	D8	Q	121	Q	Q
5E	;	073	;	;	D9	R	122	R	R
5F	~	176	~	NONE	E2	S	123	S	S
60	-	055	-	-	E3	T	124	T	T
61	/	057	/	/	E4	U	125	U	U
6B	,	054	,	,	E5	V	126	V	V
6C	%	045	%	%	E6	W	127	W	W
6D	‾	137		SHIFT-O	E7	X	130	X	X
6E	>	076	>	>	E8	Y	131	Y	Y
6F	?	077	?	?	E9	Z	132	Z	Z
7A	:	072	:	:	F0	0	060	0	0
7B	#	043	#	#	F1	1	061	1	1
7C	@	100	@	@	F2	2	062	2	2
7D	'	047	'	'	F3	3	063	3	3
7E	=	075	=	=	F4	4	064	4	4
7F	"	042	"	"	F5	5	065	5	5
81	a	141	a	NONE	F6	6	066	6	6
82	b	142	b	NONE	F7	7	067	7	7
83	c	143	c	NONE	F8	8	070	8	8
84	d	144	d	NONE	F9	9	071	9	9
85	e	145	e	NONE					

Table 6.6. EBCDIC Representations of Alphabetic Characters

| Character | EBCDIC Code | |
	Binary	Hexadecimal
A	1100 0001	C1
B	1100 0010	C2
C	1100 0011	C3
D	1100 0100	C4
E	1100 0101	C5
F	1100 0110	C6

blanks (40) on the right (note that it is not filled with binary zeros). If a variable has too little space, the characters are truncated from the left as illustrated in Tables 6.6 and 6.7.

6.4.5 Binary Coded Decimal Data

Decimal numbers can be converted to a unique representation as a string of 1's and 0's by using four bits to represent each decimal digit. If the position coefficients for each four-bit group are 8-4-2-1 (which is identical to the first four positions in a conventional binary integer) then the numer 98540_{10} can be written as:

$$1001\ 1000\ 0101\ 0100\ 0000$$

This is obviously not very efficient, since each 4-bit string is capable of representing 16 values, but there are only 10 in the decimal system. However, a large number of similar codes, some of which are shown in Table 6.8, have been used in computer-related systems—particularly in older equipment or with laboratory-oriented instrumentation. Note that a single 4-bit string will result in the same decimal number whether treated as 8421 BCD, hexadecimal, or binary. However, the same statement does NOT apply to longer bit strings.

Conversions between different coding schemes are important computer operations and are implemented by hardware and/or computer programs.

Table 6.7. Storage of EBCDIC Coded Data

| Character String | Format* | Hexadecimal Representation | | |
		Integer (1 word)	Real (2 words)	Real (3 words)
A	A1	C140	C1404040	C14040404040
AB	A2	C1C2	C1C24040	C1C240404040
ABCD	A4	C3C4	C1C2C3C4	C1C2C3C44040
ABCDEF	A6	C5C6	C3C4C5C6	C1C2C3C4C5C6

* Format specified in the FORTRAN program to read the character string specified in column 1.

Table 6.8. Some BCD Codes

Decimal	8,4,2,1	Excess-3 (8,4,2,1 code for x + 3)	2,4,2,1
0	0000	0011	0000
1	0001	0100	0001
2	0010	0101	0010
3	0011	0110	0011
4	0100	0111	0100
5	0101	1000	1011
6	0110	1001	1100
7	0111	1010	1101
8	1000	1011	1110
9	1001	1100	1111

6.5 SUMMARY

A sound understanding of how various types of information are represented in fixed-word-length binary form is essential for most computer programming and application tasks. For example, all internal computer operations are carried out using binary representations so that if the conversion from or to the binary form is incorrect, then the final result will also probably be incorrect.

Even if the conversions from, or to, the desired binary representation are carried out correctly, serious errors can arise by performing computer operations that are not consistent with the binary representation. For example: a numeric value in binary form may be mistakenly interpreted as a computer instruction; the packed ASCII code representation of the symbols 53 may be mistakenly used in an arithmetic operation instead of the integer numeric value 53, or the floating point number 53.; an output message might be sent in ASCII to a terminal device that will only accept EBCDIC; etc.

Other errors can arise by unintentionally performing "mixed mode" operations with a computer. For example, we are all used to doing arithmetic with a mixture of integer and real numbers. However, such "mixed mode" operations can lead to errors in computer applications unless care is taken to make sure that all operands involved in a particular arithmetic operation are expressed in, or converted into, a consistent binary form. For example, even the simple operation of adding A to B will give an erroneous result if A is converted to binary form as an integer and B is converted as a floating point number, BCD value, ASCII code, etc.

Probably the most difficult thing for a new computer user to do is adapt his thinking and bases for decision making so they are efficient (or "optimal") for implementation using computers. Simple examples include the logical applications discussed earlier; using powers of 2 instead of powers of 10; realizing that the floating point representation of 2. might not be equal to the integer 2 owing to round-off error; using a single bit to store a logical variable instead of a whole word, etc. Skill can be developed by experience, but a sound understanding of binary representations (and internal computer operations) is obviously a prerequisite.

An understanding of internal data representations is also necessary to make an intelligent choice of such features as word length, a hardware feature of fundamental importance for any computer system.

6.6 REFERENCES

1. Digital Equipment Corporation, *Introduction to Programming*, Small Computer Handbook Series, Digital Press, Bedford, Mass., 1969.
2. Louden, R. K., *Programming the IBM 1130 and 1800*, Prentice-Hall, Inc., Englewood Cliffs, N.J., 1967.
3. Korn, G. A., *Minicomputers for Engineers and Scientists*, McGraw-Hill, New York, 1973.

6.7 EXERCISES

1. The purpose of this exercise is to provide experience in working with binary, octal, and hexadecimal numbers and practice in converting from one to the other.
   a. Convert the following decimal numbers to binary:
      i) 27
      ii) 128.77
   b. Convert the following binary numbers to octal and decimal:
      i) 100111001
      ii) 11010$_\wedge$011 ($_\wedge$ =implied binary point)
   c. Convert the following octal numbers to binary and hexademical:
      i) 645
      ii) 2
   d. Find the sum of the two decimal numbers 31 and 16 by converting each to binary, adding the binary numbers, and converting the answer to decimal.
   e. Find the sum of the two decimal numbers 33 and −13 by converting each to binary, taking the two's complement of the negative number, and adding. Convert the answer to decimal.
   f. A computer has 12-bit words of which the 11 low-order bits are data bits and the high-order bit is a parity bit to give "ODD" parity. This computer uses the two's complement method of storing negative numbers. Give the hexadecimal representation of the following decimal numbers as they would each appear

stored in one word:
   i) 479
   ii) 604
   iii) −726
   g. Give the hexadecimal representation of the six characters CH640b (b is a blank) as they would appear stored in three consecutive words in a 16-bit computer:
      i) If the code being used is EBCDIC.
      ii) if the code being used is ASCII.

2. Fill in the following table with the appropriate numbers, codes, or symbols corresponding to the one or two hexadecimal words given at the top of the table. These words can be thought of as the contents of specific words in the main memory of the computer system.

This exercise is designed to show that the contents of any given location in computer memory is simply a string of binary ones and zeros. There is nothing included in a given word that identifies it as an instruction, integer, or alphanumeric code, etc. It is the computer program (not the hardware) and/or various conventions and standards that determine how a particular word is interpreted.

	Single Word D140	Single Word 4060	Double Word C400007E
Binary			
Decimal Integer			
Decimal Floating Point			
EBCDIC (unpacked)			
EBCDIC (packed)			
ASCII (unpacked)			
ASCII (packed)			
Instruction			
ADC input (14 bits & sign)*			
Parity Bit (odd parity)			

Notes: (1) If a double word doesn't make any sense for a given category, interpret it as two separate single words.
(2) Some conversions in the above table are not defined. Enter "illegal" in the appropriate space.

*Interpret as a percent of maximum reading. This part might best be worked after studying Chapter 10.

3. Fill in the following table by calculating the equivalent codes of the decimal numbers or characters given in the first column.

Decimal Number Or Char.	Binary	Octal	BCD[2]	EBCDIC (Hex)	ASCII (Octal)	Computer Memory Image (Hex)
0	—	—	—	FO	—	—
0.	*	*	*	*	*	—
1	—	—	—	—	—	0001
1.	*	*	*	—	—	—
.12	*	*	*	*	*	—
13	—	15	—	—	—	—
—	—	—	—	*	*	FFF3 (Integer)
13.	*	*	*	*	*	—
−13.	*	*	*	*	*	—
A	*	*	*	—	—	—
=	*	*	*	—	—	—
6.6875	*	*	*	*	*	—
1.125×10^9	*	*	*	*	*	—
—	*	*	*	*	*	C000 007E[1]

[1] Normal precision floating point number in "standard form".
[2] Assume decimal point and sign are coded separately.

7

Digital (Binary) Logic and Hardware

George P. Engelberg
Canadian National
Montreal, Quebec

James A. Howard
Department of Electrical and Computer Engineering
University of California, Santa Barbara

7.0 INTRODUCTION

In this chapter we wish to introduce and develop concepts involving digital computer logic and hardware. The term "hardware" generally refers to the electronics and other components that make up a digital computer, hence the logical circuits, the memory, the processing units, the electronic interfaces for the peripheral devices, and so on. It is usually said that the hardware supports the software, software referring to the programs that the computer runs.

Our discussion of hardware will begin with the gate. The gate is the fundamental building block of computer circuits in the same way that an atom is the fundamental building block of molecules. Different types of gates will be examined, and it will be seen how gates can be connected to form other digital elements. Most important of these is the flip-flop, the digital memory element. Gates and flip-flops in turn form the computer's adders, registers, etc. The design procedures that are usually followed in putting together such digital circuits will be indicated.

A more macroscopic view of the computer is presented in Chapter 8 where the major functional components are examined, i.e., the Memory, Arithmetic Unit, Control Unit, and Input/Output Unit. It will be seen how these components combine to form the overall architecture of the computer.

The student is probably aware that computer hardware represents a vast topic, and a rigorous treatment of it would fill many textbooks. It is not our intention to condense that subject into a single chapter. Rather, we will indicate what the important concepts are and how they can be developed. A list of materials suitable for supplemental reading has been included.

7.1 COMBINATIONAL LOGIC

The term "combinational logic" refers to digital circuits possessing outputs whose present states have a functional relationship to the present states of their inputs. As an example of a combinational logic circuit similar to one that would be found in a computer, consider the following example of a simple security lock.

Example 7.1. A Security Lock. For purposes of illustration let us assume that a badge reader is used to open a door lock which normally is closed. The lock is to open if a card inserted in the badge reader is not green, or if the card is green and contains a valid code.

The logical equivalent of this statement can be expressed compactly by letting

Z represent—the door opens (output)
A represent—the code is valid (input)
B represent—the card is green (input)

Then the following logic expression can be written:

$$Z = (\text{not } B) \text{ or } (B \text{ and } A) \qquad (7.1)$$

To make the above expression more compact, a shorthand notation has been developed. It is usual to write "not B" as $\bar{B}$ (other common symbols used are B', $\tilde{B}$, and $\sim B$). Also the word "or" is replaced with "+," and the word "and" with "·". (Another common symbol for "or" is "$\vee$" and "$\wedge$" is sometimes used for "and".) With these substitutions we obtain:

$$Z = (\bar{B}) + (B \cdot A) \qquad (7.2)$$

Note that "·" and "+" refer to the logical operations AND and OR, not to be confused with their arithmetic counterparts. Since the "·" is usually not necessary to understand the expression, it is often omitted. The parentheses in logical Equation 7.2 may also be omitted, since the expression can be read unambiguously without them. We arrive at:

$$Z = \bar{B} + BA \qquad (7.3)$$

The above expression completely describes the circuit's behavior. In logical equations, the numerical values the variables may assume are either 0 or 1. These are the only values that logical inputs and outputs may ever have. Zero usually corresponds to logical "false", and one usually corresponds to logical "true". In the use of our example, if the card were green, B would be 1; otherwise it would be 0.

Now that the operation of the circuit is described in this compact form, how would someone go about actually implementing it? The answer is that there are standard electronic building blocks, called gates, which perform the elementary functions discussed (AND, OR, NOT). In the most commonly used system, with gates a logical 0 would correspond to 0 volts (approximately) and a logical 1 to 5 volts (approximately).* Let us examine these gates

---

*What we are describing here is TTL (transistor-transistor) compatible positive logic. For some applications a designer may want to work with a negative TTL system. The logical condition would then be true when the voltage is equal to 0 volts, and false when the voltage is 5 volts. Other logic systems have been used, for example, diode-transistor logic system (DTL) which utilizes −12 (approximately) and 0 volt inputs and outputs.

Table 7.1. A Two-Input AND Gate

A	B	AB
0	0	0
0	1	0
1	0	0
1	1	1

Figure 7.1. Schematic Representation of an AND Gate

in more detail, and then implement the security lock with them.

When input variables are ANDed together, the output is 1 if and only if all the inputs are 1. The behavior of a two-input AND gate can be represented by the "truth table" given in Table 7.1. A and B are inputs to the gate, and AB is the output. It is also convenient to be able to represent gates pictorially. The symbol used for an AND gate is shown in Figure 7.1.

When inputs are ORed together, the output is 1 if one or more of the inputs is 1. Table 7.2 shows the truth table for a two-input OR gate. The symbol used for an OR gate is shown in Figure 7.2.

When the NOT or complement of a variable is taken, the output is the opposite of the

Table 7.2. A Two-Input OR Gate

A	B	A+B
0	0	0
0	1	1
1	0	1
1	1	1

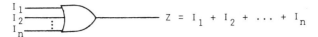

$$Z = I_1 + I_2 + \cdots + I_n$$

Figure 7.2. Schematic Representation of an OR Gate

Table 7.3. The NOT Operation

A	$\bar{A}$
0	1
1	0

Figure 7.3. Representation of the Complementation or NOT Operation

input. Table 7.3 shows the truth table, and Figure 7.3 shows the corresponding symbol.

Using the symbols just described, it is an easy procedure to represent pictorially the expression in Equation 7.3 which we have done in Figure 7.4. Note the direct correspondence between elements in the equation and Figure

7.4. A truth table, Table 7.4, can be constructed that corresponds to this figure. Figure 7.4 and Table 7.4 describe completely and unambiguously the operation of the security lock. They contain all the information a designer would need to construct the circuit.

Our example illustrates three simple logical operators—AND, OR, and NOT. Are there any others that are necessary in order to implement any required digital circuit? Surprisingly, the answer is no. It can be shown that any desired logical circuit can be built from these three operators alone. In fact, any circuit can be built using only AND and NOT, or OR and NOT. A little later other logical operators will be examined, but in each case we will see that they can be derived from AND, OR, and NOT.

7.2 BOOLEAN ALGEBRA

In the discussion of combinational logic circuits, use was made of algebraic type expres-

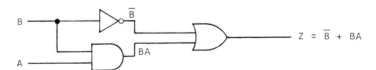

$$Z = \bar{B} + BA$$

Figure 7.4. A Logic Circuit Used to Implement the Security Lock

Table 7.4. Truth Table for the Security Lock

B	A	BA	$\bar{B}$	$\bar{B} + BA \ (=Z)$
0	0	0	1	1
0	1	0	1	1
1	0	0	0	0
1	1	1	0	1

sions, such as Equation 7.3. Expressions of this type are usually referred to as Boolean expressions. In fact, all the logical operations discussed up to this point properly belong to the branch of mathematics called Boolean algebra; the formal design of logical circuits is carried out with Boolean algebra.

In light of the ease with which the logic circuit in Figure 7.4 was developed, why is it necessary to formalize the procedure, especially since it is always possible to follow this approach to construct any desired logical circuit? The answer is that, unfortunately, the direct implementation method can be very inefficient in terms of the ultimate hardware design and the time required to build it. Expressions that are written down immediately from the verbal description often contain redundant information even though they may not appear to. By redundant information we mean that more literals (letters) than necessary are used in the algebraic expression. Since each occurrence of a literal means that a gate is necessary with that literal as an input, reducing the amount of literals will save both electronic circuitry and money. The longer a Boolean expression, the more likely it is that it can be simplified. By using the fundamental theorems of Boolean algebra, expressions can be manipulated and the redundancies eliminated.

The fundamental theorems are stated here without proof:

Uniqueness

    1a. The element 1 is unique.

    1b. The element 0 is unique.

Complementation

    2a. $A + \bar{A} = 1$

    2b. $A \cdot \bar{A} = 0$

Double Negation

    3. $\bar{\bar{A}} = A$

De Morgan's Theorems

    4a. $\overline{A + B} = \bar{A} \cdot \bar{B}$

    4b. $\overline{AB} = \bar{A} + \bar{B}$

Absorption

    5a. $A + AB = A$

    5b. $A(A + B) = A$

Sameness

    6a. $A + A = A$

    6b. $A \cdot A = A$

Union and Intersection

    7a. $A + 0 = A$

    7b. $A \cdot 1 = A$

    8a. $A + 1 = 1$

    8b. $A \cdot 0 = 0$

Commutation, Association and Distribution

    9a. $A + B = B + A$

    9b. $AB = BA$

    10a. $A + (B + C) = (A + B) + C$

    10b. $A(BC) = (AB)C$

    11a. $A + BC = (A + B) \cdot (A + C)$

    11b. $A(B + C) = AB + AC$

Of the above, de Morgan's theorems are perhaps the most important. One of their major uses is in complementing Boolean expressions. Theorems 4a and 4b demonstrate how expressions that contain only the positive literals A and B can be transformed so that the result contains only negative literals. The fact that the entire left-hand sides of 4a and 4b are complemented is no stumbling block. If the expression we wish to convert is $A + B$, for example, it can first be rewritten as $\overline{\overline{A + B}}$, and then as $\overline{\bar{A} \cdot \bar{B}}$.

Using the fundamental theorems, Equation 7.3 can be simplified as follows:

$$
\begin{aligned}
Z &= \bar{B} + BA && \text{(7.3)} \\
&= \bar{B} \cdot 1 + BA && \text{By theorem 7a} \\
&= \bar{B}(A + \bar{A}) + BA && \text{By theorem 2a} \\
&= \bar{B}A + \bar{B}\bar{A} + BA && \text{By theorem 11b} \\
&= (\bar{B}A + \bar{B}A) + \bar{B}\bar{A} + BA && \text{By theorem 6a} \\
&= (\bar{B}A + BA) + (\bar{B}A + \bar{B}\bar{A}) && \text{By theorem 10a} \\
&= (\bar{B} + B)A + \bar{B}(A + \bar{A}) && \text{By theorem 11b} \\
&= 1 \cdot A + \bar{B} \cdot 1 && \text{By theorem 2a} \\
&= A + \bar{B} && \text{By theorem 7b} \\
& && \text{(7.4)}
\end{aligned}
$$

Thus it is seen that the original expression was not in its minimal form, and it was possible to save one literal. This expression is equivalent in *every* respect to the longer original one. This can be proved by constructing the truth table for Equation 7.4 (Table 7.5).

It can be seen that the outputs of Tables 7.4 and 7.5 correspond exactly for the same input conditions B and A. A circuit diagram for the reduced expression is shown in Figure 7.5. Comparing this to Figure 7.4, it is apparent that

Table 7.5. Truth Table for the Security Lock (Alternate Realization)

B	A	$\bar{B}$	$A + \bar{B}$ (=Z)
0	0	1	1
0	1	1	1
1	0	0	0
1	1	0	1

one gate has been saved by minimizing the expression.

Applying the above theorems in the correct sequence is not easy, and for more complicated expressions it can be extremely difficult. (In fact the astute student may observe that expression 7.4 can be derived from 7.3 in only two steps!) Fortunately, there is an easier way to do the simplification. A *Karnaugh map* is a shorthand way of representing Boolean expressions pictorially. The map looks like a rectangular grid, and, when drawn correctly, terms that can be combined appear adjacent on the map. Redundant terms can be identified because they overlap other terms already entered onto the map. We will not enter into a discussion of Karnaugh maps; they are discussed in the introductory texts on Boolean algebra given at the end of this chapter. In addition to Karnaugh maps, computer programs exist that have been designed to minimize Boolean expressions.

There is one further advantage in the formal study of Boolean algebra. It is possible to manipulate Boolean expressions so that they take on a different form even if there is no savings in the count of literals or gates. This is handy for the person who must build the actual circuit. Certain electrical components have become standard (often because of cost or fabrication considerations), and it is often advantageous to be able to construct a circuit using these standard components. The two most common components are the NAND and NOR gates. A NAND gate functions as an AND gate followed by an inverter. The NOR gate functions as an OR gate followed by an inverter. The standard symbols for these are shown in Figures 7.6 and 7.7.

What advantages are gained by using NAND and NOR gates? Any Boolean expression, no matter how long or complicated, can always be expressed in terms of NAND gates only or in terms of NOR gates only. The circuit designer can then work with only one type of gate for the entire circuit. Figure 7.8 shows what Figure 7.5 looks like when implemented with two-input NAND gates, and Figure 7.9 shows the circuit with two-input NOR gates. Note that in Figure 7.9 a price is paid for using only NOR gates. One extra gate is required.

The transformation rule to obtain NAND or NOR representations is quite simple. Assume that an expression is written in the form

$$Z = A_1 + A_2 + \ldots A_m \qquad (7.5)$$

where $A_1, A_2 \ldots$ are each of the form

$$A_i = B_1 B_2 \ldots B_n \qquad (7.6)$$

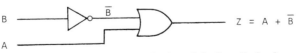

Figure 7.5. An Alternate Realization of the Security Lock

Figure 7.6. A NAND Gate

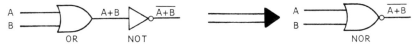

Figure 7.7. A NOR Gate

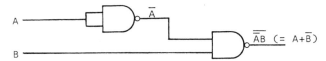

Figure 7.8. The Security Lock Implemented with NAND Gates Only

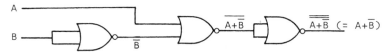

Figure 7.9. The Security Lock Implemented with NOR Gates Only

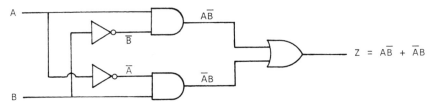

Figure 7.10. A Circuit which Performs the Exclusive-OR of Two Inputs

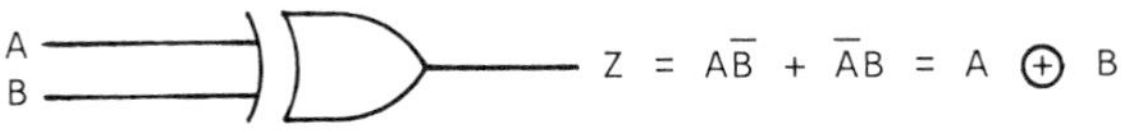

A ⊐⊐D— $Z = A\overline{B} + \overline{A}B = A \oplus B$
B

Figure 7.11. Symbolic Representation of an Exclusive-OR Circuit

Table 7.6. Truth Table for the Exclusive-OR

A	B	$A \oplus B$
0	0	0
0	1	1
1	0	1
1	1	0

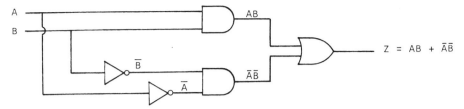

Figure 7.12. The Exclusive-NOR or Coincidence of Two Inputs

Then to obtain the NAND representation, only double negation followed by the application of de Morgan's Theorem 4a is required. For example, suppose that a truth table has described the expression

$$Z = \overline{A}\overline{B} + ABC \tag{7.7}$$

With double negation we obtain

$$Z = \overline{\overline{\overline{A}\overline{B} + ABC}} \tag{7.8}$$

Theorem 4a then gives

$$Z = \overline{\overline{\overline{A}\overline{B}} \cdot \overline{ABC}} \tag{7.9}$$

Equation 7.9 can be implemented solely with NAND gates.

The student can apply the same two transformations as above to obtain the circuit in Figure 7.8 from the original description

$$Z = A + \overline{B} \tag{7.4}$$

NOR representations are obtained by first writing the original expression in the form

$$Z = A_1 A_2 \ldots A_m \tag{7.10}$$

where $A_1, A_2, \ldots$ are each of the form

$$A_i = (B_1 + B_2 + \ldots B_n) \tag{7.11}$$

Then double negation followed by de Morgan's Theorem 4b is applied.

The NAND operation of two variables is usually represented as $\overline{I_1 I_2}$. Other representations that are used include $I_1 \uparrow I_2$, and $I_1 \wedge I_2$. The NOR operation is usually represented by $\overline{I_1 + I_2}$, but occasionally one sees $I_1 \downarrow I_2$ or $I_1 \vee I_2$.

There are two more binary operations that are very common. These are the exclusive-OR and exclusive-NOR functions. These are not elementary operations, because they are built by using a number of gates. However, they are used often enough that a shorthand notation has been developed for them.

When a circuit performs the exclusive-OR of two inputs, the output is 1 if and only if exactly one of the two inputs is 1. Figure 7.10 shows such a circuit, and Figure 7.11 shows the usual shorthand notation employed where the symbol $\oplus$ denotes the exclusive-OR operation. The truth table for the exclusive-OR is shown in Table 7.6.

A circuit is said to perform the exclusive-NOR, or the coincidence, of two inputs if and only if the output is 1 when both inputs are the same. Figure 7.12 shows the circuit, and Figure 7.13 shows the shorthand notation employed. In this case the symbol $\odot$ denotes the exclusive-NOR operation. Table 7.7 is the truth table for the exclusive-NOR.

From Tables 7.6 and 7.7, note that $\odot$ is $\overline{\oplus}$, and $\oplus$ is $\overline{\odot}$.

7.3 SEQUENTIAL LOGIC

Up until now we have been discussing circuits whose outputs are a function of the present

$$Z = \overline{A}\overline{B} + AB = A \odot B$$

Figure 7.13. Symbolic Representation of an Exclusive-NOR Circuit

Table 7.7. Truth Table for the Exclusive-NOR

A	B	A ⊙ B
0	0	1
0	1	0
1	0	0
1	1	1

inputs only. This class of circuits was referred to as combinational logic circuits. There is another class—those circuits whose behavior depends not only on the present inputs, but on past inputs as well. It is usually said that these circuits exhibit memory; these kinds of circuits are referred to as "sequential logic" circuits. To become familiar with the idea, consider another example, this time a circuit with memory.

Example 7.2. A Combination Lock. This time we consider a simple combination lock. There are three spring-loaded switches—a red one, a green one, and a blue one. The lock is to open if and only if the red button is pressed and *then* the green button is pressed. Pressing these two buttons in the wrong order does not open the lock, and pressing the blue button permanently disables the lock.

It is apparent from this description of the circuit that the output must depend on the order in which buttons are pressed; hence the circuit must have a memory to record past events.

Memory elements must possess two different stable states corresponding to a logical 0 and a logical 1. Switches, latching relays, and mag-

netic cores are elements displaying such a property. In digital circuits, an electronic element called a *flip-flop* is generally employed. Let us examine a typical flip-flop and see how it can exhibit memory. The flip-flop is usually drawn as a rectangle. The inputs R and S appear on the left. The output Q appears on the right as in Figure 7.14.

It can be seen from truth table given in Table 7.8 that when S is 1, the output is set to 1. When R is 1, the output is reset to 0. When both inputs are 0, the output remains in whatever state it was in previously—i.e., it doesn't change. This behavior allows the flip-flop to "remember" the occurrence of an event, because if the S input ever goes high, the output Q is 1, regardless of what S does afterward.

Flip-flops are really no different from the combinational logic discussed earlier. In fact, it is possible to build the flip-flop from two NOR gates as shown in Figure 7.15. (NAND gates could be used, as well, in a slightly different manner.)

The point to note from Figure 7.15 is that the use of feedback in a combinational circuit has allowed us to construct a memory element. Also note that an output Q̄ has been made available. This is simply the opposite or complement of Q. Q̄ is provided in most flip-flops as well as Q, since it can often save the circuit designer from having to use an inverter elsewhere in a circuit. For the flip-flop shown in Figure 7.15, we disallow S and R being 1 at the same time because the outputs Q and Q̄ would both be 0, and if both inputs returned to 0 simultaneously, the outputs Q and Q̄ wouldn't be predictable.

We can use the memory element just described to build the circuit for the combination lock described earlier as shown in Figure 7.16.

To understand the operation of the circuit assume that initially the outputs Q of all three flip-flops are 0. (Q̄ of FF3 is 1.) For Z to be high (i.e., for the lock to unlock), both inputs to gate 2 must be high. Now, if the blue button is ever pressed, FF3 is set and Q̄ changes from 1 to 0. This locks out the second AND gate and the output Z can never be 1. If the red button is pressed, the output Q of FF1 is set, enabling the top half of the first AND gate.

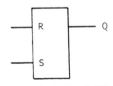

Figure 7.14. A Simple Flip-Flop

Table 7.8. Truth Table for the Flip-Flop

S	R	Q	Q_{NEXT}
0	0	0	0
0	0	1	1
0	1	0	0
0	1	1	0
1	0	0	1
1	0	1	1
1	1	0	Not allowed
1	1	1	

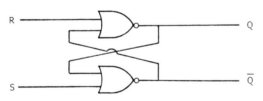

Figure 7.15. Construction of an RS Flip-Flop from Two NOR Gates

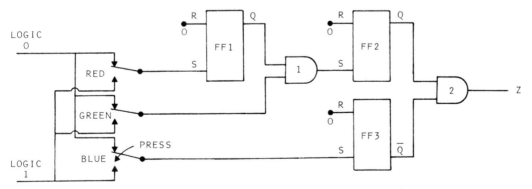

Figure 7.16. Circuit Representation of the Combination Lock

Pressing the green button can then cause the output of the first AND gate to be 1, setting FF2 and causing Z to be 1. Pressing the green button before the red one has no effect because the signal can't propagate through the first AND gate. Thus we have a simple example of how a circuit is able to "remember" events that have occurred in the past.

Most memory elements have extra features added to them in addition to those of the simple RS flip-flop discussed. The most common feature is a "clock input". The addition of

Table 7.9. Truth Table for a Clocked RS Flip-Flop

C	S	R	Q
0	0	0	Q
0	0	1	Q
0	1	0	Q
0	1	1	not allowed
1	0	0	Q
1	0	1	0
1	1	0	1
1	1	1	not allowed

a clock input allows the output to change only when the clock signal is high. The truth table for such a clocked flip-flop is shown in Table 7.9. Note that a shorthand convention has been employed in Table 7.9. Under the output column, the symbol Q is used in addition to 0 and 1. Such an appearance of Q means that the output remains in its present state given the input conditions on the left. Were the output to change to the opposite of its present state, this would be shown by entering a $\overline{Q}$.

Figure 7.17 shows how a clocked flip-flop could be constructed from combinational logic elements.

Why do we bother with a clock? The answer is that complicated circuits contain many gates, and when a circuit's inputs change, the outputs of the gates do not all change state at exactly the same instant because there is a finite time

delay or signal propogation time associated with each one, typically about 10 nanoseconds. Unpredictable behavior could arise if flip-flops were allowed to change state while the circuit was still "settling down." Thus, a single clock pulse applied to all the flip-flops after the signals had become stable throughout the circuit would insure correct behavior. Flip-flops containing a clock input are referred to as synchronous flip-flops, as opposed to asynchronous flip-flops—those that do not contain a clock.

Another feature found on many flip-flops is the direct clear and set inputs. In the discussion of the combination lock, the assumption was made that the flip-flops all had an initial state of 0. When a circuit first has power applied to it, this assumption cannot be made. Thus inputs are usually provided to clear or set

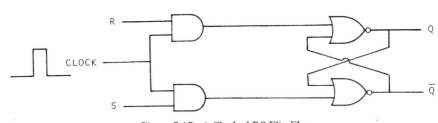

Figure 7.17. A Clocked RS Flip-Flop

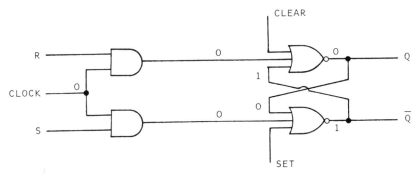

Figure 7.18. A Clocked RS Flip-Flop with Direct Set and Clear Inputs

the output without requiring the use of the clock. Depending upon the construction of the flip-flop, these inputs may be normally high (1) and perform their function when they go low (0), or vice versa. Figure 7.18 illustrates a clocked RS flip-flop with the direct clear and set feature added. In this figure, these lines are normally low (perform their function when high) Hence the circuit in Figure 7.18 may be "cleared" (reset) by placing the CLEAR and SET inputs in their logical 1 and 0 states, respectively. Setting the circuit would be accomplished by placing CLEAR and SET in their logical 0 and 1 states. Notice that these inputs override R, S, and CLOCK and, because of this property, are known as asynchronous inputs to the flip-flop. As in the case with the clock input, one direct set or clear signal can be applied to many flip-flops.

Figure 7.19 shows how such a flip-flop is usually represented in a circuit diagram. The little circles in Figure 7.19 indicate that in this

case the direct set and clear lines perform their function when they are low or at logic 0.

Other types of flip-flops exist in addition to the one described. Their operation is very similar, but the output behaves slightly differently depending on what the inputs are. The most common clocked ones are: the delay flip-flop, the toggle flip-flop, and the JK flip-flop. From the transition tables given along with Figure 7.20 we note that when the C input is high: the delay flip-flop's output Q is the same as its input D; the toggle flip-flop's output Q remains unchanged if T is 0 and is complemented if T is 1; the JK flip-flop's output Q behaves exactly as the output of a clocked RS flip-flop would except when J and K are both 1, in this case Q is simply complemented. Although the delay flip-flop is the one most commonly used, the JK flip-flop is the most versatile. In fact, all the other flip-flops may be derived from it.

Formal design procedures do exist for sequential circuits. Although it is usually possible to obtain a circuit from the verbal description, as with combinational circuits, they would often contain redundant electronic components. Various techniques make it possible to reduce the number of flip-flops necessary to construct the circuit.

7.4 ELEMENTARY BINARY SYSTEMS

The computer designer does not usually think in terms of gates and flip-flops. This is too microscopic a scale on which to work, and it is much more common to deal with modules of many gates and flip-flops which perform

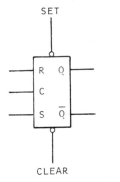

Figure 7.19. A Simplified Representation of a Synchronous RS Flip-Flop with Direct Set and Clear Inputs

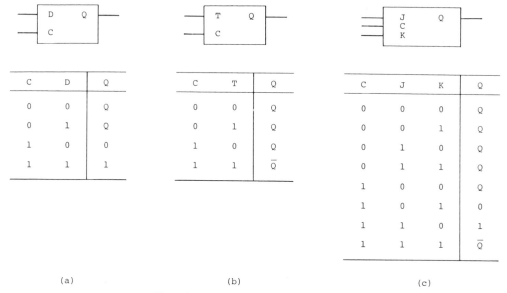

C	D	Q
0	0	Q
0	1	Q
1	0	0
1	1	1

C	T	Q
0	0	Q
0	1	Q
1	0	Q
1	1	$\overline{Q}$

C	J	K	Q
0	0	0	Q
0	0	1	Q
0	1	0	Q
0	1	1	Q
1	0	0	Q
1	0	1	0
1	1	0	1
1	1	1	$\overline{Q}$

(a)   (b)   (c)

Figure 7.20. Delay, Toggle and JK Flip-Flops

specific functions. Going back to our initial analogy, we can say we prefer to deal with molecules, not atoms. A simple example of this was shown before when the exclusive-OR and -NOR functions were introduced. Binary systems are usually broken down into two classes— combinational logic systems and sequential logic systems.

7.4.1 Combinational Logic Systems

There are many common circuits which are constructed from gates alone (as opposed to gates and flip-flops). A few examples will be considered.

Half Adder—The concepts of binary addition were discussed in Chapter 6. To accomplish addition with hardware a circuit is required which can generate the sum (S) and carry (C) for two inputs (I_1 and I_2). The truth table for the sum and carry is shown in Table 7.10. We note that the sum is 1 only when the two inputs are different. The carry is 1 only when the two inputs are both 1. Making use of the exclusive OR gate developed earlier, the circuit necessary to add two numbers can be drawn as

Table 7.10. Truth Table for a Half Adder

I_1	I_2	S	C
0	0	0	0
0	1	1	0
1	0	1	0
1	1	0	1

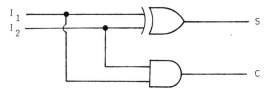

Figure 7.21. Representation of a Half Adder

in Figure 7.21. This circuit is usually referred to as a half adder.

Full Adder—The above circuit is suitable for determining only the least significant bit for the sum of two multibit binary numbers. For the other bits, there will be a carry input from the previous stage in addition to the other two inputs. This type of adder is referred to as a full adder because there are three inputs as opposed to only two for the half adder. The truth table for the full adder is shown in Table 7.11.

From the truth table, the expression for S can be written

$$S = \overline{I}_1\overline{I}_2C_{in} + \overline{I}_1I_2\overline{C}_{in} + I_1\overline{I}_2\overline{C}_{in} + I_1I_2C_{in} \tag{7.12}$$

Using the rules of Boolean algebra listed earlier in the chapter, this can be simplified to

$$S = I_1 \oplus I_2 \oplus C_{in} \tag{7.13}$$

where $\oplus$ refers to the exclusive-OR.

The expression for carry is

$$C_{out} = \overline{I}_1I_2C_{in} + I_1\overline{I}_2C_{in} + I_1I_2\overline{C}_{in} + I_1I_2C_{in} \tag{7.14}$$

which can be simplified to

$$C_{out} = I_1I_2 + I_1C_{in} + I_2C_{in} \tag{7.15}$$

Figure 7.22 shows the circuit representation of the full adder.

Using the half and full adder as modules, we can now proceed to build an adder to add two binary numbers of arbitrary length. For example, Figure 7.23 shows a circuit to add the binary numbers $A_n \ldots A_1A_0$ and $B_n \ldots B_1B_0$ to form the sum $S_n \ldots S_1S_0$.

The circuit in Figure 7.23 has the disadvantage that S_n cannot be determined until the input carry to that stage is determined, and this term is not available until all the other stages have

Table 7.11. Truth Table for the Full Adder

I_1	I_2	C_{in}	S	$\underline{C_{out}}$
0	0	0	0	0
0	0	1	1	0
0	1	0	1	0
0	1	1	0	1
1	0	0	1	0
1	0	1	0	1
1	1	0	0	1
1	1	1	1	1

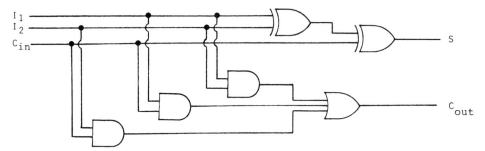

Figure 7.22. Representation of a Full Adder

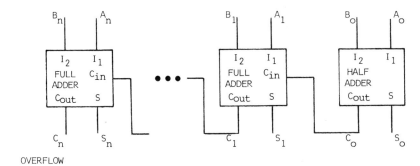

Figure 7.23. An n-Bit Full Adder with Overflow

settled down. In other words, all the carry signals must "ripple" from right to left. For many stages, the time necessary for this ripple to be completed can be undesirably long. Other faster addition schemes exist, commonly referred to as "carry look ahead" techniques. The basic ideas remain the same, however.

Decoder—Another common type of circuit is a decoder. A word of n binary bits can represent a total of 2^n different codes. For example, a two-bit word A_2A_1 has $2^2 = 4$ distinct codes, namely 00, 01, 10, 11. Often a circuit is required to determine which particular bit pattern (code) is at the inputs. Figure 7.24 shows a block diagram for such a decoder. To derive the circuit, the truth table (Table 7.12) is constructed. From the truth table, the circuit given in Figure 7.25 can be drawn. This type

of circuit could be used, for example, to decode a computer instruction. Once it was decoded, the computer processing unit could begin to execute the instruction.

Multiplexer—Often the situation is encountered where there are many lines, each containing binary information, and we would like to transfer the information on exactly one of these lines to a single output line. This type of operation is called multiplexing. Figure 7.26 shows a block diagram of a four-line input multiplexer. The lines S_1 and S_2 control which input line is transferred to the output Y. Table 7.13 shows the correspondence. Proceeding directly from Table 7.13, the necessary circuit diagram can be drawn as in Figure 7.27. If S_2S_1 is 01, for example, the reader can verify that line I_2 is connected to the output while the other lines are blocked at their respective AND gates.

The above multiplexer transfers only one bit of information at a time to the output. Many such circuits operating in parallel can be imagined, controlled by the same two lines S_2 and S_1. In this case, many bits of information could be transferred over the multiple output lines simultaneously. Figure 7.28 illustrates a four-word, N-bit multiplexer.

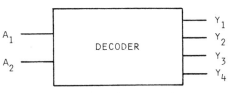

Figure 7.24. A 2 to 4 Decoder

Table 7.12. Truth Table for the Decoder

A_2	A_1	Y_1	Y_2	Y_3	Y_4
0	0	1	0	0	0
0	1	0	1	0	0
1	0	0	0	1	0
1	1	0	0	0	1

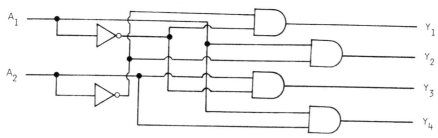

Figure 7.25. Circuit Representation of the 2 to 4 Decoder

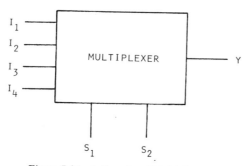

Figure 7.26. A Four-Input Multiplexer

7.4.2 Sequential Logic Systems

Systems of this type derive their name from the fact that they employ flip-flops in addition to gates. In this section a few of the more common sequential logic circuits will be examined.

Latch—One application of sequential logic is for buffering and storing data. We have seen how a flip-flop can be used to store a single bit of data. By using more than one flip-flop, many bits of data, or entire words of data, can be stored. A common circuit to perform this function is called the latch. Figure 7.29 shows a three-bit latch that employs simple unclocked RS flip-flops. The outputs Q_1, Q_2, and Q_3 follow the inputs I_1, I_2, and I_3 when the input E is high. When E is low, the outputs of the AND gates are low, and the flip-flop outputs no longer respond to the input lines. Thus, when it is desired to store the inputs I_1, I_2, I_3 the control signal E could go high for a small period of time, and then low again. We can envisage such a circuit being used, for example, in a peripheral device that has gathered data from an experiment and is waiting for the computer to access the information.

Table 7.13.   Operation of the Multiplexer

S_2	S_1	Line Transferred to Output
0	0	I_1
0	1	I_2
1	0	I_3
1	1	I_4

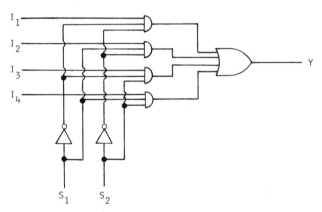

Figure 7.27. Circuit Representation of the Four-Input Multiplexer

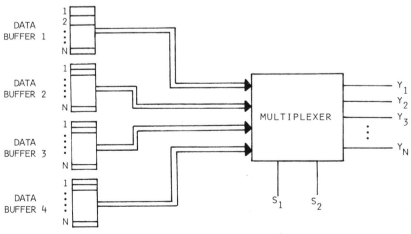

Figure 7.28. A Four-Word Multiplexer (N Bits per Word)

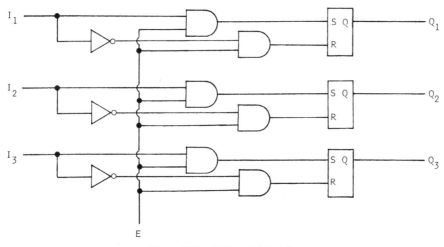

Figure 7.29. A Three-Bit Latch

Shift Registers—A shift register is similar to the data latch, but may possess other capabilities as well. In addition to being able to load information in parallel as with the latch of Figure 7.29, it can also shift binary information left or right between the flip-flops within the register. In other words, a shift register can be loaded in parallel and then send its data serially, and vice versa. This application is very common in digital logic, a simple example of which would be in receiving binary information serially from a telephone line and then sending it in parallel to a computer. Clocked flip-flops are employed in shift registers. A block diagram of a typical four-bit shift register is shown in Figure 7.30.

Figure 7.31 shows the circuit diagram of a simple four-stage shift register without a parallel load capability. This register can only right shift its one input line toward the output flip-flop. Note how the output of one stage feeds the input of the next stage. More capabilities could be added to the register, such as parallel load and bidirectional shifting, by using control gates in addition to the four flip-flops.

Monostable (One Shot) Multivibrator—The monostable multivibrator is stable in one output state only. It can be forced (triggered) into its unstable output state where it remains for a time determined by the components in its logic circuitry, and then automatically returns to the stable state. Figure 7.32 shows the generalized operation of a monostable device.

A well-designed monostable multivibrator should have the following characteristics for the time duration of the unstable state:

- It should be independent of the triggering pulse width.
- It should be adjustable.

A monostable circuit constructed from a clocked JK flip-flop and inverters is shown in Figure 7.33.

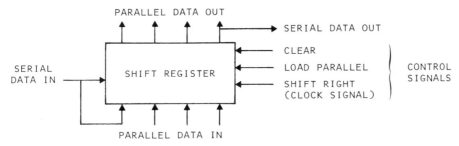

Figure 7.30. A Four-Bit Shift Register

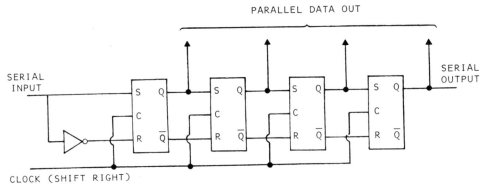

Figure 7.31. Circuit Representation of a Simple Four-Stage Shift Register

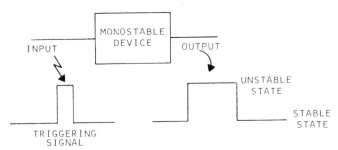

Figure 7.32. Generalized Monostable Multivibrator Operation

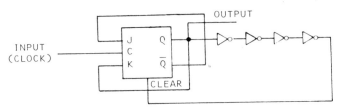

Figure 7.33. A Clocked monostable Multivibrator

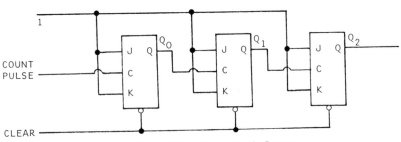

Figure 7.34. A Three-Stage Ripple Counter

For the above JK flip-flop, assume that initially $Q = 0$, which means that the output of the last inverter is also 0. Note, too, that the flip-flop clears when the clear input is high. The reader can verify that a clock pulse will set the flip-flop, and the Q output will then start rippling through the chain of inverters. The time to ripple through to the last inverter is dependent upon the individual gate delays. When the rippling is complete, the clear input becomes high, and the output returns to its stable 0 state. The time duration of the unstable state is directly controlled by the number of inverters in the feedback chain.

The above design is suitable for outputs typically up to 100 nanoseconds in duration. More versatile monostables are also available off the shelf in integrated circuit form. These require external circuitry (usually a resistor and capacitor) to adjust the output pulse width; the typical adjustment range is from a few nanoseconds to several seconds.

One application of monostable devices is in the generation of relatively long pulses (10 to 20 milliseconds, for example) which are required to drive an electromechanical device such as a reed relay or stepping motor. This requirement often arises in the interface circuitry between a small computer and external electromechanical devices driven by it. Another application is in the generation of timing pulses for computer interface control circuitry, where very accurate pulse widths may be required.

Counter—One of the most important classes of sequential logic circuits is the counter. As its name implies, this circuit can be used to record the number of times an event occurs. Counters find their way into a number of different digital applications. The register that points to the current computer instruction being executed operates as a counter. Many peripheral units must put out a specified number of pulses to control an external device. This is usually accomplished through use of a counter.

Many different types of counters exist. Some count up, some count down, some can be programmed to count up to a preset value and then reset themselves to zero, etc. Figure 7.34 illustrates a simple three-stage binary ripple counter. Note that JK flip-flops are used in this counter because their outputs change state if a clock pulse occurs and both the J and K inputs are high. In Figure 7.34, the clear line is used to initially set the output of all flip-flops to 0. Upon each clock pulse, the least significant bit, Q_0, increases by one, and the count sequence $(Q_2Q_1Q_0)$ is 000, 001, 010, 011, 100, 101, 110, and 111. The next clock pulse causes the count to be 000, since the maximum binary number that can be represented with three bits is 111. Note how the count pulse feeds only into the clock input of the first-stage flip-flop. The following stages are clocked by the transition of the output of the previous stage.

To understand that the flip-flops in the circuit respond correctly when clocked, we must expand a little on our explanation of clocked flip-flops. In the discussion of Figure 7.17, it was indicated that when the clock is high, the output may respond to the input. Although flip-flops of this type exist, it is more common for the flip-flop to be constructed so that the response to the input occurs not when the clock is high, but when the clock *changes* from high to low (or, for some types of flip-flops, from low to high). This insures that the output can change at most once during each clocking cycle. Figure 7.35 shows a timing diagram for the circuit of Figure 7.34.

The type of counter illustrated here is known as an asynchronous ripple counter because the last stage does not respond until the effects of the input clock pulse have travelled through the entire string of flip-flops. Although this delay is too small to show up in the timing diagram, it can be important for counters with many stages. To circumvent this potential problem, synchronous counters can be used which allow for simultaneous clocking of all stages. These

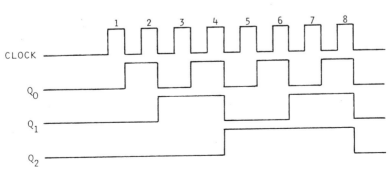

Figure 7.35. Timing Diagram for the Counter

have increased speed, but extra circuitry is added to eliminate the need for rippling.

7.5 A LOGIC DESIGN EXAMPLE

We will end our discussion of logic design with an example that illustrates many of the techniques and concepts presented in this chapter.

Example 7.3. Design of a Pulse Totalizer. Suppose we wish to be able to count the number of times a certain event occurs within a specific time interval. We would like to be able to vary this time interval in separate runs of our experiment over a range of 1 to 7 seconds. We know that our external event sensor is constructed in such a way that it generates a logic 1 level pulse lasting one millisecond at each occurrence of the observed event. It is also known that we generally can expect the event to occur 15 times or less during our maximum counting interval, but as a precaution we would like to provide some indication if it occurs more than 15 times. We also need some indication when a timing interval has elapsed, meaning that our timing device must provide a logic signal to indicate this state.

This description of the design problem is complete enough that we may now begin a solution.

The logic system we must design has two

basic requirements. First, it must be able to time an interval accurately (whose length we can specify) of from 1 to 7 seconds. Second, it must be able to count the number of event pulses occurring during the timed interval and allow us to obtain the final result when completed. Recall from Section 7.4 that we designed a device to act as a pulse counter known as a ripple counter. As shown in Figure 7.36, we can use two of these devices in designing this system.

The upper counter, labeled Time Counter in the diagram, is used in conjunction with a clock pulse generator to satisfy the first design requirement. Notice that this counter has been provided with a parallel load capability to allow specification of the timing interval. In order that we may simplify the logic design task we will require that the interval length be specified in two's complement form. This allows us to use a counter that counts up rather than down. (The counting operation amounts to adding one to the contents of the counter for each pulse counted.) Therefore, by counting pulses originating from a clock which generates one pulse every second we may detect the end of the desired interval by sensing the all zero state of the counter. We may see that this follows from a simple example. Suppose the desired interval is 2 seconds; the counter is loaded with 1110_2 $(= -2_{10})$.

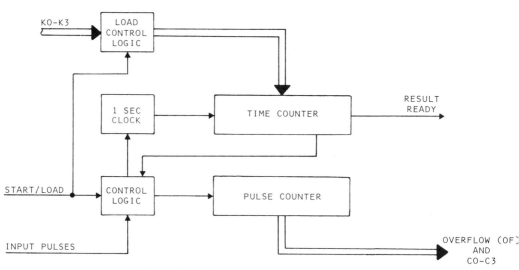

Figure 7.36. Block Diagram of the Pulse Totalizer

Counter Contents:

$$1110 \xrightarrow[\substack{\text{first} \\ \text{pulse} \\ \text{counted}}]{} 1111 \xrightarrow[\substack{\text{second} \\ \text{pulse} \\ \text{counted}}]{} 0000$$

Because the maximum time interval is 7 seconds, we use a four-stage counter in this part of the circuit. To accomplish the parallel loading of the counter we will make use of the asynchronous SET and CLEAR inputs of the JK flip-flops as shown in Figure 7.37. We may detect the overflow (all zero) state of the counter easily as a result of requiring that the desired timing interval be specified in its two's complement form. Notice that the most significant bit of a negative two's complement number is always 1. This means that K3, the fourth or most significant flip-flop in the Time Counter, may be used to detect overflow, since it will contain a 0 only when the all zero state of the counter occurs.

The second design requirement is satisfied with the Pulse Counter in Figure 7.37, a five-stage ripple counter. The Pulse Counter is constructed in such a way that the fifth stage serves to indicate the occurrence of more than 15 event pulses, i.e., an overflow.

We assure timing accuracy by using a 1-second clock with a single control input. It is assumed that the clock begins to generate pulses exactly one second after the control input is placed in the logic 1 state. The output of the fourth stage of the Time Counter is used in combination with the START/LOAD pulse to enable operation of the clock which, as a result, is synchronized to the START/LOAD pulse at the initiation of a timing cycle.

Figure 7.37 shows a diagram of all the logic required to implement the Totalizer as we have described it. Four inputs, K0–K3, allow entry of the two's complement interval constant. One input, START/LOAD, is used to load data from K0–K3 and to start the counting operation. One input, INPUT PULSES, receives event pulses. Five outputs, C0–C3 and OF, allow access to the contents of the Pulse Counter. One output, RESULT READY, is used to indicate the end of the time interval. Operation of the circuit is best illustrated by referring to the timing diagram given in Figure 7.38. The sequence of operations is initiated

by placing the two's complement representation of the desired timing interval on lines K0–K3 and applying a pulse at the START/LOAD input. The two-input AND gate shown connected to the enable input of the clock insures that pulses are not generated until START/LOAD returns to logic 0. At that point the Time Counter and Pulse Counter are enabled. Notice that pulses counted by each counter, as shown in Figure 7.38, affect the states of the counters at their negative-going or falling edge. This is a common feature of some commercially available JK flip-flops which operate in the Master–Slave mode. Each flip-flop is constructed in such a way that data appearing at the J and K inputs are transferred to the Q and $\overline{Q}$ outputs at the negative edge of the clock pulse. Completion of the totalizing operation is indicated by the $\overline{Q}$ (inverted) output of T3 going to logic 1 as shown in Figure 7.37. This output provides a "clean" edge or level transition which may be used in controlling additional circuitry. We may, for example, use this output (RESULT READY) to interrupt a computer which makes use of the Totalizer in controlling a process.

The reader should realize that there are many alternative ways of handling this design problem, some more complex, some less. For example, one approach would be to use synchronous rather than asynchronous (ripple) counters. The list of supplemental reading at the end of this chapter is designed for the person who may need to get involved in the actual design of such circuits.

7.6 SUMMARY

In this chapter we have progressed from a basic discussion of Boolean logic to the actual design of a fairly complex digital system. We have seen that if we begin with a clear, unambiguous description of a desired logical function we can arrive fairly directly at a realization of the function (if it is realizable) in the form of a digital circuit. We discovered that many times we can simplify our design by applying the theorems of Boolean Algebra to its mathematical representation.

Two classes of digital logic were described—

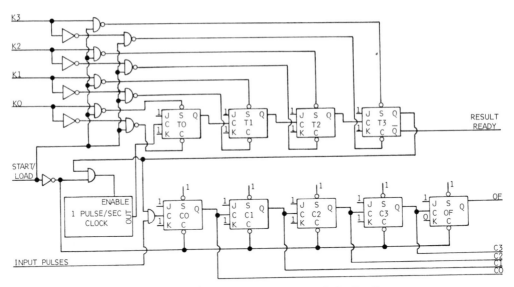

Figure 7.37. Logic Implementation of the Pulse Totalizer

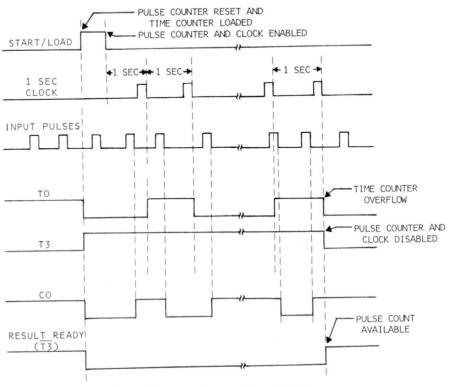

Figure 7.38. Pulse Totalizer Timing Diagram

combinational (or static) logic and sequential logic. We observed that the logical activity of a combinational circuit depends only on the present state of its inputs, while a sequential circuit makes use of past inputs as well. In order that past inputs may be used by a sequential circuit we saw that it was necessary to introduce a digital memory element—the flip-flop. Through the combined use of combinational and sequential circuits we found it possible to design such useful systems as counters and shift registers. In Chapter 8 we show that these and other similar macro-level digital elements may be used in the design of a digital computer.

Until the mid-1960s gates and flip-flops were constructed from individual transistors and resistors. Typically, about half a dozen transistors were required to construct either a gate or a flip-flop, the exact number depending upon the element's complexity. After the mid-1960s, however, it became possible to fabricate more than one transistor in the same package, and one or more complete gates or flip-flops were manufactured in a package about the size of a dime. These packages are called integrated circuits. More recently it has become possible to manufacture hundreds and even thousands of transistors in the same small package. Entire counters, shift registers, adders, even a computer CPU (microprocessor), are now obtainable in one integrated circuit. This allows complex logical systems to be designed on a functional level rather than on a gate and flip-flop level and makes possible the design of much more complex systems than would be feasible using individual gates and flip-flops.

7.7 SUPPLEMENTAL READING

1. Roth, Charles H., Jr., *Fundamentals of Logic Design*, West Publishing Co., St. Paul, Minn., 1975.

2. Kohave, Zve, *Switching and Finite Automata Theory*, McGraw-Hill Book Co., New York, 1970.
3. Peatman, John B., *The Design of Digital Systems*, McGraw-Hill Book Co., New York, 1972.
4. Hill, F. J. and Peterson, G. R., *Digital Systems: Hardware, Organization, and Design*, John Wiley & Sons, Inc., New York, 1973.
5. Dietmyer, Donald L., *Logic Design of Digital Systems*, Allyn and Bacon, Inc., Boston, 1971.
6. Brzozowski, J. A. and Yoeli, M., *Digital Networks*, Prentice-Hall, Inc., Englewood Cliffs, N.J., 1976.
7. Mowle, Frederic J., *A Systematic Approach to Digital Logic Design*, Addison-Wesley Publishing Co., Reading, Mass., 1976.
8. Hill, F. J. and Peterson, G. R., *Introduction to Switching Theory and Logical Design*, 2nd Ed., John Wiley and Sons, New York, 1974.
9. Friedman, Arthur D., *Logical Design of Digital Systems*, Computer Science Press, Inc., Rockville, Md., 1975.
10. Malmstadt, H. V. and Enke, C. G., *Digital Electronics for Scientists*, W. A. Benjamin, Inc., Menlo Park, Ca., 1969.
11. Marcovitz, Alan B. and Pugsley, James H., *An Introduction to Switching System Design*, John Wiley and Sons, Inc., New York, 1971.

7.8 EXERCISES

1. For the Boolean expression:

$$E = AC + \overline{B(C\overline{D})} + \overline{(A\overline{B} + BC)D}$$

what is the value of E when $A = B = 1$, $C = D = 0$?

2. Produce a truth table for the circuit illustrated. Why do you think this circuit is called a "Majority Voter"?

3. Using AND, OR, and Inverter (NOT) gates, realize the following function:

$$E = (A\overline{B} + CD)(\overline{\overline{\overline{AC}}} + BD)$$

4. A 4-bit True/Complement I/O gate, shown in the illustration, is to be designed. Depending on the control inputs, the gate transfers the 4-bit input, or their one's

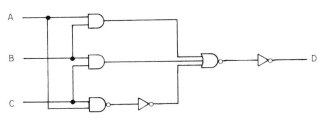

Figure for Exercise 2

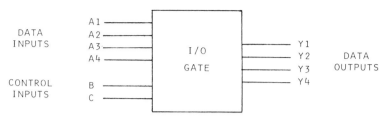

Figure for Exercise 4

complement to the output. Furthermore, the control can also set all outputs to 0 or 1. These conditions are summarized in the truth table.

Control Inputs		Outputs			
B	C	Y_1	Y_2	Y_3	Y_4
0	0	$\overline{A_1}$	$\overline{A_2}$	$\overline{A_3}$	$\overline{A_4}$
0	1	A_1	A_2	A_3	A_4
1	0	1	1	1	1
1	1	0	0	0	0

From the truth table it is apparent that the outputs Y_1–Y_4 are all the same for the various control inputs. Thus, the logic circuit of Y_1 may be designed by itself, and then replicated four times to structure the complete circuit. Create a new truth table for Y_1, expressing Y_1 specifically as a function of the inputs B, C, and A_1. Draw the resulting logic circuit for Y_1. Note that in a modular fashion all the circuitry for the

other input and output variables can be added.

5. A pair of flip-flops is connected as shown and has initial state $Q_1 = 0$, $Q_2 = 1$.
   a. If the CLEAR line is held at 1 and INPUT changes as illustrated in the figure, what is the final state of Q_1, Q_2? Assume that the flip-flops change state when their clock level changes from logical 1 to logical 0.
   b. If the CLEAR line is dropped to 0 momentarily and then the input changes as in part a, what will be the final state of Q_1, Q_2?

6. The logic circuit presented in Figure 7.33 was suitable only if the monostable's unstable output state is to last for an even number of inverter gate delays. Explain why the circuit would not function with an odd number of inverters. Make a slight modification to the circuit to permit it to produce an unstable output for an odd number of gate delays.

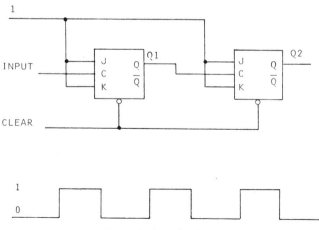

Figure for Exercise 5

Part IV
Real-Time Digital Systems Architecture

8

Digital Computer Architecture

George P. Engelberg
Canadian National
Montreal, Quebec

William R. Hughes
Department of Electrical Engineering
University of Colorado, Boulder

James A. Howard
Department of Electrical and Computer Engineering
University of California, Santa Barbara

8.0 INTRODUCTION

A brief introduction to the structure or architecture of modern small computers was presented in Chapter 2. Here we will extend this discussion in such a way that the reader may obtain a fairly detailed knowledge of computers constructed with a classical architecture. By this we mean a computer that may be functionally decomposed into four standard, interconnected subsystems—a Control Unit, an Arithmetic Unit (AU), a Main Memory Unit, and an Input/Output (I/O) Unit. Figure 8.1 illustrates such a classical decomposition. We note that the four subsystems are interconnected by four signal pathways or buses. The memory, arithmetic, and input/output buses serve to carry binary information between the various units, while the control bus carries signals that are used to properly sequence and control the operations of the AU, Memory, and I/O Unit.

The computer that we use here as an example is similar in structure to one originally proposed by Burks, Goldstine, and Von Neumann in 1946. Von Neumann et al. originated the conceptual basis for all modern computers—the *Stored Program* processor. This concept requires both instructions and data to occupy the same memory space; in fact such a computer has no way of differentiating data and instructions contained in its memory. Though this now seems an obvious storage technique to employ, it had been previously supposed that a computer should be provided with separate instruction and data memories.

The decision to store data and instructions in the same memory has a number of basic consequences on the architecture of a computer. Obviously, the amount of memory we use to represent an instruction must be compatible with the technique used to represent data so as to avoid inefficient use of available memory. In other words, if we decide always to use 18 *bits* (*binary digits*) in representing a data item, but we are able to specify all the instructions our computer may execute in 16 bits, 2 bits per instruction will be wasted if we fix the *word* (the smallest unit of uniquely accessible memory) of our memory at 18 bits. Designers of small computers have generally selected a word size that is *byte*-oriented (8 bits). Such a word size (usually 16 bits) allows efficient storage of character data—each character requires 8 bits—and has proved to be of sufficient size to specify fairly complex instruction sets.

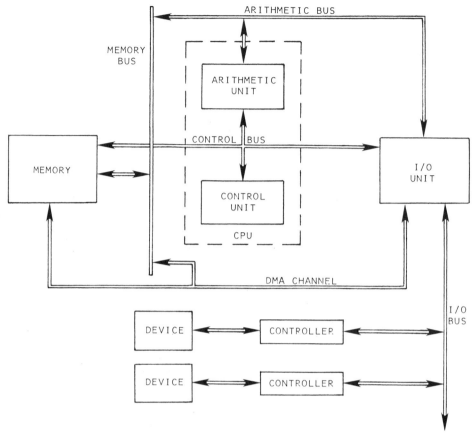

Figure 8.1. Classical Multiple Bus Computer

Fixing the size of a memory word has a general structural effect on all of the subsystems of a computer. The Control Unit—whose function it is to sequence and control the execution of an instruction—must be constructed in such a way that it can correctly interpret the meaning of an instruction word of the selected size. The Arithmetic Unit—which serves as the actual data processing element—must be designed to allow operations to be performed on data elements of the selected size. Finally, the Input/Output Unit—which is used to effect communication between the AU (or the memory) and external devices—must allow the bidirectional transmission of data words between the computer and its peripherals when required. Additionally, data paths or buses interconnecting the subsystems of the computer, as indicated in Figure 8.1, usually allow the transmission of whole words between various units. For small computers it may also be said, in general, that the width of a memory word fixes the maximum number of storage locations that may be addressed in the memory unit.

In designing a computer, after selecting the size of a data or instruction word, the designer must specify the set of instructions the computer will execute and the way in which they will be executed. The structure of each of the subsystems (AU, Memory, I/O Unit, and Control Unit) is intimately tied to the kinds of instructions that will be included in the computer's executable set. (It may also be said, in some cases, that the structure of a machine determines its instruction set.) As we will see later, instructions may specify operations on the contents of registers or memory locations, input or output between the computer and its peripheral devices (terminals, card readers, etc.), and certain internal operations (termination of instruction execution, adjustment of the contents of special processor registers, etc.). The instruction set which is selected determines, among other things, the number of operand registers or accumulators that can be included in the AU, the method and hardware to be employed in calculating any memory addresses used in executing an instruction, the structure of the instruction execution sequencing logic included in the Control Unit, and the types of signals that will be routed over the Input/ Output bus (control signals such as status testing signals, etc.).

In selecting the instructions a computer will execute, the designer would like to minimize the amount of memory required to store a particular program and maximize the *power* of individual instructions. An instruction set is said to be powerful if it allows the efficient implementation of a general class of algorithms (procedures for performing tasks) in a relatively simple way. A second consideration which must be included in the design of an instruction set is that the complexity of individual instructions not lead to excessive execution times.

In this introduction we have discussed in general terms the problems to be solved and decisions to be made in designing a computer. In succeeding sections we will treat each computer subsystem in detail and attempt to define an example machine which will aid in the understanding of the architecture and operation of a typical computer.

8.1 THE STORED PROGRAM CONCEPT

As we have mentioned, the "stored program" concept is fundamental to the operation of modern digital computers. By this we mean that the work the computer will perform is governed by the instructions, or program, contained in its memory. In addition to instructions, the memory contains data that the instructions operate on to produce desired results. Since both the instructions and data appear as binary sequences of zeros and ones in memory, it is impossible to tell just by looking at a memory location whether it contains an instruction or data. However, by virtue of having prepared the program, the programmer (or the system program) knows which words are instructions and which are data. Sequences

of operations are specified so that the computer never attempts to execute a data word as an instruction. Usually instructions are grouped together and data words are grouped together in separate sections of memory. If a data word were accidentally encountered, the computer would execute it as if it were an instruction. Although the computer would continue to function deterministically, it most likely would not execute any further instructions in the intended sequence, and the results would appear random to the programmer.

The programmer (or system software, such as a FORTRAN compiler) is responsible for breaking down the problem to be solved into a series of fundamental steps. These steps must correspond exactly to the limited class of instructions that the computer is able to understand and execute. The instructions are usually very simple, and would include basic mathematical operations, the movement of data from one part of the computer to another, and the ability to branch or transfer control to other instructions within a program. As fundamental as these instructions are, significant processing is possible because of the high speed at which they are executed. Execution of one million instructions per second is typical.

Before discussing the four major computer units, it would be worthwhile to examine the classes of instructions that a typical computer can execute. The instruction formats developed here will be used later in the chapter to illustrate the operation of the various units or subsystems.

Computer instructions generally can be classified as one of four types:

    I. Memory access
   II. Branch and subroutine jumps
 III. Arithmetic and logical
 IV. Input/Output

The specific format for each type differs widely among computers. Smaller computers may require several instructions to perform the same operation that a larger one is capable of performing with just one instruction. For our purposes, a relatively simple instruction set will be defined for the example computer we will develop and use for illustration. We will assume

Table 8.1. Four Types of Computer Instructions (Operations)

| 0 | 1 | 2 | 3 | 4 | 5 | 6 | 7 | 8 | 9 | 10 | 11 | 12 | 13 | 14 | 15 |

Single Instruction Word

Opcode

Op-Code (Bits 0 and 1)	Instruction Type
00	I. Memory access
01	II. Branching, subroutine jumps
10	III. Arithmetic and logic
11	IV. Input/Output

that the architecture or structure of the computer is meant to be general-purpose and makes use of two arithmetic registers or accumulators. The basic instruction length will be 16 bits, chosen because this corresponds to the size of the basic memory element in many modern minicomputers. Individual instructions will be described in detail as the operation of the example computer is discussed. Our computer will use a two-bit field in each instruction as an "operation code" (op-code for short) which distinguishes each of the four types of instructions listed above from the other three. Table 8.1 gives details.

In the sections to follow we will describe each instruction type in detail at such times that seem appropriate. Instructions of each type will be assigned a particular binary format and mnemonic code which will be used wherever necessary in subsequent discussions.

8.2 THE FOUR MAJOR UNITS IN A DIGITAL COMPUTER

8.2.1 Memory

We have indicated that a computer's instructions and data are stored in its memory. Memory can be thought of as consisting of sequentially-organized information storage locations; each location is referred to as a word. Figure 8.2 shows how a memory is organized for a computer with a 16-bit word size. Note that each memory word has a specific number assigned by which the computer can address it. A special register in the control unit, called the Program

Counter or Location Counter, always contains (points to) the location or address of the current instruction in memory to be executed. Computer instructions are usually executed sequentially; i.e., the contents of the Program Counter automatically increases by one every time an instruction is executed. Thus we see that the Program Counter serves somewhat the same function as a bookmark. Branch instructions can be used to change the contents of the Program Counter to any desired value, however. This would be done, for example, if, based on calculated results, the programmer wished to execute a series of instructions not immediately following the current instruction.

Most memories allow both reading (information retrieval) and writing (information storage) operations to be performed. Reading would be done to access both instructions and data. Writing would be done to store away computed results. Such a memory is referred to as a read/

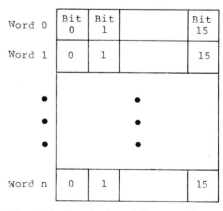

Figure 8.2. Organization of Computer Memory

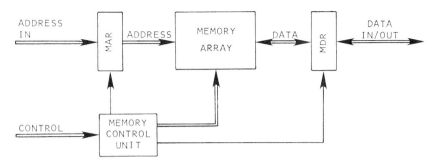

Figure 8.3. Memory Unit

write memory or Random Access Memory, abbreviated RAM.

Two registers are associated with the memory to accomplish reading and writing. These are the Memory Data Register (MDR) and the Memory Address Register (MAR). They are shown in Figure 8.3. The MAR is loaded by the Control Unit prior to each memory access, and it is set to contain the address of the desired memory location. For a read operation, the MDR receives the data from memory and sends it to either the Control Unit or the Arithmetic Unit, depending on the desired destination. For a write operation the MDR is loaded by the Control Unit with the data to be placed in memory. In each case, it should be noted, signals generated by the Control Unit are received and interpreted by the Memory Control Unit. This device causes the MDR and MAR to be loaded at appropriate times and, in addition, insures that accesses to the memory array are sequenced properly. In Figure 8.3 and following figures, pathways that would consist of only one signal are indicated by a single line; double lines are used to indicate multiple signal pathways.

Some memories do not allow alteration of their stored information; these are referred to as Read-Only Memories, or ROMs. The existence of read-only memories may be justified, to give just one example, by the fact that some computers always execute the same program. As an illustration, consider a computer in an aircraft that may be so specialized that its only function is to compute the aircraft's position, over and over again, and this is the only program that the computer ever executes. In such a case, aside from a small portion of random access memory where results will be stored, the contents of the rest of the memory need never change. Read-only memory is often used to store such a program. In addition to the fact that the memory can never accidentally be erased, read-only memory is usually faster (shorter access time), less expensive, and may consume less power than RAM. Its use is very common in real-time applications.

To give the reader an idea of the physical characteristics of common small computer memories, a few of them are mentioned here. The average minicomputer contains anywhere from 4K to 256K words of memory (where K is equal to 1024). It is possible to put 16K words or more into one module about 12 inches square where access times for a memory word range from approximately 250 nanoseconds to one microsecond. Memories may be constructed from small toroidal magnetic cores (core memory) or, more commonly, from special transistor circuits in integrated circuit packages (semiconductor memory). The use of semiconductor memory in modern mini- and microcomputers has become predominant mainly because of the high densities possible, their low cost, and high reliability.

A second, smaller memory exists in all computers. It can be as small as one word, but 4 to 16 words is more typical; some small computers even have 64. These memory words are referred to as the computer's accumulators, or registers. The contents of these registers can be loaded from or stored into main memory. In many computers, the operands for all arithmetic operations must be placed in these registers before the operation can be performed. These registers are also useful for storing temporary results from ongoing calculations and for calculating future memory addresses. In addition, they are used in transferring information to external devices. For example, a typical sequence would be to transfer a particular memory word to an accumulator, and then to transfer the contents of the accumulator to a paper tape punch. This smaller memory is not housed in the main memory, but exists in the Arithmetic Unit (as will be demonstrated in our example computer), or the Control Unit, depending upon the individual computer.

In Section 8.2.2 we will discuss the Control Unit, in particular how it is used to decode computer instructions. To complete this section, however, it would be useful to look at a set of Type I operations—those that access memory—and see how they would be *encoded*, i.e., written in machine language.* We will continue to develop our example machine, one that is not necessarily very well-designed or efficient but does illustrate the most important features of any computer. Table 8.2 illustrates a set of memory access instructions which would be typical for a two-accumulator machine. (In this case one of the accumulators can be used as an index register, i.e., a base memory address which will be defined in the sequel.)

To begin with we should note that this hypothetical machine has an "address field" nine bits long, hence it can address only 512_{10} memory locations directly (i.e., locations $0-511_{10}$). For example, the instruction to "Load Accumulator 0 with the *contents* of memory location 7" would appear as follows in machine language.

lator 1 in memory location 43_8", the appropriate machine language statement would be

$$\underline{00}\ \underline{0}\ \underline{1}\ \underline{00}\ \underline{0}\ \underline{000100011} = 010043_8$$

and would correspond to the mnemonic representation STA 1, 43.

Note that, in both these cases, when we refer to a particular memory location address, or to an accumulator, we imply the "contents of the designated storage location". Hence a Load instruction reads the contents of a particular memory cell *without* changing what is stored there and *overwrites* the contents of the accumulator with this 16-bit number. A Store reads the contents of the accumulator, again without disturbing it, and overwrites the contents of the designated memory location with this 16-bit number. Some computers have the capability of performing what is called a "load immediate" operation, e.g., "Load Accumulator 0 with the *number* 7". This sort of operation *is not* implied by the operation we have used. In our machine the only way to place the number 7 into Accumulator 0 would be to have previously stored the number in some memory location, say 255, and then perform a "load accumulator" instruction specifying location 255 (LDA 0, 255).

8.2.2 Control Unit

The Control Unit of a computer is responsible for managing the processing of each instruction

| $\underbrace{0\ 0}$ | 1 | 0 | $\underbrace{0\ 0}$ | 0 | $\underbrace{0\ 0\ 0\ 0\ 0\ 0\ 1\ 1\ 1}$ = 020007_8 |
| Memory Reference | Load | Accumulator | No Index Register | Direct Addressing | Memory Location 7 |

The mnemonic representation of this instruction would be LDA 0, 7. Similarly, for an instruction to "store the *contents* of Accumu-

and for initiating whatever action is necessary to complete execution of the instruction. It usually is the most complex part of the computer from a hardware logic point of view. Included in the Control Unit are most of the computer's timing or control sequencing circuits. In addition, it is the Control Unit that supervises

---

*We will use the term "machine language" to refer to the numerical (binary, octal, or hexadecimal) representation of instructions as they would be stored in the computer memory.

Table 8.2. Type I Operations: Memory Access Instructions

0   0

0	1	2	3	4	5	6	7	8	9	10	11	12	13	14	15

Op-code — bits 0,1
Load or store — bit 2
Accumulator — bit 3
Index — bits 4,5
Indirect — bit 6
Address — bits 7-15

Mnemonics	Bit 2	Load or Store
STA	0	Store into memory (write)
LDA	1	Read from memory (read)

	Bit 3	Designated Accumulator
0	0	Register 0 will receive the information read, or contains the information to be written
1	1	Register 1 will receive the information read, or contains the information to be written

	Bits 4,5	Index Register Addressing
	00	Don't use an index register
X	01	Use Register 1 as an index register (Indexed)
R	11	Use the Program Counter as an index register (Relative)

	Bit 6	Indirect Addressing
	0	Don't'perform indirect addressing
@	1	Perform indirect addressing

	Bits 7-15	Address
	-	The absolute memory address (0 to 511_{10}) to be written into or read out of, or a two's complement "displacement" (-256_{10} to $+255_{10}$) if an index register is used.

Format

<Opcode> $\begin{Bmatrix} X \\ R \end{Bmatrix}$ <ACC> $\begin{Bmatrix} @ \end{Bmatrix}$ $\begin{bmatrix} \text{<Address>} \\ \text{<Offset>} \end{bmatrix}$

Examples

```
STA 0,NUM     ; Store contents of AC0 in memory location NUM.
              ; (Assembler may use relative or page zero addressing depending on NUM)
LDAX 0,0      ; Load AC0 with the content of memory location
              ; indexed by AC1 plus an offset of 0 .
STAR 1,1      ; Store the contents of AC1 in the memory location
              ; indexed by PC plus an offset of 1.
STA 1,@NUM    ; Store the contents of AC1 in the memory location pointed to by
              ; the contents of memory location NUM.
```

Note: $\{\ \}$ Denotes optional alternatives, $[\ \ ]$ denotes required alternatives

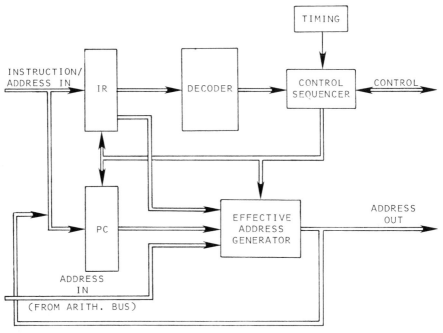

Figure 8.4. Control Unit

and controls the transfer of data and instructions from one part of the computer to another.

There are five main functions that the Control Unit performs or initiates. These are:

(1) Fetching instructions and data from memory
(2) Writing data into memory
(3) Decoding instructions
(4) Initiating action in the Arithmetic and I/O Units
(5) Performing branching

The first two functions will be described in detail when instruction and data flow are examined, i.e., when the dynamic operation of the computer is analyzed. The last three are discussed in the remainder of this section. Figure 8.4 illustrates schematically how the Control Unit is constructed. Each instruction, after it is fetched from memory, is placed in a register in the Control Unit called the Instruction Register, abbreviated IR. There it is decoded (using the first two bits, or op-code, in our computer) to determine what class the instruction belongs to, and it is then further decoded to extract other information contained

in it. For example, if a memory access instruction is being decoded, we saw that Bits 7 through 15 would be interpreted as an absolute memory address in page zero (locations $0-511_{10}$) if Bits 4, 5, and 6 are decoded and found to be all zero (direct addressing with no index register).

Table 8.3 illustrates a set of Type II instructions—those that modify program flow, i.e., initiate a branch or jump so as to cause the machine to execute statements that do not follow sequentially in memory the instruction presently being executed. Note that in Table 8.3 the number fields are decoded precisely as in Type I instructions except that Bit 3 is not used. (An accumulator does not need to be specified for this type of instruction.)

As we shall see in Sections 8.2.3 and 8.2.4, Arithmetic and Logic (Type III) instructions contain information such as what operation is to be performed (add, subtract, AND, etc.), which accumulator(s) contain the operand(s), and where the result should be placed; Input/ Output (Type IV) instructions contain information concerning what peripheral device is involved. Hence in these cases the number fields represented by Bits 2 through 15 will be

Table 8.3. Decoding of Type II Operations: Branching and Subroutine Jumps

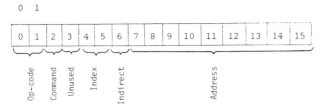

Mnemonics	Bit 2	Command
JMP	0	Jump to new address
JSR	1	Jump to new address and save (present Program Counter plus one) in Register 1 (subroutine jump)

	Bit 3	(Unused)

	Bits 4,5	Index Register Addressing
	00	Don't use an index register
X	01	Use Register 1 as an index register
R	10	Use the Program Counter as an index register

	Bit 6	Indirect Addressing
	0	Don't perform indirect addressing
@	1	Perform indirect addressing

	Bits 7-15	Address
	-	The absolute memory address (0 to 511_{10}) to which control will be transferred, or a two's component "displacement" (-256_{10} to $+255_{10}$) if an index register is used.

Format

Examples

```
JMP  0     ; Jump to location zero.
JSR 100    ; Jump to subroutine at location one hundred.
JMPR -2    ; Jump to memory location addressed by PC minus 2.
JMPX 0     ; Jump to memory location addressed by AC1 plus an offset of 0.
           ; (This instruction may be used as a return instruction
           ; in a subroutine).
JMP LOOP   ; Jump to memory location LOOP.
```

decoded in a completely different way than with Type I and II instructions.

Once the Control Unit's decoder has broken down an instruction into its important fields, it can initiate whatever action is necessary to complete execution of the instruction. To add two numbers together, it would send the operands (inputs for the operation) to the Arithmetic Unit and instruct the unit to add. The Control Unit would then retrieve the result and store it away in the designated register. For an I/O instruction, the Control Unit would inform the I/O Unit of the type of operation required and which registers were involved.

Example 8.1. Branching Within Page Zero (Memory Location 0-511)

Memory Location	Instruction or Data	Machine Language Equivalent (Octal)	Mnemonic Equivalent
.	.	.	.
.	.	.	.
.	.	.	.
30	Load Accumulator 0 from Location 32	020032	LDA 0 , 32
31	Jump to instruction in Location 40	040040	JMP 40
32	Some number, say 260_8	000260	260
33-37	Other data	.	.
		.	.
		.	.
40	Any instruction	——	——
.	.	.	.
.	.	.	.
.	.	.	.

The Control Unit is usually able to execute simple branching instructions on its own and memory access instructions require, of course, the use of the Memory Unit.

Returning to our discussion of branch instructions, we note again that they are used when it is desired to execute a sequence of instructions that does not immediately follow the current instruction. This technique is always used when a block of data follows the current instruction and it is necessary to jump over the block to get to the next instruction. Example 8.1 illustrates this technique.

We see that the machine language instruction in Location 31 is 040040_8 which, when written out in binary form and decoded according to Table 8.3, would indeed force the machine to jump over the data block in Locations 32-37 and cause the instruction stored in Location 40 to be executed next. (The op-code is 01, the command bit is 0 for a simple jump, and the address is 40_8.)

Note that our branch instruction is an unconditional branch; i.e., the contents of the registers or states of external flags do not affect the execution of the instruction. With this simple instruction set, conditional branching would have to be accomplished using arithmetic or I/O instructions in conjunction with a branch statement (as will be demonstrated in the sequel). Many computers do include conditional branching in the branch instruction type,

however. These machines allow a branching operation to be completed only if a particular condition is satisfied.

A common variation of the branch instruction saves the contents of the Program Counter "plus one" in a specified register (in our case, Accumulator 1) before performing the jump. This instruction is necessary when transferring control to a subroutine. The subroutine eventually must be able to return execution to the next location after the one from which the "call", i.e., the transfer or jump to subroutine, was made. The contents of the accumulator thus serve to inform the called subroutine of the location to which control must eventually be returned.

We will end this section with a brief description of the operation of the Control Unit from a slightly different viewpoint. If we consider the primary function of the Control Unit to be the interpretation of instructions (and their resulting execution), the function of the Control Unit may be described by what is sometimes called the Instruction Interpretation Process. This process may be presented in tabular form with the function or contents of each element in the Control Unit described (Table 8.4). Because this table is intended to represent the interpretation of any instruction which may be executed by the example computer, the descriptions of the functions of the Effective Address Generator and the Control Sequencer are

Table 8.4. Interpretation Process for One Instruction

Interpretation Step	Contents of IR	Contents of PC	Effective Address Generator	Control Sequencer
1. Fetch Cycle				
a) Instruction Fetch – Obtain the next instruction from the memory location pointed to by PC.	Before: Previous Instruction After: Current Instruction	Before & After: Current Contents	Output the contents of PC.	Generate control signals to load the MAR with the contents of PC, read instruction from memory and place it in IR.
b) PC Update – Modify the contents of PC to point to the next instruction.	Before & After: Current Instruction	Before: Current Contents After: Current Contents plus one (new contents)	Output the contents of PC plus one.	Generate control signals to cause PC to be updated.
2. Execute Cycle				
a) Operand Fetch – Calculate the address (or location) of operands needed for execution, if any, and fetch. Go to Step 2.b if no operands are needed.	Before & After: Current Instruction	Before & After: New Contents	Output the memory address of required operand calculated as specified by current instruction.	Generate control signals to load the operand address into MAR in the memory unit, read the memory and place contents into the MDR.
b) Instruction Execution – Using operand fetched in Step 2.a, if appropriate, execute instruction in IR. Return to Step 1.a.	Before & After: Current Instruction	Before: New Contents After: New Contents, or if branch instruction, address of next instruction.	For branch instructions generate address needed for next instruction.	Generate sequencing, I/O, bus and AU control signals necessary to execute the current instruction.

necessarily vague. Upon completion of this chapter the reader should be able, in general, to fill in details.

8.2.3 Arithmetic Unit

The Arithmetic Unit of any computer serves as the central arithmetic and logical processing element; it invariably contains a combinational logic network—the Arithmetic Logic Unit (ALU)—which performs arithmetic or logic operations on information contained within the machine's accumulators to produce a result.

Three types of operations normally are performed by the ALU:

(1) Arithmetic
(2) Logical
(3) Bit and byte manipulation

The standard arithmetic functions in most small computers are addition, subtraction, and negation. The other mathematical operations can be derived from these three. Multiplication and division circuitry is available for minicomputers, but often at an extra cost. Without these circuits, successive addition and subtraction are used to perform multiplication and division.

Most small computers perform at least the standard logical functions, such as AND, OR, NOT, and Exclusive OR. These operations are performed on the corresponding bits of specified operands (e.g., accumulator contents). Bit and byte manipulation is also an important part of the ALU. By bit manipulation, we mean the ability to shift the contents of a register left or right by a specified number of bits. With elaborate machine architectures it often is possible to alter the state (0 or 1) of individual bits in the ALU result, as well. Figure 8.5 shows an example of a circular left shift of one bit. This type of operation would be necessary to perform multiplication and division by the

negative or positive, etc.

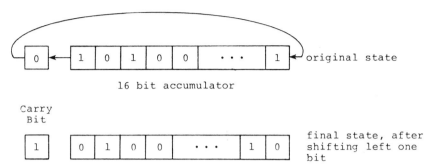

16 bit accumulator

Figure 8.5. A Left Shift Operation

techniques of successive addition and subtraction. Notice the existence of the Carry Bit in Figure 8.5. This one-bit register (sometimes called the Link Bit) is usually accessible to the programmer; i.e., it can be set to 1 or 0 or it can be tested with an associated "instruction skip" in a single instruction. It can also be used to determine when the machine's arithmetic capacity has been exceeded, if a two's complement number is negative or positive, etc.

Byte manipulation capabilities are important, especially in communications applications. One common operation performed by 16-bit computers is byte swapping (a byte is 8 bits). Figure 8.6 illustrates this operation.

Operands can enter the ALU from one of two sources, depending upon the computer. They can come from a memory location or from a register (accumulator). Similarly, the result is returned to a memory location or register.

Elements that ordinarily are found within the Arithmetic Unit (besides the ALU) are the accumulators, the Carry Bit, and, possibly, a condition-testing unit. This element might be used to test some portion of the ALU result and cause the next instruction to be skipped (or not skipped), depending on the outcome of the test.

The Arithmetic Unit for our example computer is shown in Figure 8.7. Note that it does contain the two accumulators, a Carry Bit, and a Condition Tester such as just described. In this configuration the Condition Tester is used to test the state of the ALU output (result) which includes the state of the Carry Bit as well.

Table 8.5 continues the definition of the instruction set for the example machine—Type III, Arithmetic and Logic Instructions. The particular mathematical or logical operation to be performed (i.e., add, subtract, AND, etc.) is specified by Bits 4, 5, and 6. These three bits can select one of up to eight possible operations.

The selected operation is performed on the contents of the computer's two accumulators. Bits 2 and 3 indicate the "source" and "destination" registers respectively. If our computer

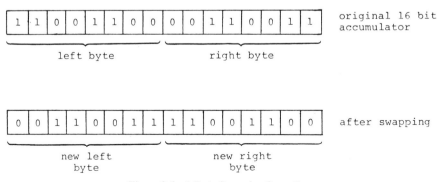

Figure 8.6. A Byte Swapping Operation

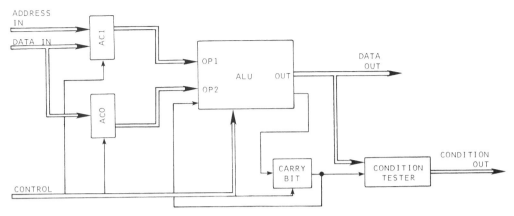

Figure 8.7. Arithmetic Unit

had more than these two general-purpose registers, more bits in the instruction would have to be allotted to specify them; i.e., if four registers existed, then the register fields would have to be two bits each instead of one. Bits 7 and 8 of the instruction specify any selected post-operation involving bit or byte manipulation, such as shifting the result of the primary operation left or right by one bit.

Bits 9, 10, and 11 specify test conditions based on whether the result of the post-operation is greater than, equal to, or less than zero, or on the state of the Carry Bit (0 or 1). In this case, if the specified condition is found to be true, the next instruction in sequence will be skipped. Thus conditional branching can be initiated after the post-operation, as shown in Example 8.2 which assumes that Accumulators 0 and 1 already contain two positive integers to be added and the result tested for an arithmetic carry from the most significant bit (Carry Bit equal to one).

Referring again to Example 8.2, the effect of executing the instruction in Location 40 is to leave the accumulators as they were but initialize the Carry Bit to 0. After execution of the instruction in Location 41 the machine will skip the instruction in Location 42 and continue in sequence only if there is no carry from the addition operation. If a carry is detected, the instruction in Location 42 will be executed, causing a jump to the instruction in Location 60. Presumably a routine located beginning there will take suitable action; also the programmer will have to prevent the instructions

in Location 43 et seq. from continuing on through Location 60 unless that is intended.

8.2.4 Input/Output (I/O) Unit

A computer's I/O circuitry allows it to send data to and receive data from all its peripheral devices. In many small computers it is difficult to point to a specific part of the hardware and say that it is the I/O Unit. More likely the I/O circuitry is distributed throughout the AU and Control Unit. The I/O circuitry consists of sequencing circuits, the I/O instruction skip logic, the direct memory access (DMA) channel, the interrupt lines from the different devices, and the interrupt acknowledgment circuitry.

Figure 8.8 gives a block diagram of the I/O unit included in our example computer. Notice that in this machine we have included a block for each function described and have assumed that the I/O Unit is an independent module, controlled by the Control Unit, and is not, in consequence, distributed throughout the computer. DMA hardware included in the I/O Unit consists entirely of a memory buffer (or isolation device) and a primitive DMA control or sequencing unit. This hardware is not sufficient by itself to allow DMA activity. Thus, to use our machine's DMA capability, a special I/O controller must be added which is designed to provide the extra control and memory elements needed for proper DMA operation (see Section 8.6.2).

In Chapter 2 the need for each peripheral device attached to the computer to have its

Table 8.5. Decoding of Type III Operations: Arithmetic and Logic Instructions

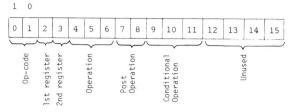

Bit 2	1st Register		Bit 3	2nd Register
0	Register 0 is the first operand		0	Register 0 is the second operand
1	Register 1 is the first operand		1	Register 1 is the second operand

(Bits 2 and 3 may specify the same register. Bit 3 specifies the destination register, where the result of the operation is placed.)

Mnemonics	Bits 4,5,6	Operation	Bits 7,8	Post-Operation	Bits 9,10,11	Conditional Operation
MOV	000	Move	00	None	000	None
ADD	001	Add	L 01	Shift left one bit	SRZ 001	Skip if result equals zero
SUB	010	Subtract	R 10	Shift right one bit	SGZ 010	Skip if result greater than zero
NEG	011	Negate	S 11	Swap bytes	SRN 011	Skip if result less than zero
AND	100	AND			SCZ 100	Skip if Carry Bit equals zero
IOR	101	OR			SCO 101	Skip if Carry Bit equals one
NOT	110	NOT			CTZ 110	Set Carry Bit to zero*
					CTO 111	Set Carry Bit to one*

*These operations are performed before the post-operation and before the skip condition is tested.

FORMAT

$$<Opcode> \begin{Bmatrix} L \\ R \\ S \end{Bmatrix} <Source\ Acc.>, <Destination\ Acc.> \begin{Bmatrix} SRZ & SCO \\ SGZ & CTZ \\ SRN & CTO \\ SCZ \end{Bmatrix}$$

Examples

MOV 0, 1	; Move the contents of AC0 to AC1
ADDL 1,1	; Add AC1 to itself then shift result one bit left, Place result in AC1
MOVL 0,0 ,CTO	; Set CARRY to 1, shift content of AC0 one bit left (with carry) and
	; place the result in AC0
IOR 1,1,SRZ	; Or AC1 with itself, place the result in AC1, and skip the next
	; instruction if this result is zero.
NEG 0,0	; Generate the two's complement of the contents of AC0, place the result
	; in AC0.
NOT 1,1	; Generate the one's complement of the contents of AC1, place the result
	; in AC1.

own controller was discussed. Although the controller is often assembled by the computer manufacturer and even though it is usually housed in the computer mainframe along with the other units, the controller usually is considered to be part of the peripheral device and not the I/O Unit itself. Regardless of which point of view is taken, however, it is important to keep in mind that the computer communicates with the device controller, and the controller in turn communicates with the actual device. Thus a controller is the means by which a computer manufacturer matches the characteristics of his computer to an actual peripheral device which may or may not be from an independent manufacturer.

Example 8.2. Conditional Branch on Overflow

Memory Location	Operation	Machine Language Equivalent (Octal)	Mnemonic Equivalent
:			
40	Set carry bit to zero, move contents of Accum. 0 to Accum. 0.	100140	MOV 0,0,CTZ
41	Add Accum. 0 to Accum. 1 (result in Accum. 1), Skip next statement if Carry Bit is 0	111100	ADD 0,1,SCZ
42	Jump to Location 60	040060	JMP 60
43	Next regular instruction (no overflow)	–	–
:			
60	Execute this instruction if overflow occurred	–	–
:			

In Chapters 1–3 it also was noted that there are two common methods of performing input/output. The first is called *Programmed I/O*, a term that may be confusing, since the other method must be programmed, too. The term Programmed I/O means that every data transfer between memory and a peripheral unit must take place via an accumulator. For example, if it were desired to initiate the operation of an analog to digital converter and to transfer the result into a particular memory location on completion of the conversion, it could be done as follows: An instruction would first be executed to start the conversion; another would

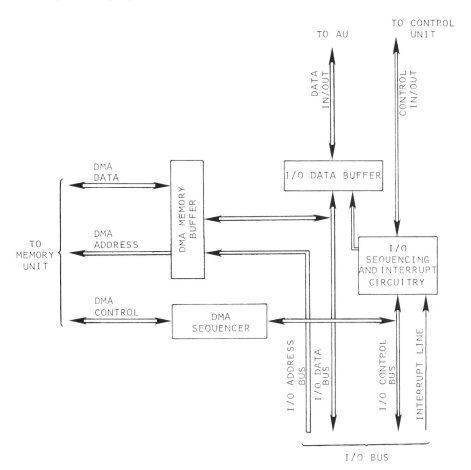

Figure 8.8. The I/O Unit

Table 8.6. Decoding of Type IV Operations: Input/Output Instructions

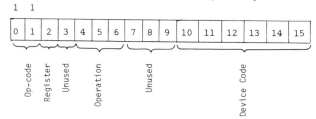

Mnemonics	Bits 4,5,6	Operation
DOA	000	Output to device controller register A
DOB	001	Output to device controller register B
DIA	010	Input from device controller register A
DIB	011	Input from device controller register B
SP1	100	Apply special pulse one to device controller
SP2	101	Apply special pulse two to device controller
SFS	110	Skip if device controller's flag is set
RCF	111	Reset the device controller's flag

	Bits 10-15	Device Addressed
	-	The device code of the peripheral unit. A device code of zero is used for special commands to the CPU. In this case Bits 4,5,6 are interpreted as follows:

	Bits 4,5,6	Special CPU Commands (Device Code = 0)
HLT	000	Halt
EIN	001	Enable interrupts
DIN	010	Disable interrupts
AIN	011	Transfer device code of interrupting device into specified register (Acknowledge Interrupt)

Format

<Opcode> { <Source or Destination Acc.>, } <Device Address>

Examples

```
DOA 0, 15    ; Output the contents of AC0 to the A register of device 15.
DIB 1,TTI    ; Input the content of the B register of device TTI to AC1.
SFS TTO      ; Skip the next instruction if device TTO's flag is set.
HLT          ; Halt program execution.
DIN          ; Disable interrupts.
```

be issued to read the converter value into an accumulator on completion of the conversion operation; a third instruction would store the contents of the accumulator into the memory location. This is a very straightforward and effective way of performing the transfer. Consider, however, the case where there are 100

analog to digital converter channels to be scanned and converted values to be placed in memory. In this case the same sequence of instructions would have to be executed 100 times, a very inefficient process as it can tie up the computer CPU (by CPU we mean the Central Processing Unit which is made up of the

Control Unit and the AU) completely during the entire operation. In addition there are some peripheral devices, such as disks, that operate at such high data rates (several hundred thousand word transfers per second) that the use of Programmed I/O techniques would be inappropriate.

To get around these problems, a second method of performing I/O has been devised called *Direct Memory Access*. Using this method the computer provides the peripheral's controller, via regular I/O instructions, with all the information necessary to define the transfer. The controller then proceeds to handle the transfer on its own. Among the items of information that must be sent to the controller are the number of words to transfer and the starting address of the block of memory locations that will be used in the transfer. The controller can access main memory with no assistance from and no interference to the Central Processing Unit except for, as will be seen later, a slight decrease in CPU speed. When the controller has transferred the number of words specified, it informs the CPU of this situation by setting a flag (an indicator testable by the Control Unit during the execution of an I/O instruction) or by generating an interrupt.

The use of programmed I/O, interrupts, and DMA are very important concepts in real-time applications; hence Section 8.6 will deal in detail with their implementation. Table 8.6 lists the Input/Output Instructions (Type IV operations) for our hypothetical machine. In this case we note that input/output instructions are used for transferring data between a computer register and an external device, such as a card reader or analog to digital converter. The fields in such an instruction specify the accumulator and the external device involved, and whether the information is transferred from the accumulator to the device or from the device to the accumulator. Input/output instructions must also be able to test the state of one or more flags which serve to indicate whether the device is ready to perform the desired operation. As in the case of arithmetic instructions, an instruction can be skipped if the flag condition tested is detected. This will be discussed in more detail in Section 8.7 which deals with

interfacing. It should be noted that the I/O instruction set defined handles DMA transfers and programmed I/O with equal ease.

8.3 SEQUENCING OF COMPUTER OPERATIONS

Now that the role of each of the major units of a computer has been examined briefly we can see how these units are joined together to form a dynamic working entity. The study of how to specifically structure a computer is broadly referred to as computer architecture. Just as the architecture of buildings can be extremely varied, so it can be with computers. In our computer model all the necessary parts are included but few of the extra features that would be found on most real computers.

Figure 8.9 illustrates the architecture of our example computer. It is capable of executing the entire instruction set defined in Tables 8.2, 8.3, 8.5, and 8.6. Initially Figure 8.9 may appear to be somewhat unwieldly. However, it is essentially a composite diagram of all the subsystems (memory, AU, etc.) we have studied in the previous sections. In this section we will refer to this drawing in summarizing the operation of the example computer as a whole.

Notice that there are four buses in our computer—the I/O Bus (data, address, and control lines), the Memory Bus (data and address lines), the Arithmetic Bus (data lines), and the Control Bus (subsystem control lines)—indicating that it is a multiple bus machine. In addition to these buses, there is one DMA channel that bypasses the Arithmetic Unit and allows the I/O Bus to access the Memory Unit directly. Since we have assumed that the computer has a 16-bit word size, then all the data buses would be 16 bits wide. The width of the Control Bus depends on the complexity of the subsystems. This can vary greatly, depending on the detailed design of the computer.

In addition to the instruction decoder and timing circuits, the Control Unit contains two registers. One register is the Program Counter, or PC. When it is time to fetch an instruction from memory, the contents of the Program Counter are transferred to the Memory Address Register. Following the fetching of the current

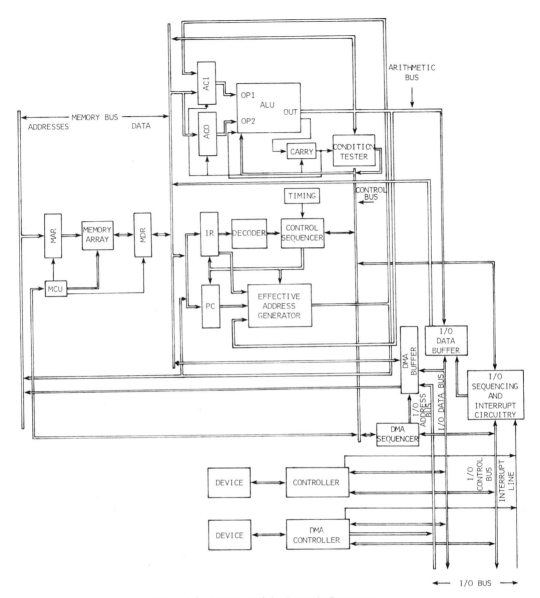

Figure 8.9. Structure of the Example Computer

instruction, the PC is incremented and in this way points to the next instruction. (Execution of branch instructions further modifies the contents of PC.)

The second register in the Control Unit is the Instruction Register (IR). When an instruction has been read from memory, it is transferred from the Memory Data Register to the Instruction Register. This register is then analyzed by the instruction decoder whose outputs are used in activating the necessary control lines and timing circuitry to execute the instruction.

So far we have discussed the transfer of instructions between the Memory and Control units, while the transfer of data has not yet been mentioned. Data transfers occur during the execution of load and store instructions. To write data into memory the contents of the address field of the Instruction Register are transferred to the MAR (through the Effective Address Generator which performs any necessary address calculation), and the contents of the specified accumulator are sent to the MDR through the ALU. The control lines to the

Memory Unit are then activated commanding that a write operation be performed. To read data, the address is transferred to the MAR in the same manner, but a read command is sent over the control lines. The memory places the word just read into the MDR, and the control lines cause the word to be sent to the designated accumulator.

Arithmetic instructions cause the Control Unit to inform the Arithmetic Logic Unit of the operations to be performed. The Control Unit then causes the result to be placed in the specified accumulator. At this time the Control Unit may need to increment the PC by one extra count if the instruction had ordered a possible skip to occur based on the Arithmetic Unit's result (e.g., subtract AC0 from AC1, skip on negative result).

When executing I/O instructions, the control lines enable the device controllers to transfer information to or from the accumulators. Those controllers which have the necessary circuitry can bypass the Arithmetic unit, when so instructed, and communicate directly with the memory (DMA). In most minicomputers the DMA channel shares the data lines to memory with the Control and Arithmetic Units. Thus, when the DMA transfer takes place, the Control Unit must wait until the lines are free before attempting to access memory. This method, called cycle stealing, is described more fully in Section 8.6.2. It slows down somewhat the rate at which program instructions are executed, since memory cycles occasionally are consumed by the DMA channel.

At this point it will be instructive to follow the register-by-register execution of a small four-word computer program (Example 8.3). This will illustrate the instruction and data flow that has been discussed.

In completing this example we will use the following notation: (1) ⟨X⟩ means "the contents of X", and (2) → means "is transferred to". Such notation is commonly used in describing information flow at what is known as the "register transfer level". This basic level of description provides a symbolic listing of actual information transfers as they occur between registers. (In this context a memory location is considered to be a register.) Assuming that the Program Counter (PC) initially contains address 100, the detailed sequence of operations for the example program is shown in Table 8.7. From the list of register transfers, it is seen that the first four operations for each instruction are the same. These are usually referred to collectively as the "fetch cycle". The remaining operations depend upon the specific instruction and are referred to in the same fashion as the "execute cycle".

8.4 OTHER COMPUTER ARCHITECTURES

The example computer that we have described in previous sections represents a very simple and straightforward architecture which is widely used. In this section we will discuss two very common alternatives to this architecture.

8.4.1 Single Bus Computers

Computer architects refer to our example computer as a multiple bus machine, since it has four separate buses, one each for memory, I/O, arithmetic, and control. This type of architecture is popular because it is simple to design, yet effective. With single bus architectures, however, only one bus is used, and all information is transferred over this bus. The most noteworthy characteristic is that every programmable register, without exception, behaves like a memory location. Figure 8.10 illustrates this structure. Accumulator 0 is programmed as if it were Memory Location 0, etc. These locations would not exist in the actual memory unit; thus there is no conflict. In an identical manner, registers in the device controllers are assigned "locations" which are also absent from

Example 8.3. Illustration of Program and Data Flow

Memory Location	Instruction	Mnemonic
100	Load AC0 with the contents of Location 200	LDA 0,20
101	Add AC0 and AC1, placing the result in AC1	ADD 0,1
102	Output AC1 to peripheral device 2, Register A	DOA 0,2
103	Halt	HLT

Table 8.7. Program and Data Flow for Example 8.3

```
100      <PC> → MAR                                                     ⎫
         READ command, on Memory Unit Control lines                     ⎬ fetch cycle
         <MDR> → IR                                                     ⎪
         <PC> + 1 → PC                                                  ⎭

         <address field of IR> → MAR                                    ⎫
         READ command, on Memory Unit control lines                     ⎬ execute
         <MDR> → AC0                                                    ⎭


101      <PC> → MAR                                                     ⎫
         READ command, on Memory Unit control lines                     ⎬ fetch
         <MDR> → IR                                                     ⎪
         <PC> + 1 → PC                                                  ⎭

         <AC0> → ALU Input 1 (OP1)                                      ⎫
         <AC1> → ALU Input 2 (OP2)                                      ⎪
         ADD command, on Arithmetic Unit control lines                  ⎬ execute
         <ALU output> → AC1                                             ⎭


102      <PC> → MAR                                                     ⎫
         READ command, on Memory Unit control lines                     ⎬ fetch
         <MDR> → IR                                                     ⎪
         <PC> + 1 → PC                                                  ⎭

         <AC1> → I/O bus data lines                                     ⎫
         Control lines order controller #2 to accept data on bus        ⎬ execute
                                                                        ⎭

103      <PC> → MAR                                                     ⎫
         READ command, on Memory Unit control lines                     ⎬ fetch
         <MDR> → IR                                                     ⎪
         <PC> + 1 → PC                                                  ⎭

         Control Unit timing circuits shut down                         ⎫ execute
                                                                        ⎭
```

the memory. The computer manufacturer informs the user which memory locations really exist and what the missing locations are actually used for. Referring to our example, the manufacturer might say that Location 200 is really the output register for the paper tape punch.

In the single bus architecture there are no separate I/O and memory reference instructions. These are replaced by a single instruction type called Move, the idea being that all the programmer is doing is moving data and instructions from one location on the bus to another. More important than the uniform instruction types, however, is the very high degree of modularity which results from the bus structure. It is very simple to add extra memory, peripheral device controllers, or other specialized hardware modules to the bus. The most important disadvantage of the single bus architecture is the increased complexity of the Control Unit. It must continually arbitrate which part of the computer may have control over the bus. Different units on the bus must wait until they are granted permission by the Control Unit to place their data on the bus. This continual arbitration requirement can slow down execution speed somewhat.

8.4.2 Microprogrammed Machines

In discussing the functions of the Control Unit in Section 8.2.2, we saw that a sequence of operations was performed. These were to fetch the instruction from memory into the MDR, to decode the op-code and the various subfields, and to initiate the necessary sequencing to execute the instruction. The logic required in the Control Unit to accomplish these functions

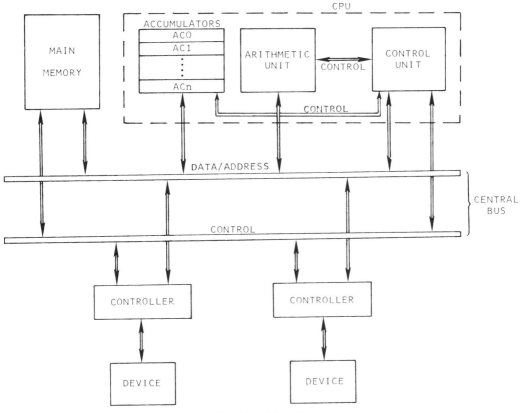

Figure 8.10. A Single Bus Computer

consists of decoders for all the fields of the instructions, plus timing circuits, address generation circuits, and internal buses. For a large instruction set the internal design of the Control Unit can easily become unwieldy. The addition of a new instruction to the instruction set might be extremely time-consuming and expensive, since, conceivably, the redesign of many parts of the Control Unit may be involved.

Many of the above problems can be solved by the use of microprogramming. With this method the Control Unit itself is a small computer. The op-code of the main computer instruction is an address to which the "microprogram" jumps (in an internal microprogram memory) in order to interpret and execute the instruction. The inner program which thus commands the operation of the main computer is often called the "control program". The architecture of the control computer is usually much different from that of the main computer, because

the control computer is optimized to interpret instructions, to route data back and forth between registers in various units, and to branch to different parts of the control memory based upon internal conditions. The control computer thus causes a particular "macro" level instruction to be executed by breaking it down into a sequence of basic, register-transfer-level operations called microinstructions.

There are two advantages to the microprogramming approach. The first is that it represents a very organized and modular way of building a Control Unit. Once all the data paths within the main computer are potentially under the control of the control computer, then any instruction can be made part of the main computer's instruction set simply by writing a microprogram or microroutine to execute that instruction. The manufacturer writes all these desired microroutines, and they are placed in read-only memory in the control memory. The second advantage of microprogramming is that, in some cases, it is possible for the user to define his or her own computer instructions. This may be done by selecting an op-code that is unused, defining the associated subfields, and then writing the microprogram to execute the new instruction. If user microprogramming is to be permitted, the control memory must not be entirely read-only memory—the user-accessible segment must be writable, in which case it is sometimes called Writable Control Store (WCS). In this instance certain main computer instructions allow the programmer to load the control memory with the appropriate microinstructions prior to executing the newly-created main computer macroinstruction.

8.5 ADDRESSING

Certain instructions, such as load, store, and jump, require an address field in the instruction word, as was illustrated by the load, store, and branch instructions that were defined for our hypothetical computer in Tables 8.2 and 8.3. There the address field was 9 bits wide, sufficient for a memory size of 512 words (2^9 words). However, since even microcomputers have larger memories than this, more sophisticated techniques are required to allow addressing of all memory locations. This section will examine the most common techniques employed to extend the memory addressing range of commercially-available computers.

8.5.1 Index Register Addressing

A field in the computer instruction for load, store, and branch instructions can indicate whether the contents of a special register, called an index register, should be added to the address field of the instruction word in calculating the desired or "effective address" to be used in executing the instruction. The resulting sum is the final memory address. Computers usually have several index registers, each programmable in the same way an accumulator is programmable. In fact, the registers of some computers do double duty as both accumulators and index registers. For our example computer, AC1 can also act as an index register. Since it is a 16-bit register, the index register can contain a memory

address as high as 64 K. The width in bits of the index register is often the determining factor for the largest allowable size of a computer memory. Table 8.2 showed that for memory access instructions, Bits 4 and 5 specify which index register to use. If these bits are 00, the final address is simply that contained in Bits 7 through 15 of the instruction word. If the bits are 01, the Effective Address Generator adds the contents of the 16-bit index register (AC1) to the 9-bit address field of the instruction to get the actual or effective memory address.

Although the above method allows higher addresses to be accessed, it is still somewhat unsatisfactory because, for all locations at a higher address than 512, it is necessary to use the index register. A further refinement that alleviates the problem is to allow the Program Counter to be used in the same way an index register is used. This would be accomplished by setting Bits 4 and 5 to 10 in our memory access or branch instructions. When the Program Counter is allowed to be used as an index register, we have what is referred to as a "floating page" of directly accessible memory locations. By this it is meant that no special calculations need be performed to specify the address, as would be necessary when deciding what number to place in AC1, if it is to be used as the index register. The size of the floating page depends, of course, upon the number of bits available in the address field of the instruction. The reason that the floating page technique works reasonably well is that most computer programs tend to reference memory locations near the address contained in the Program Counter. If the desired location is too far away from the location pointed to by the Program Counter, then an index register has to be used.

The last refinement in using index registers (including the Program Counter) involves regarding the address of the instruction field not as a positive number only, but as a two's complement number. When this is done, the range of the address field remains the same, but now locations both above and below the location pointed to by the index register or Program Counter can be accessed. This addressing mode is sometimes called *relative addressing.*

8.5.2 Indirect Addressing

Another common form of addressing allows the desired memory location to be pointed to by the contents of some other memory location accessed by a particular instruction. One bit in the instruction is necessary to indicate whether or not it is desired to perform indirect addressing. Bit 6 of the instructions that refer to memory in our example machine (Tables 8.2 and 8.3) serves this function.

As an example of how indirect addressing works, assume in executing a load instruction that it is determined the desired memory location is 1000 (using indexed addressing if required). If the indirect bit in the instruction is zero, then the contents of location 1000 are the data that will be loaded. If the indirect bit is one, however, the contents of 1000 are not the data, but represent a pointer to, or indirect address of, another location which does contain the data. If location 1000 had 2000 as its contents, then the desired data would be contained in memory location 2000.

Some computers allow chaining of indirect addresses. Bit 0 of a memory word is used to indicate if chaining should be performed. Continuing our example, if indirect addressing were specified and if Bit 0 of Location 1000 were one, then the data would not be in location 2000, but location 2000 would point to yet another location. The chaining process stops at the first memory location where Bit 0 is not one. This word then contains the last pointer. Using Bit 0 in this fashion cuts the maximum addressable memory size in half, since that bit no longer is available to indicate memory addresses.

Indirect addressing is useful for processing complicated data structures, in addition to extending the range of the address field of an instruction.

8.5.3 Page Addressing

Some computer designers prefer to divide the entire memory into segments of memory of fixed size known as pages. The high-order bits of the Program Counter then serve as the page indicator. The address field of the memory reference instructions indicates the displacement into the page. This method provides yet another way of accessing nearby information without requiring the use of an index register. As an example consider a computer in which the Program Counter is 12 bits wide, of which the three most significant bits also serve as the page indicator. We see that there can be eight memory pages, 000 to 111. Twelve bits can support a maximum memory size of 4096 words; hence each of the pages would be 512 words in length. Since the displacement into each page must come from the address field of the instruction, the address field would have to be nine bits wide to allow access to all the addresses within a page.

8.5.4 Immediate Addressing

Using immediate addressing, the desired data word is found in the memory location immediately following the instruction. When the Control Unit's decoder recognizes the "load or store immediate" instruction, it need only add one to the Program Counter to determine the necessary memory location. After execution of the instruction the Program Counter is automatically incremented one extra time so that the data word following the instruction will not be executed. The immediate addressing technique is useful not only for load and store instructions but can also be found in some computers as part of the arithmetic and I/O instruction formats.

8.5.5 Stack Addressing

The last form of addressing we will discuss involves the use of a stack. A stack is simply an area of memory in which there is a defined base, or "bottom" location. Information is added to the stack by placing it in successive memory locations above the base location. When information is removed, it is always taken off the stack starting from the top. Stack pointers (registers) are used to point to or address the bottom and the current top of the stack. This structure is referred to as a LIFO stack—Last In, First Out. Such a stack can at any time have from zero entries up to the maximum number of memory locations which have been assigned for use in the stack.

A stack is useful when the thread of an executing routine must be temporarily broken. This occurs under two conditions—(1) when calling subroutines or (2) when a program is interrupted and an interrupt service routine must be executed. For a subroutine call, information that might be "pushed" onto the stack includes: the program counter of the calling routine; the contents of all the registers, since they contain pertinent information for the suspended routine; parameters the calling routine is passing to the called routine; results from the called routine when it has completed its operation; and temporary working space for the called routine. The called routine can, if it wishes, place similar information onto the stack if it in turn calls another routine. When a routine is completed and wishes to return control to the program from which it was called, the stack is "popped", i.e., the saved information is removed from the stack and replaced in the registers and program counter.

The stack concept is important for several reasons. First, it represents an orderly procedure for managing the machine state and passing information to and from subroutines. Second, it makes subroutines reentrant, i.e., allows a subroutine, if it is structured correctly, to be in use by more than one calling routine at a time. This is important in multitasking environments. This reentrancy is made possible because, every time a subroutine is called, new working space for its variables is assigned on a stack; hence, there is no interference due to the subroutine's variables being disturbed by competing routines.

A stack can be entirely a software concept, in which the programmer himself saves all registers, manages the stack pointers, etc.; or in some small computers, special hardware instructions exist for managing the stack. This feature relieves the programmer of much work, since one instruction could cause all the machine registers to be saved, the pointers to be automatically adjusted, and other related operations performed.

8.6 INPUT/OUTPUT TRANSFERS AND INTERRUPTS

One of the main difficulties in interfacing a peripheral device to a computer is the large difference in speed between the two. Very few peripherals can respond to computer commands as rapidly as the computer can issue them; usually the speed difference is several orders of magnitude. For example, most computers can issue about one million commands every second, while a computer terminal may be able to accept only 10 characters in the same amount of time. Therefore, a major part of the interfacing problem is in providing the computer with the ability to determine when the peripheral has completed its operation. A related problem is deciding what the computer should do while waiting for the operation to finish.

We have already noted that there are two methods commonly employed in operating a peripheral device, i.e., in transferring information to or from the device. These two methods are referred to as *Programmed Input/Output* and as *Direct Memory Access.*

8.6.1 Programmed Input/Output

A programmed transfer is any transfer of data that results directly from the execution of a single I/O instruction. For each input or output instruction, one word of information would be transferred to or from a particular device (device interface). Generally, data words to be transferred are routed through one of the CPU accumulators. Table 8.6 includes a set of instructions that are intended to be used for programmed transfer of data in our example machine.

For example, the machine language command

$$11\ 0\ 0\ 000\ 000\ 000111 = 140007_8\ (DOA\ 0, 7)$$

would cause the data contained in Accumulator 0 to be transferred to Register A in the controller of peripheral device number 7. This device might be a console typewriter, and the information transferred might represent one character which is to be typed. Presumably the controller will cause the typewriter to print out the character once it has been placed in its A Register, or, as often is the case, such an operation must be initiated by the transfer of a special command from the computer to the device controller. With our example computer that might be done by resetting the device controller flag.

There are two ways of determining when a peripheral device has completed an operation, such as printing a character, with programmed I/O. In the first method a computer instruction is used to check a status flag (flip-flop) in the peripheral's controller. We refer to this technique as "status checking". In the second method the computer is informed of the completion of an operation by an interrupt from the controller of that device. We refer to this technique of sequencing as "interrupt-driven" input/output.

Status checking—We have designed the flag in each peripheral controller of our example computer to work as follows: If the flag is zero, the peripheral is still busy; if the flag is set to one, the operation has been completed. The flag can be set in one of two ways: The peripheral can set it when it has completed the operation, or the controller itself can set the flag. If the controller is to set it, then the controller must have a built-in timer, and the time to complete the operation must be fixed. This latter method is used in a computer terminal controller where it is known that the operation to output a character always takes, for example, one-tenth of a second.

Most computers have an instruction that causes the next instruction to be skipped if the

flag is set. In this way the programmer can simply test the controller flag. Table 8.8 gives an algorithm illustrating this technique both for programmed transfer into the computer, as from a terminal keyboard, and for programmed transfer from the computer, as to a terminal print unit.

We see that the computer will execute steps 2 and 3 until the flag is set. For output, the flag will set when a previously initiated output operation has been completed by the device. For input, the flag will set when the device has information ready for transfer to the computer. In either case the device controller flag must be cleared if more data words are to be transferred in or out.

The above procedure is reasonable if the computer has nothing to do but output characters to a computer terminal. Steps 2 and 3 would be executed about 50 thousand times, each, between characters. Usually, however, there are other things to do besides typing characters on a terminal. There may be computations to be performed or other devices to communicate with. One way to handle the problem would be for the program to do something else and come back later to check the flag. This method is awkward for the programmer though, especially if more than one device is involved. It can be seen that it would be helpful if the controller could automatically obtain the attention of the computer when the operation was completed. This is in fact the second method previously referred to.

Interrupt-Driven I/O—Each controller, when previously enabled to do so by a computer in-

Table 8.8. The Sequence of Programmed I/O Transfers

Output	Input
1. Load the data to be transferred into an accumulator.	1. To initiate the input operation, clear the controller flag.
2. Test the device controller flag. If set, skip 3.	2. Test the device controller flag. If set, skip 3.
3. Go to 2.	3. Go to 2.
4. Transfer the accumulator contents to the appropriate register in the device controller. Clear the device controller flag, if required.	4. Transfer the contents of the appropriate device register to an accumulator. Clear device controller flag, if required.

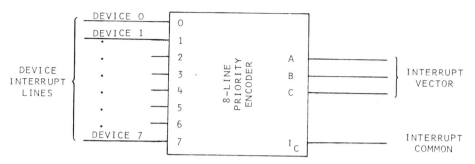

Figure 8.11. An Eight-to-Three Line Priority Encoder

struction, can place a signal on a special line that enters the CPU–the interrupt line. In many computers this line is shared among all controllers. The interrupt signal causes the computer to store away the Program Counter in a special register or memory location and to jump to a particular location in memory fixed by the computer manufacturer. In this location the programmer will have placed the first instruction of a routine that determines which device caused the interrupt and then takes whatever action is necessary to service the interrupt. Since an interrupt often catches a program in the middle of a computation, the interrupt service routine usually stores away the contents of the machine's registers in memory before attempting to decide what it must do to service the interrupt. After servicing the interrupt the last thing the service routine does is to restore the machine registers to their former state (PC is restored last). It then jumps back to the instruction following the one executed just prior to the occurrence of the interrupt. This is done effectively by executing a jump to the address specified by the pre-interrupt contents of the PC. Usually, the Control Unit will not respond to an interrupt until the currently executing instruction has been completed.

There are several different ways in which a routine can decide which peripheral caused an interrupt. It can check the flag line of each peripheral until it finds one that is set. It can issue an I/O instruction which forces the interrupting device to place its device code on the I/O bus data lines (an "interrupt acknowledge" instruction); the data lines are then gated into an accumulator. In a third method each device is assigned its own interrupt line to the CPU. This approach would require a more elaborate

set of hardware in the interrupt handling section of the CPU. A priority encoder often is used to generate an interrupt "vector" in response to a device activating its assigned interrupt line. Figure 8.11 illustrates an eight-device encoder that would generate a three-bit vector. The vector can be used to build up the address of the memory location where the computer will jump automatically to service that particular device. Using the encoder, in this case, has reduced the number of lines which need enter the CPU from eight to four.

Several different techniques can be used to handle multiple interrupts occurring at the same time or in close succession. The simplest method utilizes software to poll, i.e., check device flags, in decreasing order of priority to take care of the most important interrupt first. Other methods utilize hardware: (1) The interrupt acknowledge circuitry of all peripheral controllers may be "daisy-chained" (connected in series) such that only the controller located physically closest to the CPU in the wiring chain will respond to the acknowledge command (thus the more important it is to service a particular device quickly, the closer it is placed to the CPU in the chain). (2) The CPU may have several hardware interrupt lines with different priority levels. (3) The CPU may permit individual devices to be "masked out" under program control; i.e., a flip-flop may be set for each peripheral device which prevents it from causing an interrupt. (4) With vectored systems, only the device generating the highest-priority vector at any given time is permitted to interrupt. Figure 8.12 illustrates a vectored system with the ability to mask out any device or group of devices by loading zeros in corresponding bits of the mask register.

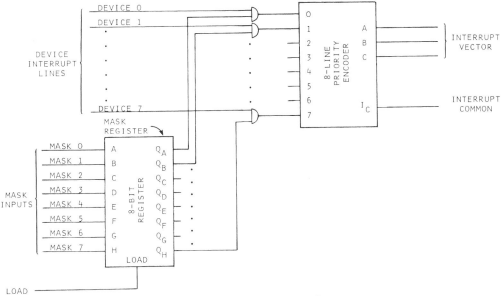

Figure 8.12. A Maskable Interrupt System

An extra feature sometimes associated with vectored machines is for the machine state (PC and all registers) to be automatically saved in predetermined memory locations (perhaps a stack), thus relieving the programmer of this chore.

Interrupts are useful for more than just servicing I/O devices. They can be used in detecting various hardware malfunctions or illegal operations. The three examples below, in addition to illustrating the use of interrupts, demonstrate some of the extra, generally optional features available on most small computers.

(a) Power Fail Detector

Should the power fail, the contents of all the computer's registers would be lost, since the registers are constructed from semiconductor circuits. Many applications cannot afford such a loss of information. Therefore, with core memory machines (or machines with semiconductor memory and emergency power supplies for memory) it may be desired to store these registers away in memory, which is nonvolatile, before the power fails completely. A circuit can be included in the computer that detects when the power is in the process of failing. Once this condition is detected, a few milliseconds exist during which operations can still

be successfully carried out. The power fail circuit requests an interrupt, and the service routine identifies this "device" as the cause of the interrupt. The registers can then be saved before power disappears.

(b) Privileged Instruction Trap

In some cases it is desirable that certain instructions be executed only by the operating system; users may not execute these privileged instructions. I/O instructions are often forbidden to the user, since the operating system will execute I/O routines for the user upon request. Allowing the user to perform I/O directly can cause costly errors, since many I/O devices are extremely complex. Therefore, the operating system can set a hardware trap before any user instructions are allowed to execute. If any user I/O instruction is attempted, the trap generates an interrupt, and the operating system will terminate the offending program.

(c) Memory Protection

When many programs reside in memory at the same time, it is desirable to restrict each program to access only the memory it has been assigned to work in. Otherwise, one program could accidentally write into another's memory area, destroying that program. In this case the operating system could load two special registers with certain memory limits that a particular program must work between. If a memory access is attempted outside these limits, an interrupt is generated.

8.6.2 DMA Transfers

In many cases the speed at which programmed transfers can occur is insufficient. An obvious way of increasing the speed of data transfers from an I/O device to main memory is to bypass the CPU entirely, i.e., to use what is known as Direct Memory Access (DMA). An interface using the DMA transfer mode must be more complex than a standard I/O interface. At a minimum a DMA memory address register and appropriate control logic are needed.

Two basic methods of performing DMA transfers exist. The first and the simplest uses what might be called "exclusive access" of the memory during transfers. During the period of time the I/O device is transferring data to memory the CPU is effectively "turned off" or in an idle state. The second method is known as "cycle stealing". Using this method the I/O device "steals" one or more memory cycles from the CPU periodically in order that it may complete its required transfers. Assuming this technique is implemented in the normal way, the CPU can never be idle for more than 50% of the memory cycles available to it. Thus, completion of both processor and I/O tasks is more readily assured with this transfer method.

Normally, DMA interfaces or channels allow for the automatic transfer of whole blocks of data rather than single data items. To provide for the transfer of blocks of data a register must be added to the DMA channel controller to hold a number specifying the number of data words to be transferred. Some computers reserve locations in main memory for both the word count and "present" or "next available" DMA storage location. This technique is slightly slower (since memory accesses are necessary to modify the word count and the storage address) but is more easily expandable. A basic DMA channel interface in block diagram form is shown in the next section.

8.7 INTERFACING

It has already been indicated that a controller acts as the interface element between the CPU and the actual peripheral device. In this section we will see just how the interaction between CPU and the controller takes place.

Figure 8.8 illustrated how controllers communicate with the CPU over the I/O bus. The signal lines of the I/O bus can be divided into six distinct groups. Figure 8.13 shows these six groups of lines along with a typical controller which would handle Programmed I/O.

(a) Data Lines

Data lines are used to pass data between the CPU and the controllers.

(b) Device Select Lines

Each controller is assigned a device code or I/O address by which it may be addressed. The device code is unique for each controller and is fixed electronically in the controller circuitry. Also, every I/O instruction has a field that contains a device code. This code is sent out on the device select lines as the instruction is executed. Only the controller whose assigned device code corresponds to the code on the device select lines will respond to the command lines. In this manner one I/O bus can be used to communicate with many devices. The controller circuitry for this function is really just a big AND gate, with certain of its inputs preceded by an inverter. The pattern of bits that can succeed in turning on the AND gate is then the device code for the peripheral.

(c) Command Lines

The command lines tell the selected controller what it should do. Commands can be classified as data transfer commands and *non*-data transfer commands. The data transfer commands can cause the controller to transmit the contents of the data lines to the peripheral, such as in outputting a character to a terminal. Alternately, an input from a device can be placed on the data lines, such as data transmitted as a result of striking a terminal key. In the latter case, the control unit would transfer the information from the data lines into an accumulator.

*Non*data transfer commands are used to initialize the controller or the peripheral in some way. These commands are necessary because many peripherals are more complicated than

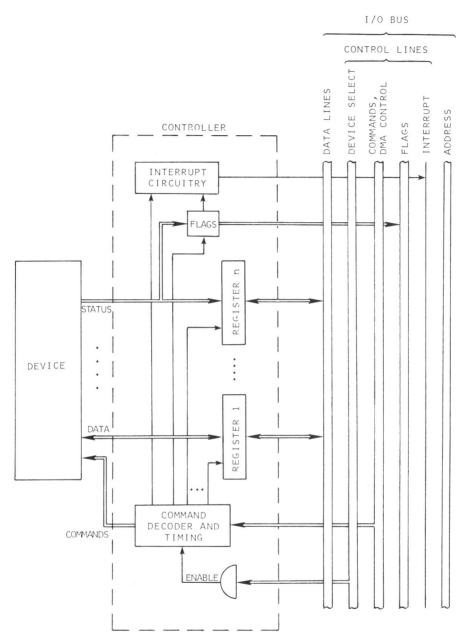

Figure 8.13. A Typical Device Controller

the simple terminal described above. For example, a magnetic tape unit might require a command to rewind the tape before any data could be transferred to it. Some *non*data transfer commands, in addition, require information to be passed over the data lines to the controller. An example of this would be in setting up a device to use the Direct Memory Access channel. It would be necessary to pro-

vide the controller with, among other things, the starting memory location for the block of memory reserved for use by the channel. Figure 8.14 illustrates the connection of a DMA Controller.

*Non*data transfer commands can also request that the controller place information describing the status of the device on the data lines. For example, the magnetic tape controller could

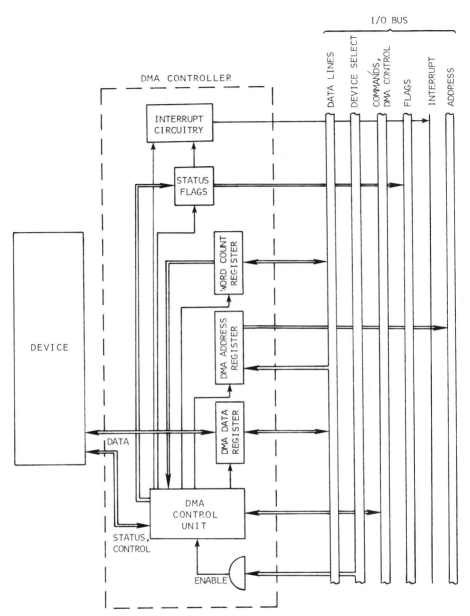

Figure 8.14. A DMA Controller

have a status register whose contents would indicate whether the unit's power was on, the reel of tape was mounted, the tape was write-protected, etc.

(d) Flag Lines

These lines can be interrogated by the CPU to determine whether the peripheral has completed a data transfer command. Earlier we described how an instruction can test a flag line and skip the next instruction if the flag is set.

The number of flag lines and their exact function vary from computer to computer. These flag lines may seem similar in purpose to the status register mentioned above. However, flag lines are used specifically to signal completion of an operation for two reasons. First, not all peripherals are complicated enough to justify including a general status register in the controller, and second, the flag circuitry normally is standard for all controllers in the computer. Thus in an interrupt polling sequence the same type of instruction can check the flags of all devices. No such instruction is available for checking a status register directly.

(e) Interrupt Line

This line is usually affected directly by the controller's flag circuitry and generates an interrupt in the computer. The flag line itself is not used to cause an interrupt, since in many applications it is desirable to disable particular devices from causing interrupts temporarily, while not disabling other devices. For our sample controllers, in Figures 8.13 and 8.14, interrupts can be disabled by issuing such a *non*data transfer command to the controller. The decoder and timing section would have to contain a flip-flop to "remember" such a command. An output line from this flip-flop then would not permit the interrupt flip-flop to set, as indicated in the figures.

(f) Address Lines

These lines are used exclusively by the DMA Channel. They permit a memory address maintained within a register in a DMA controller to be passed directly to the MAR in the Memory Unit. As a result the device controller itself maintains a pointer to the location in memory that will be accessed in a read or write cycle. Figure 8.14 illustrates the connection.

8.8 SUMMARY

This discussion of computer architecture cannot pretend to be complete. The topics are much too broad to fit comfortably within the pages of a single chapter. Nevertheless, it is still important that everyone who plans to use a computer in a real-time environment have a fundamental understanding of how a computer functions. The techniques that have been explored in this chapter correspond closely to those of the computers that are being built today. Their individual logic circuits, although invariably more complex than those presented in Chapter 7, are based on exactly the same principles. The gates, flip-flops, registers, and counters we discussed previously are the standard building blocks for even the most complicated circuits.

This chapter has attempted to describe most of the important architectural features of modern small computers. In fact, the least expensive minicomputers behave quite similarly to the example computer that was developed. More powerful minicomputers employ additional features that we have not had space to discuss; nevertheless, their fundamental behavior remains exactly the same as that of our simple model. The interested reader may consult the references which follow for a more detailed discussion of many of the topics we have presented.

8.9 SUGGESTED READING

1. Add-Alla, A. M. and Meltzer, A. C., *Principles of Digital Computer Design*, Prentice-Hall, Englewood Cliffs, N.J., 1976.
2. Bell, C. G. and Newell, A., *Computer Structures: Reading and Examples*, McGraw-Hill, New York, 1971.
3. Booth, T. L., *Digital Networks and Computer Systems*, McGraw-Hill, New York, 1972.
4. Gschwind, H. W. and McCluskey, E. J., *Design of Digital Computers*, Springer-Verlag, New York, 1975.
5. Hayes, J. P., *Computer Architecture and Organization*, McGraw-Hill, New York, 1978.
6. Hill, F. J. and Peterson, G. R., *Digital Systems: Hardware Organization and Design, 2nd Ed.*, John Wiley & Sons, New York, 1978.
7. Kline, R. M., *Digital Computer Design*, Prentice-Hall, Englewood Cliffs, N.J., 1977.
8. Klingman, E. E., *Microprocessor Systems Design*, Prentice-Hall, Englewood Cliffs, N.J., 1977.
9. Korn, G. A., *Microprocessors and Small Digital Computer Systems for Engineers and Scientists*, McGraw-Hill, New York, 1977.
10. Mano, M. M., *Computer System Architecture*, Prentice-Hall, Englewood Cliffs, N.J., 1976.
11. Peatman, J. B., *Microcomputer-Based Design*, McGraw-Hill, New York, 1977.
12. Pooch, V. W. and Chattergy, R., *Minicomputers, Hardware, Software, and Selection*, West Publishing Company, St. Paul, Minn., 1980.
13. Sloan, M. E., *Computer Hardware and Organization*, Science Research Associates, Chicago, Ill., 1976.
14. Stone, H. S., *Introduction to Computer Architecture*, Science Research Associates, Chicago, Ill., 1980.
15. Weitzman, C., *Minicomputer Systems, Structure, Implementation, and Application*, Prentice-Hall, Englewood Cliffs, N.J., 1974.

8.10 EXERCISES

1. a. Suppose the example computer described in this chapter had been equipped with four accumulators. Show how Types I–IV instructions might be modified to accommodate this feature. Retain all present "addressing modes" (i.e., indexed, indirect, etc.) in defining your new instruction formats.
   b. How many words of memory can be addressed *directly* with your instruction set?
   c. How many separate peripheral devices can be accessed by your Type IV instructions? Could more devices be accommodated in this 16-bit format? How?

2. Table 8.8 gives a standard sequence of four instructions for programmed input and output transfers. Assume that a particular peripheral device can be used for both input and output. The interface for this device is organized so that its A register is used for input to the computer, while its B register is used for output from the computer. The interface device code is 15_8.
   a. Referring to Table 8.8, write the four 16-bit words representing the machine language instructions of the output sequence in their octal and mnemonic formats. Assume that the device controller flag does not have to be cleared in Step 4 of the sequence.
   b. Assuming that the device controller flag *must* be cleared in Step 4 of the input sequence, write the appropriate machine language instructions in their octal and mnemonic formats which correspond to the listed sequence of operations.
   c. Can you determine under what conditions the controller flag would need to be cleared in the output sequence? The input sequence?

3. Suppose that a computer manufacturer has supplied you with a computer having the architecture of the example machine of this chapter, and that it included an interface consisting of a single-channel analog to digital converter (ADC). The ADC is used to convert an analog voltage input (-10 volts to +10 volts) to an 11-bit plus sign, two's complement number. Owing to an error on the part of the manufacturer, Register A of the ADC device controller (Device 21_8) has been wired so that the 12-bit result generated by the ADC is placed in the 12 least significant bits of the register. Hence, when the contents of Register A are read into an accumulator via an input instruction, the four most significant bits always are input as zeros regardless of whether the result should have been positive or negative.

Assume that "Special Pulse One" of the Type IV instruction set causes the ADC to generate a result and place it in Register A, and that the ADC's converson time is small enough that the computer can read in the generated result in the next following instruction.
   a. Write the octal and mnemonic equivalents of the instructions described above to cause an ADC conversion. Place the result in Accumulator 0.
   b. Write a "program" of machine language instructions for the example computer (octal and mnemonic equivalents with a word description of each instruction) that will determine if the number input into Accumulator 0 by the instructions of part a represents a negative or a positive value and will make the most significant four-bits all ones or leave them zero, as appropriate. If you need to have data stored for your solution, assume that you may use memory locations 6, 7, 10, 11, . . . for this purpose.
   c. If the ADC were not as fast as we described, one way to avoid problems would be to design the ADC interface so that a "Special Pulse One" command would clear the controller flag at the same time the converter is started. At the end of a conversion cycle the *converter* would set the flag as soon as it had loaded the generated result into Register A. Answer part a of this problem, again assuming that the interface and ADC operate as just described.

4. A binary "mask" often is used to extract one or more bits from a word of information in the computer. For example, 20_8 $(00 \ldots 10000_2)$ can be used as a mask to determine if Bit 11 in a particular number is 0 or 1 by "ANDing" 20_8 with the number in question and noting whether the result is zero or nonzero.

This technique is useful in testing the status of individual inputs of a digital input device which might be a part of a real-time computer system. It also can be used effectively to output 8-bit ASCII character information to a terminal when 16-bit memory words are used to store two ASCII characters each. (Bits 8–15 would contain the ASCII representation of the first character; Bits 0–7, that of the second.)

a. What mask (octal number) could be used to select only the right byte of a computer word, rejecting the left byte?

b. Write a machine language program (octal and mnemonic equivalents of the example computer instructions *and* word descriptions) that will take the contents of Memory Location 6 and output the right byte first, then the left byte, to Register A of Device 16_8 (which is some sort of terminal). The mask should be stored in Location 7, and your program should begin in Location 10_8. If a null byte (all zeros) is encountered, your program should halt immediately.

5. What will be the contents of Accumulator 0 of the example computer after the execution of the following sets of instructions:

a. NOT 1, 0
   ADD 1, 0
b. SUB 0, 0
   NOT 0, 0
   NEG 0, 0
c. SUBL 0, 0, CTZ
d. MOV 0, 1
   ADDL 0, 0, CTZ

ADDL 1, 0, CTZ
e. MOVL 0, 0, SCZ
   SUB 0, 0
   SUBR 0, 0

6. Suppose that we wished to add an unsigned integer multiplication instruction to the example computer. This instruction, when executed, would multiply the integers contained in AC0 and AC1 and leave the resulting 32-bit product in these two registers. After the instruction was executed, AC1 would contain the most significant 16 bits of the product, while AC0 would hold the least significant 16 bits.

a. To which instruction class would the multiply instruction belong? Suggest an instruction format and a mnemonic for the multiply instruction.

b. Assuming that the ALU is not capable of *directly* multiplying two 16-bit integers, what modifications would have to be made to the Arithmetic Unit and the Control Unit to allow the multiply instruction to be executed as described? Specify the algorithm to be used in the multiplication operation.

c. Assuming that each instruction requires *one unit* of time to execute, what is the *maximum* (approximate) number of time units which would be required to execute the multiplication instruction? Justify your answer.

9

Peripheral Devices and Data Communications

Walter G. Rudd
Department of Computer Science
Louisiana State University
Baton Rouge

9.0 INTRODUCTION

In Chapter 8 the architecture and operating characteristics of the computer itself—the Arithmetic and the Control Units, the Memory and Input/Output Units—were described. However, without connections to the external world to permit the computer to communicate with other devices, the most powerful processor is useless. In this chapter we consider ways to bring about the necessary interactions between a computer and its users and environment. Under the topic of peripheral devices we will also discuss the means available to enhance the total information storage capacity of a computing system. Finally, we will consider some aspects of data communications—the transfer of digital information from place to place—in this case, under computer control.

9.1 PERIPHERAL DEVICES

A peripheral device is a hardware component that is separate from the computer central processing unit and main memory. In a system block design, these devices often are lumped together and called the "input/output" system. Three components are usually involved in the connection of a peripheral device to the central system—the input/output bus, the device interface or controller, and the device itself, as illustrated in Figure 9.1. (Some high-speed Direct Memory Access devices are also connected to the memory address and memory data buses as well.) Since few, if any, peripheral devices can be plugged directly into the input/output bus, the interface or controller is necessary as an electronic intermediary between the input/output bus and the peripheral device.

Under the heading of peripheral devices, we can identify four major classes:

(1) *Interactive Devices or Terminals.* These devices enable direct communications between the human computer user or operator and the computer system. Most systems have at least one interactive console device to control the operation of the system. In many real-time systems users enter data, programs, and commands to the system via one or more interactive terminals. The system responds in real-time to these commands, thereby giving the user seemingly instantaneous feedback about the information he or she entered or about the status of the system.

(2) *Sensory and Control Devices.* These devices enable the system to communicate in real-time with the process portions of its environment. Sensory (input) devices include analog to digital converters, binary inputs, pulse counters, and the like. Control (output) devices—such as digital to analog converters, pulse and binary outputs—enable computer systems to regulate valve positions, open or close switches, and perform other external hardware control functions.

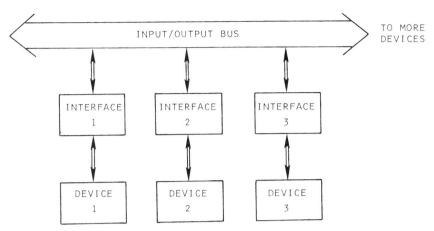

Figure 9.1. The Interconnection of Peripheral Devices to the Computer

(3) *Low- and Medium-Speed Input/Output Equipment.* This class of devices includes line printers, card readers, paper tape reader/punches, and other peripherals used for repeated input/output of large quantities of information. In some real-time systems these devices are used for program development and bulk data entry.

(4) *Auxiliary Storage Equipment.* These devices extend the power of the computing system by providing hardware access to quantities of information too large to be stored economically in the computer's main storage unit. Auxiliary storage devices include magnetic disks, drums and tapes, and bubble memories.

We shall now explore each of these classes of peripheral devices, with the exception of Class 2, in greater detail. Because of the unique place of sensory and control devices in real-time computing systems, we have consolidated discussion of computer/process interfacing in a separate chapter (Chapter 10).

9.1.1 Interactive Devices

The most important device that allows interaction (communication) between operator and computer is the terminal. A terminal—such as shown schematically in Figure 9.2—consists of a keyboard, a display which may produce output on paper, a CRT or video display screen or other medium, hardware to encode and decode character data, and circuitry to control communications on the line or cable to which the terminal is attached. Before investigating the various kinds of terminals which exist, we shall look at how the data are formatted.

Data Formats. A fundamental principle of interactive computing is that human beings are slow in comparison with most of the things that go on inside a computer system. The computer is capable of executing several hundred thousand or even millions of instructions during the time it might take the user of an interactive terminal to decide which key to strike next. The point is that the arrival of a data character from the keyboard of an interactive device is a rare event from the standpoint of the computer. Virtually all interactive devices transmit and receive data one character at a time.

Characters are represented and transmitted in the form of binary codes. When one strikes a given key on the keyboard of a terminal, a mechanical or electronic encoder within the terminal produces a binary code corresponding to the character that was entered. This binary-coded character is transmitted as a series of electrical pulses to the computer. (We provide details of the transmission process in Section 9.2.1 under the more general topic "Data Communications".)

The code used almost universally for character-by-character data communications is the

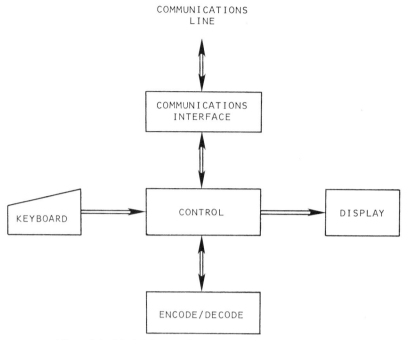

Figure 9.2. Block Diagram of a Generalized Computer Terminal

7-bit ASCII Code,* also called full ASCII, extended ASCII, or USASCII Code. The exception to this rule is IBM's adherence to their Extended Binary Coded Decimal Interchange Code (EBCDIC). The code used for communications is usually the same as that used for internal character representation in the computer system.

In Chapter 6, we presented both the 8-bit EBCDIC Code and the 7-bit ASCII Code, the latter of which we will discuss here. In using the ASCII code, devices actually transmit a total of 8 data bits. The last 7 bits are the character code as in Table 6.5. The first, or most significant, bit is transmitted either as a binary 0 (7-bit ASCII), a 1 (8-bit ASCII), or it can be used as a parity check bit. If the last alternative is used, we have a choice of whether to use even or odd parity in specifying the first (or parity) bit. If we choose even parity checking, the first bit is chosen so that the total number of logic 1 bits, including the parity bit, is even. If we choose odd parity checking, the parity bit will be set to make the total number

of 1 bits including the parity bit odd. For example, the 7-bit ASCII Code for "A" is 1000001_2. Using odd parity the data transmitted would be 11000001_2. Similarly, using even parity the character "S" would be transmitted as 01010011_2.

Parity checking permits detection of an odd number of errors in the transmission of bits within a single character. The transmitter computes the proper parity value and sends it with the character. The receiving end checks to see that the 8 bits received have either an odd or even number of 1's, whichever is appropriate. If the parity condition is not satisfied, the receiving end must request a retransmission of the character or flag the error in some way.

The choice between even, odd, or no parity checking is made at the time hardware is selected and installed. Since few manufacturers charge extra for parity checking hardware these days, there is no reason to do without it.

Now that we know something about the form that data characters take during transmission, let us look into the various kinds of terminals available.

One way to categorize terminals is by the display medium that is used. The two most

*ASCII stands for American Standard Code for Information Interchange.

popular media are paper and CRT (Cathode Ray Tube) display. Terminals that produce their output on paper are called *hard-copy* terminals. Terminals that display their output on CRT screens are often called Video Display Units (VDUs).

Hard-Copy Terminals. The precursor of all terminals is the Teletype* machine. While use of Teletypes is declining in the face of technological improvements, the slow model ASR 33 (10 characters per second) is still in use in many laboratories and other locations where inexpensive paper tape and keyboard I/O are required.

The Teletype printer is an example of an impact printer, so named because the output is formed by striking a ribbon against paper with an outline of the desired character. The character font of an impact terminal can be arranged in several ways—on a cylinder as in the Teletype, on a type ball as in IBM typewriters, on a daisy wheel, or on a bar or a chain. Terminals that include type balls and daisy wheels ordinarily produce "letter quality" output comparable in appearance to material from an electric typewriter. These terminals are frequently found in "word-processing" applications, a relatively new and burgeoning field of on-line computing.

An alternative impact printing technique is the dot-matrix method, in which characters are printed by a mechanism that presses the appropriate wires in a 5×7 or 7×9 matrix onto the paper. This latter technique appears to have the upper hand in the market because of speed and reliability advantages that the method offers.

One of the key considerations in selecting a printing terminal is speed. Impact terminal printers that are not of the dot-matrix variety do not exceed 55 characters per second in printing speed. Dot-matrix terminals that print 180 characters per second are not uncommon.

The most popular nonimpact printer is the thermal printer, which prints by heating points on special heat-sensitive paper. The advantages of thermal printing are that the printers are lighter in weight, quieter, and tend to be cheaper than impact printers of the same speed. These advantages are largely offset by the expense of the special paper they require and the fact that they cannot be used to make multiple copies.

Almost all terminals can be purchased without keyboard in an RO (Receive Only) configuration. These terminals often serve as attractive replacements for line printers. As an example, one can purchase a 120-character-per-second dot-matrix printer which is the functional equivalent of a line printer. It prints between 55 and 720 lines per minute, depending upon the lengths of the lines.

CRT Display Terminals. CRT display terminals display their output on TV screens. They are virtually noiseless and can operate at much higher speeds (most run up to 960 characters/second) than hard copy terminals. They also tend to be less expensive than hard copy devices.

In real-time computing applications, CRT display terminals are often used for data entry and for reporting process data. In a data entry application a form is displayed on the screen and the operator simply fills in the blanks with the information requested. Similarly, large amounts of information can be displayed rapidly in response to an inquiry. This makes CRTs appropriate for displaying process data to plant operators. Because CRTs typically display only 24 lines of output at any one time, they are not very convenient for scanning large programs for one particular program element. On the other hand, they are quite convenient and widely used in inputting original programs to the computer and in making minor changes to programs, as in editing and debugging.

Each data character is formed in a character generator circuit as depicted in Figure 9.3 and displayed in 5×7 or 7×9 dot-matrix on the screen. A typical screen holds 24 lines, 80 characters per line.

Some of the older terminals use a storage tube for display, in which data stay on the screen until the entire screen is erased. Modern terminals use a storage-refresh system, in which a screenful of data is stored in a special buffer

---

*Teletype is a registered trademark of the Teletype Corporation.

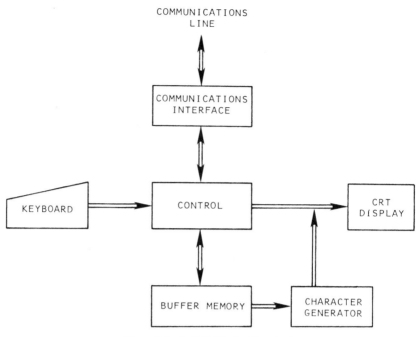

Figure 9.3. A CRT Display Terminal

memory built into the terminal. The entire screenful of data is rewritten onto the screen 60 times per second, providing an image that appears to be permanent. The advantage of storage refresh is that individual characters can be changed without having the computer erase and rewrite the entire screen.

CRT terminals operate at speeds between 10 and 1920 characters per second. A "dumb" terminal, i.e., one without built-in electronics to permit local editing, is the least expensive interactive device. "Smarter" terminals offer sophisticated editing capabilities that permit the construction of one or more screensful of data at a time. The operator can insert and delete individual characters and lines and can therefore visually check and debug the data before they are transmitted to the computer. The result is increased efficiency in data or program entry, since the chance of transmitting erroneous information is reduced. Many terminals permit the attachment of a character printer, a hard-copy unit, thereby permitting the generation of a permanent copy of the displayed data when desired. Other features available include dual-intensity display, blinking fields, protected fields that cannot be erased by

the operator, and buffers than can hold more than one screenful of data at a time.

Other Display Media. A few terminals are available with LED (Light Emitting Diode) or plasma displays primarily for use in situations that require large displays that must be visible to others besides the operator. Point-of-sale terminals often seen in supermarkets and drug stores use these displays. These media have not been accepted for use in general-purpose interactive terminals.

Intelligent Terminals. Some modern terminals contain at least one microprocessor for character translation, buffer management, screen editing, and other functions. The user neither knows nor cares whether or not that internal processor is there in the sense that it is preprogrammed by the vendor. A few terminals include a user-programmable microcomputer and memory. Thus the terminal becomes a computer in its own right and is called an *intelligent terminal.*

Figure 9.4 shows the structure of a typical intelligent terminal. The read-only control memory contains the firmware drivers (a program or software stored in some form of nonvolatile, read-only memory is known as

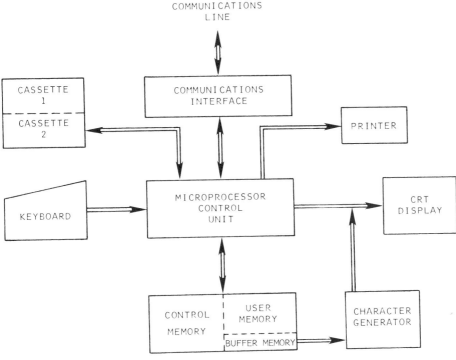

Figure 9.4. Structure of an Intelligent Terminal

firmware) for all the attached devices and possibly a compiler or language processor. The user programs and temporary data reside in the user memory.

A typical application of an intelligent terminal is in store-and-forward data entry. The user preprograms the machine to display a data entry format on the CRT display. The program requests data and checks the data entered for errors before writing them onto one of the cassettes.* At the end of the day, the terminal calls up a remote computer over a telephone line and transmits the collected data from the cassette to the remote computer.

Several intelligent terminal systems include a graphics display unit and a high-level language interpreter. Since these types of systems really are rather powerful tools for scientific compu-tation and for computer graphics applications, the distinction between terminal and computer vanishes when we consider them.

The new magnetic bubble technology (a non-volatile memory technology) provides a sub-stitute for cassettes or floppy disks in intelligent terminals. One now can purchase a thermal printing terminal with 20 kilobytes (20,480 bytes) of bubble memory in which data can be accumulated, edited, deleted, and otherwise manipulated off-line. Additional memory can be purchased up to a maximum of 80 kilobytes (about one third the capacity of a cassette). Once the data are acceptable, the terminal can transmit them to a computer in one sequence.

Graphics Terminals. A graphics terminal is a CRT display terminal with the added capability of being able to plot points at arbitrary locations on the screen. If the graphics terminal contains a *vector generator*, the computer to which the terminal is attached need specify only the end points of lines to be plotted, and the terminal plots the line and continually refreshes the image. Without a vector generator, the attached computer must supply the x-y coordinates of every point to be plotted, a task that, depend-ing on the available display resolution, often becomes prohibitive in terms of computer time

---

*Auxiliary storage devices such as cassettes and floppy disks are discussed in Section 9.1.3.

required and the amount of display memory which must be included in the terminal.

In several commercially available graphics systems a mini- or microcomputer takes on the character generation, vector generation, and storage-refresh responsibilities. Some of these units offer language translators and enough memory to function as programmable, stand-alone systems in addition to fulfilling their roles as terminals.

Several manufacturers offer the capability of displaying in multiple colors, a feature that is especially attractive in process control applications, since several trend plots in various colors can be superimposed to give a much better picture of what is happening than would be available from separate plots.

In simpler systems, the user controls the display on the screen through the use of the terminal keyboard or by altering the program in the computer to which the terminal is attached. Several devices exist that provide a more direct means of interaction between the user and the system. One such device is the light pen, which can detect the presence or absence of the CRT electron beam when held close to the screen. The attached computer then can determine the x-y coordinates of the light pen and respond accordingly. This feature gives the operator a way to "draw" on the screen and to select options that depend upon their position on the screen. For example, a graphics console in one computer process control installation displays a schematic of the entire plant at the highest "level". By striking a key on the console keyboard and pointing the light pen at a given unit, the operator can cause the graphics system to "explode" the view of that section of the screen, allowing the operator to observe valve settings, switch positions, set-points, and sensory readings from various devices in the unit. Using the keyboard and light pen, the operator can request more detailed information—trend data from a given set of sensors, for example— or can cause the system to change a switch position or set-point at a given control point.

Other devices that serve to provide an x-y coordinate information are the "joy stick" and the "mouse". The joy stick works on the principle of an airplane attitude control stick.

The operator moves the joy stick to a desired position; A/D converters transform electrical voltages proportional to the x-y coordinates indicated by the joy stick to digital form. The mouse works on the same principle, but the x-y coordinates are physically determined by rolling the mouse over a flat surface to the desired position.

Choosing a Terminal. There is a bewildering array of terminals on the market, and terminal selection can be a real headache. Fortunately, in most cases the specifics of the application help eliminate large sections of the market early in the selection process. Procedures similar to the following can help a system designer make a final selection.

First, determine whether or not hard-copy is required. If not, then the economic, speed, and noise-level advantages of CRT display terminals dictate the selection of one of those. If a small amount of hard-copy is required, but not a record of every character entered and displayed, investigate the possibility of attaching a character printer to the terminal.

If a hard-copy terminal is needed, decide whether or not multiple copies of the output are necessary or desirable. If so, select an impact printing terminal. Then decide whether or not letter-quality output is necessary before selecting a printing technique.

If multiple copies are not required, the slightly lower purchase price, higher degree of portability, and relatively quiet operation of a thermal printer suggest that a thermal printing terminal may be chosen over an impact printer. The user must be careful here, however, because the increased cost of special paper for a thermal printer could eventually counteract any initial economic advantage of this type of terminal. An estimate of current and future throughput requirements, while sometimes difficult to obtain, can help one to make this decision and to select the proper display speed as well.

The amount of intelligence required of the terminal further narrows the field of possible candidates.

The next most important criterion is the manufacturer. Select three or four of the top manufacturers in terms of volume and make the final selection from among them. There are several reasons for this, among which are that maintenance will be more readily available, delivery earlier, response from the company better, and support for the terminal more secure if a well-established vendor is chosen.

If the above guidelines are followed, the end decision can be based on price and operator and human engineering factors. In the latter group are such criteria as the appearance of the letters displayed, the feel of the keyboard, keyboard layout including separate numeric keypads and special function keys (if needed), and other similar features. A representative assortment of interactive devices is presented in Table 9.1.

Table 9.1. Representative Interactive Devices

Interactive Device	Specifications	Approximate Relative Cost
1. Thermal Printing Terminal	80 columns, 30 char/sec, requires special paper	1
2. Dot-matrix Printing Terminal	a) 30 char/sec, 132 characters per line	1
	b) 120 char/sec, 132 characters per line	2
3. Letter-quality Printing Terminal	Up to 55 char/sec, daisy wheel	2-4
4. CRT Terminal -- Nonintelligent	19,200 baud* (bit-serial) transmission and reception of ASCII information. Received information is typically displayed on a 80 char/line, 24-line CRT. Normally the data format utilized may be switch-selected as well as full and half duplex terminal operation. Some terminals also allow special cursor and display control.	0.5-2
5. CRT Terminal -- Intelligent	Normally includes the same features as 3. above plus additional capabilities which may include: offline editing, a language processor, block data transmission, etc. Device intelligence is usually provided by a microprocessor.	2-4
6. Graphics Terminal	a) Terminal designed especially to allow graphical display of information on a CRT. These devices usually have some of the capabilities listed for nonintelligent and intelligent terminals in addition to graphics generation features. Display/resolution: 10-4096 points/inch. Display size: 4 in x 4 in to 20 in x 20 in	2-10 Prices for these devices vary greatly depending on display size and resolution.
	b) Similar to the device described in (a) with the addition of a mini- or microcomputer to provide enhanced editing and computational capabilities.	5-15

* Data transmission terminology is discussed in Section 9.2. Individual terms are defined in a glossary contained at the end of this chapter.

9.1.2 Low- and Medium-Speed Input/Output Devices

While card readers, line printers, and similar devices are never used directly as real-time peripheral devices, they nevertheless form an important part of many real-time systems because of the need to develop software and to process bulk data in batch operations.

The trend in modern computing systems is away from the use of physical media such as punched paper tape or cards. Programs and data are entered directly into the system via a terminal and stored on a disk or other auxiliary storage device as described in Section 9.1.3. We therefore present in the following a brief overview of the characteristics of paper tape and card devices, with the caveat that current trends are away from the use of such devices.

Paper Tape Readers and Punches. Standard paper tape is one inch wide (Figure 9.5). A set of sprocket holes is flanked by three rows of data on one side and five rows on the other side. Each 8-bit frame punched across the tape contains either one character code (7-bit ASCII code with a parity bit, for example) or 8 bits (a byte) of binary data. A hole represents a logic 1 at that bit position.

Tape readers are usually photoelectric devices (but may be mechanical as is the case with a Teletype). The tape is pulled over eight photo sensors (or nine if the sprocket hole is not used to drive the tape through the reader), one per track, which detect light passed through the holes from above. Speeds available at a reasonable price range from 150 to 600 frames per second. Most computer vendors offer optical readers that read 300 frames per second.

Paper tape punches are slower, typically running 50 to 150 frames per second. Several small-computer vendors offer a reader/punch combination with a single interface card—reading at 300 characters per second, punching at 150 cps—at a cost somewhat below that of separate reader and punch. Also available at much lower prices are the "micro" tape readers and punches, which operate at about 30 characters per second.

Card Readers and Punches. For compatibility with large systems and for processing large programs and/or sets of data, card readers are more appropriate than paper tape equipment. As is the case with other peripherals, the small-computer market has forced the price of card equipment down while improving price/performance ratios. Speeds available range from 100 cards per minute, in a reasonable price range, to over 1000 cards per minute, at considerably higher prices. A primary concern in choosing a card reader is how "finicky" it is. Does it require frequent, meticulous adjustments? Does it read cards with bent corners, or does the deck have to be in perfect condition? One manufacturer advertises that his reader can read the same deck 1400 times without missing a character. This is the kind of performance to look for.

Card punches are and will probably continue to be one of the most expensive peripherals in computer systems. Nobody has yet discovered a way to punch a hole in a card without going through a lot of mechanical manipulations, which means high prices and reliability problems. A 300 card-per-minute punch for a small computer costs four to five times as much as an equivalent card reader. (Manual card punches or "Keypunches" are similarly expensive.) This suggests that if program and/or data portability is necessary, paper tape, cassette tape, floppy disk, and cartridge tape are all preferrable in terms of initial hardware expense.

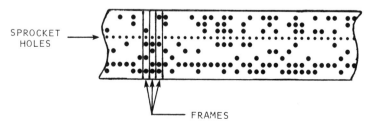

Figure 9.5. Paper Tape Data Format

Printers. In Section 9.1.1 we noted that high-speed character printers offer an economical way to obtain hard-copy output. When greater throughput is required than character printers permit, line printers may offer appropriate operating characteristics. Typical impact line printers come with line widths of from 80 to 132 columns and print at speeds between 100 and 1400 lines per minute.

Some high-speed alternatives to impact line printers are available. An example of such a device is a thermal printer/plotter, a device that includes character-generating facilities as well as 100-points-per-inch resolution in plotting. This very expensive device is capable of speeds up to 4800 lines per minute of character data. When it is used as a plotter, the slight sacrifice in resolution in comparison with a conventional mechanical plotter is more than offset by the several orders of magnitude increase in plotting speed.

Plotters. For hard-copy graphics output there is a wide range of plotting devices available. Analog X–Y plotters require voltage-level x–y coordinates, meaning that the computer plotting the data must contain digital to analog converters. Digital plotters accept digital inputs and contain their own D/A converters. Electrostatic or thermal plotters accept x–y coordinates in digital form and plot dots at the indicated points. As we have mentioned previously, some graphics terminals allow hard-copy output of displayed information. This means that we could effectively use a graphics terminal as a plotter if we are willing to settle for a plot resolution that is limited by our display device (CRT). Several other varieties of plotting devices are available at higher costs, including drum plotters, digital flat-bed plotters, etc.

A listing of typical low- and medium-speed I/O devices is given in Table 9.2. Notice that in many cases a range of devices of various capabilities is listed under a single item to provide an idea of the variety of available devices.

9.1.3 Auxiliary Storage Devices

Many real-time applications require storage capacities far beyond those economically feasible through the use of main-frame semiconductor or core memory. Auxiliary storage devices, including magnetic tapes, disks, drums, and others, provide the necessary means for bulk storage of information including program libraries, historical data, and files containing parameters, quantities, and schedules for process control. Two such devices often are used so that one can provide back-up storage to permit recovery from errors involving the primary bulk storage device.

Selection of the type of auxiliary device to use depends upon several factors, including the quantity of information to be stored, the frequency of use of the stored information, and the time available for access to specific units of information.

Magnetic Tape Transports. The least expensive bulk storage medium now available is magnetic tape. A typical 2400-foot reel of 9-track tape costs less than $15 and has a capacity close to 40 million 8-bit characters. Also available, are 600- and 1200-foot reels. Reels are mounted on peripheral devices called tape transports or drives for storage or recovery of information on the tapes. Data transfer rates on the order of 15,000 to 800,000 bytes per second are typical.

These characteristics make magnetic tapes the choice for compact, inexpensive, long-term storage of information that is not often or speedily needed. Tapes are often used in real-time computing for back-up storage of program libraries, process data, and tables that are normally stored on-line on disks or drums. Information lost from disks or drums because of hardware or software failure can then be recovered from back-up tapes. Tapes also are used for bulk storage of historical data. For example, process data might be accumulated on a daily basis and stored on disks in a process control system. At the end of each day, that day's data are copied onto tape for long-term storage. Similarly, many on-line accounting, inventory management, and other data processing systems use tapes both as back-up for on-line bulk storage and for long-term storage of historical data.

One final advantage of tapes is that, until quite recently, they were the only reasonable means of storing data in a form readable by

Table 9.2. Representative Low- and Medium-Speed I/O Devices

Device	Specifications	Approximate Relative Cost
1. Paper Tape Reader	Reads 8-level (8 holes/frame) one inch wide paper or Mylar tape at 400 characters (frames)/sec.	1
2. Paper Tape Punch	Punches 8-level characters (and sprocket hole) in one inch paper or Mylar tape. Characters are punched 10 per inch at 63.3 or 100 characters/sec.	1.5
3. Punched Card Reader	Reads standard punched cards at rates of 100 to 1000 cards/min.	1-3
4. Card Punch	Punches standard cards at a rate of 300 cards/min.	4
5. Line Printer	Prints 125-300 lines (80 or 132 columns/line) of character information per minute. Notice that information is printed a line at a time rather than one character at a time. Uses standard width computer paper and in some cases will allow the use of special size forms.	1.5-6
6. Serial Printer	Same characteristics as the line printer in 5 except printing is accomplished one character at a time. This is slower but also less expensive. Print speed: 30-180 char/sec	0.75-2
7. X-Y Plotter	This device plots graphical information in an X-Y coordinate system on flat sheets of paper or sometimes Mylar. Many plotters of this type generate graphic information using ball point or fountain-style pens though some are arranged in much the same way as a serial printer; i.e., plots are printed using a print head and ribbon. It should be noted that these devices usually possess little or no intelligence and in consequence generate plots under the direction of a computer.	1-6 Prices are a function of plot size possible and speed of plotting.

other machines. For example, most minicomputer tapes are compatible with IBM tape equipment, so that data acquired on-line and copied onto tape by a small computer can be processed using a larger, more powerful machine.

Most magnetic tape devices process data on seven or nine parallel tracks that run the length of the ($\frac{1}{2}$-inch-wide) tape as shown in Figure 9.6. On a nine-track tape each frame written across the tape contains eight data bits and one parity bit. Recording density is generally either 800, 1600, or 6250 bits (frames) per inch.

The parity bit is set or reset to give overall odd or even parity, depending on which is in use, for that frame. Data are written in blocks, usually several hundred 8-bit bytes long. Each block is preceded by a header, which marks the start of the block and gives some information about the remaining contents of the block. After the data block come one or two cyclic check (CC) bytes, which are used to check for and isolate errors not detectable through use of the transverse parity bits. Each block is separated from the following one by an inter-record gap (IRG). Blocks are written sequentially down the tape in the order the processor sends them to the tape unit (Figure 9.6).

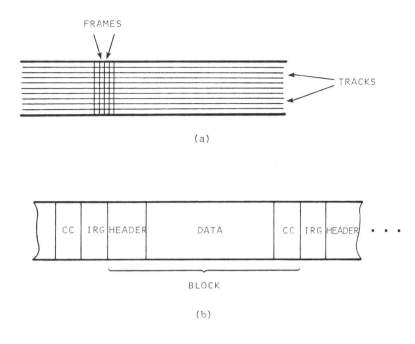

CC: CYCLIC CHECK BYTE
IRG: INTER-RECORD GAP

Figure 9.6. Magnetic Tape Data Format a. Detail b. Block Format

Several files, each containing one or more blocks, can be written on the same tape. A file usually begins with a special block that identifies the file and is separated from its predecessor and successor by a special end-of-file mark.

The primary limitation in using magnetic tapes, particularly in real-time environments, arises from the fact that tape is a sequential storage medium. Access of a desired data item requires moving the tape past the reading heads until the item is located. This process can take several minutes. For bulk storage of data and programs that must be accessed more quickly, we turn to magnetic disks or drums.

Cassette Tape Units. Cassette tapes fulfill two basic objectives: They serve as low-cost alternatives to magnetic tapes for auxiliary storage of small amounts of data in applications where rapid access is not necessary. And they have proved to be successful replacements for paper tape and punched cards in software development and data preparation.

Computer cassettes are almost identical with high fidelity audio cassettes. They consist of about 300 feet of 40 mm wide Mylar* tape on two spools inside a plastic package that measures about $2\frac{1}{2}$ by 4 inches. The cassette is loaded by simply placing it in the tape drive just as with the audio versions. A cassette holds about 250,000 bytes of data, including extra parity check bits. Data formats are essentially the same as with full-size magnetic tapes, with the exception that the data are written in serial fashion on a single track which is usually written in parallel with a separate timing track. On many units blocks must all be of the same length, with 256 bytes/block typical. Data transfer rates range from 10 to 2000 bytes per second. A "fast forward" feature that permits high-speed searches for the beginning of files is nice to have. Many recorders can be connected directly to data communications hardware for transmission of recorded data over telephone lines. These recorders serve as the storage medium for store-and-forward terminals.

---

*Mylar is a registered trademark of E.I. du Pont de Nemours & Co., Inc.

Cartridge Tape Units. Cartridge tapes offer price, performance, and capacity intermediate between full-size magnetic tapes and cassettes and are currently not as popular as either of the others. Some cartridge tape recorders offer recording density and speeds that rival those of full-size drives, but with a lower capacity.

The chief advantage of cartridge tapes is that error rates tend to be significantly lower than those of cassette tapes, primarily because in cartridge tape systems the read/write heads do not come into contact with the recording medium as they do in cassette systems.

At least one manufacturer offers a Winchester disk drive (see below) with a built-in cartridge tape as a back-up storage device. Cartridge drives for these systems are much simpler than those for general-purpose use because they do not require the complex logic needed to start and stop the tape drive on block boundaries. Instead, the data from the disk are recorded as a single unbroken stream of data from the beginning of the tape to the end.

Magnetic Disks. In comparison with magnetic tapes, rotating memory devices offer faster access to data with the disadvantages of more expensive and bulky storage media and higher cost for the mechanical and electronic mechanisms. Disks are *random access* or *direct access* devices, in which a recorded data item can be retrieved without searching through previously recorded data.

A magnetic disk storage module (disk *pack* or *cartridge*) consists of one or more thin disks. These are made of aluminum coated with a magnetic oxide (Figure 9.7). The disks are mounted on a spindle, and a motor rotates the entire pack at about 3000 rpm. Data are recorded magnetically on the disk surfaces and retrieved from them via read/write heads as the surface on which the desired data are written passes beneath the heads.

Data are recorded in concentric circles called *tracks.* A track is the strip of the disk surface that pases directly beneath a read/write head in a single revolution. An index mark indicates the "beginning" of a track. In a "hard-sectored" disk system each track is divided into several pie-shaped *sectors*, as shown in Figure 9.8. Typical systems might have 8 or 16 512-byte sectors per track. Data must then be written 512 bytes at a time, since the block length is fixed. Because each track contains the same total number of data bytes, data are at a higher density on the inner tracks than on the outer tracks.

To locate a given piece of information we must specify the surface, the track, and the sector on which the data are recorded. Thus a disk address would consist of three numbers— the surface number, the track number, and the sector number—which would be a unique combination. Of course, if the system has more than one disk drive, we must also indicate which drive contains the data.

On a "soft-sectored" disk, there are in effect no sector boundaries. Data blocks of any length can be written, and a disk address consists of surface, track, and block number. Soft-sectored disks are more expensive than hard-sectored disks, but eliminate the wasted space and the system overhead involved in dealing with fixed-length data blocks.

In order to read or write a data item on a given track there must be a read/write head over the track. There are two physical ways of arranging the read/write heads with respect to the disk surfaces. *Fixed-head* or head-per-track disks (Figure 9.9) have one immovable head for each track. Thus a fixed-head disk with four surfaces and 100 tracks on each surface requires 400 read/write heads. *Moving-head* disks have one or two heads per surface (Figure 9.10).

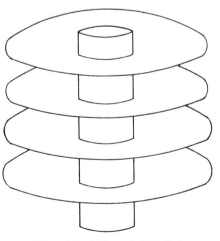

Figure 9.7. A Magnetic Disk Pack

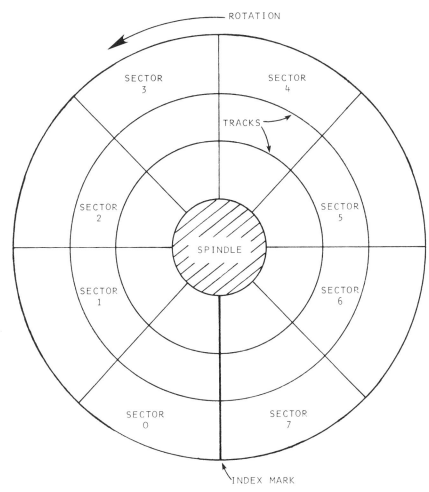

Figure 9.8. Layout of the Disk Surface in Sectors and Tracks

In the conventional configuration, the heads are mounted in a comblike arrangement on a movable arm which is moved in and out by a linear motor and feedback servo-mechanism to select the desired track.

In conventional moving-head drives, the read/write heads are tiny magnetic coils encased in flat disks of plastic or ceramic material about $\frac{3}{4}$ inch in diameter and $\frac{1}{8}$ inch thick. A spring-loaded arm attempts to force the head down onto the disk surface; however, the disk surface moves at speeds between 50 and 150 miles per hour, and, just as with any body moving through air, the disk surface has a thin layer of

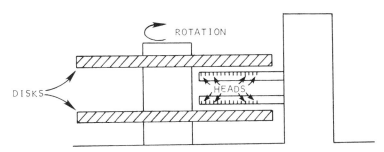

Figure 9.9. Arrangement of Read Write Heads in a Fixed-Head Disk

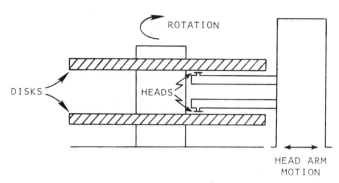

Figure 9.10. Arrangement of Read/Write Heads in a Moving-Head Disk

air above it that rotates with it. Thus the force exerted by the spring is counteracted by the tendency of the head to float on the air layer, and the head actually flies over the disk surface.

Figure 9.11 illustrates the structure of a *Winchester* disk drive. These drives use a relatively new technique in which the read/write heads are mounted on an arm that rotates about an axis to position the heads over the appropriate tracks. The motion of the arm is similar to the motion of the cocking lever in a Winchester rifle. These drives offer higher storage densities at a lower cost than conventional drives, but access times are longer and data transfer rates are slower than those of cartridge or disk pack drives.

The distance between the head and the disk surface is only about 100 microns (10^{-4} inch). Thus a smoke particle of 250 microns diameter is a large impediment to the disk head. Figure 9.12 gives a rough idea of the magnitudes involved. The scales and speeds involved work out to be about the same as those of a Boeing 747 flying 2 feet off the ground at 240,000

miles per hour! Thus if a disk head encounter even the smallest speck of dirt, a catastroph known as a "head crash" occurs, in which th head or particle scrapes on the disk surface destroying the surface and the head. Dis drives have sophisticated self-cleaning an ventilating systems to help prevent this. Neve theless, it is a good idea to keep compute rooms and storage areas for disk packs an cartridges as clean as possible.

To write to or read from the disk, an I/C driver program, usually part of the operatin system, sends the disk address—surface, track and sector (or block) number—to the disk inte face, i.e., the disk controller. What happen then depends upon whether the disk is moving-head disk or a fixed-head disk. If it is fixed-head disk, the controller turns on th read/write head for the selected surface an track. This head selection takes a negligibl amount of time. The unit must then wait unt the chosen sector (or block) on the disk rotate to a position beneath the read/write head, s that the data will be transferred at the prope position on the track. Since the position sough is, on the average, halfway around the disl when the input/output is requested, the averag rotational delay is half the time it takes for th disk to make a complete revolution. Therefor rotational delays are on the order of 10 to 2(milliseconds.

If the disk is of the moving-head variety, a additional delay is involved, since the head must be positioned over the track addressed On modern disks the time it takes to move th heads, the *seek time*, is between 15 and 15(milliseconds (depending upon the number o tracks the heads travel across and the disl

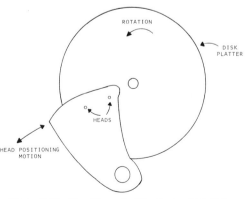

Figure 9.11. Top View of a Winchester Disk Drive

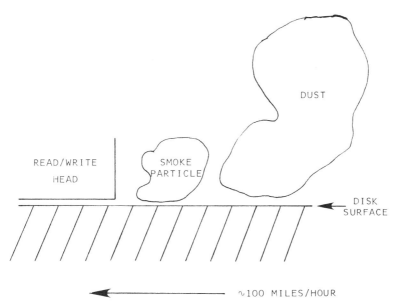

Figure 9.12. The Relative Size of Impediments to Disk Read/Write Head Operation

model). The average is typically 30 to 50 milliseconds.

Thus a moving-head disk takes significantly longer than a fixed-head disk on the average to access a given data item. Once the data are beneath the read/write head, the disk can transfer the data to or from the input/output bus of the computer. Data transfer rates vary between 300,000 and 2,000,000 bytes/second.

The advantage in access time the fixed-head disk offers is offset by the cost of the disk drive. Fixed-head disk units rarely offer more than one megabyte of storage, and drives designed for small computer systems are not inexpensive. Between 50 and 100 megabytes of moving-head disk storage can be purchased for an equivalent price. A further advantage offered by most moving-head disks is that at least part of the disk storage medium can be used to transport programs and data. Removable disk storage media also allow some added degree of flexibility in back-up storage of data and programs.

There are two kinds of removable-medium disk devices. *Cartridge* disks allow the removal of a single platter within a thin plastic cartridge. A popular configuration using this technique is one that contains two platters, one of which is removable, the other mounted permanently on the spindle. The removable cartridge can be used for back-up storage of the information on the nonremovable platter. *Disk pack* devices have removable packs containing several platters.

In Winchester drives, the platter and the entire drive mechanism is a hermetically sealed unit, so that removal of the platter is impossible. This presents one of the biggests handicaps in the use of Winchester drives; it is hard to provide an inexpensive yet adequate back-up storage system for them.

Floppy Disks. Just as cassette tapes answer the demand for low-cost sequential data storage, the flexible or floppy disk fulfills the need for low-cost direct-access storage.

The principle behind the floppy disk is the same as that used in full-sized moving-head disks. A read/write head moves in or out to select a track for data transfer; but here the storage medium is a magnetic oxide coated disk of heavy gauge Mylar whose flexibility gives rise to the name floppy disk. The disk, which is 8 inches in diameter, is encased in a flexible plastic case much like a record jacket. The entire package is inserted into the drive, and the disk rotates inside the plastic case. A smaller "mini-floppy" unit is available with a disk diameter of $5\frac{1}{4}$ inches.

Head positioning on early floppy disks and

some of the less expensive current versions is accomplished either via a worm drive mechanism or by a grooved spiral wheel. Seek times between 100 and 500 milliseconds are typical. Some of the newer drives use the voice-coil system of their larger counterparts, and average seek times on the order of 50 milliseconds are now feasible. In floppy disks the heads are in constant contact with the disk surfaces. This leads to high error rates and short lifetimes for the disks, in comparison with those of conventional "hard" disks.

A typical floppy disk has a storage capacity of 256 K bytes and a data transfer rate of about 80,000 bytes per second. Mini-floppies have capacities on the order of 100 K bytes, and transfer data at about 12,000 bytes per second. Ordinary floppies use only one side of the disk for data storage. Dual-sided floppies use both sides and have double the storage capacity of the single-sided versions. Capacities of over one megabyte with transfer rates on the order of 200,000 bytes per second are available on dual-sided, double-density drives.

Data Structures. Direct-access devices offer an extra dimension in the patterns in which data can be stored by contrast with sequential devices, such as magnetic tapes, where data must be stored and processed sequentially, i.e., in the same order in which they are written.

There are three fundamental logical data structures of primary interest in real-time systems. (1) *Sequential files* are used for storage of programs and data that are inherently sequential, such as historical process data recorded in chronological order. (2) *Indexed files* permit the easy access of data in some order other than the usual sequential order. (3) *Library* or *partitioned files* are used for on line storage of collections of program or data files. The individual files with each library or partition are sequential or indexed files in their own right. We now examine each of these structures in greater detail.

With tape, cards, or other sequential media, a sequential file is stored physically in its logical order, the order in which it must be processed. On a direct-access device it is possible to store data sequentially, but this is not usually done. Instead, a sequential file is usually stored as a linked list, in which each block of the file contains some data items and also a pointer to the next logical block in the file (Figure 9.13). The pointers are usually disk addresses; thus in processing a sequential file stored as a linked list, the system first obtains the address of the first block from a table of contents (or directory) on the disk. After accessing the data from the first block, the system uses the pointer in the first block to locate the next block. This procedure is repeated until a special end-of-file mark is encountered in the last block.

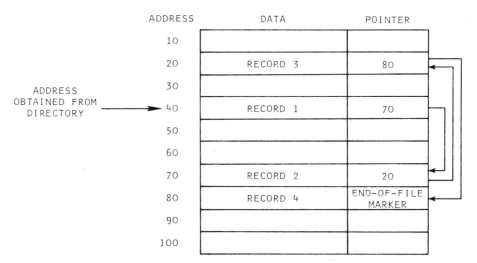

Figure 9.13. Linked List Disk Format for a Sequential File

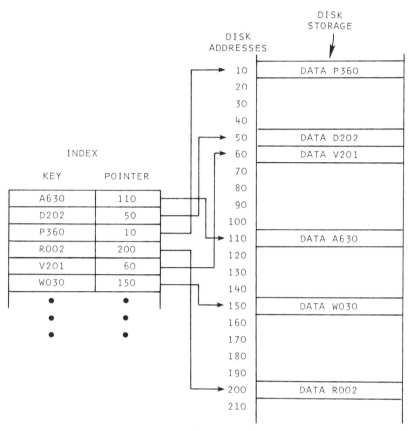

Figure 9.14. An Indexed File Structure

In indexed file systems (Figure 9.14) we generally assume that *each* data item or record has associated with it a unique key or label that can be used to identify the record. An index is a table that the system can use to determine the disk address of a record, given its key. These index tables often have a much more complex structure than a simple directory, but space does not permit a detailed treatment here. To locate a given record, the user program presents the key to the system program that handles indexed files. This program looks up the key in the index, determines the disk address of the record being sought, reads the desired record from the disk, and passes the data back to the user program. Note that it is *system* software that does all this; hence the vendor should supply software and a disk system to handle indexed files without a great deal of programming effort by the user.

A library structure (Figure 9.15) consists of, in part, a table of contents, called a catalogue or directory, in which each entry contains the name of a file and the address of the first data block of the file. When a user program requests access to a given file within the library, the system first searches the directory for the desired file name. Once found, the system obtains the disk address of the first block and proceeds to process the file as usual.

All three disk file structures discussed above are commonly found in real-time computing systems. Sequential files are used for on-line storage of symbolic and machine language programs and data to be processed in sequential order. Groups of related sequential and indexed files are often combined into libraries. For example, all the programs having to do with a single application might be in one library, while all the systems programs might be kept in another, and all the language translators in a third. An example of an indexed file might be found in the process control computer of a pharmaceutical house, in which each record

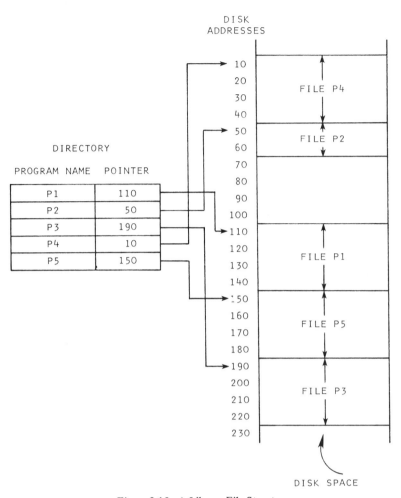

Figure 9.15. A Library File Structure

contains the control sequence necessary to produce a batch of a particular one of several thousand products. Another real-time control system for a refinery might keep a back-up copy of current sensor inputs, set-points and other control parameters in indexed files on disk so that the system can recover automatically from short-lived hardware malfunctions or power failures.

System Design Considerations. A large number of auxiliary storage devices presently are on the market, but the choices are narrowed considerably if one adheres to the same vendor for storage devices as for the other components of the computer system. Unless there are overwhelming reasons not to do so, the system designer should try to purchase all the major

components from the same vendor. Otherwise when something malfunctions, the user often will be presented with a dispute among vendor maintenance people, each company blaming the other's hardware and/or software for the malfunction and no one wishing to fix it. A typical vendor offers a spectrum of storage devices including the ones shown in Table 9.3.

While the detailed mix of these devices in a real-time system is strongly applications dependent, there are some general guidelines that the designer should follow:

First, each class of device is better suited for some functions than for others. In most medium-to-large-scale real-time systems, cassette tapes and/or floppy disks are used only for transporting small amounts of data or programs or as a "bootstrap" device to restart the system

Table 9.3. Representative Bulk Storage Peripherals

Storage Peripheral	Storage Density	Access Time (msec, average)	Transfer Rate (Bytes/sec)	Approximate Relative Cost
1. Cassette Tape	800 bits/inch 250 K bytes/tape	Variable	600	1
2. Magnetic Tape (reel)	800 bits/inch 40 M bytes/2400' reel	Variable	60,000	4
3. Bubble Memory	20,000 bytes	50	5,000	0.2
4. Mini-floppy Disk	100 K bytes	300	12,000	0.5
5. Floppy Disk	256 K bytes/disk	100	31,000	1
6. Moving-head Cartridge Disk	10 M bytes/drive	35	300,000	5
7. Moving-head Cartridge Disk	20 M bytes/drive	35	600,000	6
8. Fixed-head Disk	256 K bytes/drive	8.5	250,000	3
9. Winchester Drive	12.5 M bytes/drive	70	900,000	3
10. Disk Pack Drive	50 M bytes/drive	43	800,000	11
11. Disk Pack Drive	96 M bytes/drive	43	800,000	13
12. Disk Pack Drive	190 M bytes/drive	43	800,000	16
13. Disk Pack Drive	277 M bytes/drive	43	1,200,000	19

If full-size magnetic tape can be justified for the system, then it too can be used for the same purpose, making a floppy disk or cassette unnecessary. Because of their relatively long data retrieval time, full-size tape units are rarely used in normal activities in real-time systems. However, if back-up storage for disk files is important, and/or if data must be transferred to or from another computer, as is often the case in data acquisition systems, magnetic tape is often the most economical medium to use.

There is now available a separate magnetic core memory unit that replaces some fixed-head disks at prices comparable to fixed-head disk prices. The fixed-head disk or its core memory replacement often serves as a "swapping disk" in many real-time systems. Most computers in a multitasking real-time environment (see Chapter 13 for a discussion of multitasking systems) do not have main memories large enough to hold all the programs or tasks being used at any given time. The operating system keeps a copy of each program or task on the swapping disk when the task is not currently active. When an interrupt or other event occurs that requires the reactivation of a task that is not in main memory, the system copies the executing tasks onto the swapping disk from memory to make room for the new task, which is then loaded into main memory. In some applications—those in which rapid task swapping is required—the speed at which this swapping activity occurs is the limiting factor in system throughput. In such systems, a separate, fast swapping disk can sometimes be justified.

The number and sizes of moving-head disks can usually be determined from estimates of on-line storage requirements. When in doubt one, should purchase extra space. Disturbing a production system to add more disk storage can be expensive in terms of necessary software modifications and lost production while the computer system is down. As a general rule, specify at least twice as much disk space as is actually needed.

The Future. Some experts feel, at the time of this writing, that the dominance of rotating memory devices—disks and drums—for rapid-access storage is near an end. We have noted that one manufacturer now offers a terminal with built-in "bubble" memory for bulk storage of data. Other bulk storage technologies such as charge-coupled devices (CCDs) are just beginning to emerge. Predictions are that within the next five to ten years these and similar devices will offer alternatives to disks and drums that will be two or three orders of magnitude faster and cost the same per byte of storage.

9.2 DATA COMMUNICATIONS

In the broadest sense, communication is the transmission of information from place to place. This information may be in any one of a number of forms, such as newsprint, spoken voice, analog signals, and digital data. As the digital computer increasingly becomes a central feature in many endeavors, we find that digital communications techniques are rapidly replacing traditional methods in a wide variety of applications. In particular, real-time computing systems almost always involve some on-line communications between the computer and remote devices or among several computers.

For our purposes, data communications will be defined as the transfer of *digital* data from one point to another. Current technology dictates that digital data—whether it in fact represents coded characters or numerical information, computer instructions, or spoken voice equivalents—be transmitted as binary numbers.

Conceptually the problem is, given a set of N binary digits, how do we transfer this information from Point A to Point B? The simplest answer is to connect Point A and B together using N wires and to provide a mechanism for A to put the appropriate voltage levels onto the lines, as in Figure 9.16. B can then sense the voltage and process the information it receives. Clearly, there are severe limitations to the practicality of this method. A typical message might consist of hundreds of bits. Connecting a computer in Los Angeles to one in New York would be prohibitively expensive.

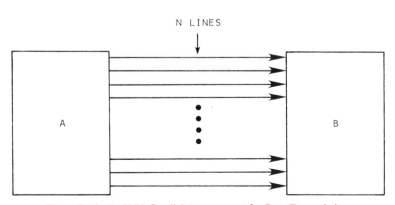

Figure 9.16. An N-Bit Parallel Arrangement for Data Transmission

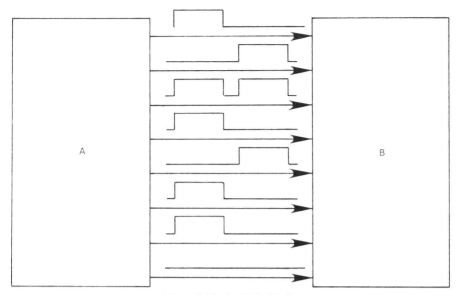

Figure 9.17. An Eight-Bit Bus

Furthermore, the internal architecture of computers and other digital devices precludes effective use of such broad, parallel communications techniques. Because of the same economic considerations, computers move information internally from point to point and to and from peripheral devices over narrow data paths called buses. A typical bus consists of a small number, often 8, of parallel lines. With a given data byte, all 8 bits are placed onto the bus simultaneously, one bit per line (as in Figure 9.17). To transmit more than 8 bits of data, we would split the data file into bytes and then send the bytes down the bus one after another.

This technique adds another level of complexity to the problem. How does B know when A is transmitting data, and how does B know when each individual byte arrives? How does B know whether the data byte it receives is correct? We shall consider the answers to these questions in detail later.

There is a virtually limitless number of ways in which buses can be designed to connect computer components and their peripheral devices together. In the past, each manufacturer has tended to use exclusive techniques, which means that it is often difficult or impossible to connect peripherals made by one manufacturer to computers made by another. Recently,

however, several vendors, particularly those in the real-time data acquisition and process control market, have agreed to adhere to a set of connection standards called the IEEE 488 system. This set of rules specifies how to design processor buses, peripheral devices, and electronic signal patterns so that peripherals can be more easily interchanged among processors.

A similar set of standards called the S100 bus system has been developed for 8-bit microcomputers. The S100 bus specifications indicate how processor components—CPU, memory, and peripheral device interfaces—should be interconnected in order to make interchanging components as easy as possible.

Use of 8-bit or wider buses is economically feasible for high-speed transfer of data between a processor and its attached on-site peripherals. But 8-bit parallel communications is too expensive for most general data communications applications because of the high costs of communications lines. Therefore, the common practice is to transmit each byte as a serial string of binary digits. Timing and error checking information are also transmitted in this fashion. Using "bit-serial" communications, we can transmit messages of arbitrary length over a single communications line. The trade-off here is that bit-serial communications are much slower than parallel communications.

The largest existing network suitable for long-distance communications at reasonable costs is the telephone system. While many local communications problems, such as in-store inventory and accounting systems or in-plant process control systems, do not require the use of the commercial telephone system, the basic techniques used in such local systems are the same as those that are used with commercial telephone lines.

9.2.1 Data Transmission

With the exception of some ultra-high-speed special-purpose systems and within computers themselves, data communications are most often done in bit-serial fashion using single communications lines. Most terminals and other communications equipment offer two electronic means for digital signal transmission. The simplest is the current loop, in which the communications circuit forms a closed loop. The direction of current flow within the loop determines the logic value, 0 or 1, being transmitted. Current loops can be used to transmit data over standard telephone lines for distances up to several miles. The actual maximum distance for a given connection depends upon the condition of the line, how many switching points the line passes through, and the electrical noise environment—motors, power lines, and so on—near the line. Keep in mind the fact that the path a telephone signal takes between neighboring buildings can be several miles long; in any case the telephone company's communications office should be consulted before attempting to use a current loop for transmission over phone lines.

The other major transmission method is the EIA RS 232C standard established by the Electronics Industries Association. The standard defines the characteristics of the interface between the data terminal and the data communications equipment to which the terminal is attached. It specifies the electrical signal characteristics (–12 V and 12 V for the logic levels) and the details of the 25-pin cable connection between the terminal and the communications equipment.

Do not confuse the RS 232 "interface" with the communications device controller/interface mentioned in the introduction to this chapter. The former is simply a cable connection, while the latter is generally a moderately complex piece of electronic circuitry. RS 232 standard equipment is used in communications in which the distances involved and/or line conditions prevent use of the current loop approach.

The major problem in transmitting digital signals over long distances is the attenuative properties of "voice grade" lines. Telephone lines are designed to handle frequencies in the voice range, between roughly 300 and 3000 cycles per second. When transmitted over a phone line, the high frequency components of a digital square wave are rapidly filtered out, so that a digital signal is difficult or impossible to detect after it has traveled a long distance over a voice grade line (Figure 9.18). The RS 232 standard was implemented to standardize connections to devices that remedy this problem.

A hardware device called a *modem*, short for *mo*dulator–*dem*odulator, extends the transmission range by transforming digital signals to a form more suitable for the phone lines. For example, most low-speed modems translate input data so that a logic 1 input produces a low frequency output from the modem, while a logic 0 input leads to a high frequency output, as illustrated in Figure 9.19. Typical frequen-

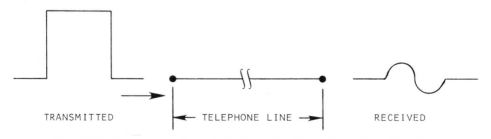

TRANSMITTED   ⟵ TELEPHONE LINE ⟶   RECEIVED

Figure 9.18. Distortion of a Square Pulse Transmitted Over a Long, Voice-Grade Line

Figure 9.19. Frequency Modulation of a Square-Wave Pulse by a Modem

cies are 1200 cycles per second for a 1 bit, a *mark*, and 2200 cycles per second for logic 0 bit, a *space*.

A complete telecommunications link requires two modems, one at each end of the line. As shown in Figure 9.20, in transmitting data from Point A to Point B, Device A outputs the serial bit string to the modem. The modem modulates the signal and places it onto the transmission line. Once the signal is modulated, it can be transmitted for indefinite distances over long-distance telephone lines and overseas via satellite links. At the receiving end the B modem demodulates the signal, converting it back to its original bit-serial string form. Device B can then input the transmitted data and respond appropriately.

If the system is equipped for two-way communications, the reverse process, transmission from B to A, occurs in the same way. B outputs the digital data to the modem; the B modem modulates the signal; the A modem demodulates the signal; Device A inputs the data from the modem.

There are three transmission modes available—simplex, half-duplex (HDX), and full-duplex (FDX). A simplex system can carry data in only one direction. A typical example is a remote character printer or receive-only (RO) terminal. A half-duplex link can carry data in either direction, but in only one direction at a time. A full-duplex data communications line carries data in both directions at the same time.

Which of the three modes is used determines to some extent the kinds of modems and telephones that will be required. Many modern modems include a switch that determines whether HDX or FDX mode is in use. Others require a minor hardware alteration (a "wire strap") to select the mode. Still others can operate in only one of the modes.

At least two wires are necessary for any electronic communications link. The telephone company standard is a two-wire, twisted pair line which supports simplex, half-duplex, and low-speed full-duplex operation. The phone company calls a two-wire line half-duplex even though it might be in use in full-duplex mode. A full-duplex line consisting of four wires, two for each direction, is necessary for medium- or high-speed full-duplex operations. Low-speed full-duplex mode is accomplished over two wires by using two sets of frequencies simultaneously, one set for each direction. For example, the Bell System 103-type modems use the following frequencies:

Originating end:
    Mark (logic 1)   2025 cycles per second
    Space (logic 0)   2225 cycles per second
Answering end:
    Mark   1070 cycles per second
    Space   1270 cycles per second

Here the originating end is understood to be the end of the line that initiates the data transmission sequence. Virtually all modern terminals are equipped with a half/full-duplex switch to offer the alternative of full-duplex operation as described above. When in the full-duplex mode, typing a character simply sends the character to the computer; the computer must "echo" the character, i.e., send it back to the terminal, if the character is to appear on the screen or paper output of the terminal. The software I/O driver program in the computer operating sys-

Figure 9.20. Modems in a Communications Link

tem is responsible for echoing the input from a terminal in most systems.

A common measurement unit used to describe digital data transmission speeds is the *baud*. One baud is one increment of data (not necessarily a data bit) per second in a binary signal train. For binary signals, the baud rate is equivalent to the transfer rate in bits per second, but, as we shall see, not all bits transmitted constitute useful data or information. Also, some rarely used transmission techniques exist in which the baud rate does not measure bits per second, but some fraction or multiple of the bit rate. Thus one must be careful in using baud rate to designate data transmission speed. Some authors suggest that for this reason one should never use the baud unit for a measure of speed, but, since the transmission rates of modems are always given in bauds, it is impossible to follow this suggestion.

9.2.2 Asynchronous Communications

Having seen how individual bits can be transmitted over long distances, let us now look into the question of synchronizing the devices at the two ends of the line so that larger units of data can be transferred.

Implicit in data communications practice is the necessity to establish a timing relationship between the transmitter and the receiver so that the receiver knows when the data flow starts and how it should split the data into bits and bytes. There are two ways to do this—asynchronous and synchronous techniques.

In asynchronous communications, each end has its own timer. The transmitter outputs serial bit strings at a rate determined by its clock. The receiving end inputs bits at a rate determined by its own clock which must operate at virtually the same frequency as the

clock at the transmitting end. Data are sent byte by byte, and transmission of individual bytes is random in time. That is, given that one byte has been received, there is no way for the receiver to predict when the next byte will arrive.

Since the two ends have separate clocks, extra control bits are sent with each byte to establish synchronization during the transmission of the actual byte of information. A *start bit* precedes each data byte. The receiver starts its clock upon receipt of the leading edge of the start bit. The receiver then waits for one and one-half "bit-times", after which it shifts the data bits in from the line, continuing to shift in bits, one each bit-time until all the bits in the byte have arrived.

The transmitter appends a *stop bit* (two in a 110-baud link) to the end of each data byte. The stop bits serve to help the receiver detect errors due to noise and to prepare for the next byte. Figure 9.21 illustrates how an 8-bit data byte is embedded within a start bit "1" and two stop bits "00".

The asynchronous technique is necessary when data are to be transmitted at random times. For example, there is no way for a device to anticipate when the user of a nonbuffered terminal will send the next character. Since data rates are usually low in such applications, asynchronous techniques are used for low-volume communications at speeds that rarely exceed 2400 baud.

At 110 baud eleven bits are required for each character—a start bit, eight data bits and two stop bits. So 110 bits per second equals 10 characters per second. Faster asynchronous devices use only one stop bit, so that the data rate in characters per second is one-tenth the data rate in bits per second. A 1200-baud link then carries 120 characters per second.

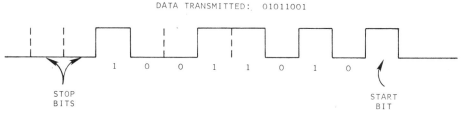

Figure 9.21. Asynchronous Data Transmission

9.2.3 Synchronous Communications

The synchronous (also called BISYNC or 2780 protocol) technique eliminates the need for extra bits to be transmitted with each byte, thus offering more efficient use of communications lines at the expense of somewhat more expensive hardware and software for communications. In contrast to the asynchronous technique (where clocks at both ends play a role in single-directional transmission), in synchronous transmission the transmitter's clock determines the timing at both transmission and receiving ends.

In the absence of data to transmit, the transmitter sends a square wave timing signal to the other end. Data, when transmitted, are superimposed on the timing signal in such a way that the receiver continues to receive the timing signal along with the data. Data are transmitted in *blocks* consisting of continuous strings of bits, with no extra bits or gaps between bytes. In a single transmission sequence, one or more blocks is processed. Each block contains a fixed number of bytes, usually on the order of 500 or so.

The transmitter is responsible for establishing the timing signal and for encoding the information transmitted. The receiver must be able to distinguish real information from noise and from the idle-line timing signal, must know when to start reading in bits as data, and must be able to respond to the transmitter to indicate whether or not the data were successfully received. The additional information transmitted between the two ends to perform these functions is called the communications *protocol*. The protocol consists of extra control bytes sent with messages or data and of "handshaking" messages between the transmitter and receiver.

Figure 9.22 shows a much oversimplified summary of the events that occur in the transmission of a block of data from A (the sender) to B. All messages, whether part of the handshaking or data, are preceded by at least two "SYNC" characters. SYNC is a special control character that informs the receiver that the bit string following the SYNC characters is to be treated as a message or as data. Anything the receiver detects that is not preceded by SYNCs is ignored.

In our simplified system, A starts the handshaking process by sending two SYNC characters followed by an "EOT" (End Of Transmission) character. B understands this message to be an inquiry as to B's ability to receive data. B then assumes the role of transmitter and responds in one of three ways:

(1) *SYNC-SYNC-ACK-EOT*. The ACK (ACKnowledge) response here indicates that B is ready to receive the data

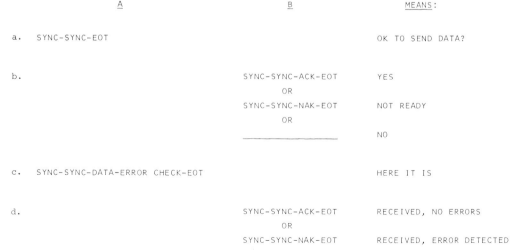

Figure 9.22. A Synchronous Communications Protocol

(2) *SYNC-SYNC-NAK-EOT*. The Not Ac-Knowledge response here indicates that B is functioning but is not prepared to receive data at this time

(3) *Nothing*. B is not operating

In Cases 2 and 3, A tries again several times before concluding that the transmission cannot be completed. After an ACK response from B, A sends SYNC-SYNC, the data block, error-check information, and EOT, in that order. B uses the error-check information to check the data and sends a message to A to indicate the result. An ACK (Case 1 above) means that no errors were detected that could not be corrected. A NAK message means that B detected an error that could not be corrected. A then retries the transmission, repeating the entire process from the beginning.

At each end of the line, checks for proper formatting of the messages and of the transmitted data are made. Additional error checking takes the form of an automatic "time-out" condition if one end or the other does not respond within a certain time after a response is due.

The error checking information sent with the data takes several forms. It is possible to use one bit from each byte as a parity bit for the byte. These *vertical* or *transverse* parity bits enable the detection of odd numbers of errors in bit values within a single byte. Vertical checking alone does not permit any error correction; the data must be retransmitted if an error is detected.

If more error checking information is sent with the data, it becomes possible for the system to correct some minor errors in the data after they have been received, without retransmitting any of the data. In one such technique, the transmitter appends an extra *longitudinal* parity check byte to the end of the data block. If we imagine the data to be arranged in a matrix in which each byte forms a column, the vertical parity bits check for errors in columns. Each bit of the longitudinal parity byte is a parity bit for the row in which the bit lies (Figure 9.23). For example, Bit 3 of the longitudinal parity byte is set to indicate the parity of all the Bit 3's in the data. The receiver computes the value the longitudinal parity byte should have and compares the computed value with the value received. If the computed and received bytes differ in a given bit position, the receiver knows that an odd number of bits in that row were altered in the transmission process. If only one bit is in error, the combination of vertical and longitudinal parity permits the receiver to locate the error and correct it. This combination also permits detection of most multiple bit errors.

An alternative to longitudinal checking is the cyclic check, which is a one- or two-byte number appended to the end of the data. The

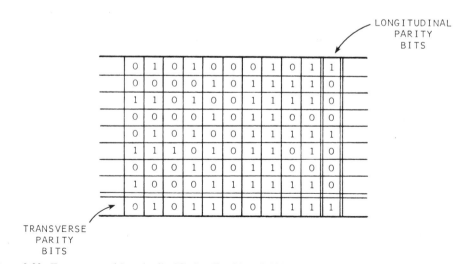

Figure 9.23. Transverse and Longitudinal Parity Checking (Odd Parity is Used for Both in this Example)

number is the sum of the data bytes with some constant, usually 2^n (where n is the number of bits used to transmit the cyclic check), as a base or modulo. The cyclic check offers basically the same capabilities as the longitudinal check.

Some more sophisticated techniques, such as the Hamming Code and the BCH code, offer more powerful error checking capabilities at the expense of large numbers of check bits, more complicated algorithms to compute the error check data, and, consequently, more expensive hardware and software for error checking.

A two-wire link supports synchronous transmission in only one direction at at time; two-wire synchronous links are of necessity half-duplex. In such a system, when one end finishes transmitting and requires a response from the other end, the former receiver becomes the transmitter and vice versa. It takes a relatively long time, called the "turnaround" time, to effect this transformation. Turnaround times are usually on the order of 150 milliseconds. Four-wire synchronous communications links operate in full-duplex mode, thereby avoiding the delays for turnarounds.

Synchronous hardware and software cost considerably more than the equivalent support for asynchronous communications. Medium- and high-speed communications—at 4800 baud and faster—use synchronous techniques. Typical applications are to link two processors together or to connect a processor to a remote job entry terminal consisting of a line printer and card reader.

At the time of this writing, a new IBM communications technique, called SDLC for Synchronous Data Link Control, is just emerging. IBM hopes SDLC will become *the* communications system in the future; however, with several minor exceptions, its applications to the real-time computing field have had little impact.

As is the case with hardware interconnections, a set of standards called the X.25 communications protocol has been developed to allow easier communication between devices of mixed brands and types. The X.25 standard specifies, for various communication techniques, the message formats and protocols that are to be used.

9.2.4 Communications Hardware

Modems. As we have seen, the function of a modem or data set is to convert the digital data from a sending device to a form suitable for transmission and to undo the conversion at the receiving end of the line. Since modems are vital components of most data communications systems, we need to discuss them in greater detail.

The Bell System was, until recently, virtually the only supplier of modems. Therefore their numbering system has become a de facto industry standard. We refer to classes of modems from other manufacturers by descriptions such as "Bell System 103A-type" or "Bell 202T-compatible."

The three major categories of modems are:

100 series—low speed: 75–300 bits per second

200 series—medium speed: 300–9600 bits per second

300 series—high speed: 9600 bits per second and faster

The 100 series is used exclusively for slow terminals using asynchronous techniques at 10 to 30 characters per second.

There are two main categories of 200 series modems. The 201-type modems all use synchronous techniques in the 4800–9600 baud range. The asynchronous 202 types run between 300 and 2400 baud and are used primarily to connect the faster character printers and CRT terminals to remote computer systems. The 300 series modems are used almost exclusively in high-speed processor-to-processor communications.

Synchronous communications generally use *dedicated* lines that are leased from the telephone company to form a permanent connection between the two ends. In asynchronous links one can lease the line; or one can use the ordinary switched phone lines in which a link is established only when one end of the line calls the other. If the latter alternative is selected, then we must distinguish between the originating end of the line, which does the calling, and the answering end, which answers the call. If the originating end always does the calling, then an *originate-only* modem is appropriate. The answering end requires an *auto-answer* or "dial-up" modem if connections are to be made at the answering end without human intervention. Auto-answer modems usually include a built-in telephone receiver in addition to the data conversion circuitry. *Auto-dial* modems permit computers to dial the phone numbers of other computers and transmit and receive information to and from each other without human intervention.

Most modems are wired directly to the telephone line. An alternative for low- and medium-speed communications is to use an *acoustic coupler,* in which a telephone receiver is placed in a pair of rubber "ears" in the coupler instead of using a more permanent, wired connection. The advantages in using an acoustic coupler are increased portability and flexibility: the terminal can be used anywhere that there is a telephone; and the terminal can be connected via a regular phone call (perhaps long-distance) to any computer that has a dial-up modem operating at the proper speed. Many terminals include an optional built-in acoustic coupler; however, it usually is cheaper to buy the terminal without the coupler and to purchase the coupler separately. Having a coupler separate from the terminal is less convenient, since there is one more component to carry around, but, on the other hand, the coupler can be used with more than one terminal.

Communications Interfaces. Terminals are largely self-contained: the RS 232C interface is connected directly to a modem to establish a communications link. In connecting a modem to a computer at the other end, it is necessary to include a hardware interface (as in Figure 9.24) between the line or modem and the computer Input/Output Bus. These interfaces are often called multiplexers, not to be confused with the communications multiplexers described below or the analog input multiplexers described in Chapter 10.

A typical interface performs the following functions:

(1) *Parallel-Serial Conversion.* The interface receives data from the Input/Output Bus one byte at a time on eight parallel lines. Each data byte is placed into a shift register and shifted out to the modem one bit at a time at a rate appropriate for data transmission. Conversely, received serial data bits coming in from the modem are shifted into a register until a full byte is accumulated. When a byte has been transferred one way or the other, the interface informs the computer by setting a READY bit (or turning off a BUSY bit), perhaps requesting an interrupt in the process.

(2) *Level Conversion.* The interface contains circuitry to convert internal logic levels (0 and 5 V at 1 ma, for example) to levels appropriate for the line or modem (-12 and 12 V at several ma, for example), and vice versa on input.

(3) *Timing and Formatting.* In asynchronous communications, the interface is responsible for shifting the data in and out at the proper bit rate, e.g., 300 bits per second on a 300-baud line. Some synchronous modems expect the interface to do the same. In others, the modem sends a signal to the interface that the interface uses to determine when to shift data bits to the modem.

Asynchronous interfaces attach the Start and Stop bits for output and take the extra bits off on input. Synchronous protocol and handshaking usually demand the intelligence of a computer for message and data formatting and error checking.

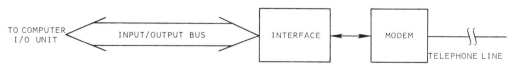

Figure 9.24. A Computer Communications Interface

(4) *Error Detection and Status Monitoring.* Most interfaces recognize parity errors and inform the computer by setting an error status flag when an error occurs. The interface also determines whether or not the modem is "on" and receiving a signal from the other end, and subsequently sets status bits to indicate the condition of the transmission system to the computer.

Most modern asynchronous communications devices use a single LSI integrated circuit, called a UART—*U*niversal *A*synchronous *R*eceive *T*ransmit—that provides all the above functions with the exception of Item b. The UART allows one to select from a wide range of data transmission rates (110-9600 baud), character lengths (6, 7, or 8 bits), parity checking (none, even, or odd), and number of stop bits (1 or 2). An analogous USRT—*U*niversal *S*ynchronous *R*eceive *T*ransmit—integrated circuit is also available for synchronous communications.

Communications interfaces are usually purchased from the computer vendor, since the interfaces must be connected to the computer Input/Output Bus.

Multiplexers. A digital or communications multiplexer, as shown in Figure 9.25, is a hardware device that permits several communications terminals to share the same communications line, often at a large savings in line rental costs in comparison with the expense of leasing one telephone line for each device. Two multiplexers are necessary, one at each end of the line. In a typical multiplexer system several remote terminals share a common telephone link. The data transmitted from the terminals are multiplexed onto the phone line for transmission to the host computer. The multiplexer at the host computer end demultiplexes the incoming data so that the operation of all these devices is transparent—it appears to the host computer as if each remote terminal has its own separate line.

Two multiplexing techniques are in common use. Frequency-division multiplexers (FDMs) split the communications frequency band into several smaller bands. Each device then has its own band. For example, one device might operate in the 1000-1500 cycles per second frequency range, while another operates in the 2000-2500 cycles per second range.

Time-division multiplexers (TDMs) use a single frequency band for all devices. Used almost exclusively with asynchronous communications, a TDM takes a data character from each device in turn. Each time a cycle is completed, the TDM uses a simple synchronous technique to transmit the data to the other end, where the multiplexer there splits up the data into individual characters and distributes them to the various lines into the computer. Of course, the same process takes place for data transmitted from the computer to the remote devices.

Statistical multiplexers use a modified TDM technique in which the cyclic polling and transfer of information from each terminal is replaced by a first come–first served protocol.

The decision whether or not to use a multi-

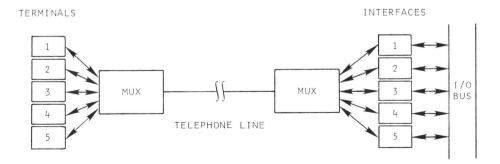

MUX = MULTIPLEXER

Figure 9.25. A Digital Multiplexer

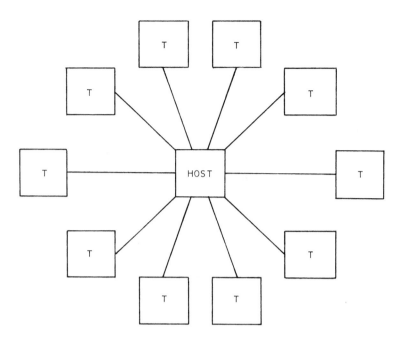

T :   TERMINAL

Figure 9.26. Structure of a Simple Time-Sharing Network

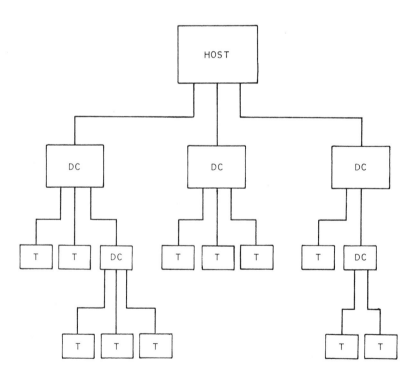

DC = DATA CONCENTRATOR

T = TERMINAL

Figure 9.27. Structure of a Hierarchial Communications Network

plexer is usually a matter of comparing telephone line rental costs with multiplexer prices.

9.2.5 Communications Networks

Network Structures. Communications networks come in all shapes and sizes, ranging from two terminals connected to a minicomputer at 30 characters per second, up to the ARPA (Advanced Research Projects Agency) network which links about 30 large-scale computers (IBM 370's, CDC 6600's, etc.) via high-speed transmission lines. The simplest and currently dominant network structure is shown in Figure 9.26. Many time-sharing systems have a host computer as the central node, which is connected to the outer or terminal nodes.

A more complex network has a tree structure as shown in Figure 9.27, where one or more of the nodes connected to the host computer is also connected to more remote peripherals. The intermediate nodes are often called *data concentrators.* A data concentrator's primary function is to control the flow of information between more remote devices and devices closer to the host. The simplest data concentrator is a digital multiplexer, which simply combines information from the remote sites into a form suitable for transmission over a single communications link. A growing trend is to use small computers as data concentrators. These can then buffer messages from remote sites, preprocess the data, and transmit messages in larger units to the host machine, using higher-speed communications.

The primary advantage of the use of data concentrators is, of course, reduced line costs. In a typical airlines reservations system, many terminals that are geographically close to each other communicate asynchronously with a local minicomputer which serves as a data concentrator. The minicomputer buffers the characters it receives and transmits data to the host, which may be across the country, only when it has filled a buffer from the terminals. The data concentrator–host link is usually synchronous at medium speeds.

The hierarchical network structure of Figure 9.27 is often found in process control systems. The top-level computer is a large-scale machine that runs optimization programs to compute production schedules for the plants and units at the lower level. At the next level down, the computers, usually medium-size minicomputers, serve as data concentrators for the top-level machines and also have some control and scheduling responsibilities. Until recently, hierarchical control systems have included at most three levels, in which the bottom level includes the operator's console devices, displays, and sensory and control units for process inputs and outputs. With the advent of the microcomputer, we are beginning to see a new level emerge between the bottom level and the minicomputer level.

The hierarchy structure presumes that some processors are inherently more important than others. Some of the more powerful networks, such as the ARPA network and Michigan's MERIT network, consist of several large host computers connected by serial links. Their structures are an extension of the "multidrop" structure (Figure 9.28) in which messages pass from node to node (drop to drop) in a daisy-chain fashion until they reach their destination. Each node can and usually does serve as a local data concentrator.

The primary advantages of such networks result from economies of scale. One feature is the capability to share data bases and peripherals, so that data required by all nodes need be stored at only one of them. This *resource-sharing* capability implies that each host can specialize in a certain area; the results are available to all other network nodes. In a way, the whole is greater than the sum of the parts. In addition, *load-sharing* is possible, so that if one node finds itself overloaded, it simply sends its extra work to another site. That processor performs the necessary work and sends the results back to the source computer.

As a point of fact, the ARPA and MERIT networks and others, as well, have cyclic structures, as illustrated in Figure 9.29. The advantages of cyclic structures include increased communications flexibility (more than one possible path between two nodes) and higher reliability (loss of a link or node leaves the network intact, although operating in a suboptimal mode). We emphasize that in these networks, the hosts have a great deal of autonomy; each is a

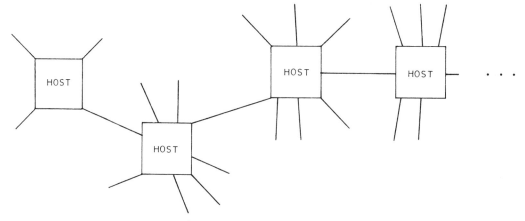

Figure 9.28. A Multidrop Network

powerful computer in its own right. There is no "boss" or primary network controller. Any host can initiate communications with any other host, provided the line locally is not busy.

Such *distributed* processing networks also have a place in process control. Figure 9.30 shows an example of the kind of network structure many feel represents the future direction for process control communications systems. All communications are handled asynchronously over inexpensive phone lines. Each node can initiate communications in any of several formats with any other node. Communications priorities are based on real-time conditions. For example, in an alarm situation all extraneous communications would be routed in such a manner that those nodes necessary to process the alarm would have the freest paths. Notice that, in this structure, loss of a few lines still leaves the network intact, although communications might proceed in a less than optimal manner.

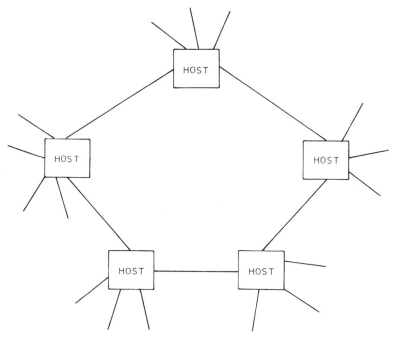

Figure 9.29. A Cyclic Network

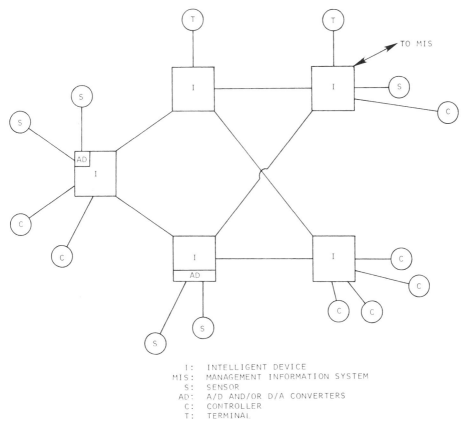

```
I:    INTELLIGENT DEVICE
MIS:  MANAGEMENT INFORMATION SYSTEM
S:    SENSOR
AD:   A/D AND/OR D/A CONVERTERS
C:    CONTROLLER
T:    TERMINAL
```

Figure 9.30. A Distributed Process Communications System

9.2.6 The Man–Machine Interface in Process Control

The trend in computer-based automatic control systems today is to take advantage of the new input/output and communications technology in communication between the operator and the process. Trend recorders, strip chart recorders, switches, gauges, knobs, and other equipment usually associated with control rooms are difficult to find in many new control rooms. Instead, one finds two easy chairs in front of two graphics terminals. The operator activates a screen with a light pen to display data, trends, and set-points and initiate high-level supervisory control and scheduling activities.

What is happening here is that on-line computers equipped with bulk storage devices and sophisticated I/O equipment are replacing all the old control display facilities. This does not mean that the operator's control power and ability to obtain information are decreased. He has more information available than with the older techniques. But he and/or the communications and data acquisitions systems decide what information is relevant and needs displaying for human consideration at a given time. If the operator wishes to display trends, he so indicates via his console. The data, perhaps several superimposed trend plots, are displayed in the form of lines on the screen. If an alarm condition arises, the computer system reacts and informs the operator, displaying the pertinent information. Data logging, report generation, and other tasks that are not time-sensitive are carried out in the background automatically using bulk storage devices to record the data.

9.3 SUMMARY

In this chapter we have attempted to describe the characteristics and applications of peripheral devices found typically on all small-computer systems, real-time or otherwise. In doing this we grouped the devices according to function and discussed each device within a class in terms of other peripheral equipment which might serve a similar purpose. (The difficult decision facing the systems designer usually is not *whether* to utilize an auxiliary storage unit but rather *what type* of auxiliary storage unit to use.)

The specification of peripheral equipment for a computer system requires major tradeoffs between two important areas—(1) convenience of using the system and (2) cost. With respect to the first category it is important to note that, while design characteristics of a particular central processor may determine whether a real-time application can be carried out (or carried out efficiently), the software and peripheral equipment complement has more to do with convenience. For example, a real-time system without a high-level operating system that supports an auxiliary storage unit might be fully adequate for a fixed and unchanging data acquisition assignment. The same system might be totally inadequate (in terms of convenience) if the data acquisition program has to be revised often using this system.

The second area is important because a major cost of almost any computer system is represented by the peripheral hardware; often the combined cost of input/output and auxiliary storage equipment is many times the cost of the processor itself. Hence it is important to pay careful attention to detail in specifying the peripheral equipment complement for large or elaborate systems, if total system costs are to be held down.

With any system the object should be to try to obtain the maximum convenience at the minimum price.

Inflation and increasing market size will have effects on peripheral prices which may cancel each other somewhat. We should note that many of the devices discussed in this chapter contain mechanical components that have not been subject to major price reductions as a result of technology changes such as have occurred with electronics components through changing solid state technology. However, we can expect that mechanical devices gradually will be replaced by electronic equivalents, wherever possible. Major price changes may be expected when this occurs; for example, the replacement of fixed-head disks by bulk electronic memory based on magnetic bubble or equivalent technology should eventually bring down the cost of purchasing and maintaining fast-access auxiliary storage units.

9.4 READING LIST

Books

Asten, K. J., *Data Communications for Business Information Systems*, The Macmillan Company, New York, 1973.

Booth, T. L., *Digital Networks and Computer Systems*, John Wiley & Sons, New York, 1971.

Martin, James, *Design of Real-Time Systems*, Prentice Hall, Inc., Englewood Cliffs, N.J., 1974.

Martin, James, *Telecommunications and the Computer*, 2nd Ed., Prentice Hall, Inc., Englewood Cliffs, N.J., 1976.

McNamara, John E., *Technical Aspects of Data Communication*, Digital Equipment Corporation, Bedford, Mass., 1978.

Pooch, U. and Chattergy, R., *Minicomputers*, West Publishing Co., St Paul, 1980.

Soucek, B., *Minicomputers in Data Processing and Simulation*, John Wiley & Sons, Inc., (Interscience Division), New York, 1972.

Tebbs, D. and Collins, G., *Real-Time Systems*, McGraw-Hill, London, 1977.

Weitzman, Cay, *Minicomputer Systems*, Prentice Hall, Inc., Englewood Cliffs, N.J., 1974.

Yourdon, Edward, *Design of On-Line Computer Systems*, Prentice Hall Inc., Englewood Cliffs, N.J., 1972.

Articles

Leibson, S., "What are NS232C and IEEE 488?," *Inst. Cont. Sys.*, 47–53, January 1980.

Pomzin, L. and Zimmermann, H., "A Tutorial on Protocols," *Proc. IEEE, 66*, 1346–1370, 1978.

Technical Journals

Computer, IEEE Computer Society, Long Beach, Cal.

Computer Decisions, Hayden Publishing Co., New York.

Computer Design, Computer Design Publishing Corp., Littleton, Mass.

Datamation, Thompson Publications, Chicago, Ill.

Digital Design, Benwill Publishing Corp., Boston, Mass.

EDN, Cahners Publishing Co., Boston, Mass.

Electronic Design, Hayden Publishing Co., New York.

Electronics, McGraw-Hill Publishing Co., Albany, N.Y.

Instruments and Control Systems, Chilton Press, New York.

Mini-micro Systems, Modern Data Services, Hudson, Mass.

9.5 APPENDIX: A COMMUNICATIONS GLOSSARY

Acoustic Coupler—A modem that is connected to the telephone lines by plugging in a telephone receiver.

Asynchronous Transmission—A data transmission technique in which data are transmitted one character at a time with variable times between characters. Using this technique the receiver and transmitter operate in independent or unsynchronized time references.

Auto-Answer—Describes a communications device that is capable of responding automatically to an incoming call.

Auto-Dial—Describes a communications device that is capable of self-dialing a telephone.

Baud—A measure of data transmission speed. One baud is one transmission unit per second. If the data transmitted are serial binary data, one baud is one bit per second.

Bisynchronous Transmission—IBM's name for their form of synchronous transmission.

Bit-Serial—A method of data transmission in which data are sent one bit at a time over a single communications link.

Communications—Transfer of information from one place to another.

Current Loop—A means of data transmission in which the direction of current flow in a closed loop circuit determines the logic value transmitted.

Data Communications—Transfer of *digital* information from one place to another.

Data Set—See Modem.

Dedicated Line—See Leased Line.

Dial-Up—See Auto-Answer.

EIA RS 232—A set of standards that describe how a digital device should be connected to communications equipment. (EIA stands for Electronic Industries Association.)

Firmware—Programs and data stored in non-volatile, read-only memory.

Full-Duplex (FDX)—A communications link in which data can be transmitted in both directions simultaneously.

Half-Duplex (HDX)—A communications link in which data can be transmitted in either direction, but in only one direction at a time.

High-Speed Communications—9600 bits per second and faster.

IEEE 488—A set of standards that specify how peripherals are to be attached to processors.

Leased Line—A telephone connection rented from the telephone company in which the two ends are connected at all times; no dialing or switching is necessary, since the ends are "permanently" connected.

Longitudinal Parity Check—A parity technique in which a parity bit is obtained for all bits in a given bit position for all the characters in the data string.

Low-Speed Communications—75–300 bits per second.

Mark—In data communications, a binary signal with logic value 1.

Medium-Speed Communications—300–9600 bits per second.

Modem—Modulator–demodulator; a device that converts digital signals to signals suitable for transmission over telephone lines and vice versa.

Originate-Only—Describes a communications device that cannot answer an incoming call.

Parity Check—An error-checking technique in which the number of 1's in an array of binary data is computed and used to determine the value of a distinct parity bit.

Private Line—See Leased Line.

Protocol—Supplementary information transmitted in synchronous communications to establish synchronization and facilitate error detection and handling.

RS 232—See EIA RS 232.

S100 Bus—A standard bus configuration for interconnecting processor components.

Space—In data communications, a binary signal with logic value 0.

Simplex—A communications line in which data can be transferred in one direction only.

Start Bit—In asynchronous communications, an extra bit appended to the beginning of each character to enable the receiver to synchronize its clock with that of the transmitter.

Stop Bit—In asynchronous communications, an extra bit appended to the end of each data character to enable the receiver to detect the next start bit.

Synchronous Transmission – A data transmission technique in which data are transmitted continuously with special characters used to indicate the beginning and end of data and messages. Using this technique the transmitter continuously sends synchronizing information in addition to data to provide a common time reference between sender and receiver.

Transverse Parity Check – A parity check technique that is based on the parity of individual characters.

Turnaround Time – In a half-duplex link, the time required to reverse the direction of data transmission.

UART – Universal Asynchronous Receive Transmit; a large-scale integrated circuit that provides interfacing functions for asynchronous transmission.

USRT – Universal Synchronous Receive Transmit; a large-scale integrated circuit that provides interface functions for synchronous transmission.

Vertical Parity Check – See transverse parity check.

X.25 Communications Protocol – A set of standards that specifies the sequence in which data and control information are transmitted in data communications.

9.6 EXERCISES

1. Table 9.2 indicates that several character printers with speeds on the order of 120 cps can be purchased for the same price as one line printer. Discuss the tradeoffs involved between these two possibilities for providing hard-copy output. How do you calculate the break-even point at which n character printers provide throughput equivalent to that of a line printer? How does price then enter into the picture? What about reliability? Flexibility of use of the peripherals? Operational problems? Software complexity?

2. A travel agency wants to develop a complete on-line system to automate their reservation and billing system. The system is to provide stand-alone billing, payroll, and financial accounting for up to 10,000 customers and is to be linked to a remote airlines reservation system at a large airlines headquarters.

There are ten employees, four of whom are responsible for dealing with clients and making reservations. The agency processes the data and accounting for about 20,000 tickets per month, including data entry, ticket printing, and billing. Make your own estimates of disk space and printing and data input equipment needs, and specify a complete set of peripherals for this system.

3. Describe the files and data structures that would be appropriate for the system in Exercise 2.

4. Estimate the following:
   a. The time it takes to print ten pages (66 lines per page, average line length 30 characters), using
      (1) A terminal operating at 300 baud
      (2) A 30 line-per-minute line printer
      (3) A 120 cps character printer
   b. The time it takes to fill a CRT screen (24 lines at 80 characters per line) operating at 19,200 baud.
   c. The time it takes to transmit ten pages of data over a 4800-baud synchronous communications link.

5. Compute the number of bytes of data that can be written on a 2400-foot reel of magnetic tape at a recording density of 800 bytes per inch. Assume that record gaps are $\frac{3}{4}$ inch long. Do the computations for 80-byte records and 8000-byte records.

6. If the tape drive of Problem 5 moves at 75 inches per second, how long does it take, on the average, to locate a given 80-byte record on the tape?

7. A certain disk rotates at 3000 rpm and has four surfaces, with 200 tracks per surface, each of which holds 24 512-byte sectors.
   a. What is the disk capacity?
   b. If the time it takes to move the heads from one track to another averages 30 milliseconds, how long does it take, on the average, to locate a specific record if the file occupies the entire disk, and there are no indexes for the file?
   c. How long, on the average, does it take to locate a record on the disk if the disk address of the record is given?
   d. What is the data transfer rate for this disk?

10

Digital Computer/Process Interfacing

William R. Hughes
Department of Electrical Engineering
University of Colorado, Boulder

10.0 INTRODUCTION

For a digital computer to be effectively employed as the monitoring and control agent in a real-time environment it must be provided with the ability to easily sense parameters defining the state of a particular process and, normally, to effect some control over the sensed state. This task is accomplished by providing an interconnection of the computer and process through a device known as an interface. The purpose of this chapter will be both to discuss interfacing philosophy and to describe various interface components and techniques, at least the most important interfacing methods and associated building blocks. Whenever appropriate, examples will be employed to aid in clarifying important ideas and to assist in their practical application. The reader must remember throughout this chapter that our purpose is not to provide an exhaustive study of computer/process interfacing, but rather to generate enough understanding to deal with the basic complications and decisions involved in specifying or operating a real-time data acquisition and control system.

10.1 AN INTERFACING OVERVIEW

Initially it should be recognized that process interfaces are composed of two subsystems. One part, which was discussed in Chapter 8, is the interface control hardware. The structure of this section of the interface is intimately tied to the characteristics of the computer with which it is to be used. Functionally, this hardware serves to decode commands originating from the computer, to control interface opera-

tion, to provide status information to the computer through its I/O system, and to transmit information to and from the devices that make up the identifiable process interface. The second part of the process interface consists, then, of the easily identified peripheral devices that communicate with the process—the process-related peripherals discussed in an introductory way in Chapters 1 and 2.

Process-related components of the interface could be thought of as the interpreters of a real-time data acquisition and control system. Just as human beings are employed to translate spoken languages, these interface components are used to translate the "languages" of the computer and the process so as to provide a mutually understandable system of communication. Two common examples of such components are Analog to Digital and Digital to Analog converters (ADCs and DACs) which can be used together to provide two-way data/voltage translation. ADCs translate an analog voltage (continuous rather than quantized voltage) to digital information that represents the original voltage in quantized form. DACs translate coded digital information into an analog voltage. Obviously an ADC/DAC pair could be used to provide a two-way communication link between a computer and a process that "speaks" and "understands" only analog voltages. It is not important at this time that the internal operation of an ADC or a DAC be understood. It is important, however, for the reader to know about these devices which can be viewed and studied independently of both computers and processes and which are an integral part of most process interfaces.

The interface must, therefore, completely

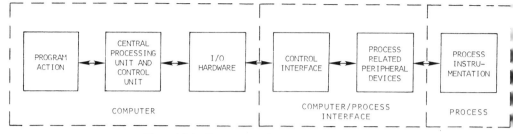

Figure 10.1. Computer to Process Communication

characterize the operating parameters of the process for the computer and at the same time operate as a transducer for communication from the computer to the process. Figure 10.1 illustrates a general block diagram of this interface communication link. As indicated in the figure, computer-to-process communication may be a combination of hardware and software (program) components. As an example, program action in the computer might cause it to issue appropriate control signals to transfer a digital word of data from a computer register to a process DAC via the I/O hardware and the DAC control interface.

Perhaps it would be useful at this point to furnish a definition:

Digital Computer/Process Interface—All the devices that serve to connect a digital computer to a process in such a way that computer/process communication can be effected in one or both directions, as required.

This definition leaves a great deal of latitude in deciding what components are included in the process interface for a very good reason. The actual makeup of the interface varies widely with the level of complexity of both the computer and the process. An attempt will be made, in the sections that follow, to discuss many of the devices that make up the interface. Since we have already discussed at least the general principles of peripheral device controllers in Chapter 8, we will restrict our discussion in this chapter to process-related peripheral devices themselves and mention the control interface only where some special feature or problem deserves comment or clarification.

10.2 PROCESS/COMPUTER INTERFACE COMPONENTS AND TECHNIQUES

In this section we will discuss, in a fair amount of detail, important process-related peripheral components. Our presentation for each device will consist of an explanation of its operation, an illustration of alternate implementations, if relevant, and finally, in most cases, an application example utilizing the device in question. We will assume, in some cases, that the reader possesses a minimal knowledge of electronics.

We will not include in our discussion devices that simply perform binary data encoding or decoding, since this class of devices was described in Chapters 7 and 8. Instead we will describe devices that have a more direct relationship to the process such as A/D and D/A converters, pulse generators, timers, etc. In presenting descriptions of these devices a value judgment has been made in that only a limited number of devices can be discussed. These were chosen primarily for their wide applicability and general usefulness.

10.2.1 Digital (Binary) Input/Output

In Chapter 2 we observed that in many cases information is transferred between computer and process in the form of two discrete voltage (or current) levels. Such two-state (i.e., binary or digital) information may be used to represent the closure of switches or relays tied to the process. For example, operator inputs, variables out of limits, etc., may represent inputs to the computer. Similarly, digital information may be sent back to the process to open or close valves, or start and stop motors, to name just a few examples. Such transfers are not, of course

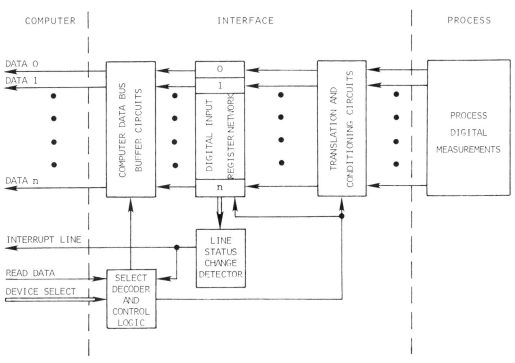

Figure 10.2. A Generalized Digital Input Interface

strictly limited to the exchange of control and process state information. Coded or uncoded data could just as easily be moved between pairs of computers or other complex digital devices using similar techniques. In either case, it will soon become apparent that binary interfaces usually consist of nothing more than intermediate memory elements (registers), an assortment of digital combinational logic devices (gates, etc.), and a variety of signal translation (level shifting) and conditioning devices. Hence they usually represent the least complex form of computer/process interconnection devices, in effect an indirect connection of the computer's I/O bus to process inputs or outputs. In other words, information transferred from an accumulator, internal to the computer, to the I/O bus is received by the interface and transmitted to the process, usually as a single word (e.g., 16 or 32 bits) without any intermediate processing. We will point this out clearly later in this chapter.

Input. Figure 10.2 illustrates the general structure of a Digital Input Interface. The interface includes all components necessary to connect it to the computer's I/O bus and to interpret commands (the Select Decoder, Control Logic, and Data Bus Buffers), all components necessary for its connection to the process (Translation and Conditioning Circuits), and the central input storage and status change hardware (Data Input Register Network and Line Status Change Detector). The Select Decoder serves to sense when the interface is to furnish information to the CPU. For example, to input an n-bit binary word stored in the Digital Input Register Network, the computer would execute a single I/O instruction addressing that particular interface device. As described in Chapter 8, the Select Decoder would then enable the Data Bus Buffer to transmit the required word to the I/O bus. Control Logic simply acts to provide necessary control of interface operations that must be sequenced, such as loading and clearing of registers in the Input Register Network. The Data Bus Buffers in this case would consist of signal translation devices designed to insure that the computer and device are able to share the I/O bus without mutual interference.

The Translation and Conditioning Circuits

usually consist of voltage-level translators (e.g., translate a 0-10 volt signal to a 0-5 volt signal), current flow detectors (e.g., detect a current flow through a switch contact), or level change detectors (e.g., detect a $0 \to 5$ volt or $5 \to 0$ volt signal level change). In addition a limited amount of digital (e.g., switch input debouncing hardware) or analog noise filtering hardware might be included. The Digital Input Register Network may be as simple as a single n-bit latch or as complex as a multi-register structure used in concert with Status Change Detector to detect the change in state of any input signal line. Figure 10.2 i intended to picture the input interface in it most general form; hence we have included Line Status Change Detector. This devic makes use of a record of previous line configur ations (stored in the Register Network) t generate a computer interrupt whenever an input line changes state. This feature is no strictly necessary, but does allow interfac

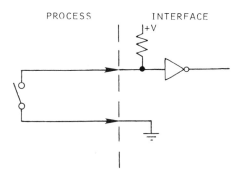

(a) Contact Closure Sensing

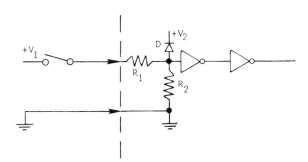

(b) Contact Closure Sensing with an External Voltage

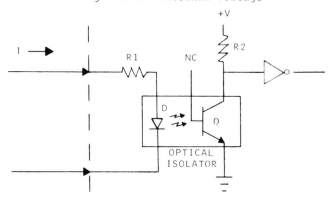

(c) Current Flow Sensing

Figure 10.3. Digital Input Types and Sensing Techniques

operation to be more automatic and eliminates computer overhead in checking the status of digital inputs periodically.

As mentioned above, Digital Input Interfaces are designed to sense and record changes in voltage and current levels on lines originating from a process. In this discussion we will consider translation and conditioning circuits that provide logic-level-compatibility (e.g., TTL or transistor-transistor logic where "0" = 0 volts, "1" = 5 volts) with input lines.

Currently available Digital Interfaces must in general have the capability of sensing the three types of process signals illustrated in the left-hand portion of Figure 10.3. In the right-hand section of this drawing we have included examples of circuits representative of those that might be used to sense *DC* digital inputs. This, however, does not exclude the possibility of allowing *AC* inputs as well, by use of different sensing circuits. Figure 10.3(a) represents the most common type of process input, a switch (or relay) contact closure, and the manner in which an input may be sensed. In this instance the interface must supply a current to flow through a circuit completed by the contacts. If the switch contact is connected as shown, its closure will result in the generation of a logic "1" (5 volts) at the output of the TTL inverter and a "0" otherwise. Resistor R in this circuit supplies current needed to sense the status of the line.

Figure 10.3(b) illustrates another variation of a contact closure input. In this case an external voltage appears on the line only when the contact is closed; otherwise an open circuit condition exists. The main problem here is insuring that the open circuit line status is interpreted correctly. Here we have shown one simple method of doing this although in practice more complex and more general solutions would be used. In this circuit resistor R_2 is selected in such a way that the voltage appearing at the input to the inverter when the switch is open is at logic "0" level, while a switch closure generates a logic "1". Notice that the input of the inverter is connected directly, through a resistor, to $+V_1$ when the switch contact is closed. Because of this we must insure that the voltage that appears at the input of the inverter does not exceed its allowable input range. This may be accomplished by "clamping" the voltage at a second lower level using a resistor and diode as indicated. In effect, D insures that the inverter's input is never larger than $+V_2$ ($+V_1 > +V_2$).

The third input type is shown in Figure 10.3(c) where the presence or absence of current flow in the signal line determines the logic output. This circuit makes use of a component called an optical isolator. When there is no current flow, transistor Q is turned off, since D, a Light Emitting Diode, is also off. Because Q is off, the voltage at the input to the inverter is V+ which corresponds to a logic "1" and results in the generation of logic "0" at the gate output. When a sufficiently large current flow exists, D emits enough light to turn Q on, placing a logic "0" on the gate input and resulting in the generation of a logic "1" at the output.

We should note that all three of these circuits have been designed as illustrations and are, therefore, simplified. Additionally, other input arrangements are possible; for example, the input type illustrated in Figure 10.3(b) might generate a signal that switches between two different voltage levels rather than between a single level and an open circuit condition. Actual commercial interfaces are available that can accept digital inputs over a range of voltages using either direct or alternating current (DC or AC voltages). Special-purpose interfaces may be restricted to single input type and/or levels, however.

Figure 10.4 shows schematically how an 8-bit (8-line) Digital Input Interface might be structured. This interface accepts inputs from any one of the sensing circuits pictured in Figure 10.3, detects status changes in these inputs, and reports their occurrence through generation of computer interrupts. Two latches (see Chapter 7) are required in the Digital Input Register Network; one latch contains the current line configuration (Sense Latch), and another stores the line configuration existing before the last line status change (Compare Latch). A program that makes use of this particular interface must maintain a record of the line configuration since the reception of the last interrupt, i.e., the contents of the Compare Latch. (The interface is made simpler by not allowing direct

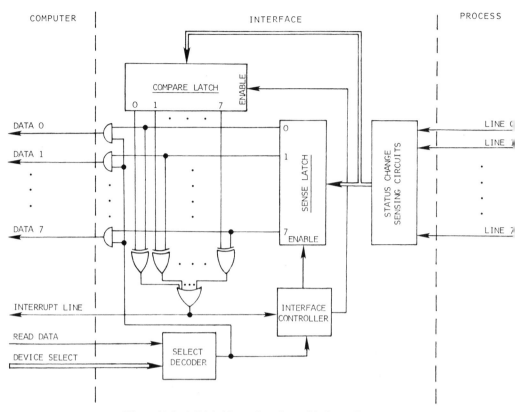

Figure 10.4. A Digital Input Interface with Status Detection

computer access to the contents of this latch.) Initially, after momentarily enabling the Compare Latch, the contents of the Sense and Compare Latches will be identical, indicating no change in line status. Notice that corresponding output lines of the two latches are used to sense a change in an input line through an array of exclusive-OR gates (Chapter 7). If the states of any two lines differ, as would be the case when an input change occurred, the corresponding exclusive-OR gate would cause an interrupt to be generated through the 8-input inclusive-OR gate and disable the Sense Latch until its contents could be read. A running program would, upon detection of an interrupt, read the contents of the Sense Latch and compare this information with its record of the previous contents of the latch to determine which line (or lines) changed state. The simple process of reading the contents of the Sense Latch might cause the Interface Controller to store the current input line status in the Compare Latch, thus clearing the interrupt and readying the interface for the next line change.

This interface is, of course, designed to be used in applications where response time to *changing* digital inputs is important. Many applications exist where a much simpler interface structure might be utilized. For example, if all we need do is occasionally input the state or setting of an array of toggle switches, this situation involves the input of digital signals of the type shown in Figure 10.3(a). In this case an array of switches could be included as part of a control panel to define to a running program the function or functions it is to perform; hence the interface need only consist of a set of gates to act as a buffer between the digital input and computer's I/O bus. No storage is needed. A variety of other possibilities exist, some of which are included in one or more commercial units available from various manufacturers.

Output. The general characteristics of Digital Output Interfaces can be illustrated in much the same way as for Input Interfaces. Figure 10.5 gives a schematic representation. Again

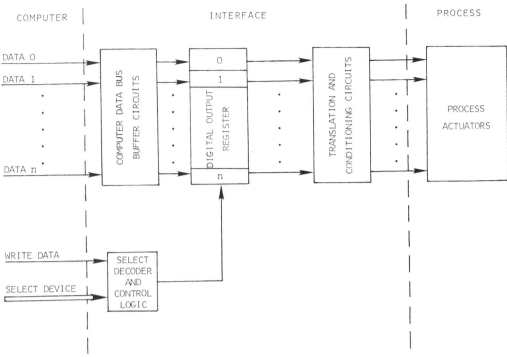

Figure 10.5. A Generalized Digital Output Interface

the interface of Figure 10.5 includes a Select Decoder. Three other subsystems are identifiable: the Computer Data Bus Buffer Circuits, the Output Register (or Latch), and a set of output Translation and Conditioning Circuits. The Bus Buffer Circuits serve to isolate the interface from the computer's I/O bus, while the Output Register simply serves as a memory for a desired digital output. These devices represent nothing new, since they were also included in the input interface which was described previously. Because the function of the Output Interface is to turn external devices off and on, problems, arise in choosing an appropriate Translation or Conditioning Circuit which allows connection of a line from the Output Register to the process. In general, we can identify four types of outputs, each associated with a particular driving signal:

(1) Sustained AC or DC voltage
(2) Pulsed DC voltage
(3) Relay contact
(4) TTL or logic

Figure 10.6 illustrates each of these types. Sustained AC or DC outputs (Figure 10.6(a))

may be implemented using relay contacts or more commonly by making use of solid-state devices which serve the same function (triacs for AC, transistors for DC). These outputs, in general, are meant to drive lamps, solenoids, relays, or similar elements. Pulsed DC outputs normally make use of solid state output components (as shown in Figure 10.6(b)) but may instead include relays. Pulsed outputs allow a DC pulse of fixed width (usually manually adjustable) to be generated on each computer output command for use in driving devices such as stepping or latching relays. The same effect may be had by turning a sustained output on and off as required; however, the advantage of the pulsed type output is that is reduces program overhead, since a pulse may be generated in one step rather than two. Also, a pulsed output always generates pulses of repeatable width, something that might not always be possible in a system under program control.

The third output type, the relay contact (Figure 10.6(c)), is primarily useful because of its high voltage and (especially) current-carrying capabilities. In many cases, the current requirements for a device to be driven from an output interface cannot be met by available solid state

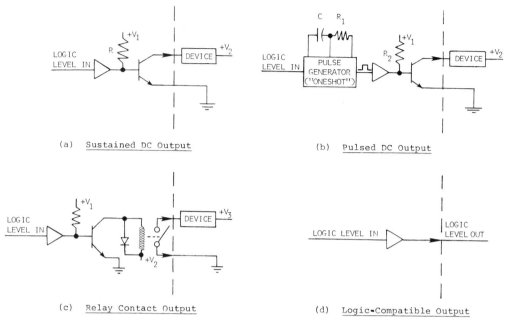

(a)   Sustained DC Output   (b)   Pulsed DC Output

(c)   Relay Contact Output   (d)   Logic-Compatible Output

Figure 10.6. Types of Digital DC Outputs

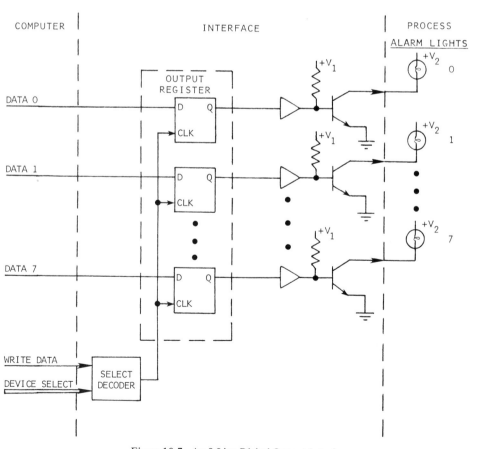

Figure 10.7. An 8-Line Digital Output Interface

components. In consequence, a set of relay contacts must be used instead. Additionally, electromechanical relays, unlike solid state "relays", have no open circuit leakage. Solid state circuits, even when off, allow small currents to flow through output components, an effect that may not be desirable.

TTL or logic outputs (Figure 10.6(d)) are useful mainly in driving compatible instruments or custom-designed interface circuitry. For example, a logic-compatible output interface could be used to transmit several bits in parallel to a receiving logic-compatible input interface on another computer. Other uses include the driving of logic level indicators (e.g., Light Emitting Diodes), controlling special I/O devices, etc.

Figure 10.7 shows how an actual sustained DC output interface might be constructed. In this particular example the interface is to be used to light various alarm lights on a large control panel. Many interfaces of this type allow the user to select the driving voltage to be used by leaving the output of the driver transistor open for connection of the desired DC voltage.

We have now completed a basic discussion of the techniques and hardware used in Digital I/O. We should, however, note the existence of one additional interface component in widespread use in Digital I/O—the optical isolator. This is the same device used in the current-sensing circuit of Figure 10.3(c). The greatest hazard to I/O interfaces in general is in their incorrect wiring to external processes. Suppose that we accidentally connect 115 VAC to a logic-compatible input or output. At a minimum the interface itself will be destroyed; worse yet the I/O bus and the CPU of the computer itself might be badly damaged. The most obvious way to prevent such an occurrence is to isolate all external interface inputs and outputs in such a way that application of a high voltage destroys nothing but the isolator. As noted in Chapter 5, optical isolators provide such protection. These devices consist of a light source and a light sensor in close proximity and are available in a variety of types, e.g., with logic-compatible inputs and outputs. Figure 10.8 illustrates how a logic-compatible input/ output interface might be fitted with optical isolators to prevent accidental damage.

We end this section with a brief listing, in Table 10.1, of examples of currently available commercial Digital I/O Interfaces. These interfaces can be purchased as individual circuit boards which may be placed directly in the main chassis of a computer or as parts of large-scale I/O subsystems. A complete I/O subsystem is an independent device which may contain a variety of data acquisition and control interfaces including Digital I/O Interfaces. Such subsystems are connected to a host computer, usually by way of a single intercon-

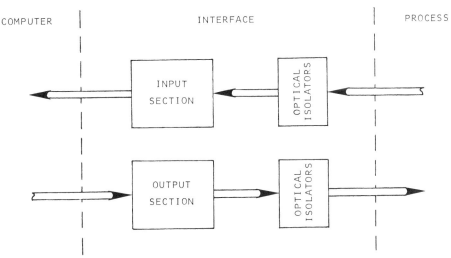

Figure 10.8. A Logic Compatible Digital I/O Interface with Optical Isolators

Table 10.1. Typical Digital I/O Interface Hardware

Device	Description
1. Digital I/O interface circuit board.	May be installed directly in main computer chassis. Includes: a) 16 Logic-compatible digital inputs. b) 16 Logic-compatible digital outputs. c) General purpose I/O connector.

2. I/O Subsystem with digital input and output interfaces.	Includes:
	
	a) I/O subsystem chassis to accept up to 16 digital interfaces.
	
	b) Digital input interface with
	  -- 16 lines for
	    6 to 55 VDC max. input
	    or
	    15 to 125 VAC max. input
	  -- Line status change detector
	  -- Optically-isolated inputs
	
	c) Digital output interface with
	  -- 16 sustained DC lines with
	    55 VDC maximum per line
	    1.5 amps maximum current
	  -- Optically-isolated outputs
	
	d) Interconnection cables and primary computer interface.

3. I/O Subsystem with digital input and output interfaces.	Includes:
	
	a) I/O subsystem chassis to accept up to 16 digital interfaces.
	
	b) Digital input interface with
	  -- 16 input lines with a choice of input types.
	  -- Optional contact bounce eliminators or filters.
	
	c) Digital output interface containing
	  -- 16 output lines with a choice of:
	    Logic-compatible outputs
	    Sustained or pulsed DC outputs

4. Digital Input Interface for Device 3 above.	8-line AC input interface. Accepts 117 VAC inputs and includes line status change detection circuitry
5. Digital Output Interface for Device 3 above.	8-line AC output interface with 117 VAC Triac outputs

nection cable. The individual interfaces may be accessed using either assigned computer interface-select codes or some special addressing protocol (such as selection of interface with one word, transmission of data with the next). In the following pages we will describe other types of interfaces that may be installed in an I/O subsystem.

10.2.2 Digital to Analog Conversion

Frequently it is necessary to translate binary-coded information into a current or voltage that has a direct or proportional relation to the original binary number. The resulting electrical quantities are known as continuous or analog voltages and currents. Circuits that perform such translations are called Digital to Analog Converters or DACs. Rather than begin with a description of these converters we will provide some necessary preliminary information.

As discussed in Chapter 6, several representations for binary information are possible. Although the selection of one of these coding systems has some influence on the total structure of a DAC, a basic converter may be designed by considering the unsigned binary representation only. Such a design then may be easily modified to allow for any desired input coding scheme. In designing an example DAC we may think of each bit in a binary number as representing the inclusion (1) or exclusion (0) in a sum of weighted amounts of an electrical unit (voltage or current) whose magnitude is directly proportional to the positional value of the bit under consideration. In this case the magnitude or size of the quantum assigned to the least significant bit of our binary word determines the *resolution* in that system of representation. When considering an uncoded binary input (i.e., not BCD or a similar coding), the magnitude of the amount of the electrical unit assigned to Bit i, q_i, must have the following relation to the size of the amount assigned to Bit i - 1:

$$q_i = 2q_{i-1} \text{ for } n \geq i \geq 1$$

where n defines the number of bits in the binary word used as an input to the DAC.

A digital to analog conversion can be per-

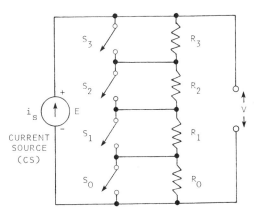

Figure 10.9. A Simple Digital to Analog Converter

formed simply using the circuit shown in Figure 10.9. In this circuit CS is a current source which supplies a constant current i_s. The voltage, V, appearing at the output of the circuit is determined by the setting of switches S0–S3 and is equal to E when all switches are open (as shown). If we interpret the positions of S0 to S3 as defining a four-bit binary input to the circuit where 1 = Open, 0 = Closed, we would like to choose R0 to R3 so that V reflects in analog form the value of the digital input. This may be accomplished by requiring that

$$R_0 = \tfrac{1}{2}R_1 = \tfrac{1}{4}R_2 = \tfrac{1}{8}R_3$$

Therefore E is determined by

$$E = i_s(R_0 + R_1 + R_2 + R_3) = i_s \, 15R_0$$

Suppose that we want to force V to be the analog equivalent of the binary number 1011 (11_{10}). We can do this by leaving open S0, S1, and S3 and closing S2. V is then given by

$$\begin{aligned} V &= i_s(R_3 + R_1 + R_0) \\ &= i_s(8R_0 + 2R_0 + R_0) \\ &= i_s \, 11R_0 \end{aligned}$$

and

$$i_s = \frac{E}{15R_0}$$

Therefore:

$$\begin{aligned} V &= (E/15R_0)11R_0 \\ &= \tfrac{11}{15}E \end{aligned}$$

which is the result we require. We can see readily that the resolution of this DAC is E/15, corresponding to the change in output voltage that may be generated by opening or closing S0.

It is obvious that switches S0 to S3 could be replaced with a 4-bit storage register and suitable additional circuitry to allow input of a binary number directly from a computer. We probably would not want to do this, however, since this simple DAC suffers from at least one disadvantage; i.e., in its present form any load placed on the output terminals will have a radical effect on V. It would be possible of course, to buffer the output with a suitable analog amplifier, but we will instead present several more realistic converter designs.

The first design, commonly known as a *weighted-resistor* DAC, is illustrated in Figure 10.10. The operation of this circuit differs from the last example in that it performs the required digital to analog conversion through the use of a current-summing technique. Currents generated by the four input resistors are summed at point Σ and generate an amplifier output according to the relation

$$V \cong -i_{in} R_f$$

where i_{in} is the magnitude of the input current at Σ, and R_f is the value of the feedback resistor R. This arrangement is known as a current follower and results in an output voltage whose polarity is opposite that of its input voltage.

The value of each resistor in the input section

of the DAC is selected so that it contributes a quantity of current at Σ that maintains the linearity of the binary to voltage translation. I the current associated with the least significan bit of the input is I_{LSB}, then each resistor con tributes $I_{LSB}2^m$ amperes where m is an intege equal to the number associated with the posi tion of the bit in the input word (bits in the input word are numbered as usual in ascending order from zero, right to left). Notice that we could just as easily have assigned BCD (Binary Coded Decimal) weights, or any other weighting corresponding to a particular binary coding scheme, to the input resistors of the DAC.

This DAC is, of course, an improvement over the original example, but it too has at least one major disadvantage. The problem inherent in the design is not readily apparent at this point because of the limited size of the input word. However, for a DAC allowing a 10-bit input we require ten *precision* resistors (the accuracy of the DAC depends largely on the accuracy of the resistor values), whose values must range from 2R to 1024R. If R is 5000 ohms, this mean that we must have available precision resistors over the range 5K to 5120K (K = 10^3) ohms In general, it is costly to manufacture resistors over such a large range that are not only precise but also have stable temperature and aging characteristics. The next DAC design provides a solution to this problem.

An obvious way of overcoming the difficulty encountered in the last example is to attempt to design an input network that uses only a few resistor values. The most popular way of accomplishing this is illustrated by the DAC

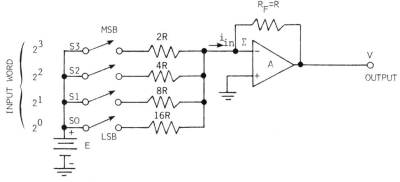

Figure 10.10. A Weighted Resistor DAC

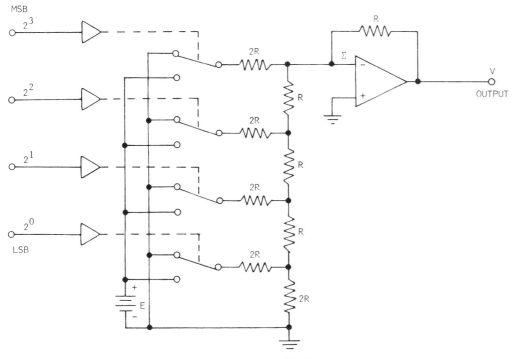

Figure 10.11. A Ladder DAC

circuit in Figure 10.11. This particular design makes use of an input network resembling a ladder and is for this reason sometimes referred to as a *ladder* DAC. This circuit requires only two different values of precision resistors in its input network. Notice that in this design we have introduced the use of a set of electronic switches. These switches are commonly-used components in modern digital/analog interface circuitry because they allow simple, totally solid state, implementation of circuits that would otherwise require the use of either electromechanical relays or a large number of discrete components. The switches that we employ in this design are controlled by a binary input through the use of digital buffers included as part of each switch. Switches in Figure 10.11 are shown in the position determined by a 0 input to their control buffers.

The ladder network in this circuit is arranged so that the current entering the operational amplifier at point Σ is equivalent to currents observed at the same point in the last example for any setting of the four input switches. To see this, suppose that the binary number 0001 is applied at the inputs to the circuit. We can show that the full-scale current at Σ is given by

$$I_{max} = E/R$$

The current that flows from the circuit ground or common to the positive reference can be calculated if we assume that point Σ is at the same potential as common (a good approximation) and draw the equivalent circuit for this situation. The total resistance for this circuit is found to be

$$R_T = \frac{128}{43} R$$

By analyzing currents flowing in the network we determine that the total current (i_T) flowing as a result of closing switch S_1 is $43/128\ I_{max}$ and the current entering Σ is

$$i_\Sigma = \frac{8}{43} i_T = \frac{1}{16} I_{max}$$

which is the desired result.

We mentioned at the beginning of this section that we might consider the use of any one of several different input coding schemes in designing DACs, yet our designs have been

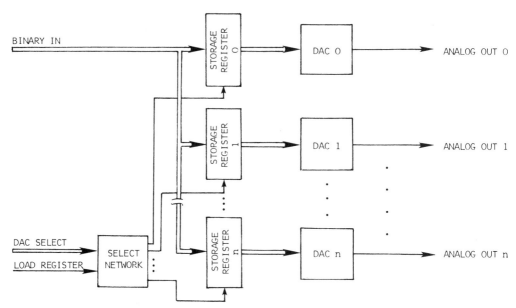

Figure 10.12. An n-Channel Digital to Analog Converter.

Table 10.2. Typical DAC Hardware

	Description	Resolution (Bits)	Settling Time	Output Range
1.	Single channel monolithic integrated circuit DAC. Bipolar or unipolar operation possible.	6	3 μsec	Bipolar ±5 Volts or ±10 Volts Unipolar 0-10 Volts
2.	Single channel hybrid integrated circuit DAC with binary input register.	12	3 μsec	0-5 Volts
3.	Single channel modular current output DAC. Bipolar or unipolar operation possible.	10	0.3 μsec	±1 mA or 0-2 mA
4.	Four channel DAC controller circuit board designed to be inserted directly in computer chassis. 64 ADC channels may also be included on the same circuit board at extra cost.	12	4 μsec	±10 Volts
5.	I/O subsystem including a four channel DAC module. The subsystem chassis can accommodate up to 16 digital and/or analog I/O modules. May be connected directly to I/O bus of computer or to a data channel controller.	12	5 μsec	±10 Volts

oriented toward unsigned binary inputs only; in fact, the last two designs even inverted the polarity of the analog output! Modifications to these designs to accommodate different input codes are relatively minor. In the case of two's complement coding for the second design (weighted resistors) we have only to use a negative reference as an input to the resistor corresponding to the most significant bit of the input and to double the value of the feedback resistor on the operational amplifier. Another way of solving the same problem would be to modify the operational amplifier circuit to provide for an offset in output. A noninverted output could be obtained by adding a second stage of amplification to simply invert the output of the first amplifier.

In ending this discussion we should point out that it is never necessary for the user to generate a design for a DAC, since many good, inexpensive commercial devices are available and usable in most systems. Consequently, information presented in this section dealing with the internal workings of DACs is intended to be of value as background information in selecting or understanding the operation of devices used in a specific application.

Manufacturers now produce self-contained DAC modules (integrated circuits and circuit boards) which require some degree of interfacing expertise to be used effectively in a particular real-time application. Also, many manufacturers offer "multi-channel" DACs (where each DAC comprises a separate channel) which may be accessed by a computer as a single I/O device. This hardware arrangement is illustrated in Figure 10.12. These units and I/O subsystems offer the greatest convenience to the system designer, since they are, in most cases, offered in configurations that may be directly connected to a wide variety of existing computers.

Table 10.2 lists a number of typical DAC units including general specifications. The list includes units consisting of arrays of individual DACs housed in a common mainframe.

In selecting DACs the most difficult task consists of understanding DAC specifications and deciding which particular unit fits the requirements of the intended application. As an aid in

this process, Table 10.3 presents, a summary of the terminology generally used to describe the characteristics of DACs. This table, of course, contains only a partial listing of terms used to describe DAC characteristics, but it includes those used most often by various manufacturers.

In specifying DACs or DAC systems, the most important questions to be answered include:

- What output voltage (or current) range is required? In general, DACs including high voltage (greater than 10 volts) or high current outputs are more expensive than DACs with low or medium-range outputs. DACs are offered with unipolar and/or bipolar outputs.
- What kind of input coding will be required? DACs are offered that will accept a number of input codes including two's complement and BCD.
- What output accuracy and resolution is required? DACs with high accuracy and resolution are normally expensive so it is advantageous to specify only that which is actually needed.
- In what kind of environment will the DAC be operated? Inexpensive DACs may be unusually temperature-sensitive, but may be suitable in applications where temperatures are relatively constant.
- What is the maximum DAC response time that may be tolerated? Extremely fast DACs are usually very expensive.

In concluding this section we present a simple example of an application requiring the use of a DAC. The DAC to be used is similar in structure to the device illustrated in Figure 10.11. It has a full-scale output of 10.0 volts with an 8-bit input network (0.039-volt resolution). The DAC is represented here as a block with labeled inputs and outputs only, since we have no need to discuss the internal workings of the converter any further.

In this example we wish to regulate the intensity of a light source in an indoor hydroponic garden to simulate the time-varying energy flux that could be expected during a normal period of natural sunlight. To ac-

Table 10.3. DAC Terminology

<u>Accuracy</u> -- A measure of the maximum error which may be expected in representing a
digital input as an analog voltage. An accuracy of 0.1% of full-scale is typical.

<u>Resolution</u> -- The size of the change in DAC output which is produced by changing the
least significant bit (LSB) of its input. For a 10-bit (input) DAC we would say
that its resolution is 1 part in 2^{10}. Generally a DAC's resolution is specified
simply by listing the number of bits required as input.

<u>Linearity (or Nonlinearity)</u> -- The degree to which there is a linear relation between
the binary inputs to a DAC and the analog voltages which it produces. This
quantity normally is expressed as percentage of the device's full-scale output.

<u>Gain Temperature Coefficient (TC)</u> -- A measure of the effect of change in temperature on
the accuracy of the output. This characteristic normally is expressed in parts
per million per degree centigrade (ppm/°C) and describes the amount of variation
in the DAC's output which can be expected per degree of temperature change.

<u>Settling Time</u> -- Specifies the length of the time interval required between a change in
digital input and the generation of an accurate analog representation of the
input at the DAC output. This quantity may be interpreted as a measure of the
"speed" of the DAC. Settling times range from a fraction of a microsecond to
tens of microseconds depending on the design and cost of the converter.

<u>Offset</u> -- Voltage present at DAC's output with a digital input of zero. When present,
this output effect may be eliminated if required through the use of some minor
adjustments to the DAC circuitry.

<u>Output Voltage (or Current) Range</u> -- Range over which the DAC output may vary with ·
appropriate digital inputs. DAC outputs may be capable of either unipolar
(+ <u>or</u> - voltage or current) and/or bipolar (+ <u>and</u> - voltage or current) operation
as required. Some manufacturers also offer DACs allowing high power output
operation.

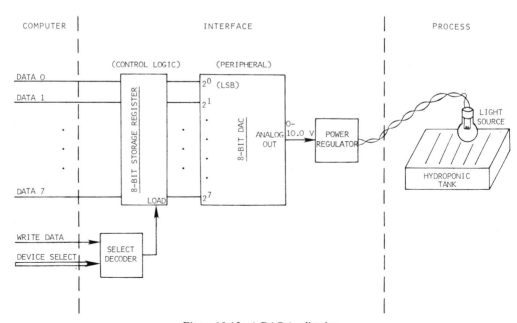

Figure 10.13. A DAC Application

complish this we use the output of the DAC to drive the input of a solid state power regulator that serves as the energy supply for the light source. The power regulator is designed in such a way that it requires a 10-volt input to control the light source at its maximum intensity. For an input of 0.5 volt the source is completely off. All logic related to timekeeping and correlation of light source intensity to the time of day is to be included in a program executed periodically by the computer to which these devices are connected. Figure 10.13 illustrates one hardware arrangement that might be considered in such an application. Notice that required hardware consists of little more than the DAC, a data-holding register, and the power regulator. Many commercially-available DACs include a built-in data storage register.

10.2.3 Analog to Digital Conversion

The most important interfacing component in data acquisition applications is the Analog to Digital Converter (ADC). This device performs a function that is the inverse of that performed by the DAC discussed in the previous section; i.e., it converts an analog voltage, applied to its input, to a scaled binary number whose magnitude is proportional to that of the analog voltage. Quite often some sort of normalizing, prescaling, or offsetting of the analog voltage occurs before it reaches the input of the ADC. This allows standardized ADC modules to be used in a variety of applications.

In this section we will describe three types of ADCs:

(1) Integrating
(2) Servo or Feedback
(3) Parallel

In designing any type of ADC we are faced with the problem of representing, in quantized form, the analog input to our converter. For these devices the conversion *resolution* is determined by the number of bits in the binary output. Each change in the least significant bit (LSB) of the output must correspond to one quantum change in input voltage. We would like to have these changes occur at half-quantum levels to provide for the generation of a conversion curve that is as close as possible to the ideal over the entire range. In other words, the output should ideally change at each intermediate voltage as indicated in Figure 10.14.

In general (except for parallel converters which represent a special case), ADCs may be placed conceptually in one of two separate

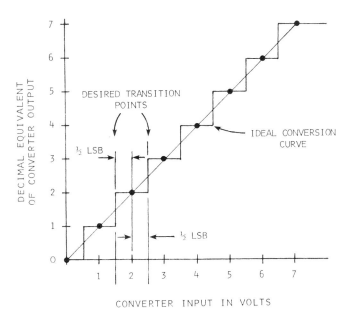

Figure 10.14. Conversion Curve for a Three-Bit Converter

classes differentiated by the basic method used to measure an analog voltage. One class relies on the measurement of electrical charge, while the other uses feedback techniques to perform required conversions. The charge-measuring devices that we discuss first are usually referred to as *Integrating ADCs*.

Charge-measurement ADCs measure input voltages indirectly by counting the number of units of an accumulated charge whose magnitude is known to be proportional to the unknown voltage. Two related charge-measurement conversion techniques have found wide use. The first of these is used in what is generally known as a *Ramp* or *Single Slope* ADC. Figure 10.15 illustrates the structure of an 8-bit Ramp ADC.

The precision oscillator and the integrator circuit composed of the operational amplifier, resistor R, and capacitor C are basic to the operation of the ADC. The frequency of the oscillator is fixed in such a way that the charge accumulated on C during one period of oscillation is related to the smallest unit of voltage that may be resolved by the ADC. Therefore, by applying an input voltage at V_{IN} and counting oscillator periods until the voltage across C equals or just exceeds V_{IN}, a measure of V_{IN} may be obtained. This converter gets its name from the shape of the curve representing the time-varying voltage across C during the conversion process.

The operation of the converter, pictured in Figure 10.15, may be described as follows. In the quiescent state both flip-flops and the counter are reset (counter output equals zero). When a pulse is applied at the Convert input, Flip-flops A and B act to synchronize the start of the conversion with the output of the precision oscillator. When Flip-flop A sets, it removes the short circuit across C and allows it to begin charging through R. At the same time, count pulses are enabled for the 8-bit counter. Counting continues until the voltage appearing across C exceeds the input voltage at V_{IN} as determined by a voltage comparator. This device maintains a logic 1 voltage level at its output, while the voltage at its minus input is less than that at its plus input. When the voltage at the comparator minus input exceeds V_{IN}, one last oscillator pulse is allowed to increment the counter, and the converter is returned to its quiescent state. During the conversion process the status of the circuit is indicated at all times by the converter Done output. When a conversion has been completed, this output is placed in its logic 1 state, and the contents of the counter indicate the magnitude of the converted voltage.

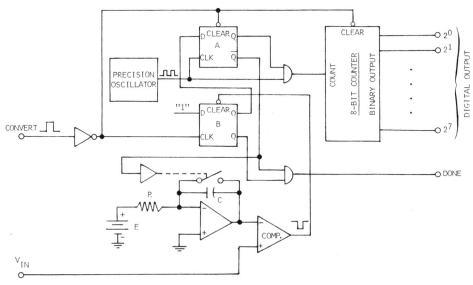

Figure 10.15. A Ramp Analog to Digital Converter

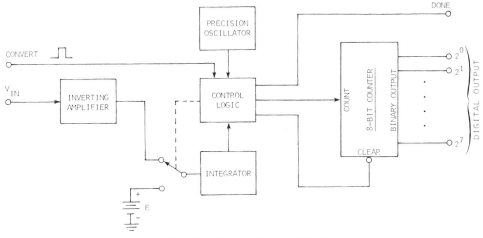

Figure 10.16. A Dual Slope Analog to Digital Converter

The main problem with this type of converter is that of accuracy. Because the conversion process depends on proper matching of the oscillator frequency to the time constant of R and C, accuracy depends on all three of these components. Therefore problems of mismatching, as well as all of the usual aging and stability problems, are the major error sources in the system. The next converter to be discussed decreases the effect of these error sources somewhat through a special technique.

The operation of a *Dual Slope* ADC is similar in concept to the Ramp converter. Here the capacitor in a voltage integrator is allowed to charge toward the input voltage V_{IN} for a fixed time interval which is an integral number of periods of a precision oscillator. If this integration time is given by T_I, then the charge accumulated on C is proportional to the average value of V_{IN} during T_I. At this time we connect the input of the integrator to the reference source E and reset the counter. The charge on C will be lost in a time interval, T_m, proportional to the value of the input voltage V_{IN}. Because the counter has been allowed to count oscillator pulses during the discharge of C, it will contain a binary number representing V_{IN} when the voltage across C reaches zero.

Figure 10.16 is a representation of a Dual Slope ADC in block diagram form. We can see that this converter gets its name from the shape of the curve (Figure 10.17) that depicts the voltage across C at any time during a conversion.

The Dual Slope technique obviously is slower and requires more hardware for its implementation; however, as we mentioned earlier, conversion accuracy obtainable with the Dual Slope

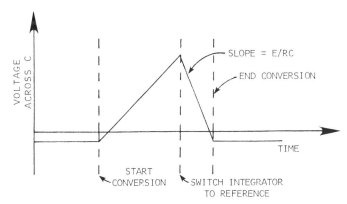

Figure 10.17. Dual Slope Waveform

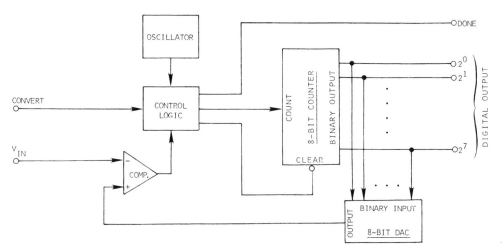

Figure 10.18. A Staircase Analog to Digital Converter

method is superior, since it utilizes the same components, particularly the timing oscillator, in charging and discharging the integrating capacitor. This has the effect of cancelling errors introduced by these components in the charging and the discharging phases. Dual Slope converters are widely used in instruments requiring accurate results but not requiring extremely short conversion times.

We now turn to a discussion of the second conceptually different type of ADC, known generally as *Feedback* or *Servo* ADCs. One way an analog to digital conversion could be made is to make a guess of the probable digital value of the input and then subsequently adjust this estimate based on a comparison of the guess with the actual value of the input. Such a technique is used by ADCs having a Feedback or Servo structure. The simplest example of this type of conversion is represented by the *Staircase* ADC illustrated in Figure 10.18. The operation of this circuit may be described as follows: The 8-bit Conversion Counter initially contains zero. A voltage comparator is included to operate in such a way that it continuously compares the value of the input voltage to the output from a high-speed DAC whose input is the binary number residing in the Conversion Counter. When the converter is started, the comparator allows the counter to be incremented until the output of the DAC exceeds the input V_{IN}. This conversion procedure obviously suffers from the disadvantage

that the required conversion time is directly proportional to the magnitude of the analog input. Therefore, if a counter clocking frequency of 1 MHz is assumed, the conversion would require 255 microseconds to convert the maximum input. A higher clock rate could, of course, be used, but this approach is limited by the conversion time of the internal DAC.

We note that conversion times may be reduced if a faster search method can be devised to determine the correct value for the digital output. One such method, called interval halving or bisection, is illustrated in Figure 10.19. The true value of some quantity known to be within an interval of range A, B can easily be found to lie within a particular one of the smallest resolvable intervals. In performing the search we initially guess that the value we are searching for lies either above or below the center of the interval A, B at Point 1. We find that the true value is below this point. Consequently we narrow the search by dividing the

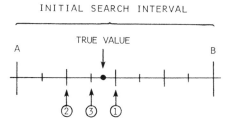

Figure 10.19. Estimation of a Value by Interval Bisection

interval in the order shown and determine in three guesses in which subinterval the true value must be located. Because interval bisection is easily implemented using binary logic, it naturally came to be included in a converter type known as a *Successive Approximation* ADC.

An implementation of a 4-bit unsigned binary Successive Approximation ADC is given in Figure 10.20. Note that the conversion time for this type of ADC is determined by the number of bits in its digital output only.

Operation of this converter is completely controlled by the Sequencer module which generates control signals and makes the logical tests necessary to obtain the correct conversion of an analog input. Conversion proceeds in the following steps after the detection of a Convert pulse by the Sequencer:

(1) All flip-flops in the Conversion Register are reset to generate a digital output of zero.

(2) The Sequencer counter, Count, is set to N where N equals the number of bits in the output word.

(3) The flip-flop specified by Count is set to 1.

(4) The output of the DAC settles to the voltage specified at its input in Step 3.

(5) The output of the Comparator is examined. If its output indicates that the present contents of the Conversion Register exceed the required value of the output, the flip-flop specified by Count is reset.

(6) Count is decremented. If Count is not zero, the sequence returns to Step 3.

(7) Done is set to 1.

At the conclusion of this procedure the scaled binary equivalent of the analog input will be found in the Conversion Register. This converter always requires N converter clock periods (where N is the number of bits in the result) to complete a conversion regardless of the magnitude of the analog input. The major problem with this converter (and all others described previously) is that is requires the analog input to be held constant for the entire conversion cycle. Converter systems, as a consequence, include what is known as a Sample

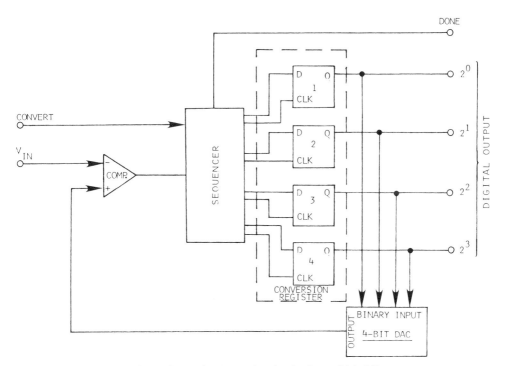

Figure 10.20. A Successive Approximation Analog to Digital Converter

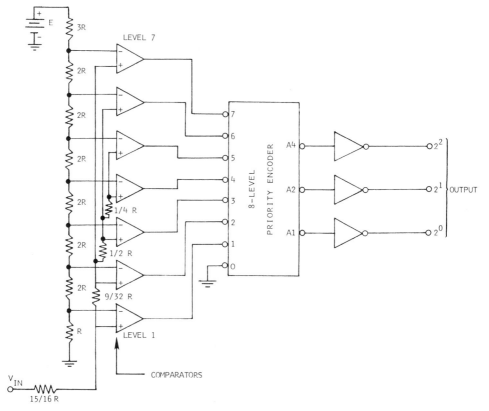

Figure 10.21. A Parallel Analog to Digital Converter

and Hold circuit—specifically designed as an analog voltage sampler. This device enables the storage of an analog voltage for use as a converter input for a period of time that is sufficiently long to allow completion of the conversion. Sample and Hold circuits are discussed in more detail in Section 10.2.4.

The third converter type to be discussed in this section is a *parallel* or "flash" ADC. The description of this ADC will be brief because of its limited applicability and generally wasteful hardware implementation for output word lengths of useful size.

All other converters presented in this section use some form of sequential comparison to determine the appropriate output for a particular analog input. Parallel converters, on the other hand, perform all required level comparisons simultaneously. Figure 10.21 illustrates an implementation for a 3-bit parallel ADC. Notice that a comparator is required for each quantized voltage level to be checked. Such a circuit is very complex for an output word length much

larger than three. A second problem encountered when designing an ADC of this type is that generally the comparator outputs must be encoded in some way to allow for generation of easily-used data. We have chosen in this design to perform the required encoding through the use of an integrated circuit module known as a priority encoder. This device is normally used in computer interrupt systems to aid in the generation of interrupt vectors (as described in Chapter 8). Each input of the encoder is assigned a relative priority. The binary code appearing at the module's outputs, if more than one input is activated, is determined by the highest priority active input. Thus, if inputs 7 and 5 are simultaneously active, 000 will appear at the encoder outputs rather than 010 (encoder outputs are inverted).

Conversion times for such ADCs can be as short as a few tens of nanoseconds, depending on the speeds of both the comparators and the encoder used. For example, an integrated circuit implementation of such a converter can perform one 8-bit conversion every 50 nano-seconds. A converter having this speed would most likely find application in situations involving signal processing at radio frequency rates. As we mentioned earlier, however, parallel converters find infrequent usage because of the expense of the hardware required for their implementation; and applications in which their use may be justified are fairly rare.

ADC circuits discussed in this section are obviously extendable to any output word size and may be operated over any input voltage range with proper scaling. Use of any of these circuits with any of the bipolar output codes (such as two's complement) is possible, usually simply by offsetting the analog input (through the use of additional circuitry) by half of the converter's full-scale range. We should note that there are several other ADC designs in common usage. Here we have presented only those designs that are most often encountered; other designs are similar enough that understanding their operation should not be difficult.

Several commercial ADCs are available as standard modules that utilize one or more of the conversion techniques we have discussed. The kinds of ADC modules and systems available follow the same sort of pattern discussed for DACs. In other words, ADCs may be purchased as integrated circuit chips, as circuit boards, or as part of complex I/O subsystems.

Data acquisition systems often are designed to allow a single ADC to simulate, functionally, the operation of several ADCs. Figure 10.22 illustrates a hardware arrangement that might be used in a basic system of this type. Notice that any one of several analog channels (normally 4 to 64) may be selected for connection, through an analog multiplexer (treated in Section 10.2.5), to the ADC included in this system. More complex acquisition systems might also include the capability of sequentially scanning and converting each of the voltages on all, or a selected set, of the input channels.

Several manufacturers offer converter building blocks such as successive approximation registers and DAC modules. High-performance units are also available in practically any configuration imaginable as modules that are comprised of several integrated circuits or circuit boards. Table 10.4 lists several currently available ADC modules and a typical complete data acquisition system.

In selecting an ADC it is helpful to be familiar with the performance terms used by designers. A short summary, similar to that presented in the last section for DACs, is given in Table 10.5.

A user is never forced to design a custom ADC because of the wide range of hardware available. However, it is wise to use a similar set of guidelines in selecting an ADC as was given for DACs. Use of such a check list cover-

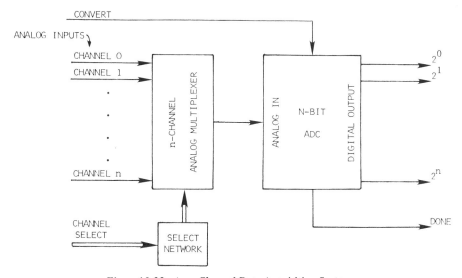

Figure 10.22. An n-Channel Data Acquisition System

Table 10.4. Typical ADC Hardware

Description	Resolution (bits)	Conversion Time	Input Range
1. A single-channel, integrated circuit, integrating ADC. Bipolar or unipolar input is possible.	8	1.25 msec	±5 Volts or 0-10 Volts
2. A single-channel hybrid, successive approximation ADC with built-in sample and hold circuit. Bipolar or unipolar operation possible.	12	8 µsec	0-5 Volts or 0-10 Volts or ±2.5 Volts or ±5.0 Volts or ±10.0 Volts
3. A single-channel, modular, dual-slope ADC. Sign-magnitude binary output.	10	1.25 msec	±1, ±5 or ±10 Volts
4. I/O subsystem including a 16-channel ADC data acquisition system. This system can also accommodate a 4-channel DAC and up to 15 more digital or analog I/O modules. ADC includes sample and hold.	12	20 µsec	±10 Volts
5. A 64-channel ADC controller circuit board designed to be inserted directly in a computer chassis. Four DAC channels may also be included on the same circuit board at extra cost.	12	20 µsec	±10 Volts

ing resolution, accuracy, etc., tends to narrow choices quickly and minimize costs. One additional consideration should be added to the check list, however:

- Does the intended application require the use of a Sample and Hold circuit at the input to the ADC? All of the ADC types discussed here require that the converter's analog input be held constant during the conversion process.

Finally, in selecting a particular ADC it should be remembered that, as a general rule, *resolution* and *conversion time* have the greatest effect on the cost of the device.

We end this section with another brief example (Figure 10.23), this time illustrating an ADC application. We assume the use of a 10-bit ADC and require no particular minimum conversion interval, since the manually-controlled process does not change rapidly. For this example we wish to monitor the temperature of a liquid in a stirred tank. The tank receives an input flow of liquid with varying temperature and releases an output flow at a temperature to be measured. The temperature of the liquid within the tank can be adjusted by means of a heating element contained in the tank. The temperature of the liquid leaving the tank is measured by a thermocouple in the output flow line. In this case the low-level signal

Table 10.5. ADC Terminology

Accuracy -- A measure of the error which may be expected in representing an analog
   voltage as a binary number. An accuracy of 0.01% to 0.8% of full-scale is
   typical.

Resolution -- Analogous to the definition of DAC resolution: specifies the size of the
   input voltage change required to produce a one unit change in the digital output.

Linearity -- The degree to which there is a linear relation between the analog inputs to
   an ADC and the digital outputs which it produces. Expressed as a percentage of
   the device's full-scale output or fraction of LSB.

Gain Temperature Coefficient -- Analogous to the definition given for DACs.

Conversion Time -- Specifies the length of the time interval required to perform an
   analog to digital conversion. This time may range from a few microseconds to a
   few milliseconds depending on the converter type and cost. Time required to
   activate an associated Sample/Hold unit may be included for ADC systems with S/H.

Input Voltage Range -- Voltage range over which an ADC's analog input may vary. ADCs
   may be designed to accept either unipolar or bipolar input.

Monotonicity -- A specification of the degree to which a continuously increasing analog
   input produces a continuously increasing digital output.

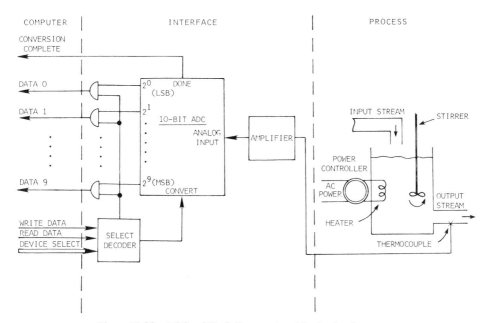

Figure 10.23. A Stirred Tank Temperature Monitoring System

generated by the thermocouple is amplified by an instrumentation amplifier before it is input to the ADC.

Normally when a computer uses such an interface to monitor the associated process, it does so through execution of a monitoring program. The only special requirement is that the program wait, after commanding a conversion, before sampling the outputs of the converter. The end of the wait interval will be signaled by a change in state of the ADC Done output, which might notify the computer that the conversion is complete by an interrupt.

10.2.4 Sample and Hold Devices

As we noted in the previous section, it often is necessary to sample an analog signal in order to maintain its value for a period of time, for example, to insure that the input to a successive approximation ADC is held constant during the entire conversion cycle. A device designed to perform such a function is known as a Sample and Hold circuit. Basically, a Sample and Hold Circuit may be considered to be an analog memory with a data input, a control input, and one output. Figure 10.24 shows a widely used S/H circuit consisting of two operational amplifiers, a resistor, a capacitor, and an analog switch. When the analog switch is closed (Sample = 1), the voltage across C and the output voltage V track the input V_{IN}. To hold a sample we have only to open the analog switch and allow the "plus" input of the second operational amplifier to float at the voltage appearing across C. This action results in maintaining the circuit's output, V, at the sampled voltage level.

Sample and hold circuits have a number of possible error sources. We simply list them here

with a brief definition or explanation as required:

(1) *Droop*–A decrease in V as a result of a loss of charge from C following the taking of a sample. This problem may be caused by leakage in the capacitor C, amplifier input current, or leakage through the solid state switch.
(2) *Offset*–Basically a problem involving the operational amplifiers in the circuit. The addition of a small voltage offset at the input to either amplifier results in an offset of V.
(3) *Gain Error*–Improper representation of the input voltage can result if the gain* for both amplifiers is not exactly equal to one.
(4) *Switching Offset*–If changing the analog switch from Sample to Hold mode has any effect on the gain of the second amplifier or on the quantity of charge on C, representation of the sampled voltage V will be in error.

In using a sample and hold circuit, several timing factors must be considered to insure that V represents the sampled V_{IN} at a particular time. Because the Sample and Hold circuit is a physical device, it has an inherent signal propagation delay associated with its operation. With the sampling switch closed, a change in voltage at V_{IN} produces a change in the output

---

*The *gain* of an amplifier may be defined as the ratio of the amplifier's output voltage (or current) to its input voltage (or current). Therefore, an amplifier with a gain of two, for example, produces an output whose magnitude is twice that of its input.

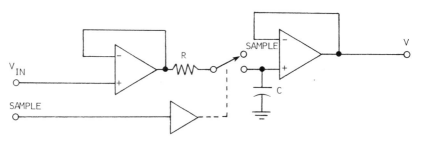

Figure 10.24. A Basic Sample and Hold Circuit

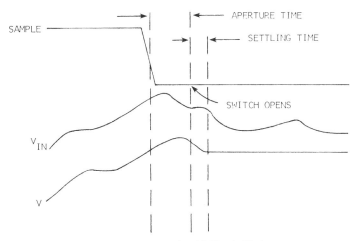

SAMPLE

APERTURE TIME

SETTLING TIME

SWITCH OPENS

V_{IN}

V

Figure 10.25. Sample and Hold Circuit Timing

V after a time interval commonly called the circuit's *settling time*. A second built-in time delay—this one associated with the sampling operation—is known as the circuit's *aperture time*. This delay is produced by the noninstantaneous operation of the sampling switch. That is, after a sample and hold cycle is commanded, the sampling switch requires a certain amount of time to open, thus allowing V to track V_{IN} for this delay period. Figure 10.25 illustrates the relationships among these factors. It is evident that we should not rely on the accuracy of the output V after issuing a hold command for at least a period of time given by

$$t = \text{aperture time} + \text{settling time}$$

This time interval is usually referred to as the *acquisition time* of the circuit. It can be seen that when such a circuit is used in conjunction with an ADC, the maximum conversion rate of the composite system is determined by the sum of the acquisition time of the sample and hold circuit and the conversion time of the ADC. Table 10.6 gives a brief listing of currently available sample and hold units. In many cases, commercial ADCs are available in configurations including sample and hold circuits.

10.2.5 Analog and Digital Multiplexing

Many times in process control applications there is need to have access to more data and control paths than the computer can accommodate. It also may be desirable to minimize duplication of expensive items of hardware anyplace where it seems that a single item might be shared. A technique known as *multiplexing* (also discussed in Chapter 5) allows for economical time sharing of system data paths

Table 10.6. Sample and Hold Circuits

Description	Aperture Time	Acquisition Time	Accuracy	Input Range
1. Monolithic integrated circuit sample and hold. Requires an external hold capacitor.	100 nsec	6 µsec	0.01%	±10 Volts
2. Modular sample and hold. Includes an internal hold capacitor.	10 nsec	100 nsec	0.1%	±10 Volts
3. Hybrid sample and hold. Includes an internal hold capacitor.	20 nsec	1 µsec	0.01%	±10 Volts

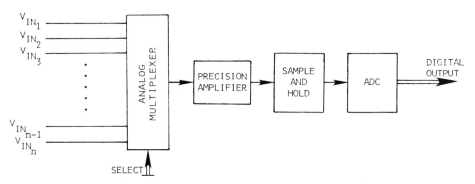

Figure 10.26(a). Multiplexing the Use of an Amplifier S/H and ADC

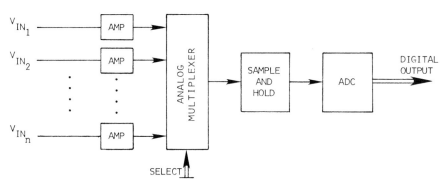

Figure 10.26(b). Multiplexing Remotely Amplified Low-Level Inputs

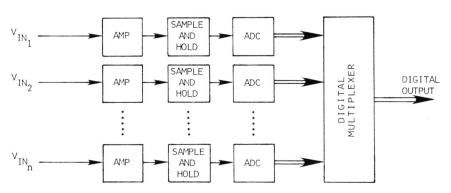

Figure 10.26(c). Multiplexing the Use of a Digital Signal Pathway

and hardware resources. We may, for example, wish to share a fast, precision (expensive) ADC among several different analog inputs. This section will discuss in a conceptual way several examples of commonly used multiplexing techniques.

Multiplexing may occur at several levels in an interfacing system, depending on the resources and signal paths whose use is to be optimized.

Figure 10.26 illustrates three ways in which multiplexing might be utilized in a multi-channel A/D conversion interface. In Figure 10.26(a) we have indicated that sharing might be desired in a system using a precision input amplifier and high-speed ADC. In this situation we would want to share the expensive hardware (amplifier and ADC) among all n inputs. This approach requires the use of a device known as an *analog multiplexer*, little more than an array of analog switches or relays with a digital decoding and driving network much like the example shown in Figure 10.27. Here we might obtain any one of a variety of

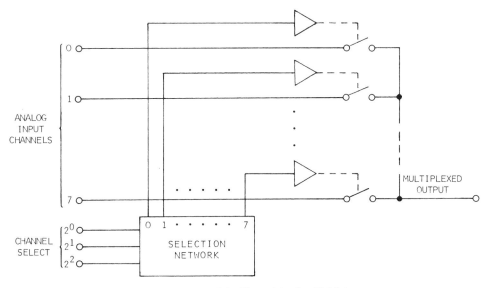

Figure 10.27. An Eight-Channel Analog Multiplexer

commercial units which will perform the same function.

In Figure 10.26(b) we have assumed that either the signal amplifiers are fairly inexpensive and not worth sharing, or that the characteristics of the signals we are dealing with are such that we must locate an individual amplifier remotely at each signal source. This last situation is a common occurrence in systems utilizing fairly low-level (less than 100-milivolt) signals such as are obtained from thermocouples. Signals at these levels should be amplified as near as possible to their source to prevent their corruption by incidental electrical noise. It may be noted that we have also multiplexed a sample and hold circuit in both of these examples.

Figure 10.26(c) introduces a somewhat different type of multiplexing. In the previous situations our goal was to share expensive hardware; in this application, however, our desire is to multiplex the use of another potentially expensive resource, a digital signal pathway. Devices that perform this sharing function are known as *digital multiplexers*. Internal signal switching in a digital multiplexer is performed by complex logical networks and is controlled by a selection network similar to that in an analog multiplexer. A digital multiplexer might be used in a variety of ways: providing the capability to input a large number of digital signals, to

individually output the contents of several registers to a shared data bus, etc.

An example of analog and digital multiplexing is shown in Figure 10.28 in what is commonly called an *analog data acquisition system* (though several of the systems previously discussed might be given the same name) or *analog front end*. Data gathering in this system is fairly automatic once initiated. This interface has the ability to transfer acquired data to the computer's main memory by way of a DMA channel. Initially, the computer must provide the acquisition system with several pieces of information to specify its mode of operation—as depicted, a channel scanning range and the starting address for the DMA storage buffer in main memory are required. Storage for this information, provided in the interface, is accessed through use of reserved I/O instruction codes. A channel scanning range is defined by two numbers—one for the first or lowest numbered channel in the range, the second specifying the last channel to be scanned. A single channel may be specified by defining the beginning and ending range number to be the desired channel. Additionally hardware must be included to determine and control the status of the interface, usually by making a control and status word available to the computer. In this interface we have reserved one instruction code to allow access to a single-word register (part of

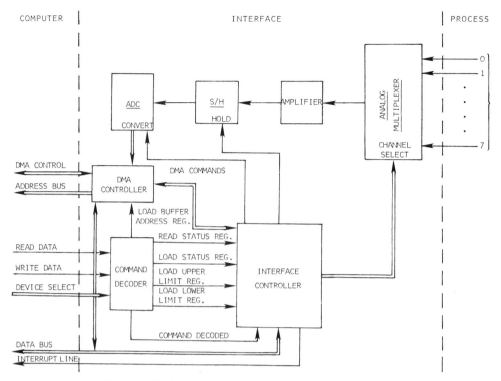

Figure 10.28. An Eight-Channel Analog Data Acquisition System

the interface controller). By accessing and modifying or checking this word we may initiate or stop the data collection operation or determine the present status of the interface.

This system employs both analog and digital multiplexing, the analog portion obviously through the analog multiplexer. Where digital multiplexing is used in this system is not quite as obvious; however, notice that the computer Data Bus is shared by the DMA Controller *and* the Interface Controller. This data pathway must carry information between the DMA Controller and the computer's main memory and between the system's status and control registers (status register, upper and lower limit register, etc.) and the computer's accumulators. All operations occurring within this system are initiated, synchronized, and controlled by the Interface Controller, an element that might consist of a fairly complex digital network or possibly a microprocessor (see Section 10.4).

A Data Acquisition System similar to the one just described may be used in a variety of ways. Suppose we wish to measure the temperature profile of a pipeline carrying a hot fluid, per-

haps a crude oil pipeline. Eight temperature sensors are embedded in the pipe at intervals, and, as shown in Figure 10.29, each sensor has been provided with its own amplifier (located remotely for reasons discussed previously). Assume that it is necessary to generate a profile of the pipeline covered by sensors 1–5 only. The data acquisition process would proceed as in Table 10.7.

Data acquisition systems much like the one we have described are available as commercial units. Most of these allow for the scanning of a larger number of analog input channels (some have provisions for 256 or more channels), and some include a number of DAC output channels as well. Because the Data Acquisition System is essentially self-sufficient once started, such systems are sometimes located remotely, close to the process, and communicate with the control computer by way of a digital serial or parallel data link. This sort of arrangement eliminates, of course, the need for a DMA Controller (at least one contained within the Data Acquisition System itself).

The reader should be aware that even though

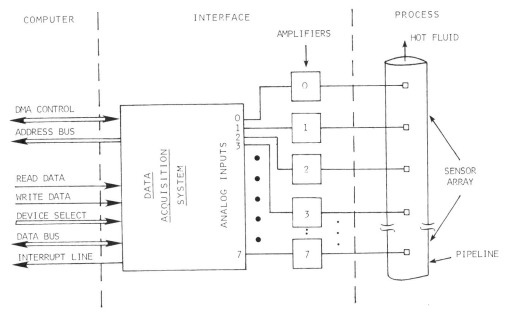

Figure 10.29. A Temperature Profile Monitoring System

multiplexing of control and data signals usually results in a less expensive solution to the problem of transferring data in a multichannel system, other problems may be encountered. As might be expected, the major problem with multiplexed systems is that resources must be time-shared, and this situation rarely allows a data collection system to operate at its maximum speed. Hence we must sometimes pay the price of including extra hardware in order to insure that a required data transfer rate is in fact achieved.

10.2.6 Pulse Generators and Counters

Several transducers that are used with real-time systems either require a pulsed input or generate a pulsed output. In this section we will dis-

Table 10.7. Operation of the Data Acquisition System

Under Program Control

1. Load the DMA Buffer Address Register with the starting address of the DMA buffer in main memory.

2. Load the Upper Limit Register with 5.

3. Load the Lower Limit Register with 1.

4. Load the Status Register with a code commanding a single scan cycle (the Interface Controller will generate an interrupt at the end of this cycle).

5. Wait for an interrupt.

Interface Controller

1. Save the contents of the Lower Limit Register in the Current Channel Register.

2. Select the analog channel specified by the contents of the Current Channel Register, sample the voltage (S/H) and convert it.

3. Initiate operation of the DMA Controller to write data from the ADC into memory at the address defined by the contents of the Buffer Address Register (Register incremented following each write).

4. Increment the Current Channel Register. If the result is less than or equal to the contents of the Upper Limit Register return to Step 2; otherwise generate an interrupt.

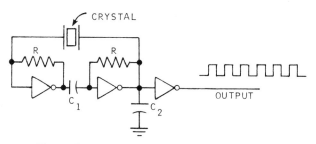

Figure 10.30(a). A Crystal Controlled Ring Oscillator

cuss the operation and uses of a *pulse source* or *generator* and of a *pulse accumulator* or *counter*.

Pulse Generators. A pulse generator (as we will define the term) is nothing more than a circuit whose output consists of a train of one or more digital pulses. Figure 10.30(a) illustrates a circuit, called a ring oscillator, that can be used to generate a continuous or free-running train of pulses at a frequency determined by the resonant frequency of the crystal included in the circuit. Normally, such oscillators generate pulse trains at a frequency much too high to be of use in most real-time applications. To eliminate this problem we can connect the output of the pulse oscillator to an N-bit binary counter circuit as shown in Figure 10.30(b). The effect of this counter is to divide the original oscillator frequency by 2^N. In other words, the counter will produce one output pulse for every 2^N input pulses. Thus, we can obtain a pulse train of practically any frequency we wish through the use of a suitable counter.

To complete the design of the Pulse Generator we need to add additional circuitry that will both help generate pulses of a desired width and allow us to control when the generator begins generating pulses. Both functions can be accomplished through the use of a single monostable multivibrator or "one-shot" (Chapter 7). This device, when enabled, generates a single output pulse for each input pulse, with pulse width determined by associated timing components (R and C). (We should note that this pulse generator design is intended as an illustration only; many other approaches exist for pulse generator designs, including one making use of a single one-shot and a few logic gates, and a second that uses the computer itself in generating pulses through programmed manipulation of digital output lines).

Pulse generators of a similar sort may be utilized in a variety of ways. A common example of a device requiring a pulsed input is an electric stepping motor. This type of electric motor is constructed so that its output shaft rotates a precise angular distance for each pulse received (the motor actually requires a multiphase drive signal, but we assume that motor drive electronics are included with the motor). The pulse generator discussed in the previous paragraphs may be used in combination with a binary counter and a small number of gates in building an interface that, for example, controls the position of a motor operated valve as indicated in Figure 10.31. To

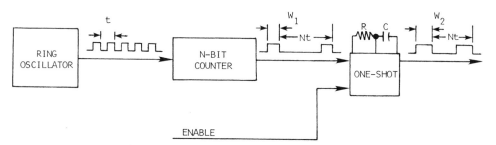

Figure 10.30(b). A Pulse Generator with Output Enable

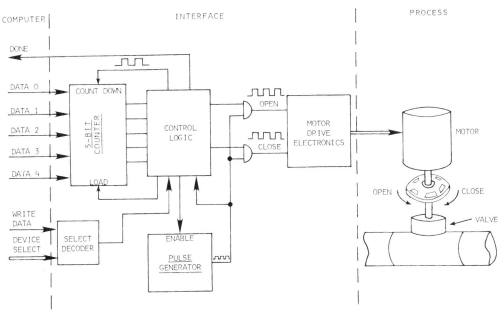

Figure 10.31. Stepping Motor Valve Control

operate the valve we simply specify the number of incremental units of valve position to move and the direction of change. This specification may be given in the form of a 5-bit sign magnitude number. For example, a positive position number might specify opening of the valve and a negative one specify closing of the valve. When a number is loaded into the Position Counter, a sequence of events is initiated (by the interface control logic) to command the stepping motor to move one increment each time the pulse generator causes the counter

to be decremented (by way of the control logic) by one. Because this system does not provide for feedback, to indicate the actual position of the valve, the program used to drive the hardware must maintain its own positional information in the computer's main memory.

Pulse generators have also been widely used to transmit control information by a technique known as pulse modulation. Suppose we want to control the voltage applied to a remotely-located heating unit as illustrated in Figure 10.32. Included with the heating unit is a re-

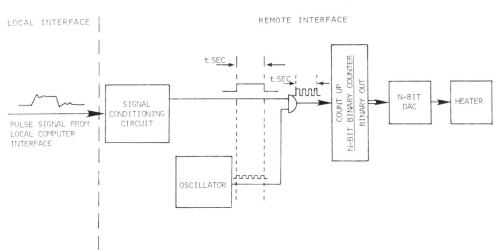

Figure 10.32. Pulse Modulation Level Control for a Remote Heater Interface

ceiving interface that contains a device that may be called a pulse to analog converter. This device integrates received pulses to generate an analog voltage that is then used as input to a power regulator driving the heating element. Thus to specify the desired operating level or incremental increase in level we have only to transmit a pulse whose width is proportional to the level change we want to effect. The integrating hardware is arranged in such a way that its operation is cyclic. That is, if the heater is operating at maximum power, a short input pulse will cause the heater control voltage level to return to zero (minimum heater power). In effect this allows the decrease or increase of heater power using a single pulse input from the computer. Generally in real applications, some form of analog feedback has been used to report the true value of the controlled variable (voltage).

Pulse Counters. Several velocity and liquid flow rate transducers exist that generate a pulse train whose frequency is proportional to the measured parameter. One way of translating these pulse trains to meaningful data is illustrated in Figure 10.33. In this example we want to measure the speed in revolutions per minute of an operating DC motor. It will be assumed that a continuous indication of the motor's speed is not needed. The motor includes an integral electromagnetic pulse-generating system consisting of a magnetic sensor with an associated magnet embedded in the motor shaft so that a pulse is generated each time the shaft magnet passes beneath the sensor. The motor's speed of rotation then can be measured by counting in a Pulse Counter (i.e. by integrating), pulses received from the motor for a fixed period of time. The interface is arranged so that when a measurement is initiated, the Pulse Counter immediately is reset and then incremented by pulses received from the motor for the fixed period of time. At the completion of this operation the interface provides an indication (Done) of data readiness. The reader will notice that the Pulse Counter in this circuit is nothing more than a binary counter and some control logic. This counter could have also been constructed in such a way that it provided a BCD (Binary Coded Decimal) output, a commonly-used alternative. If required, the computer itself can be used as a pulse counter through its digital input system. Assuming that the computer can afford the

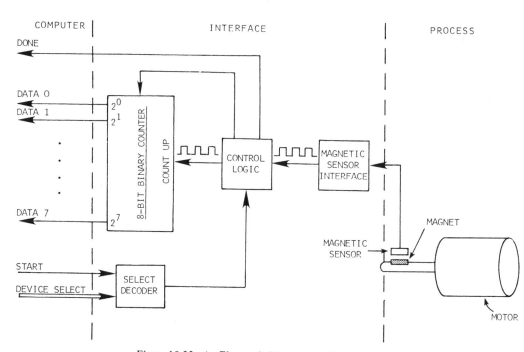

Figure 10.33. An Electronic Motor Speed Measuring System

extra computational load, a pulse input line could be allowed to generate a series of interrupts which then must be accumulated under program control.

10.2.7 Hardware Timers

In the last section we observed one use for a digital pulse counter. By changing our point of view slightly we can "manufacture" a new interface component from this counter known as a Hardware Timer. To make a Hardware Timer from a Pulse Counter we have only to force the counter to count *down* from a preset value at an exact, fixed rate. If the countdown rate is fixed at one count per millisecond (or any other division of a second) we can use the Timer to time real-time events that are expected to last for some multiple of this time unit. Normally Hardware Timers operate in the following way: The Timer is loaded with a binary number specifying the number of time units in the desired interval and started. The computer may now return to other processing tasks; when the Timer has been decremented to zero, its associated control circuitry will cause an interrupt to be issued notifying the CPU that the timer has "timed out", i.e., that the interval has elapsed. Upon recognition of the interrupt the computer can perform whatever processing task was scheduled in response to the Timer's notification.

Hardware Timers normally derive their timing or clock signal from a continuously running, highly accurate pulse generator, which might provide multiple outputs at intervals of thousandths, hundredths, tenths, and whole seconds. The user selects the clock rate, either by external switch or under program control. Timers much like the one described here may be purchased as commercial units in a variety of packages, for example, as a single unit or as an array of several timers. Hardware Timers may be used to time maximum-allowable interrogation-to-response delays, control the duration of pulsed outputs (when using the computer as a pulse generator), and provide for the execution of some real-time task periodically, among numerous other applications.

10.2.8 Programmable Gain Amplifiers

Because output voltage ranges of transducers used in data acquisition and control applications vary to such an extent, it is sometimes necessary to change the transducer output in such a way that it comes within a standard range (possibly the input range of an ADC). For example, suppose that we know that several input transducers generate voltages in the ranges 0–0.1 volt, 0–0.125 volt, and 0–0.5 volt. Additionally suppose that it is required that the system data acquisition hardware, which includes an 11-bit, 0–10-volt ADC, be able to resolve at least a 1% (full scale) change in any one of the transducer voltages. This means that the system must be able to detect a 0.001-volt change, a 0.00125-volt change, and a 0.005-volt change in the outputs of each of the three types of transducers to be used. The 0–0.5-volt transducer presents no problem, since the resolution of our ADC is 4.9 millivolts. However, it would be possible to resolve only a 4.9% or a 3.9% change in the voltages produced by the 0–0.1-volt and 0–0.125-volt transducers. One solution to this problem is to add a Programmable Gain Amplifier, shown in Figure 10.34, to the input of the system ADC. This amplifier may be externally programmed to have a gain of 1 to 15. To specify the desired gain the appropriate analog switches are caused to be closed, thus effectively selecting from a network of resistors the value of the feedback resistor in the first stage of the amplifier. Notice that the input to each switch buffer includes an inversion "bubble" to indicate that a "0" must be applied to the gain programming inputs whose associated switches must be closed.

For example, to obtain a signal gain of 3, switches S2 and S3 would be closed (Gain = 0011) leaving the R and 2R resistors in the amplifier feedback loop. To allow for the required resolution of the output voltages of the three types of transducers, the outputs of the 0–0.1-volt, 0–0.125-volt and 0–0.5-volt transducers should be amplified by factors of five, four, and one, respectively. A computer program controlling this process would simply route the appropriate analog input channel (through an analog multiplexer) to the pro-

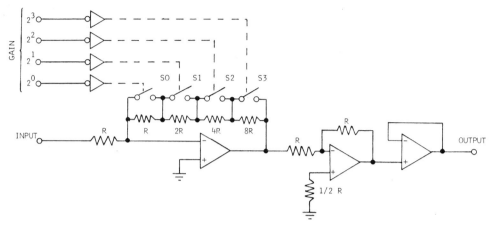

Figure 10.34. A Programmable Gain Amplifier

grammable amplifier, determine the required gain for that particular channel, output it to storage in the interface controlling the amplifier gain, and finally command the required conversion to take place.

Amplifiers much like this one are included in many currently-available analog data acquisition systems. Several of these systems include ADCs with sample and hold and a programmable gain amplifier, in addition to several DACs and possibly even a limited digital I/O capability.

10.3 MICROPROCESSORS—THE COMPUTER AS AN INTERFACE ELEMENT

At various points in Section 10.2 we spoke of an interface control unit or sequencer. Most of the interfaces that were used as examples were simple, and their control units were not required to perform very complicated functions. If real interfaces are considered, however, we will find that an interface control unit can be very complex and contribute a large part of the total cost of the interface. In addition, another problem often is faced—inflexibility of the control unit. Even simple changes in the operating procedure of the interface may require drastic and expensive changes in the hardware that comprises the control unit.

The advent of the microprocessor provided a solution to the problem of interface complexity and inflexibility. A microprocessor is a stored-program processing element that may be compared in a general way to a conventional computer. Most devices available currently have a limited instruction and data word size (4, 8, or 16 bits) and usually require external program memory to begin operation. Several microprocessors now in wide use also include both RAM and ROM memory on the processor chip (integrated circuit). In most cases processors are contained within a single or a few integrated circuit packages—requiring only external memory, clock generator circuitry, signal line buffers, and hardware to facilitate I/O. Instruction sets for many microprocessors are comparable in the number and kinds of operation that are available in instruction sets of many minicomputers.

There are, in general, three types of microprocessors available at the present time. The first and the most widely used, is what might be called a general-purpose processor. Presently most of these processors operate on an 8- or 16-bit word and have a fairly general instruction set. The second variety of processor, usually a 4- or 8-bit machine, is basically control-oriented i.e., it has an instruction set in which most emphasis is placed on control or I/O-related activities. The third type of processor is generally known as a bit-slice microprocessor, getting its name from the fact that it is built up from a number of basic modules, each of which handles only 2 or 4 bits, so that they may be thought of as slices of a machine with an arbitrarily large instruction word. These machines have no instruction set that could be compared to the other two kinds of processors; operations are controlled by their instruction sets at a much more basic level. However, groups of instructions for these machines (microinstructions) may be used to construct instruction sets (macroinstructions) that are equivalent to and in many cases faster-executing than instructions possessed by the other two processor types.

Microprocessors are used in data acquisition and control interfaces to act as control units and to provide capability of preprocessing (arithmetically or logically) input data, the so-called front end. Microprocessors may also be used to distribute real-time functions remotely, as will be discussed in the next section. In both ways we allow our central computer to perform supervisory tasks and free it from the responsibility of actually gathering process data and outputting control data. All that is required in this case is that the control computer be able to communicate with the distributed (micro-) processors and specify exactly what their functions are to be.

As a specific example of the application of a microprocessor consider the use of an 8-bit machine as the Interface Controller in the Data Acquisition System discussed in Section 10.2.5 (Figure 10.28). We might implement the controller as shown in Figure 10.35. In this case the controller consists of the microprocessor, program memory (ROM/RAM), interface registers and latches, bus buffer circuits, and a clock oscillator. The Upper and Lower Limit Registers and the Current Channel Register (see Section 10.2.5) could be implemented as memory locations in the system RAM or as accumulators in the processor. All controller functions are built into a computer program, stored in the system ROM, which is continuously executed by the processor. It is this program that gives the interface its "identity"; i.e., basic changes in interface operation may be made simply by changing the structure of the program. For example, suppose that it is desired that channels sometimes be scanned in nonsequential order. The interface control program could be modified to scan only those channels stored in

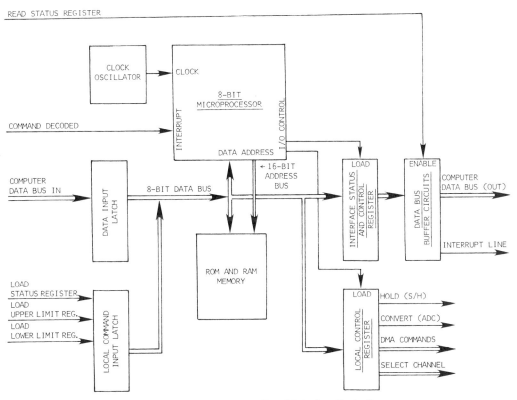

Figure 10.35. A Microprocessor-Based Interface Controller

a list in RAM as part of the scanning start-up procedure. In fact, the choice of sequential or nonsequential (random) scanning might be specified through the Status and Control Register.

Notice that the Interface Controller processor initiates interface functions on an interrupt-driven basis. Each I/O instruction addressed by the host computer to the Data Acquisition System Interface causes an interrupt to be generated in the microprocessor by way of the Command Decoded signal line. This is a common method used to drive the operation of a microprocessor-based system. An interrupt in some other type of interface could just as easily be generated through the actuation of a pushbutton or toggle switch. In fact, for a more complicated "configurable" interface we could add a command keyboard which would allow the user to define, in a simple way, the operation mode or configuration the interface is to assume.

As mentioned above, microprocessors can be used as arithmetic preprocessors or front ends in some applications to relieve the main computer of some of the more mundane data conversion tasks. For example, in the Data Acquisition Interface of Figure 10.28, a microprocessor employed as a DMA Controller might be used not only for interface control functions, but also as a preprocessor of data received from the ADC. Such processing could consist of anything from simple scaling operations to more complex integer-to-floating-point number conversions.

10.4 DISTRIBUTED CONTROL INSTRUMENTATION SYSTEMS

One relatively recent development in real-time computing involves the wholesale application of microprocessor-based systems to the manufacturing and process control fields. The approach has not been the traditional one, i.e., to start with a central process control computer and design downward, but rather to begin with the bottom-level units in the system, i.e., the analog control instrumentation and/or digital logic units, and design upward. The basic intent is to replace these traditional units, on virtually a one-for-one functional basis, with more flexible digital systems. Although the capabilities of high-level computers often exist in such systems, most commercial digital instrumentation systems are not oriented toward taking advantage of such features.

We should note that real-time data acquisition and control equipment can be distributed both geographically and functionally. Geographical distribution, as we have noted, permits the equipment to be placed near the process. Data concentration and serial digital data transmission used in such systems usually bring about a large reduction in wiring costs, a big advantage in many commercial facilities. Functional distribution allows the equipment to be partitioned in a logical way at the component level. Hence separate units can be supplied that are designed specifically for data acquisition, multiple single-loop control, batch process sequencing, data transmission, operator display, etc. The major advantages of functional hardware distribution are flexibility in system design, ease of expansion, and reliability.

A representative installation of a distributed instrumentation system is depicted in Figure 10.36. Such a system might consist of one or more of the following elements interfaced to the process:

(1) *Local Control Unit (LCU).* Typically, this unit can implement 8 to 16 individual control loops, with 16 to 32 analog input lines (a multiplexed ADC), 8 to 16 output signals (DACs with 4–20 ma current output, digital to Triac converters, or pulse outputs), and a limited amount of digital input, output, and internal logic capability.

(2) *Data Acquisition Unit.* Generally, this unit contains two to three times as many analog input channels as the LCU. This device is an intelligent remote multiplexer of the type discussed in Section 10.3.

(3) *Batch Sequencing Unit.* Typically, this unit contains a number of external event or timing counters, arbitrary function generators (e.g., for preprogrammed

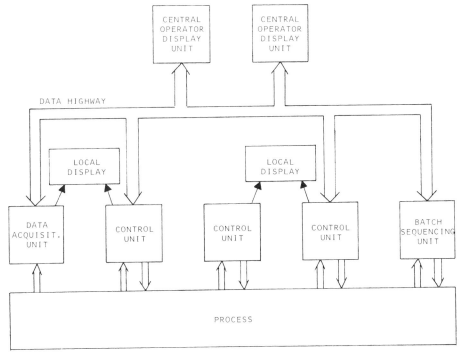

Figure 10.36. A Distributed Digital Instrumentation System

logical control of process elements), and considerable digital input/output and internal combinational/sequential logic capability.

(4) *Local Display Unit.* This device generally will contain only analog display stations and, perhaps, analog trend recorders. Alternatively, a CRT display may be used for printout of selected variables.

An instrumentation system might be purchased that contains only one or two elements interfaced to the process along with a local display unit. A big advantage of these systems is that the user can start out at just such a low level of investment. However, the true benefits of the distributed hardware accrue when a Data Highway is used to link the local process units with a Central Operator Display and, perhaps, a high-level computer. The data highway, a serial digital data transmission link, may consist of just two wires or a coaxial cable, depending on transmission speeds. Some commercial systems allow for redundant data highways, to reduce

the possibility of a loss of data to the display and operational elements higher in the hierarchy. However, one advantage of this type of architecture is that complete loss of the data highway will not cause a complete loss of system capabilities. Often local units will continue operation with no significant loss of function over moderate or extended periods of time.

The Central Operator Display typically will contain one or more consoles for operator communication with the entire system, multiple CRT units for video display, graphic display units, and hard-copy output devices. It should be noted that the high-level computer shown in Figure 10.36 accesses the process only through lower-level units, not directly. Consequently, while individual control loop set-points can be changed and batch sequences can be modified, it may not be possible to implement particular, user-designed, control functions or algorithms as with a completely general-purpose system.

In implementing distributed instrumentation systems, vendors make copious use of microprocessors, generally at least one in each separate unit or module. Since each module is designed to perform a reasonably small subset of all system functions and to deal with a small subset of process inputs and outputs, a table-driven programming approach (see Chapter 16) can be used efficiently to implement system programs within the lower-level microprocessor-based modules. These programs typically are not accessible to the user; i.e., they represent "firmware" stored permanently in read-only memory that is used only by the local microprocessor.

The design philosophy employed in structuring such systems is to supply the user with many of the benefits of digitally-based hardware without requiring the user to know how real-time digital systems operate. Their commercial success is in fact a function of how transparent the internal workings are to the operator and to the process engineer. Nevertheless, the most successful use of digital instrumentation systems requires the user to understand both their significant advantages and their obvious limitations.

10.5 INTERFACE BACK-UP OR REDUNDANCY

In some data acquisition and control applications, the integrity of connections between a computer and a process through the associated interfaces is of such importance that failure of these links could have catastrophic effects. In such cases duplicate (and perhaps triplicate) process interfaces might be employed to insure uninterrupted communications. This technique is commonly known as interface back-up or redundancy.

Figure 10.37 illustrates the interconnection of a singly-redundant control system. In this system a third device called a Monitor is used to oversee the operation of the two duplicate control interfaces and to route the output of the primary unit through the switching network. The monitoring function could be accomplished, for example, by comparing control voltage outputs generated by each unit. Wide disparity in these voltages would be interpreted as a sign of a malfunction in one of the units and would cause the Monitor to determine the location of the fault. If the fault occurred in the primary unit, the Monitor would cause the secondary unit to assume control. Regardless of the location of the fault, the Monitor should notify the control computer, and possibly an operator, of the occurrence. Notice that an inherent problem in the configuration pictured in Figure 10.36 is that the system is not completely redundant, since the Monitor and Switching Network are not duplicated. Failures in these components might go undetected with unwanted results.

Interface redundancy need not involve auto

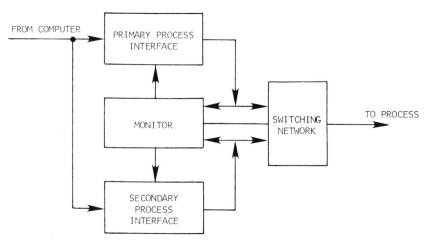

Figure 10.37. A Redundant Process Interface

matic switching between interfaces; in many cases notification of an operator might be sufficient. The operator, in this case, would perform the actual fault location and switching decision functions in response to an alarm originating in an interface or computer. Surprisingly, however, this might not be the most economical solution (even assuming that the process interface hardware is expensive), since the cost of paying the human fault locator and the cost of lost time in process operation might overshadow the cost of providing redundant interfaces in all critical computer/process links.

10.6 SUMMARY

In this chapter we have seen that a process interface may be discussed independently of the computer with which it is used. We progressed from a basic discussion of interfacing philosophy to a detailed treatment of each of the most important interfacing or peripheral components. The reader should have obtained from this discussion a general feeling about interface structures that might be used in particular applications. The goal here has been not to teach interface design, but instead to impart a broad knowledge of the techniques and hardware available to the system user. Hopefully, such knowledge will avoid any duplication of effort in developing already-existing hardware and will, in addition, help remove the aura of "magic" from specification and operation of the process interface.

10.7 SUGGESTED READING

1. *Analog–Digital Conversion Handbook*, Analog Devices, Norwood, Mass., 1977.
2. Bibbero, R. J., *Microprocessors in Instruments and Control*, John Wiley & Sons, New York, 1977.
3. Block, W. W., *An Introduction to On-Line Computers*, Gordon and Breach, New York, 1971.
4. Clulley, J. C., *Computer Interfacing and On-Line Operation*, Crane, Russak and Co., New York, 1975.
5. Harrison, T. J., *Minicomputers in Industrial Control*, Instrument Society of America, Pittsburgh, Pa., 1978.
6. Hnatek, E. R., *A User's Handbook of D/A and A/D Converters*, John Wiley & Sons, New York, 1976.
7. Johnson, C. D., *Process Control Instrumentation Technology*, John Wiley & Sons, New York, 1977.
8. Korn, G. A., *Microprocessors and Small Digital Computer Systems for Engineers and Scientists*, McGraw-Hill, New York, 1977.
9. Leventhal, L. A., *Introduction to Microprocessors: Software, Hardware, Programming*, Prentice-Hall, Englewood Cliffs, N.J., 1978.
10. Malmstadt, H. V. et al., *Digital and Analog Data Conversions*, W. A. Benjamin, Inc., Menlo Park, Cal., 1973.
11. Millman, J. and Halkias, C., *Integrated Electronics: Analog and Digital Circuits and Systems*, McGraw-Hill, New York, 1972.

10.8 EXERCISES

1. Two different approaches can be used in bringing analog signals into a real-time computer:
   a. Place an ADC near the analog signal source (process) and transmit binary results from the converter to the computer.
   b. Place an ADC near the computer and transmit analog signals from the process to the computer.
   Describe circumstances under which method a might have advantages over method b and vice versa.
2. Show that the full-scale current I_{max} at the point marked Σ in the Ladder DAC of Figure 10.11 is E/R.
3. Suppose that you have been assigned the task of interfacing an optical scanning system to a small digital computer. The scanner is used to encode information from a photograph, measuring the "degree of whiteness" at each of a number of points and storing this information in the computer's memory for later reconstruction of the photograph. The scanning system has the following characteristics:
   a. Two stepping motors are used to drive the single light sensing device, one for the x direction and one for the y direction. A single pulse applied to a motor input causes the light sensor to be advanced one step on the selected axis. A pulse on a separate reset line provided for each motor causes the selected

motor to return the sensor to the left-most or bottom position in the scanning area.

b. A normal step requires 125 milliseconds to complete on either axis. A reset operation requires 650 milliseconds, again for either axis (both motors may be reset simultaneously).

c. The light-sensing device, used to measure the degree of whiteness, continuously outputs an analog voltage in the range 0–10 volts (0 volts = black, 10 volts = white).

d. A fixed black frame is placed around the photograph to be scanned, and the sensor is calibrated so that it outputs 0 volts if it is positioned over the frame.

   i) Describe how you would interface the scanning system to a computer. Pay particular attention to the specification of the types of interfacing components used and their characteristics.

   ii) If it is required that measurements of degree of whiteness be made using a grid that is 1024 points by 1024 points, what is the minimum scanning time required? Assume that 1024 steps is just sufficient to allow a complete scan of the area within the frame. Explicitly state any additional assumptions which you make.

   iii) Discuss sources of error in the scanning system and its interface, and suggest how these errors could be eliminated or, at a minimum, detected.

   iv) Would there be any advantage in designing a special interface for the scanning system rather than using the standard interfacing components discussed in this chapter? Discuss.

4. A device known as a Voltage Controlled Oscillator (VCO) operates in such a way that it continuously produces at its output a logic-compatible square wave signal (sequence of pulses) at a frequency governed by an analog voltage (V) applied to its single input. The frequency of oscillation (f) for the VCO is given by the relation

$$f = V/(1.0 \times 10^{-2})$$

where f is given in Hertz (cycles/sec) and V in volts ($0 \leqslant V \leqslant 10.0$ volts). A simple Analog to Digital Converter may be constructed using this VCO, a hardware or software binary counter, and a small amount of control logic (implemented in hardware or software).

a. Assuming that an ADC is implemented as described above (using the VCO), what is the conversion time for such a device if a resolution of 0.039 volt is required?

b. If an ADC is implemented using a VCO, what is its resolution as a function of conversion time?

c. Draw a suitably-labeled diagram showing how the VCO could be interfaced to a computer. Describe the operation of a program that could make use of the VCO in measuring an analog voltage.

d. What type of ADC has been described in this question (i.e., integrating, feedback, or parallel)? Explain.

5. Suppose that the following hardware components are obtained and connected as shown in the drawing below:

a. A 13-bit two's complement ADC (input range $-10.0 \rightarrow +10.0$ volts).

b. A four-channel analog multiplexer whose output is connected to the input of the 13-bit ADC.

c. A Programmable Gain Amplifier or PGA (gains available range from 1 to 16). This device has been configured in such a way that the voltage appearing at its output C is the difference between the voltages at its two inputs, A and B, multiplied by the programmed gain.

d. A 13-bit two's complement DAC (output range $-10.0 \rightarrow +10.0$ volts).

   i) What is the resolution of the ADC and DAC in volts? If the maximum conversion error produced by the ADC (DAC) is one-half the voltage represented by a one in the least significant bit of its output (input), what is this error in volts?

   ii) Suppose that it is desired that voltage measurements be made to 16-bit accuracy, what voltage resolution is required? Propose a method that makes use of the hardware described above in obtaining voltage measurements to 16-bit accuracy.

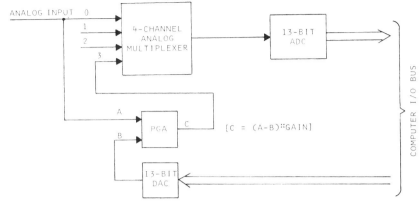

Figure for Exercise 5

You should *completely* describe the way in which each measurement is made.

iii) Using the voltage measurement technique proposed in part (ii), what is the smallest resolvable voltage increment? State any assumptions which you make.

Note: Assume that the values obtained from the ADC (or sent to the DAC) are left-justified on the 16-bit I/O Data Bus (i.e., the three least significant bits in the 16-bit word obtained by reading the ADC output are always 0).

6. Suppose that you wish to use a real-time computer system to control the flow of a liquid in three separate pipelines. Each pipeline includes a turbine flowmeter to measure liquid flow. This device operates in such a way that it generates a sequence of binary pulses (a pulse is a $0 \rightarrow 1 \rightarrow 0$ transition sequence) whose frequency of occurrence is proportional to the average value of the flow in its pipeline.

a. What kind of interface may be used to connect the flowmeters to the computer system? Justify your answer.

b. How could the frequency of a pulse sequence originating from a particular flowmeter be measured?

7. Suppose that you made use of a 12-bit, two's complement output, successive approximation ADC with a conversion cycle time of 10 microseconds in measuring an analog voltage originating from a process.

a. What effect would the variation of the

voltage with time have on the measurement produced by the ADC? Explain.

b. If the ADC has an input range of -10 volts $\rightarrow +10$ volts, what is the maximum *rate* of variation in this input that could be tolerated if the usual successive approximation measurement accuracy (i.e., the accuracy obtainable from commercially available ADCs) is to be maintained?

c. How may the effect that you described in part a be minimized or eliminated?

8. A small 16-bit computer is located 1000 meters from a manufacturing process that it must monitor and control. The process generates the following signals:

- 10 thermocouple signals in the range 0–10 millivolts which provide measurements of temperatures within a furnace.
- 10 pressure transducer signals in the range 0–10 volts from transducers located in high-speed liquid flow lines.
- 10 binary alarm signals where 0 volts = no alarm and 5 volts = an alarm situation.

The computer is required to return 20 analog signals and 10 binary signals to the process, whose values are based on inputs received from the process (i.e., those listed above). Because of the relatively large distance between the process and the computer assume that the cost of transmitting each signal is high, but that the environment is electrically quiet (i.e., low electrical noise levels).

a. Describe how the process should be connected to the computer. Specifically, discuss the characteristics of the computer interface to be used, the signal transmission technique(s) employed, and the number of wires necessary.

b. If it was determined that the wires connecting the process to the computer existed in an electrically noisy environment, what changes, if any, would you make in your answer for part a?

9. If the Weighted Register DAC in Figure 10.10 were to have a BCD coded input, what changes would have to be made in the circuit?

10. How could you test the monotonicity of an ADC?

Part V
Real-Time Systems Software

11

Assembly Language Programming

D. Grant Fisher
Department of Chemical Engineering
University of Alberta

11.0 INTRODUCTION

One of the main reasons for the value and versatility of general-purpose digital computers is that they utilize the concept of a stored program of instructions. What the computer does is determined by the sequence of instructions that is "programmed" for each specific application. These instructions are stored in the memory of the computer in exactly the same way as any information or data which are processed. Therefore the instructions (programs) can be changed as easily and as rapidly as one set of data can be substituted for another.

The stored program approach gives tremendous power and flexibility, but it also means that the program must be written or created by someone. The digital computer hardware by itself has tremendous potential, but until it is combined with software (programs), it is of little use to anyone. This chapter looks specifically at the general concepts and procedures that apply to the development of a computer program so that a given digital computer will do the desired job.

Unfortunately, at the most basic level, programming depends both on the details of the computer hardware on which the program is to be executed and on the "systems software" that is available as part of the overall computer system. And since there are hundreds of different types of computer systems, each of which uses its own uniquely tailored set of executable instructions, it is impossible to write a completely generalized guide to assembly language programming. This chapter will therefore try to describe and illustrate the basic concepts of assembly language programming so that the reader can get a sound overall idea of what programming involves and how it interrelates with other aspects of computer systems. However, in writing an actual program the reader will have to refer to the documentation for the specific computer system he will be using.

We are going to be discussing computer programming at a very basic level—at the machine language and the assembly language level; hence we need first to define some general terms. For this purpose we can take a brief look at the operation of calculators. Modern electronic pocket and desk calculators are actually special-purpose digital computers, and, since they are so common, they provide an excellent means of illustrating hardware features and the basic concepts of programming. The key features are:

Basic Instruction Set. Calculators differ in capability, but the basic operations for any particular calculator are the functions that can be done in "one step", i.e., add, subtract, multiply, divide, ... If the calculator has square root capability, then the operator can use it where required; if not, he must go to another calculator which does have square root capability or develop his own sequence of basic operations which the calculator can perform that will produce an exact or approximate square root.

Data Input. Operands, i.e., numbers to be used in arithmetic calculations, are entered via a numeric keyboard and transferred to the "central processing unit" (CPU) of the calculator when the *Enter* key is pushed or, in some cases, when an operational key is pushed.

Data Output. The results of operations done by the CPU are usually displayed automatically, but keys are often provided to display specific pieces of information or to control the display, e.g., to display the contents of specific registers, or to fix the number of decimal places which will be displayed.

Storage or Memory. More powerful calculators have registers that can be used to store data (e.g., a frequently-used constant) or to store intermediate results. Pressing the *Store* key will put information into the memory, and the *Recall* key will bring information from memory to the central processing unit. (Some calculators use magnetic cards to store relatively large amounts of information.)

Programming. A calculator, although it has great potential, will do nothing until the user provides instructions (by pressing the appropriate operation keys) and data (i.e., operands). The sequence of operations or instructions that the user performs can be called a *program.* The program may originate with the user, or be provided by the manufacturer. Note that the program might be unique in the sense that it performs a specific, original task, but each step or instruction in the program must come from the basic instruction set of the calculator; i.e., the program can be original, but the instructions are fixed by the design and construction of the hardware.

If programs can be put into "machine-readable form", for example, on a magnetic card, then it is possible to use a calculator equipped with a magnetic card reader to read the desired program from the library of programs supplied by the vendor or prepared previously by the user and stored on magnetic cards or in memory chips. Obviously, if a program is to be used repeatedly, then it is worthwhile to use a (magnetic card) programmable calculator rather than punch the same sequence of operation keys over and over again on a manual calculator. In some cases, for example, statistical calculations, the pro-

grams might be so widely used that the vendor includes them as part of the "hardware" (perhaps in read-only memory), and they can be "called" by pushing one or more special keys. Data can also be prepared in machine-readable form, and magnetic cards are obviously a convenient means of handling large data sets that will be used repeatedly.

11.1 DESIGN BASIS FOR PROGRAMMING

In Chapter 8, the hardware and architecture of a general-purpose computer was described in detail, and an example computer was used to illustrate specific points. Unfortunately the basic points of system design are by no means standardized from one computer to the next and different manufacturers use different nomenclature as well. In this chapter we wish to discuss assembly language programming *generally*; hence we start with a brief review of basic computer concepts and as few assumptions as possible which might restrict generality. In some sections comments will be included to indicate alternative nomenclature and/or features.

(1) It is assumed that we are dealing with a general-purpose, stored program computer with internal fixed-word-length, binary data representation, and parallel transfers and operations. (A variation would be an 8-bit parallel machine that could operate on 1, 2, 3, or 4-byte data.)

(2) The operations performed by the digital computer are determined by a program stored in the memory of the computer— i.e., a sequence of instructions specified by the programmer, all of which are from the "basic instruction set" that can be executed by the computer hardware. This chapter deals with how to write such a program.

(3) As stressed in Chapter 6 on internal data representation, all information—including the executable instructions to the computer—must be in binary, fixed-word length form. Some computers allow

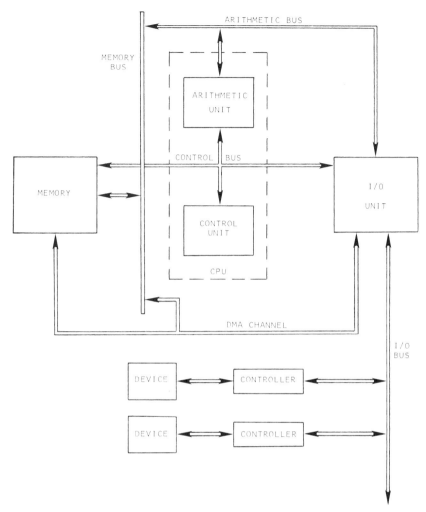

Figure 11.1. Basic Computer Architecture

mixtures of one- and two-word instructions.

(4) Discussion in this chapter will assume a basic system architecture similar to that in Figure 11.1. The main components are listed in the following paragraphs.

Central Processing Unit (*CPU*). This is the general name given to that part of the computer which

- Decodes and executes instructions
- Performs arithmetic and logical operations
- Loads/stores information from/to memory
- Reads/writes data from/to the Input/Output (I/O) bus

- Controls information transfers and system operation

The CPU normally contains one or more registers for each of the following operations:

- Arithmetic and logic operations (often called accumulators)
- Calculating the "effective address" of memory access operations
- Pointing to the next executable instruction (program counter)
- Malfunction and status indicators, e.g., "carry" or "overflow" from an arithmetic operation.

Notice that Figure 11.1 illustrates the decomposition of the CPU hardware into an Arithmetic Unit and a Control Unit. In general, the Control Unit regulates the execution of instructions and provides control signals needed in all portions of the computer, while the Arithmetic Unit manipulates the contents of internal registers to produce desired arithmetic and logical results.

Memory. All program instructions and data at time of "execution" (i.e., use by the CPU) must be stored in fixed-word-length binary form in main memory and are referred to by location— i.e., contents of the location defined by the "effective address". In simple computers the address is a number between zero and the total number of words in memory.

General arithmetic and logical operations are not usually performed in the memory. Information is "loaded" from memory or "stored" into memory under the control of instructions executed by the CPU (or by I/O controllers associated with external peripheral devices).

Information is usually loaded from or stored into memory one word at a time from/to the storage location specified in a memory address register (MAR). The address is supplied to the memory address register from the program instruction counter (PC) or the CPU operational registers.

Input/Output Unit. The I/O Unit contains hardware that serves to interface the CPU to the I/O Bus. If the word length of the computer is N bits, the I/O bus consists of N parallel paths to carry information between external peripheral devices and various parts of the computer system, plus a number of parallel lines for addressing and control. A computer may have one bus for both input and output; separate buses for input and output; separate buses for each peripheral or group of peripherals. Obviously the hardware architecture will determine the details of the instructions needed to control the input/output operations.

In some computer systems the memory unit(s) and the CPU(s) themselves are attached to a single bus in a manner that is logically equivalent to other major peripherals. Thus all operations

consist of a transfer from one device to another via the bus, and there is no real distinction between memory access operations and I/O operations.

In the simplest and most common mode of operation, usually referred to as "programmed I/O", information from an external peripheral passes—one word at a time—from the I/O bus, through the CPU, to the memory as directed by instructions executed by the CPU. In an alternative mode known as "direct memory access" (DMA), information is communicated directly between external peripheral devices and memory, bypassing the CPU. If an "input/output controller" or "I/O Channel" is available for use (as described below), direct memory access can occur simultaneously with regular CPU operations.

An important part of most I/O Units, the hardware interrupt system, allows one or more binary (i.e., on/off) signals (internal or external to the computer system) to be transmitted directly to the CPU, at the time they originate. An "interrupt" normally results in a "hardware-forced branch" instruction that directs the computer to a program provided to service that particular interrupt. A multi-level hardware priority interrupt system will let an interrupt through only if it is "higher priority" than the current status of the CPU. (Provision might exist to enable, disable, record, simulate, etc., interrupts.)

Input/Output Controllers. In order to reduce the load on the CPU and/or to permit more efficient (asynchronous, overlapped) I/O operations, some computers include I/O controllers. These are essentially special-purpose mini- or microcomputers dedicated to the control of I/O operations between memory and the I/O bus(es). After the I/O controllers are initialized by instruction(s) from the CPU, they control the transfer of blocks of data to one or more peripherals without disturbing the contents of the CPU registers. Except for "cycle stealing" by multiplexer type I/O channels (discussed later), CPU operations can proceed simultaneously and asynchronously with several I/O operations.

11.2 PROGRAM INSTRUCTIONS

11.2.1 Information Specified by Instructions

Instructions usually contain the following types of information explicitly, or it is implied by context or convention:

(1) To whom the instruction is directed.
(2) What the basic instruction is.
(3) Where the result is to be put.
(4) The operand(s) or information to be processed.
(5) Special options or variations of the basic instruction.

(1) *Whom?* In small computer systems, instructions are almost always directed to the instruction decoding and control unit of the central processing unit. Exceptions occur in more powerful computers that have I/O controllers with (limited) capability to interpret I/O instructions. Even when there are multiple CPUs in one computer system, the question of which instruction is executed by which CPU is decided by convention, e.g., where it is stored in memory, or by a special hardware or software "executive system". We will assume that all instructions are automatically directed to the correct part of the CPU and hence there is no need for the programmer to specify this explicitly as part of each instruction.

(2) *What?* The basic operations that a computer can perform are determined by the hardware design and construction and are specified in the vendor's documentation for that particular computer system. Instruction sets are not standardized and are often designed for particular types of applications. Table 11.1 lists various classes of instructions that will be discussed more fully later. In most cases the user must simply accept the list of assembly instructions implemented in his particular computer as the starting point for programming. (Microprogrammable computers, discussed later, are one exception.) Deciding which instruction set or computer is best for a given application is a difficult problem that requires a lot of experience.

(3) *Where?* Some instructions, such as STORE, by their very nature define the "destination" of the instruction result. In arithmetic or logical operations it is almost always assumed that the result stays in the CPU accumulator. In computers with multiple accumulators it is necessary to identify the accumulator that is to contain the result. Output instructions may appear to be an exception, but in later discussion we will show that they are really a transfer of information to a single destination—i.e., the I/O bus. However, the identity of the device that is to "pick up" the information from the bus must be specified, and this is one of the factors that make input/output operations more complicated than, say, arithmetic instructions.

(4) *Operand(s).* Most instructions involve one or two operands, i.e., data, numbers, or pieces of information on which the operation is to be performed; but some may involved "strings" of information or "blocks" of data as in input/output operations.

The LOAD, STORE, and TRANSFER instructions generally involve one operand that is to be transferred, say, from memory to a register in the CPU. When writing a program step-by-step, the programmer may know that the information to be transferred is the numeric value, 7, or he may refer to it by the symbol, SEVN. Internally, however, computers *always* refer to information in memory by the address or the location (word) in which it is stored. Thus if the variable "SEVN", which has the current numeric value 7 and is stored in memory location 129, is to be "LOADED" into the CPU, the instruction is equivalent to:

"LOAD the current contents of memory
     location 129 into CPU Accumulator #1".

As we will see later this is normally shortened to something like "LDA 129" and there are other, more powerful ways of specifying the "effective address" than writing the "absolute" value 129.

Some computers have only one register in the CPU that is capable of performing the full range of arithmetic, logical, etc., operations. It is normally called the accumulator and is specified implicitly. Thus an instruction like "SHIFT"

Table 11.1. Typical Digital Computer Instructions

1) ARITHMETIC (fixed point and floating point)

- Add
- Subtract
- Multiply
- Divide

2) LOAD, STORE OR TRANSFER

- Load register (from memory)
- Store register (to memory)
- Transfer from one register to another

3) REGISTER OPERATIONS

- Shift register right or left
- Shift and count or test overflow
- Rotate (shift) register plus extension
- Clear register

4) LOGICAL

- Logical AND
- Logical OR
- Logical EOR
- Compare
- Complement

5) PROGRAM CONTROL AND SEQUENCING

- Pause or Halt
- Branch (conditional or unconditional)
- Skip (conditional or unconditional)
- Branch and store Program Counter (subroutine branch)

6) INPUT/OUTPUT (and control of peripheral units)

- I/O using CPU registers
- I/O using direct memory access (data channels)
- Set/test/clear device status indicators
- Simulate interrupt

7) SYSTEM CONTROL AND COMMUNICATION

- Set Program Counter
- Supervisor Call
- Check/set status flags

8) BIT, BYTE or HALFWORD MANIPULATION

- Load Byte (or bits, or halfword, etc.)
- Logical OR byte

9) USER-DEFINED INSTRUCTION

- Microprogamming

10) TEXT OR STACK MANIPULATION

- Add/remove from top/bottom of stack (list)

or "COMPLEMENT" always refers to the accumulator, and hence there is no need to explicitly identify the single operand, since it is always "the current contents of the accumulator". For example, a complete instruction could be simply: "SR" (meaning Shift the accumulator one bit to the Right). When there are multiple accumulators in the CPU, then it is necessary to specify which one is involved. Note, however, that the number of accumulators is typically between 1 and 8, whereas there are thousands of memory locations, so that it takes less space in the instruction word (i.e., bits) to identify the accumulator than it does to fully specify a memory location.*

Some instructions such as "PAUSE" or "HALT" (which cause the computer to "pause" or "halt" rather than execute the next instruction in the program) have no operand in the usual sense.

Arithmetic and logical instructions generally involve two operands. One operand is almost always implicitly assumed to be in the accumulator. The second, depending on the design philosophy and hardware of the computer, will be either: (1) in a specified location in memory; (2) in a specified CPU register (accumulator).

Consider Examples 11.1 and 11.2 which ADD two integer numbers (stored in memory at Locations 121 and 122) and leave the result in the accumulator. Note that the actual format of the instructions will be different for different computers and will be considerably abbreviated, e.g., "LDA 121", when written as an assembly language statement.

With Computer B there is an additional LOAD instruction because it was assumed that all operands have to be in the CPU, whereas with Computer A one operand was specified by its location in memory. Thus, as written, the second case (B) would (probably) take longer to execute and requires a "longer program". Note that the difference is negligible when only one instruction is involved, but in a typical programming application this difference might arise thousands of times and hence be very significant. Seemingly trivial differences such as this can have a significant effect on the ease of writing programs, the time it takes to execute the program, and/or the space it takes to store the program in memory, and hence are an important consideration when evaluating different computers for a specific application. However, it is *not* possible to conclude that Computer A is better than B for all applications because other factors must also be taken into account, and it is not hard to think of an example where B would be "better" than A—e.g., when the two values were already in Accumulators #1 and #2 as a result of previous calculations and hence the first two instructions in Example 11.2 could be eliminated.

These examples point out how the programming of one computer can differ from that of another computer even when the task to be performed is the same, and how programming—at its most basic level—involves details of hardware and internal operations.

(5) *Special Options.* A small computer may have from 16 to 32 basic instructions, but options or variations of these basic instructions may increase the total number of different instructions to several hundred. The number of options available and how they are specified is a very important factor in evaluating and using computers, but it is hardware-dependent and hence difficult to generalize.

---

*We will follow the convention that a machine language instruction is written in the binary (or octal or hexadecimal) representation that is used when the instruction is stored in the computer memory. Any symbolic or mnemonic representation of a machine language instruction will be referred to as its assembly language equivalent. Hence, there ordinarily is a one-to-one correspondence between machine language and assembly language, the latter being by far easier for humans to read, understand, and manipulate. (More powerful assembly languages will, however, contain instructions that require several basic machine code operations to implement. These "macro instructions" must be broken down to the instruction level, usually by the Assembler programs discussed later.)

Example 11.1. Adding Two Integer Numbers With a Single Accumulator

```
                 | LOAD the contents of Memory Location 121 into the accumulator
  Computer A     |
                 | ADD the contents of Memory Location 122 to the accumulator
```

Example 11.2. Adding Two Integer Numbers with Multiple Accumulators

Computer B
{
LOAD the contents of Memory Location 121 into Accumulator #1

LOAD the contents of Memory Location 122 into Accumulator #2

ADD the contents of Accumulator #2 to #1
}

A JUMP or BRANCH instruction is used in computer programs so that instead of using the word in memory that immediately follows the location of the current instruction, the computer will "JUMP" or "BRANCH" to the specified location in memory to find the next instruction. This JUMP, BRANCH, or "GO TO" instruction may be unconditional, in which case the computer will always take its next instruction from the location specified in the JUMP instruction, e.g., "JUMP to Location 129" or "JUMP 129". Alternatively, the instruction might be conditional:

IF   1) the content of the accumulator is >0

2) the content of the accumulator is $=0$

3) the content of the accumulator is <0

4) the content of the accumulator is logically TRUE

5) the content of the accumulator is logically FALSE

6) the overflow indicator is "ON"

etc.

THEN: JUMP to specified location.

ELSE: execute the next instruction in sequence.

The particular condition upon which a JUMP is to take place must be specified in the instruction. The "otherwise" is implied. (The reader who is used to programming in FORTRAN should recognize these as the basic operations involved in the "GO TO" or "IF" instructions.) Once again the actual assembly language format used for writing a conditional branch instruction varies with different computer systems. For example,

JMP {>0} 129   or   JMG 129

or   SGZ 129

could mean "jump (skip) to Location 129 if the content of the accumulator is >0".

11.2.2 Binary Instruction Format

As discussed in Chapter 8, at the time of execution all instructions must be stored in memory in binary form. Let us start with the simplest case where:

(1) The instruction is implicitly directed to the CPU.

(2) There are N_i basic instructions defined by the manufacturer.

(3) The result of the operation stays in the accumulator.

(4) Only one operand is specified explicitly in the instruction.

(5) There are no options or variations.

Then a single instruction could have the binary format

$$0 \; 1 \ldots n_i \qquad \ldots n$$

op.code	address

where

n—implies each instruction word has $n + 1$ bits (typically 16)

n_i—implies $n_i + 1$ bits are used to uniquely identify the N_i different instructions (typically $N_i = 32$, $n_i = 4$)

op.code—is the binary operation code that uniquely identifies the N_i instructions

address—is the address of the operand in memory, or the specification of which accumulators are involved.

Thus in a 16-bit computer with 32 basic operation codes (of which the LOAD instruction is designated as number 3) a binary instruction to load the contents of Memory Location 37 into the accumulator could look like

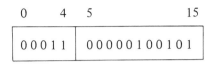

Obviously this is not a very convenient way to write a program, and a moment's thought will show that the 11 bits available for the address will only permit the unique specification of up to 2048 memory locations, which is inadequate for most computer applications.

To accommodate various options and special features a more general binary format can be thought of as

where the op.code and address are as defined previously, and the bits "x . . . x" (and in some cases the address bits) are used for one or more of the following:

(a) ≤1 bit to indicate whether one word or two words are used to define this instruction. Using two words of memory to define each instruction provides 2 × (n + 1) bits (typically 32) which is usually enough to conveniently specify the op.code, address, and various options. However, not all instructions require two words, and this approach obviously uses more memory. Hence, as discussed later, special addressing schemes and instruction formats have been developed to permit use of single-word instructions only. Some computers permit mixing of one- and two-word instructions; hence these formats require one bit to specify the instruction length.

(b) ≤4 bits to indicate what other registers, and/or operations, are to be used in calculating the effective address of the instruction.

(c) ≤4 bits to indicate which accumulator(s) in multi-register CPUs are involved in this instruction.

(d) One or more bits to indicate options in

instructions, e.g., the conditions on which a JUMP is to occur.

(e) One bit to indicate direct or indirect addressing (defined later).

(f) One or more bits to specify what optional operations are to be included as part of this particular instruction.

Different manufacturers may have other options and formats that are unique to their particular line of computers, and hence it is impossible to generalize. Specific examples are given in Tables 8.2, 8.3, 8.5, and 8.6 and later in this chapter. Also, it is obvious that not all of the above operations are required in every instruction. The following statements will summarize the situation with respect to instruction formats:

(1) *There is no standard instruction format.* Formats differ from one manufacturer to the next and sometimes even for different models of computers in the same series.

(2) *The instruction format may vary depending on all, or part, of the basic operation code.* Not all options or variations apply to every instruction so the same bit locations can be used for different purposes in different instructions; e.g., in nonmemory-reference instructions the "address bits" can be used to specify the accumulator(s) involved, additional or optional suboperations, conditions, etc.

(3) *Assembly languages and high-level languages make programming much easier than working with binary (machine language) instructions.* Programmers do not need to know the details of binary instruction formats, etc., in order to write most application programs. However, such knowledge gives a better understanding of how the computer works, why programs are written in particular ways, how program efficiency can be improved, etc. It is essential for most "systems work" and detailed "debugging" of programs. "Driver programs" for most user-constructed process interfaces and many computer peripherals supplied by vendors other

than the computer manufacturer are written in assembly language. These programs are converted to machine language by another program, the Assembler, which we discuss in Sections 11.2.4 and 11.6.

11.2.3 Binary (Machine) Coded Program Example

Let us assume that a digital computer is available, and it is desired to write a binary computer program that will:

(a) ADD one number to another
(b) Save the result
(c) STOP or PAUSE

With the hypothetical computer defined in Chapter 8, the resulting program as it would appear in memory just prior to execution might be as in Table 11.2 (each line represents a successive word in memory).

Each word is used to store either an instruction or a piece of data. When a word contains an instruction, the first two bits represent the operation to be performed (as defined in Tables 8.2 to 8.6), and the last nine bits, with some exceptions, represent the address of the operand involved in the operation.

Table 11.2. Program in Binary "Machine Code"

MEMORY WORD ADDRESS	MEMORY WORD CONTENTS
0	0010000000000100
1	0011000000000101
2	1010001000000000
3	0100000000000111
4	0000000000000110
5	0000000000000010
6	0000000000000000
7	0000000000000110
8	1100000000000000

Addressing schemes for computers are many and varied, but, in this example, the nine-bit address is interpreted simply as the absolute position of a word in memory. If this program were stored sequentially in memory starting at Location 0, and the computer instructed to start executing the first instruction, the sequence of operations performed by the computer would be those illustrated in Table 11.3. The computer executes each operation in sequence until instructed to do otherwise, as illustrated by the JUMP instruction in Memory Word 3; the JUMP instruction is of course necessary to prevent the computer from trying to interpret the data in Location 4, 5, and 6 as instructions.

When the computer stops executing, Memory Word 6 will contain the binary number 0 000 000 000 001 000 which is equivalent to the decimal integer 8. The computer has thus carried out the required task of adding the integer constants previously stored (in binary form) in Memory Locations 4 and 5, and storing the sum in Memory Word 6.

11.2.4 Assembly Language Example

The computer program listed in Table 11.2 is usually referred to as being in "machine language". It is quite apparent that writing a program of several hundred instructions in binary format would be very time-consuming and conducive to error, and that if potential users had no other choice, there would be very few computers in use today. To overcome this, most computer manufacturers supply a computer program, called an Assembler, along with the computing hardware. The Assembler program makes it possible to write our example computer program "symbolically" as shown in Table 11.4 rather than as a string of binary words. In Table 11.4 it will be noted that symbols or mnemonic codes have been used to represent:

(1) Operation codes, e.g., LDA, JMP
(2) The location of the operand, e.g., I, J, K, AC0, AC1
(3) The memory location or "statement label", e.g., SAVE

Table 11.3. Program Written in Expanded Form

MEMORY WORD ADD.	MEMORY WORD CONTENTS	TRANSLATION OP CODE	ADDR	EXPLANATION OF OPERATION CARRIED OUT
0	0010000000000100	LDA 0	4	Load AC0 with contents of word 4 (Table 8.2)
1	0011000000000101	LDA 1	5	Load AC1 with contents of word 5 (Table 8.2)
2	1010001000000000	ADD 1,0		Add AC1 to AC0 (Table 8.5)
3	0100000000000111	JMP	7	Jump to word 7 (Table 8.3)
4	0000000000000110			Integer constant (6 in base 10)
5	0000000000000010			Integer constant (2 in base 10)
6	0000000000000000			Space reserved for result
7	0000000000000110	STA 0	6	Store AC0 in word 6 (Table 8.2)
8	1100000000000000	HLT		Halt (Table 8.6)

Note that the right-hand part of each line can be used, if desired, for comments that explain each step. It should also be noted that the constants 6 and 2 are specified directly in decimal form. This particular hypothetical Assembler program will convert them to the appropriate binary form.

The "source code" version of the user's program, written as illustrated in Table 11.4, is read into the computer as input information and processed by the Assembler program. The output from the Assembler includes a listing of the program, similar to Table 11.4; a list of the corresponding executable binary instructions ("object code" as in Table 11.2); plus other options that will be discussed later. The ease and efficiency of programming in assembly language depend a great deal on the Assembler program as well as on the basic computer hardware.

11.3 ADDRESSING OPTIONS

This section describes various common options for calculating the *effective address* of operands in memory reference instructions. The beginning programmer invariably has more difficulty specifying the addresses of operands than he does specifying operations such as ADD, MULT, etc., because the latter are familiar mathematical operations.

In previous sections it was assumed that the address of the operand was included in the instruction as an "absolute" numeric value that

Table 11.4. An Assembly Language Program with Symbolic References

Address Label	Op-Code and Operands	Explanatory Comments
	LDA 0,I	Load AC0 with contents of location labelled I
	LDA 1,J	Load AC1 with contents of location labelled J
	ADD 1,0	Add AC1 + AC0 (result in AC0)
	JMP SAVE	Jump to instruction in location labelled SAVE
I	6.	Integer decimal constant labelled I
J	2.	Integer decimal constant labelled J
K	0	Space reserved to store result, labelled K
SAVE	STA 0,K	Store AC0 in location labelled K
	HLT	

identified the memory location that contained the operand. If two words are used for each instruction, then there are usually enough bits available to address every word absolutely, since one 16-bit word can identify 64K locations. However, if a suitable addressing scheme can be found, then one-word instructions would be much more efficient. Other addressing schemes are also desirable to make it easier to alter the order of instruction execution (e.g., JUMP instructions) and to make it easier to process "lists" or "strings" of data. The main methods of addressing discussed below are:

(1) Absolute
(2) Relative
(3) Indirect
(4) Calculated by user program

It is suggested that initially the reader think in terms of absolute addressing or the simpler versions of relative addressing and leave the more complex methods of specifying "effective addresses" until he has gained some programming experience.

11.3.1 Absolute Addressing

The address is completely defined by the numeric value stored explicitly within the instruction. This was discussed in Section 11.2.

11.3.2 Relative Addressing

Absolute addressing can be thought of as specifying the location in memory "relative" to zero. In relative addressing the number contained in the "address" part of an instruction is assumed to be relative to the contents of another register, R_x. Thus,

Effective address = Displacement
$$+ \text{ contents of Register } R_x$$

where displacement is the "address" part of the instruction word, and Register R_x may be:

(1) The instruction register (Program Counter)
(2) Register(s) or accumulator(s) designed for address calculations

Minicomputers frequently allocate about 9 of the 16 bits in each one-word instruction to define the address or displacement (a total of $2^9 = 512$ possibilities). When two's complement values are used, the displacement may be any value from -256 to $+255$.

If the instruction register (program counter) contains the memory location of the instruction currently being executed, then the operands of the instructions in Tables 11.3 and 11.4 can be specified using "relative addressing" as shown in Table 11.5. Note that since all the addressing is relative, this program can be loaded and executed starting at any location in memory. (When absolute addresses are used, they must be modified every time the program is "relocated".)

If it is desired to store the constants in our example program in a data area that starts at Location 4000, instead of in the middle of the executable code, then this could be accomplished by using an index register, R_k (in the computer of Chapter 8 this is one of the accumulators) as shown in Table 11.6.

Table 11.5. Addressing Relative to the Program Counter (PC) Register

Note: "R" at the end of the op-code implies Relative addressing

LDAR 0,4	Load AC0 with the contents of the location with address PC+4 (i.e with 6)
LDAR 1,4	Load AC1 with the contents of the location with address PC+4 (i.e. with 2)
ADD 1,0	Add contents of AC1 to contents of AC0
JMPR 4	Jump to instruction with address PC+4
6	Integer value 6
2	Integer value 2
0	Storage space for sum
STAR 0,-1	Store AC0 in address PC-1 (i.e. in the location preceding this instruction)
HLT	

Table 11.6. Addressing Relative to the Index Register, R_k

assume location 4000 contains the value 6 base 10

assume location 4001 contains the value 2 base 10

assume location 4002 is reserved to store the sum

assume the starting address of the data area, i.e. 4000, is stored in the
  location labelled TABAD.

"X" at the end of the op-code implies indexed addressing using AC1.

LDA 1, TABAD	Load AC1 with data table address
LDAX 0,0	Load AC0 with contents of location in index register AC1 offset by 0 (i.e. value 6 from 4000)
LDAX 1,1	Load AC1 with contents of location identified by index register AC1 offset by 1 (i.e. value 2 from 4001)
ADD 1,0	Add AC1 + AC0 (result is in AC0)
LDA 1, TABAD	Load AC1 with data table address again (since AC1 is used both as an index register and in the arithmetic operations)
STAX 0,2	Store AC0 in location indexed by AC1 offset by 2 (i.e. store into location 4002)
HLT	
TABAD A[TABLE]	The memory location with the label TABAD contains the numeric value 4000 which is the starting address of the data table. The Assembler is responsible for determining that A[TABLE] = 4000.

Note that if the hypothetical computer of Chapter 8 had a separate index register in addition to the accumulators AC0 and AC1, then the program in Table 11.6 could be written as (where I following the op-code implies indexed addressing using the special Index register rather than one of the accumulators):

```
LDAI    0,0
LDAI    1,1
ADD     1,0
STAI    0,2
HLT
```

The comparison with Table 11.3–11.5 is now more direct. (Note that it was assumed that the index register was already loaded with the beginning address of a three-word data area containing 6, 2, and SUM respectively.) This illustrates how the hardware architecture can affect the programming of a given function.

Relative addressing of memory is often called "relative paging" ("floating page") because the contents of the index register can be thought of as the page number and the displacement can be thought of as the position of the desired word on that page. In a truly "paged" system, the most significant bits of the index register *only* would be taken as the page number with the displacement from the instruction taken as the position of the desired word on the page. Page ZERO can be specified directly on many computers (e.g., does not require loading an index register with zero) and is used for frequently-referenced data and system values.

A typical way of specifying relative addressing in binary form is to use two bits of each memory reference instruction as follows:

00 → page zero or absolute addressing
01 → relative to Index Register #1
10 → relative to Index Register #2
11 → relative to Instruction Register

Executing an instruction using addressing relative to an index register usually takes slightly longer than absolute addressing, but permits the operands to be anywhere in memory. Addressing relative to the instruction register permits one-word instructions to be used and does not tie up any extra registers, but means that JUMP instructions must be included to make sure that data are not interpreted as instructions (cf. Table 11.4). The real advantages of index registers lie in advanced applications where they can be incremented or set by other (sub)programs.

A program that calculates the sum of 100 numbers stored in memory is shown in flow diagram form in Figure 11.2 and as an assembly program listing in Table 11.7. The program assumes there are 100 numbers to be added and that the starting address of the data table has the label TABAD.

Note: For someone who has not programmed in assembly language before, it can easily take an hour to carefully work through the example in Table 11.7a. It is suggested that the reader construct a blank table with the same column headings as Table 11.7b. The blank table can then be filled in line by line as the reader "mentally executes" each instruction in Table 11.7a in its proper sequence. For purposes of preparing Table 11.7b it was assumed that the 100 data values were {200, 199, 198 . . . 101} stored in memory locations {A + 99, A + 98 . . . A + 1, A} respectively; a question mark means the value is unknown; a dash means no change.

11.3.3 Indirect Addressing

In indirect addressing the location specified by any of the previously described addressing options is assumed to contain the *address*, rather than the actual value, of the desired operand:

Effective address
$$= \text{Contents of } \{\text{displacement} + R_x\}$$

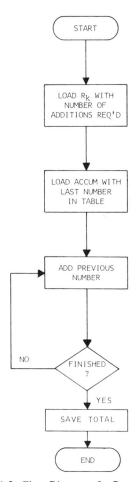

Figure 11.2. Flow Diagram of a Program to Calculate the Sum of a Table of Integer Constants

One use of this option is as follows: Assume it is desired to JUMP from Program A to Program B, and that after Program B is loaded into the computer its starting address, 1234, is stored in Memory Location 2. The the statement in Program A:

JUMP 2 (indirect)

will cause the computer to jump to Program B. The advantage of this approach over absolute or relative addressing is that if Program B is moved to another location in memory, then it is only necessary to put the new starting address in Location 2; NO CHANGE IS REQUIRED IN PROGRAM A or B.

Different computer systems have different options that can add to the flexibility and

Table 11.7a. An Assembly Language Program to Calculate the Sum of 100 Integer Constants

Instruction No.	Label	Op-Code	Comments
1	START	SUB 0,0	Subtract AC0 from AC0 (i.e. zero AC0)
2		STA 0, TOTAL	Store AC0 in location TOTAL (i.e. zero TOTAL)
3		LDA 0, HUND	Load AC0 with 100 (i.e. number of values to be added)
4	LOOP	STA 0, COUNT	Store AC0 in COUNT (i.e. store the number of values remaining to be added)
5		LDA 1, TABAD	Load AC1 with the address of the beginning of the area reserved to store the values to be added, TABAD
6		ADD 0,1	Add AC0 + AC1 to get A[TABLE] + COUNT
7		LDAX 0,-1	Load AC0 indexed by AC1 and offset by -1 (i.e. load the contents of (A[TABLE] + COUNT-1) which is the next value to be added)
8		LDA 1, TOTAL	Load AC1 with TOTAL
9		ADD 1,0	Add AC1 to AC0 (i.e. add existing TOTAL in AC1 and next value from AC0)
10		STA 0, TOTAL	Store AC0 in TOTAL (i.e. store new sum)
11		LDA 0, COUNT	Load AC0 with COUNT
12		LDA 1, ONE	Load AC1 with 1
13		SUB 1,0,SRZ	Subtract AC1-AC0
			Skip next instruction if Result is Zero (i.e. when 100 values have been summed)
14		JMP LOOP	Jump unconditionally to LOOP
15		HLT	Halt
	TABAD	A[TABLE]	Store ADDRESS of data storage area TABLE
	HUND	100.	Define constant 100 base 10
	ONE	1.	Define constant 1
	COUNT	0	Reserve storage space for COUNT
	TOTAL	0	Reserve storage space for TOTAL
	TABLE	(Reserve 100 words to store values to be summed)	

(Instructions 11, 12, 13 are bracketed with the note: subtract one from count)

complexity of programming. For example, some computers will allow multiple indirect addressing (i.e., the contents of the addressed location is the address that contains the address of the operand . . .). Also "indexing" (i.e., the use of an index register) may be allowed before and/or after the indirect addressing reference; for example:

$$\text{Effective address} = \text{Contents of } \{\text{displacement} + R_1\} + R_2$$

If $R_2 = 0$, the "pre-indexing" is intended; if $R_1 = 0$, then "post-indexing".

To illustrate indirect addressing, the program in Table 11.3 is rewritten as shown in Table 11.8.

Indirect addressing allows the numeric values to be stored anywhere in memory as long as the addresses are stored in locations 4, 5, and 6. More complete examples of indirect addressing, and combined indexed-plus-indirect addressing, are included later in this chapter.

Table 11.7b. Contents of Registers and Memory Locations after each Instruction

Instruction	Registers		Memory Locations	
	ACO	AC1	COUNT	TOTAL
0	?	?	?	?
1	0	–	–	–
2	–	–	–	0
3	100	–	–	–
4	–	–	100	–
5	–	A	–	–
6	–	A+100	–	–
7	200	–	–	–
8	–	0(TOTAL)	–	–
9	200	–	–	–
10	–	–	–	200
11	100(COUNT)	–	–	–
12	–	1	–	–
13	99 skip if zero	–	–	–
4	–	–	99	–
5	–	A	–	–
6	–	A+99	–	–
7	199	–	–	–
8	–	200(TOTAL)	–	–
9	399	–	–	–
10	–	–	–	399
11	99(COUNT)	–	–	–
12	–	1	–	–
13	98 skip if zero	–	–	–

Continue until all 100 values have been added at which time ACO → 0 after execution of instruction 13.   This causes the next instruction (#14) to be skipped and instruction #15 executed which HALTS program execution.

11.3.4 Calculated Addresses

We have seen that an address is simply a binary number.   Therefore any series of calculations that can be used to calculate a binary number can be used to calculate an address.   The calculated number can be loaded into an index register and used for relative addressing or put in a storage location as an indirect address. These techniques are relatively straightforward.

The calculated number also can be used to modify the displacement or address portion of an instruction by loading the instruction into an accumulator, arithmetically or logically adding the desired address, and storing it back in memory for later execution.   In some machines the calculated address may be put in the program counter to serve as the location of the next instruction; or, in some cases, put into the addressing hardware directly.   These

Table 11.8. A Program Using Indirect Addressing

assume location 4 contains the ADDRESS of the value 6 base 10

assume location 5 contains the ADDRESS of the value 2 base 10

assume location 6 contains the ADDRESS of the sum

the symbol "@" at the end of the op-code implies indirect addressing.

LDA@ 0, 4	Load ACO with the contents of the location whose address is in location 4 (i.e. with value 6)
LDA@ 1, 5	Load AC1 with the contents of the location whose address is in location 5 (i.e. with value 2)
ADD 1, 0	Add AC1 + AC0 (i.e. add 6+2)
STA@ 0, 6	Store AC0 (the sum 6+2) in the location whose address is in location 6.
HLT	

Note: The memory references in the above instructions were made using the "absolute" values 4,5,6 similar to Table 11.3. If desired, the symbolic labels I,J,K, could be used as was done in Table 11.4.

techniques of programmed or dynamic address calculation are very powerful but should be used only in special circumstances, as they are easy to misapply and hard to debug. The beginning programmer should avoid them altogether and utilize only the simplest addressing techniques.

11.3.5 Example of Addressing Options and Use of JUMP Instructions

Assume that it is desired to write a subprogram to calculate the sum of N integer constants previously stored in locations DATA to DATA + N - 1, and that this subprogram is to be called from another program called MAIN. Assume

Table 11.9. A "Mainline" Program that "Calls" a Subroutine

	· · ·		Executable code in program MAIN
	JSR	SUBP	This "Jump to Subroutine" will store the address of this instruction (i.e. the program counter's current value) plus one in AC1 and cause the program to JUMP to the address SUBP
	JMP	NEXT	Jump unconditionally to NEXT (i.e. jump over following DATA)
	A[DATA]		Stores the Address of DATA
	A[N]		Stores the Address of N
	A[SUM]		Stores the Address of SUM
NEXT	· · ·		Next executable statement in MAIN
	· · ·		More code
	HLT		Halt (end of MAIN program)
SUM	0		Reserve storage for SUM
N	n		Value preset by user or program
DATA	(reserve a block of words ≥ largest N)		

Table 11.10. An Assembly Language Subroutine to Sum N Values

		It is assumed that the address of the next executable instruction in the calling program (i.e. the PC register) is stored in AC1 at time of entry to the subroutine
SUBP	STA 1, RETN	Store AC1 (return address) in RETN
	SUB 0,0	Subtract AC0 from AC0 (i.e. zero AC0)
	STA 0, SUM1	Store AC0 in SUM1 (i.e. zero SUM1)
	LDAX@ 0,2	Load AC0 using indirect addressing indexed by AC1 and offset by 2. (i.e. load the value of N from Table 11.9)
LOOP	STA 0, NUM	Store AC0 in NUM (number of characters to be summed)
	LDA 1, RETN	Load AC1 with return address, RETN
	LDAX 1,1	Load AC1 indexed by AC1 and offset +1 (i.e. get address of DATA from Table 11.9)
	ADD 0,1	Add AC0 + AC1 to obtain A[DATA] + NUM. The result stays in AC1
	LDAX 0,-1	Load AC0 indexed by AC1 offset by -1 (i.e. load the DATA value contained in location A[DATA] + NUM - 1)
	LDA 1, SUM1	Load AC1 with current sum, SUM1
	ADD 1,0	Add AC1 + AC0 (i.e. add previous sum in AC1 and next data value in AC0)
	STA 0, SUM1	Store AC0 (new sum) in SUM1
	LDA 1, NUM	Load AC1 with number of values still to be added
	LDA 0, ONE	Load AC0 with numeric value 1
	SUB 1,0,SRZ	Subtract AC1 - AC0 (i.e. subtract 1 from NUM 1 and Skip the next instruction if the Result is Zero. Result is left in AC0)
	JMP LOOP	Jump to instruction labelled LOOP
	LDA 0, SUM1	Load AC0 with final sum, SUM1
	LDA 1, RETN	Load AC1 with return address
	STAX@ 0,3	Store AC0 using indirect addressing via AC1 offset by 3 (i.e. get the address of SUM from Table 11.9 and then store the new sum in this location)
	JMPX 0	Jump to instruction in address indexed by AC1 and offset by 0 (i.e. to the "JUMP NEXT" instruction in Table 11.9)
ONE	1.	Define the constant ONE
RETN	0	Reserve storage for return address
NUM	0	Reserve storage for number of values
SUM1	0	Reserve storage for accumulating sum

that the data values and the value of N have already been stored in the proper locations and that the starting address of the subprogram is SUBP. The relevant part of the MAIN program is given in Table 11.9 for the computer of Chapter 8.

The subprogram to sum the N values stored in locations DATA to DATA + N - 1 and store the result in the location labelled SUM is given in Table 11.10.

The following points should be noted about this example:

(1) There are many different ways of writing a program to accomplish the same result. This program was written to illustrate certain basic concepts of assembly language programming.

(2) Different computers have different instruction sets. For example, many computers can zero the accumulator with one instruction, and others have (indexing) registers that are decremented by one "automatically" each time they are used.

(3) The format of some of the statements in this program does NOT follow the actual conventions required by any specific Assembler but is illustrative of what must be specified in each instruction.

(4) This program does not include checks for error conditions (e.g., accumulator overflow) that are desirable in most practical applications; nor does it store the registers at the beginning of the subprogram and restore them just before it returns to the calling program (MAIN). The latter action is a good practice to adopt because it ensures that the register contents used by the calling program are the same before and after the subroutine executes.

11.4 INTERRUPT SERVICING

Everything that a computer does must be directed by program statements written by the user and/or others such as the computer vendor. The preceding examples illustrated typical code for performing arithmetic operations. The type and sequence of instructions are familiar to most technically-trained people because they are similar to ordinary mathematics. Other computer operations are less familiar to users and hence more difficult to program. Therefore, this section and the next discuss the programming support needed for interrupt servicing and input/output operations—two of the most important and difficult areas for new programmers.

Interrupts furnish a way of initiating the execution of a specified program and/or changing the sequence of program execution. Along with sensor-based input/output and the ability to synchronize internal and external operations to a real-time clock, user-accessible interrupts represent one of the most important differences between real-time and data processing applications.   For example, a hardware interrupt system makes it possible for a computer to handle tasks in accordance with user-defined priorities coupled to real-time events rather than simply in the sequence fixed *a priori* or in the order in which the tasks arise. Using interrupts a computer can stop processing a low-priority "background" job in order to service a high-priority interrupt that signals an emergency in the real-time application.

A flow diagram of a generalized interrupt handler is shown in Figure 11.3. The following discussion starts with the simplest care and then indicates how additional complications arise to make up the general situation of Figure 11.3. Note that an interrupt might be caused by an "internal" condition such as a parity error, or by an "external" condition such as the closure of a contact in a relay connected to the user's process equipment.

In the simplest case, the computer is idle (i.e., the hardware is operable but awaiting direction as to which instruction should be executed next), and upon receipt of an interrupt, the computer does a "hardware-forced" BRANCH to a program that will take appropriate action. This program is usually called the interrupt service (sub)routine or program.

In more complicated applications several devices or external interrupt sources could be connected to a single interrupt level.   Then when the interrupt level is "tripped", the first thing that the interrupt service program must do is to determine which particular device activated the interrupt and then transfer to the particular program or routines associated with that device.   Every system is different, but a typical sequence of operations is:

(1) *INTERRUPT*: Cause a BRANCH to the appropriate set of instructions.

(2) *IDENTIFY INTERRUPT*: For systems where individual hardware interrupt levels are not available for each device or interrupt condition, it is necessary to connect several interrupt sources to one level.   Depending on the hardware fea-

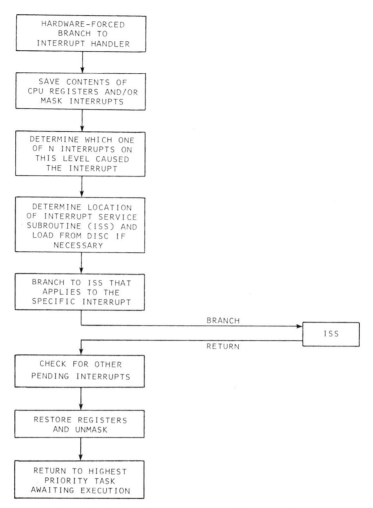

Figure 11.3. Flow Diagram of a General Interrupt Handler

tures available, the interrupt handler may use one of several different techniques to identify the interrupt source. Examples are:

a. Read an "interrupt level status word" and check to see which bit(s) is on in that word,

b. Interrogate the status registers of all devices connected to that level,

c. Check "flags" set by other programs.

(3) *BRANCH*: Branch to the interrupt service subroutine provided by the vendor or by the user for the particular interrupt that has just occurred.

(4) *RETURN*: Return to the main interrupt service program, reset the interrupt hardware if necessary (or desired), and check

to see if any other points on that level have been tripped. If not—return to "idle" state.

Note that some of the above operations may be done automatically by hardware, whereas others may require explicit instructions.

If any interrupt occurs while a lower-priority program is executing (rather than when the computer is in an idle state as assumed in the above example), then before transferring control to the appropriate interrupt service program it is necessary to SAVE the contents of the CPU registers by storing them in an area of memory reserved for that purpose. At the conclusion of the interrupt service program the CPU registers are reset to the values that they had at the moment of interrupt, and execution of the original program resumes at the exact point where it was interrupted. In some computers these SAVE/RESTORE REGISTER operations are done automatically by special hardware or microcoded subroutines. It is important to realize that the data and instructions that constitute the low-priority program remain stored in memory, and the contents of the CPU registers completely and uniquely define the state of that program. Thus the interrupted program can be restarted, at the place it was interrupted, simply by restoring the registers.

On systems that have multiple, say M, levels of hardware interrupt it is possible to have M "nested interrupts"; i.e., Program 2 interrupts Program 1, 3 interrupts 2, 4 interrupts 3, . . . M interrupts M-1, at which point program M will execute to completion because—by definition—it is the highest-priority program in the system. The interrupted programs are then completed, in order of priority, but subject to additional interrupts.

Computer systems also have provision for "masking" interrupts, i.e., turning them off, and for initiating them by a program call. For example, in the case of the M nested interrupts described above, when servicing an interrupt on Level M it might be desirable to do part of the interrupt servicing at Level M and the rest at Level 4. In this case, the most important part of the Level M program is completed, and the remainder is "set up" as a pending interrupt job to be done at Level 4.

An additional complication occurs when all of the interrupt service programs are not stored in memory. With a disk-based operating system it might be necessary to:

- Acknowledge the interrupt
- Transfer a low-priority program from memory to disk
- Transfer the interrupt service program to memory from disk
- Execute the interrupt service program
- Restore the low-priority program to memory and resume execution

A key factor in interrupt servicing is the response time, i.e., the time that elapses between the tripping of the interrupt level and the execution of the first instructions that deal specifically with that interrupt. Note that checking for which specific device tripped the interrupt, saving and restoring registers, disk operations, higher-priority interrupts, etc., all extend the response time. It can vary from a couple of microseconds to several milliseconds.

Figure 11.3 showed the main functional parts of an interrupt handler. A handler of this type is usually supplied by the computer vendor (in software and/or hardware), and hence the user is normally concerned only with writing the interrupt service subroutines associated with particular peripheral devices unique to his application.

As a final point in this section, it is important to distinguish between *interrupt-driven* and *polled* input/output operations. An "interrupt-driven computer operation" is initiated in response to a specific external or internal interrupt; i.e., the initiation of the computer operation is determined by the event that generates the interrupt (e.g., the closing of an external switch). A "polled input/output operation" is initiated by the computer, i.e., at a certain timer interval or when the computer comes to a particular point in a sequence of operations. Consider a single computer, dedicated to the task of executing a program whenever an external switch is closed. If the external switch is wired to an interrupt level, then when the switch is closed the computer will start executing the desired program. Between closings of the switch the computer is idle (i.e., completely free for other applications). If the external switch is wired to a digital input instead of an interrupt, then in order to guarantee a response time of τ seconds the computer must execute a program every τ seconds (as determined by the real-time clock) that: reads the digital input, checks to see if it is on, and if so, transfers to the desired program. Note that even during the period between closures of the external switch, the computer spends a significant amount of time polling the digital input. The choice between interrupts and polling is based on many factors including cost, response time, flexibility, etc. Some large-scale computers rely on polling techniques, and many microcomputer applications utilize polling as well.

Example 11.3. Example of Simple Programmed Output

| (1) | LOAD | Load accumulator with word to be output |
| (2) | WRITE | Execute an I/O instruction which specifies the WRITE operation, identifies the peripheral device, and transfers the data to the I/O bus. |

Although the details of programming an interrupt handler are not obvious from this brief overview, it should be clear that interrupt handling is important and more complicated than most arithmetic calculations. Some features of interrupt handling are illustrated by the example in the following section.

11.5 INPUT/OUTPUT OPERATIONS

Input/output operations, like interrupt handling, are less familiar to the programmer and hence are more difficult to program than most arithmetic operations. Unfortunately they also differ more from one computer system to another than do arithmetic capabilities, and hence are difficult to generalize. The main concepts and types of I/O situations will be illustrated by the following discussion; the reader will have to consult the operating manuals of his own computer system for details.

11.5.1 Single-Word I/O Via Accumulators (Programmed I/O)

Output from memory to a peripheral device is relatively simple if only one program is involved. Input is similarly trivial, as the essential operations in Example 11.3 demonstrate.

If more than one word is to be output, then these steps must be repeated. However, since the computer typically is much faster than the peripheral device (e.g., 10^6 operations/sec vs.

30 characters/sec printed on a terminal), it is necessary to "synchronize" the operation of the computer with that of the peripheral. This can be done in two ways:

(1) By testing device status flags (Busy/Free, etc.)
(2) By using hardware interrupts

An example of a computer output operation using the status test approach is given in Example 11.4. This general approach achieves the desired result, but most of the computer time is spent in the TEST "loop" e.g., the computer would ask a terminal "Are you busy"? several hundred thousand times before it got a "no" answer and could output the next character! If there is other work to be done, this procedure is very inefficient. Input/output using interrupts eliminates the "waiting" in a TEST loop—but requires additional programming to service the interrupt. The major programming steps are illustrated in Example 11.5.

In this example an interrupt is serviced for each piece (word, record) of data output, but the computer is free for other calculations; i.e., it is not tied up "polling" busy flags, between each I/O record. Note that, in the general case, all I/O operations would normally start with a TEST to determine if the peripheral device was on-line and ready to handle I/O. In some cases it might be necessary to send some preliminary "control" instructions to the device

Example 11.4. Computer Output Using Status Testing

(1)	LOAD	Load data from memory to accumulator
(2)	WRITE	Write data to I/O bus
(3)	---	Decide if more data are to be output (If yes, continue, otherwise exit)
(4)	TEST	"Poll" busy indicator of peripheral to determine when first operation is complete and device is "ready" to receive another word, i.e. iteratively test until the peripheral is not busy
(5)	JUMP	Branch back to the load instruction with suitable modification of the data address to get the next word (e.g. increment the index register used in the relative addressing of the LOAD instruction).

Example 11.5. Computer Output Using Interrupts

(1)   START OUTPUT   i.e. LOAD accumulator preparatory to performing the WRITE

(2)   ENABLE INTERRUPTS   i.e. enable hardware + store appropriate addresses (start of
                             ISS, etc.)

(3)   WRITE first data word to device

(4)   Continue with execution of any program that does not involve output to the same device

(5)   When the peripheral device has finished with the first request and requires more
      data (or servicing) it interrupts the computer and control is transferred to the
      interrupt service program

(6)   The interrupt service program transfers control to the "output program"

(7)   This output program decides whether more data are to be output (in which case
      the cycle is repeated) or whether the output is complete (which means transmission
      is terminated and the program peripheral device is "released" for use by other
      programs)

Table 11.11. A Simple Assembly Language Program to Read a Block of Characters
from a Paper Tape Reader

START	SP1 PTR	Send pulse to the Paper Tape Reader to turn it on
LOOP	STA 0, NCHAR	Store the number of characters still to be read from AC0 to memory location labelled NCHAR
	LDA 1, AINP	Load AC1 with the starting address of the area of memory reserved to store the INPUT
	ADD 0,1	Add AC0 + AC1 to get address where first character will be stored (A[INPUT] + NCHAR)
	DIA 0, PTR	Digital Input to AC0 from Paper Tape Reader (i.e. digitally-coded equivalent of first character)
	STAX, 0,-1	Store AC0 in memory address contained in AC1 but offset by -1 (indexed address)
	LDA 0, MIONE	Load AC0 with MIONE ⎫
	LDA 1, NCHAR	Load AC1 with NCHAR ⎬ subtract one from NCHAR
	ADD,1,0,SRZ	Add AC1 + AC0 ⎭ Skip next instruction if Result is zero (i.e. if all characters have been read)
	JMP LOOP	Jump to instruction labelled LOOP
	SP2 PTR	Send Pulse to Paper Tape Reader to turn it off
	HLT	Halt
MIONE	-1	
NCHAR	100	
AINP	A[INPUT]	
INPUT	(reserve 100 words)	

Notes:

(1)   It is assumed that the number of characters to be read from the paper tape
      reader is stored in accumulator AC0 at the start of execution of this program.

(2)   PTR identifies an output channel or pulse I/O interface and hence is different
      from a variable name or address label.

(3)   It is assumed that the DIA (digital input to accumulator) instruction reads
      only one character each time the paper tape reader status indicator changes
      from not-ready to ready, i.e. each time a new character is under the read
      head.

before actual data could be handled. For example, before output of alphanumeric data to a printing device it is usually necessary to issue "format control" instructions such as "line feed", "skip to new page", etc. A very simple driver program for a paper tape reader is shown in Table 11.11. This program uses single-word input via the accumulator and assumes that no other programs are active in the computer system.

On some computer systems it will be necessary to replace the single DIA instruction by a set of instructions that

(1) Test the "ready" status indicated on the reader.
(2) Detect when it changes from not-ready to ready.
(3) Issue an input command to read one character.
(4) Wait for next character.

The computer is so fast relative to the paper tape reader that if no checking were done, it would take dozens of readings during the time that a single punched character was under the read head.

If the controller for the paper tape reader has been designed to generate an interrupt each time a new character moves under the read head and is ready to be read, then the instruction DIA in Table 11.11 could be replaced by a set of instructions that:

(1) Cause the computer to wait in an idle state.
(2) Service the interrupt by branching to instructions that read a single input from the interrupting device.

Once again the assembly language programmer must know the details of operation of the computer and the peripheral device before a satisfactory "driver" program can be written.

11.5.2 High-Speed I/O of Larger Data Blocks (Data Channel or Direct Memory Access)

For the output of large blocks of data to a high-speed device such as a disk, programmed I/O via

the accumulator would leave very little CPU time available for handling other tasks. Therefore, data channel controllers are frequently added to the computer system to handle high-speed I/O. Data channel operations are initiated and terminated by instructions executed in the CPU, but while the transfer is in progress, the CPU is free to do other calculations. In some systems the data channels "cycle steal" one computer memory cycle for each word transferred between memory and the I/O bus. This slows operation of the CPU slightly, but does not interfere with the CPU operations in any other way such as by altering the contents of any registers. To initiate a data channel operation the CPU must check to see if the channel is already busy, supply the device address, specify the starting address of the data in main memory, specify the number of words to be transferred or the last address of the data, and (in some cases) specify the action to be taken at the completion of the transfer—e.g., interrupt the CPU. Note that once the I/O operation is started, it proceeds at a rate determined by the peripheral and is not related to what instructions might be executing simultaneously in the CPU; i.e., the two operations are asynchronous, and I/O is overlapped with program execution in the CPU. This means that the CPU is always available for the execution of high-priority interrupt programs even when large amounts of I/O are being handled via a data channel.

11.5.3 I/O in Multiprogrammed Systems

In multiprogrammed systems or in real-time applications where several "nested interrupts" can occur, it is possible to have a number of programs requesting I/O from the same peripheral device. The programmer must make sure that the computer is not slowed down to the speed of the peripheral (e.g., by programming so that the I/O from the first program must be finished before the next program can start execution) and/or that a garbled mixture of output from several different programs isn't directed to a single peripheral. The programmer must also be familiar with the idiosyncrasies of output devices that must be programmed at a basic level. For example, with some terminals the keyboard and the printer part of a single unit appear as two separate devices to the computer. If a character is printed when a key is struck, it is because the computer has serviced the interrupt from the keyboard, checked the transmission, perhaps done code conversion, and output the character to the printer part of the terminal! This will give some indication of computer response time.

Most real-time, sensor-based computer systems are "I/O bound" rather than "CPU bound" in the sense that the total number of tasks that can be handled is determined by the ability of the system to handle I/O; hence the CPU computational capability is not usually the limiting factor. Therefore, despite their complexity, I/O hardware capability and program efficiency must be given very careful consideration, by both the designer and the user.

11.6 FEATURES OF THE ASSEMBLER

As we have seen, at execution time the computer requires binary-coded data and instructions as illustrated by Table 11.2, but the programmer would prefer to work with a "higher-level", symbolic language such as illustrated in Table 11.4. An *Assembler* is a computer program that accepts the "assembly language" produced by the programmer and converts it into the binary code or machine language required by a particular computer. From previous discussion it is obvious that the Assembler must do at least the following:

(1) Convert the operation code mnemonics into binary code (in the simplest case this can be done by a one-to-one table lookup procedure, cf. Table 11.3).

(2) Assign a memory location (address) to every symbol (data constant or variable) used in the program and substitute the binary memory address whenever a symbol is used to specify an operand. This is easy with straight absolute addressing, but relative and indirect addressing require a more sophisticated Assembler. The "symbol table" produced by the Assembler showing the correspondence between the symbols used in the program and the assigned memory locations is printed out if requested. In most computers the symbol table is destroyed after the assembly operation is complete and is *not* available at program execution time. Thus it is not normally possible to type-in symbol names and obtain the numeric values because the table of symbols is not available.

(3) Assign a memory location to each instruction. If the instruction has a statement label, then the Assembler must build a "label table" and substitute the actual memory address whenever a label is used, e.g., in a JUMP SAVE instruction. Note that labels can be handled by the Assembler in the same way as symbolic references to variables; hence a single table may be used for both symbols and labels.

(4) Perform code conversions so the user can specify integers, real numbers, floating point numbers, text strings, etc., in the familiar user's form rather than in binary (cf. Table 6.1).

(5) Set up "linkage tables" so that the user's program can BRANCH to other programs or subroutines and return.

(6) Supply the binary addresses and appropriate code so that I/O operations can be specified symbolically.

(7) Do all the necessary bookkeeping and ensure that the appropriate information such as program length, external references (symbols or labels identified as being in separate programs), etc., is available to other system programs such as the "loader program". On demand these tables and/or the program listing can usually be printed by the Assembler.

(8) Diagnose "all" the common programming errors e.g., JUMP SAVE where SAVE has not been defined as a statement label, and print an appropriate message.

(9) Perform "utility functions" such as LIST, STORE (e.g., on disk), PUNCH (e.g., paper tape) copies of the source or object code. (As noted previously, "source" code refers to the programmer's original assembly language listing of the program,

"object" code to the Assembler's machine language output listing.)

The efficiency of program execution is strongly influenced by the "power" of the basic instruction set, but the ease of programming is often more dependent on the "power" of the Assembler program. The more sophisticated Assemblers include capabilities such as the following:

(1) MACRO: Sets of instructions that are used over and over again can be defined as "macro instructions" and referred to symbolically. (The Assembler substitutes the appropriate set of basic instructions for the Macro instruction.)

(2) 1:N Assembly: Operations are defined by the manufacturer that can be specified as a single instruction, but actually require several basic machine instructions to implement. An example would be floating point arithmetic operations on a computer system where the floating point hardware is optional. The programming is done the same whether the hardware is installed or not. If hardware is not available, the instruction is "assembled" as a set of basic instructions (cf. Macros). If the hardware is installed later, there is no change in the user's source program. It just has to be reassembled with a message to the Assembler that the hardware is available.

(3) A wider range of permissible data forms, e.g., double precision numbers, complex numbers, logical variables, literals, etc.

(4) Automatic "off-page" memory references, i.e., defining the index register contents, etc., for relative addressing rather than simply using absolute addressing or addressing relative to the program counter.

(5) Easier specification of block data transfers, data channel operations, interrupt servicing, etc.

(6) "Disk-overlays" so that parts of the Assembler program, user's program, and/or symbol table can be on disk rather than requiring them all to be in memory at the same time.

(7) "Conditional Assembly Capability", mean-

ing that a general program can be written that will handle several options, e.g., several I/O devices. When this program is assembled, instructions can be given to the Assembler to include only specific options, e.g., output to terminal #3. (Instructions like IF . . . , ELSE . . . , DO . . . are directed to the Assembler.) This saves having several almost-identical versions of the source code program file.

(8) "Pseudo-operations", which direct the Assembler to define symbols in certain special ways; data or information structures—e.g., a "block" of memory to store a matrix; instructions to set hardware registers—e.g., the program counter, or software indicators; list-processing operations. In general, pseudo-operations do not result directly in executable code, but assist the programmer in defining and generating a program. An example of a pseudo-operation was given in Table 11.9 where the address of a particular symbol needed to be calculated and placed in memory.

11.7 PROGRAMMING AND SYSTEM CONCEPTS

Once the programmer has learned how to write each instruction individually, he must learn how to combine them into programs that will efficiently accomplish the desired task within a reasonable amount of program development time. Learning how to code assembly language instructions takes a few days; learning to develop simple application programs takes a few weeks; but learning to create good systems programs requires months or years of full-time experience.

Most programming applications can be thought of in the following stages although they are not always done sequentially:

(1) Problem definition and organization
(2) Data gathering and quantifying of performance criteria
(3) Program design, e.g., flow diagrams, system interactions, etc.
(4) Coding
(5) Maintenance

The above steps are part of "software engineering" which is an increasingly important field. In most computer applications the software engineering costs, in terms of both time and money, exceed those of "hardware engineering". Furthermore, "coding" is only 10 to 20% of the software engineering task, so, although coding is important and necessary, it is only a small part of the total computer application.

Coding a single instruction such as "LOAD I" takes only a few seconds, but a good "rule of thumb" for completing all of the above steps in a practical commercial application is ONE HOUR PER INSTRUCTION IN THE FINAL PROGRAM, and this does not include any hardware development or interfacing that may be required. Most people don't believe this until they've experienced a few applications that took them ten times longer than they expected when they started. Provision must also be made for revisions, updates, additions, and long-term debugging. The following comments will give the reader a glimpse of some typical programming techniques.

11.7.1 Relocatable Programs

In their executable form all programs include binary numeric references to specific memory locations. However, since programs are usually used in groups rather than individually, it is not always possible to know where in memory a specific program will be located—hence *a priori* address specification is difficult. Some "Assembler" and "Relocatable Loader" programs are designed to handle this problem. However, another approach is to make the program independent of its starting address. A little thought should lead to the conclusion that, in principle, if all addresses were specified relative to the program counter, then the program could be loaded and executed anywhere in memory. A second alternative would be to calculate all addresses relative to a specified index register and include statements at the start of the program to set it to the proper reference value.

11.7.2 Reentrant Programs

It is possible to encounter a series of interrupts so that, at a specific point in time, a number of programs might be only partially executed. Assume there are M "nested" interrupts; that the M programs associated with the interrupts all call the same subprogram, A; that each of the M interrupts occurred at the point where the computer was in the middle of executing the subprogram A. If any data, or intermediate results, are stored within the subprogram, then obviously they will be modified every time the subprogram is restarted. Thus it will not be possible to resume execution of the lower-priority programs at the point where they were interrupted. One solution is to provide M copies of the subprogram, one for each interrupt level, but this is wasteful of memory. Another alternative is to store all the data, flags, counters, intermediate results, etc., associated with the subprogram in a "work area" of memory provided for each interrupt level and write the subprogram instructions so they are unmodified by execution. Then the subprogram is "reentrant" and can be in use by all M interrupt service programs "simultaneously". One way to allocate this "work area" efficiently is to use a "stack", i.e., a section of memory that expands and contracts (within strict bounds) as data and intermediate results are added to or removed from the stack.

11.7.3 Table-Oriented Programs

In real-time applications it is common for several programs to need access to the same data, e.g., to measurements taken from process equipment. In this case it is best to store the information in a table that is independent of any one program. Once the table layout is defined, any program can read or write data in the table by using addressing relative to the starting address of the table. (By convention this address can be stored in a specific location on page zero so it is "available" at all times.) Alternatively, a program could be written to "manage" the data base, and all requests to read, write, or edit data would be routed through this one program. Care must be taken to prevent mistakes due to interrupted programs that only partially update the table before a second program references it, or due to programs getting out of synchronization, e.g., reading data from the table before the data are written in. This can be done through a carefully devised system of software "flags" (or busy indicators) and/or "interrupt masking".

11.7.4 "Optimized" Programs

Once a program has been written, it can be modified to optimize factors such as response time, memory required, execution time required, number of disk transfers or I/O operations required, etc. The result is rarely truly "optimal", but it is frequently possible to make significant improvements with a moderate effort. Optimizing compilers and/or assemblers are available for some computer systems, but they are not yet in general use. However, when available, they should be considered for all real-time applications that are executed repeatedly and/or where time is critical. The extra time required to optimize a program (up to ten times longer than an "ordinary" compiler) is more than made up by the improved efficiency at execution time.

11.7.5 Cross Assemblers

In most minicomputer applications an application program is assembled using the same computer on which it will be executed, or on a "development computer" from the same hardware series. However, the Assembler is simply a program and can be written to run on any computer. A number of "cross assemblers" are available which run on large, interactive, data processing computers, such as an IBM 370 series, and produce object code that can be executed on a mini- or microcomputer. This possibility should always be investigated because the larger computer may be more readily accessible, have more convenient peripherals for program development, have a cross assembler that is more powerful than the assembler for the actual minicomputer, etc. In some cases the user can also debug and/or "execute" his application program on the large computer using "simulators" or "emulators". (All of these ideas are discussed more fully in Chapter 12.)

11.7.6 Micro-programming

The two most common levels of application programming are "high-level languages" such as FORTRAN, and assembly or machine-level languages such as discussed in this chapter. However, there are an increasing number of (mini) computers being marketed that are user-programmable at the hardware, "micro-code" level. Conceptually this can be thought of as simply a third, more basic level of programming.

Assume that a single, high-level language instruction can be replaced by a *set* of assembly language instructions (i.e., a "MACRO") that performs the same function. In the simplest case the assembly language op codes are simply mnemonic representations of binary codes that can be interpreted by computer hardware. However, in a microprogrammed computer the assembly language instructions are really just subroutine, or MACRO, calls to sets of instructions implemented in "micro-code". The micro-code is usually written by the computer vendor and is invisible to the user. However, some manufacturers are now providing micro-assemblers, editors, loaders, etc., so that sophisticated users can write some or all of their programs in micro-code. The advantages of using micro-code rather than assembler-level instructions are analogous to the advantages of assembler-level over high-level languages: more efficient programs, faster execution, more direct access to basic hardware capabilities, etc. The disadvantages are also analogous: it is more tedious and time-consuming to program, requires more knowledge of basic machine functions, is harder to debug, etc.

It is the author's opinion that *nobody should try to micro-code a computer until he has had extensive experience developing, executing, and debugging assembly language programs on the same computer.* The micro-coded part of a computer system functions in many ways like a separate computer that controls and/or emulates most of the functions seen by the assembly programmer. That micro-coded part of the computer has its own separate architecture, registers, memory, instruction set, etc., and can differ significantly from the system that the assembly programmer sees. However, most of the concepts of programming in assembly language can be extended to micro-coding; hence the recommendation that micro-coding be left until assembly language programming is mastered.

11.7.7 Firmware

Most of the discussion in this chapter assumes that the program instructions and data are stored in random access memory (RAM) and can be changed or modified by the user at any time. However, increasing use is being made of read-only memory (ROM). As the name implies, the contents of ROM can be read at any time but cannot be modified. The advantages of ROM are usually faster access speeds and safety; i.e., the contents, which might include important system or application programs, are protected from unintentional modification. Some computer manufactures supply special hardware and software so that users can take any program, "burn" it into ROM (i.e., store the program in the ROM), and then install the ROM as part of the regular system memory (or of the microcode memory). This use of ROMs does not fit the classical definitions of "hardware" or "software" and hence has been called "firmware". However, note that except for the extra steps of "burning" and installing the ROM, the programming concepts are the same as illustrated in this chapter. It also means that copies of programs can be supplied on a ROM "chip", instead of the familiar cassette tape or "deck of cards"!

11.8 SUMMARY

This chapter has outlined the basic concepts and principles involved in programming at the assembly language level. The reader may feel that it has focused too much on the operation of the computer hardware and system software and not enough on the details of writing the actual assembly language instructions. However, it is the author's experience that the most conceptually difficult part of the assembly language programming is the definition of what specific *functional* steps must be implemented to accomplish a desired task. The results of this stage of programming are often conveniently expressed as a flow diagram (cf. Figure 11.2). The coding stage, which involves translating the flow diagram into a set of instructions that the computer system will accept, is a relatively straightforward if often tedious task. Coding is also strongly system-dependent and will require frequent reference to the documentation that describes the assembler, hardware, and system software. However, a general-purpose computer is useless without a program, and hence programming is a necessary, but also challenging and creative part of system implementation.

11.9 SUPPLEMENTAL READING

Assembly language programming is strongly system-dependent. Thus it is *essential* that the reader consult the reference books for the particular computer that he intends to use. All the major computer vendors supply books on programming, and these are often available from the local sales office at a nominal cost.

1. Flynn, M. J. and Rosin, R. F., "Microprogramming: An Introduction and a Viewpoint", *IEEE Trans. Computers, C20*, 727–731, 1971.
2. Gear, William, *Computer Organization and Programming*, McGraw-Hill, New York, 1974.
3. Knuth, Donald E., *Fundamental Algorithms*, Vol. 1, Addison-Wesley, Reading, Mass., 1969.
4. Ledgard, Henry F., *Programming Proverbs*, Hayden Book Co., New York, 1975.
5. Martin, J., *Programming Real-Time Systems*, Prentice Hall, Englewood Cliffs, N.J., 1965.
6. Soucek, Branko, *Minicomputers in Data Processing and Simulation*, John Wiley and Sons, New York, 1972.
7. Stone, H. S., *Introduction to Computer Architecture*, Science Research Associates, Chicago, 1975.

Supplemental Reading on Software Engineering
1. Jackson, M., *Principles of Program Design*, Academic Press, London, 1975.
2. Jensen, R. W. and Tonies, C. C., *Software Engineering*, Prentice Hall, Englewood Cliffs, N.J., 1979.
3. Myers, G. L., *Composite/Structured Design*, Van Nostrand Reinhold Co., New York, 1978.
4. Weinberg, G. M., *The Psychology of Computer Programming*, Van Nostrand Reinhold, New York, 1971.

5. Yourdon, E. and Constantine, L. L., *Structured Design*, Yourdon, Inc., New York, 1975, and Prentice Hall, Englewood Cliffs, N.J., 1979.

11.10 EXERCISES

It is strongly recommended that the reader become proficient in designing and implementing programs in a high-level language such as PASCAL or FORTRAN before attempting to write a major program in Assembly Language. (Some high-level language compilers will produce a listing of an "assembly language equivalent" of the original source program. Reading this listing is a good way to begin and/or to compare assembly vs. high-level languages.)

Examples suitable for assembly language programming are scattered throughout this book and the introductory programming books available from computer vendors. It is recommended that the user start with very short, very simple programs and progress, as indicated below, to more complicated problems. Note that the ease of programming depends strongly on the system hardware, software, and program libraries available for the particular computer to be used. For example, $A = 6.2 \times 3.1$ is easy to program if floating point support is available for multiplication. It is more difficult on a microcomputer that performs only integer addition.

1. *Arithmetic Operations*
   a. With integers

   $$J = 3$$
   $$K = 8$$
   $$K = (6 - 2 + J - K) \times 3$$

   b. With floating point, double precision, or complex numbers

   $$B = 3.14159$$
   $$A = 6.2 \times 3.1/0.02 + B$$

   c. "Mixed-mode"

   $$A = 6.2 \times B + 3 \times K$$

2. *Logical Operations*
   a. If $(J + K$ is greater than $M)$
      THEN $L = M$
      OTHERWISE $L = J + K$
   b. Bit manipulation using logical operations

   i) turn on Bit 5 in Word K
   ii) count how many bits in a given word are "on," i.e., equal to one
   c. Sort a group of values into numerical order
   i) positive integers
   ii) negative or positive, integer and/or floating point
3. *Character Manipulation*
   a. Count the number of blanks in an "input record" that has already been stored in memory.
   b. Change some characters stored in memory from EBCDIC to ASCII code (or any other codes).
   c. Convert an integer to floating point and vice versa.
4. *Input/Output Operations*
   a. Write a message to a terminal

   THE RESULT IS K = 7

   b. Write a message to prompt the user from the terminal

   ENTER VALUE OF J

   and then read the value from some input device, e.g., the terminal keyboard
   c. Read/write values via the process interface:
   i) digital inputs/outputs
   ii) analog inputs/outputs
   iii) pulse inputs/outputs
   d. Read/write information from/to a peripheral that requires "control" as well as input/output information, e.g.,
   i) line printer (page control)
   ii) magnetic tape (rewind)
   iii) a process instrument that must be "set up" before it is used (e.g., a programmable gain amplifier and multiplier that must be set before reading analog input)
5. *Interrupt Servicing and Polling*
   a. External interrupts
   i) When a process switch is turned on by the user, write the time on the terminal printer.
   ii) Count the number of times a switch is turned on and off. (Note: make sure you count only one for each switch actuation.)
   b. Internal interrupts (or checks via polling) Assume some system device (e.g., line

printer) enters an error condition (e.g., runs out of paper). Diagnose the condition and print a message.

   c. Do some operation every N seconds, e.g. Read a digital input every 2 seconds for a total of 5 minutes.

6. *Interactions with the Operating System*

   a. Call an existing subroutine, e.g. to get the time or system status

   b. Schedule an existing program, then abort it.

   c. Multitasking

      Schedule several different programs to run asynchronously; then, when particular operations are complete, combine the result produced by each separate program into some overall action or report (cf. Chapter 13).

12

Utility and Systems Software

James Wm. White
Modular Mining Systems
Tucson, Arizona

Joseph D. Wright
Xerox Research Centre
Mississauga, Ontario

12.0 INTRODUCTION

Modern small computer systems are sold in a seemingly endless variety of configurations. Central processor capabilities may be enhanced further by adding hardware arithmetic options (integer-only or integer and real) or extended memory addressing capability. Most Input/Output control sections may be expanded to support the multitude of peripheral equipment that floods the marketplace today. These additions increase the complexity and size of the operating system and may, in turn, require very specialized software which is only mentioned briefly below.

Note that special utility routines must be provided to simulate the action of multiply, divide, and other functions (e.g., exponentiation) for systems without any arithmetic hardware. For systems with an integer-only hardware arithmetic option, floating point computation software must be provided to access this hardware and perform the desired floating point operation. Use of such software is always much more costly in time than the use of hardware, and because of this, timing considerations are usually stated for each version offered. For computer systems whose market is in a specific area (e.g., on-line data entry), software packages designed for a specific application are common. These specialized application-oriented packages frequently are geared to specific lines of terminals or peripheral equipment whose characteristics are exploited

through software and generally do not transfer well into existing operations. As another example, statistical packages are usually available for scientific and engineering users. These come either as additions to the FORTRAN or BASIC language libraries or as independent programs which themselves rely on an existing high-level language library. Statistical packages generally are available only for medium-size computer systems. Computer graphics utility routines, however, are finding considerably wider application as graphics terminal costs decrease. These software packages usually are provided as language extensions or as subprogram additions to existing libraries. They include packages for on-line or off-line plotting as well as CRT-oriented high-speed graphics and can be expensive. Furthermore, their implementation can appreciably increase the size of programs using them.

Now the rather general phrase "utility and systems software" usually refers not to such specialized applications packages but rather to vendor-supplied programs whose function is to facilitate effective use of a particular small computer system. We must recognize, of course, the key importance of the operating system itself for coordinating the use of all other programs whether user-written or vendor-supplied. Chapter 13 discusses real-time multitasking operating systems in detail. We should note in passing that some operating systems support background as well as foreground (real-time) operations. In this case the operating system may time-slice the background or support several background areas for multiple concurrent program development activities. Other operating systems will support but one background program development activity concurrent with real-time operation. Some systems allow overlap of background and foreground memory areas, whereas others do not distinguish between background and foreground areas at all but rather allocate memory and other system resources based on priority, need, and system-manager-established limits. Utility routines primarily serve housekeeping functions and make life with any given operating system more enjoyable. Here we deal with those major classes of utility software that accompany all but the most primitive small computer configurations. Table 12.1 summarizes the major systems and utility software components with a brief description of each.

In addition to the major software classes summarized in Table 12.1, there are literally hundreds of other utility and service routines (e.g., a tape "header" program for systems equipped with a paper tape punch) available either from the vendor or through a vendor-supported user group.

It is virtually impossible to compare the various features of these utilities by vendor or, indeed, by product line for a given vendor. Such detailed considerations would themselves fill a book equal in size to this one. Rather, we choose to describe utilities in their most general qualitative terms using a few specific examples. We believe that this approach is sufficient for an overall grasp of the concepts involved. However, detailed understanding and instructions for use of specific systems and utility software must be obtained from the appropriate reference manuals for a specific computer system.

Note that such software may be "bundled", i.e., provided "free" on purchase of the appropriate hardware, or, as is usually the case with small computer systems, it may be "unbundled" which means simply that hardware and software are sold separately.

12.1 INPUT/OUTPUT SUBSYSTEMS

A well-designed and implemented Input/Output Subsystem (IOSS) is a key part of any real-time operating system. This modular system component greatly simplifies I/O programming in a real-time environment. The user can call for an IOSS function from either an assembler- or a compiler-level program. Usually, the function to be performed, the location and size of the associated memory area, and the peripheral unit are specified. Other information related to proper processing of I/O requests is contained in parameter tables called *Device Control Blocks* (DCBs) discussed in detail below. Each physical unit in the system must have an associated DCB. Furthermore, there must be some type of device driver and interrupt handler for each peripheral device. A separate and distinct device driver can be supplied for each required device and configured into the system. Alternately, a single reentrant general-purpose device driver can be used. The IOSS allows specification of an I/O operation to be independent of the steps actually needed for a particular device. The general sequence is: OPEN (e.g., punch leader on paper tape), one or more READs or WRITEs, and CLOSE (e.g., punch trailer on paper tape). We can distinguish between three types of IOSS operations which may occur on an I/O bus (for systems with programmed I/O):

- *Programmed I/O* occurs whenever an instruction in the I/O group is executed by the I/O driver routines.
- *Program Interrupts* allow a peripheral device controller to interrupt a running program when it requires service, e.g., to signal completion of I/O. This priority interrupt system is used by the interrupt processing routines in the IOSS.
- *Direct Memory Access* allows the I/O driver to initiate the transfer of a block of data which, once initiated, runs to completion under supervision of the peripheral controller.

System input/output is initiated by user programs or tasks (cf. Chapter 3). One objective of the system is to maximize the throughput of any peripheral device and to minimize the time spent waiting for data transfers in the event some higher-priority task is ready to execute.

Table 12.1 Utility and Systems Software Components

Software	Comments
Operating Systems	Coordinate use of other system processors (programs) and utilities; the key software element. (cf. Chapter 13)
Input/Output Subsystems	Coordinate all device-independent input or output to or from executing programs.
Text Editors	Allow creation and/or modification of source code; a good editor is absolutely necessary in a noncard environment.
Assemblers	Translate assembly source code to relocatable machine code for input to a linking loader or to absolute machine code for input to an absolute loader. (cf. Chapter 11)
FORTRAN Compilers	Translate FORTRAN source code to relocatable machine code for input to a linking loader; several versions are usually available. (cf. Chapter 15)
BASIC Interpreters	Translate and execute BASIC statements as encountered and usually operate as a single background program or in a time-sharing mode under a real-time executive. (cf. Chapter 14)
Loaders	Load programs in relocatable or absolute format from any device supported by the resident input/output subsystem.
Memory Load Builders	Accept one assembled or compiled user mainline routine and any necessary user support routines, append any required system service routines and build a relocatable or absolute memory load; overlay load capability is usually provided with bulk systems.
Foreground Job Processors	Accept the linked relocatable output from a Memory Load Builder and generate an absolute memory load suitable for execution under control of the Foreground Executive.
Debuggers	Permit on-line examination and modification of memory contents (and disk sector contents in bulk systems).
Execution Monitors	Supervise execution of foreground programs under control of the Background Executive; some systems also output program execution time histograms.
Map Sorters	Produce a load map with references sorted by label and/or sorted by loading address; especially useful for large applications.
Concordance Generators	Produce cross-reference listings of computer programs.
File Comparators	Allow byte-by-byte comparison of two input files to ensure that they are exact duplicates.
Backup Processors	Provide system backup in bulk systems by duplicating tapes or disks in use onto some suitable storage medium.
Cross-Assemblers	Assemble small computer assembly source code on a suitable host machine and output relocatable or absolute machine code acceptable to the loader of the target computer.
Cross-Compilers	Compile small computer FORTRAN source code on a suitable host machine and output relocatable or absolute machine code acceptable to the loader of the target (usually small) computer.
Simulators	Allow time-precise register-by-register simulation of small computer program execution on a larger host machine; especially useful in dedicated machine applications with limited peripherals.

In general the device handlers are initially called by the Macro I/O commands or, in FORTRAN, by subroutine calls. Note that a WRITE command in FORTRAN is really a subroutine call to a device such as a line-printer. A system call to a device initialization routine is the first step in initiating I/O. First of all, the device is checked for availability, since there is no guarantee that some other task is not already using the device. If the device is available, the I/O operation is initiated, i.e., the request for data transfer is executed. If the device is not available, the I/O request must be stacked or placed in a queue of waiting requests for the device. Once the request for an I/O operation has been made, the task from which the request originated is usually suspended until the operation is complete (cf. Chapter 13).

Given that an I/O operation has been initiated, the operating system can expect an interrupt at some future time dependent upon the speed of the device and its hardware priority relative to other devices in the system. The priority control program, which acknowledges all interrupts, determines which device caused the interrupt and returns control to the interrupt servicing section of the device handler once an interrupt has occurred. At this point, the device must be disabled from the interrupt mode, and the handler becomes a task competing with all other tasks for CPU time to execute. Generally handlers have a higher priority than user tasks, and are ordered according to a prearranged hierarchy of devices. For example, a terminal always has lower priority than a disk, and, in most cases, powerfail has the highest system priority. Note that the CPU time spent servicing a device is very small relative to the time required for computation in user programs. Hence, the CPU time spent in this function does not normally affect the user tasks and is considered to be part of the system overhead.

Finally, within each device control block there must be a data section for parameters associated with the system I/O command. Such parameters include the number of characters or data blocks to be moved, the format of the data, the memory locations to or from which data are to be moved, error return addresses, and addresses of interrupt data which must be stored by the interrupt routine. Note that some of these functions are provided by hardware and/or software depending on the particular computer being used. The interrupt data area allows for reentrancy of the interrupt handling program with the possible restriction that once a particular device has generated an interrupt, it is prevented from interrupting again until the interrupt request has been completely serviced.

Proper use of an IOSS under a disk-based multitasking real-time operating system can be tricky. On some systems, failure to identify I/O-oriented tasks properly to the foreground

executive can even cause a low-priority task waiting for operator input to block a higher-priority task needing the same area of memory. I/O operations should be buffered for certain slow-speed peripheral devices operating in a real-time system. For example, buffered I/O eliminates the possibility of a background area of memory not being available for use by foreground programs, while a background program waits for I/O to or from a slow-speed peripheral device. Buffered I/O is generally used with keyboard input devices and paper tape readers and punches. The buffer, or character storage area, is usually equal in size to one line on the output device and is located within the device DCB, which itself is part of the resident monitor.

While the vendor normally supplies ASCII source and relocatable object code for DCBs of common peripherals, the nonstandard DCBs must be developed in assembly language by the user following "skeletons" or "templates" provided by the vendor. A typical buffered DCB appears in Table 12.2. Developing these DCBs and integrating them into the operating system can be a nontrivial task. The DCB in Table 12.2 contains permanent peripheral unit information (e.g., parameters L0, L5, L6, L7, C0, C2, C4, D0, E2, E3, E4, E5) and also holds status data for the current operation (e.g., the remaining parameters). In general, a given handler or device driver is reentrant and, via the individual DCB tables, can service several peripherals of the same type.

An Input/Output Spooling Program can also prove most useful in a real-time environment, especially where multiple tasks output reports to a line-printer. Such a program, operating under an IOSS, is usually designed so that output from selected drivers can be "spooled" to a high-speed bulk storage device without modifying the driver source code. Basically, the spooling program sends all output from a given program to disk until the program CLOSEs the spooled device. The stored data are then output to the device, as a unit, as soon as the device is available and not needed by a higher-priority task. No other data can be output to the same physical device while these stored data are being output. This permits information to

Table 12.2. Typical Device Control Block (DCB)

```
*BUFFERED DCB AND INTERRUPT PROCESSOR
*FOR NONSTANDARD
*UNIVERSAL ASYNCHRONOUS RECEIVER/TRANSMITTER (UART)
*
          DEF       C$T100,C$T1A
          DEF       I$T1A
          REF       D$AMP,I$AMPA,H$CHR
MXBYTE    EQU       80
          PSECT
          DC        B1
          DC        0,0,0
C$¢100    EQU       $
C$T1A     DC        'T1'          L0    PHYSICAL UNIT NAME
          DC        0             L1    TYPE
          DC        0             L2    FUNCTION
          DC        0             L3    OPCOP
          DC        0             L4    AREA
          DC        X'2201'       L5    FLAGS
          DC        D$AMP         L6    DRIVER ADDRESS
          DC        X'71'         L7    INTERRUPT ADDRESS
          DC        I$T1A         C0    INTERRUPT RESPONSE ROUTINE
          DC        0             C1    OPTIONAL PSEUDO INFORMATION
          DC        X'AE8D'       C2    OPTIONS, TERM CHARACTER
          DC        0             C3    CHARACTER INPUT STORAGE
          DC        X'C083'       C4    TTY INDICATOR
          DC        0             C5    STATUS, PAD COUNT
          DC        0             C6    CURRENT BYTE COUNT RESIDUE
          DC        0             C7    CURRENT BUFFER POINTER
          DC        H$CHR         D0    HANDLER ADDRESS
          DC        0             D1    INPUT INTERRUPT POINTER
          DC        0             D2    OUTPUT INTERRUPT POINTER
          DC        0             D3    STATUS INTERRUPT POINTER
          DC        0             D4    BUFFER TRANSFER COUNT
          DC        0             D5    TOTAL TRANSFER COUNT
          DC        0             D6    GENERAL PURPOSE COUNTER
          DC        0             D7    SUBR RET SAV INDICATOR
          DC        0             E0    UNSOLICITED EXIT ADDRESS
          DC        0             E1    CONTROL CHARACTERS
          CTRL      2,X'19'       E2    RECEIVE W/O ECHO SET
          CTRL      4,X'19'       E3    RECEIVE W ECHO SET
          DTIR      A,X'19'       E4    I/O INSTRUCTION
          CTRL      0,X'19'       E5    TRANSMIT MODE SET
*DCB BUFFER STORAGE FOR 80 BYTES
*
B1        DC        MXBYTE
B2        DC        $-$
B3        DC        $-$
B4        DC        $-$
B5        DS        (MXBYTE+1)/2+2
*
I$T1A     IENT      9                   INTERRUPT ENTRY
          LDV       X,C$T1A
          JMP       I$AMPA
          END
```

be output from one program without having data from other concurrent programs interleaved. A several-fold increase in throughput and system utilization is a further advantage of running I/O-bound tasks under control of a spooling program. Unfortunately such spooling programs increase the size of the resident monitor by up to $7FF_{16}$ words. Typically, spooled devices may be assigned and deassigned by console commands to the resident monitor. Operation of the spooling program is transparent to the user.

In addition, some sort of Historical File Management or Data Base Management package operating in conjunction with the IOSS can prove most useful in real-time applications, especially where large bases of plant data are to be generated during normal production. Such data bases can then be used for daily, shift, weekly, and monthly reporting and equipment monitoring as well as model and correlation development for certain process control applications. File management routines may be memory-resident as part of the monitor or built into the application program by a memory load builder (cf. §12.5). High-level language calls are used to insert, delete, retrieve, or update records of a user-defined random access data base. In addition, some systems provide complete file search, merge, sort, and report capability. Additional memory required can range from $16FF_{16}$ words for a basic system with its required I/O buffers to $1AFF_{16}$ for an expanded system. Although sequential and random file access capability is usually available for higher-level languages supported by the operating system, such conventional file access for on-line data storage has nowhere near the power and capability of the typical data base management system. The tradeoff is obvious— lots of memory vs. lots of programming effort. Each situation must be studied carefully and decisions reached in the context of time and resources available.

12.2 TEXT EDITORS

A source code text editor is used to prepare and/or modify files consisting of character strings and is absolutely essential in cases where paper tape, casettes, diskettes, or magnetic tape are the only I/O media. Even in systems equipped with a card reader/punch, the source text editor is frequently the preferred means for file creation or modification. A line or groups of lines may be inserted, moved, or deleted, and characters in individual lines may be modified using suitable editing commands. The text editor is definitely the preferred means for ASCII file creation or modification in most dedicated real-time systems. A truly powerful text editor should include extended search capabilities, text justification, intra- and interline character replacement, and file manipulation features. Note that editors can be restricted by the system on which they are implemented. Thus, nonbulk text editors can only operate on the amount of data that can be placed in a main memory working buffer of fixed size. Bulk-storage-oriented editors make use of one or more temporary bulk storage files for intermediate working storage, as well. Bulk-based working storage permits the bulk editor to process forward and backward refer-

ences as though all source code were in working memory. Some bulk systems even support very powerful full-screen text editors. Here a page of text is displayed on a CRT. As the user alters screen contents, the associated source file is automatically updated.

12.2.1 Nonbulk Systems

Nonbulk system source editors can only operate on the amount of data that can be placed into the working buffer of fixed size in memory. If a line is referenced that is not in memory, the current contents of the buffer must be output to a suitable peripheral and the next block from the input file read into the buffer. Such editors are limited to sequential processing of each buffer with subsequent disposition of the edited buffer contents to an output device. This may require several passes over a large file when many distributed changes are necessary. Furthermore, available editing commands are usually quite limited. These editors are most frequently found in paper tape systems operating under a simple resident monitor. Although magnetic tape systems speed up operation somewhat, they too do not perform well on large files. Usually, the editor uses but one input and one output device. Extensive editing on nonbulk systems invariably proves most frustrating.

12.2.2 Bulk Systems

Bulk system source editors also operate on data placed in a working memory buffer. These editors, however, use one or two disk scratch files for intermediate working storage in order to process forward and backward source file references as though all of the source file code were memory-resident. The bulk system editors input from a Source Input (SI) device and output to a Source Output (SO) device. These logical units may be assigned dynamically using executive commands. SI may be any input peripheral—e.g., paper tape reader, card reader, tape cartridge, etc.—or it may be a named data file residing on a bulk storage device. Source output may also be a physical unit or named bulk data file, or it may be a

Working Storage (WS) scratch file; alternately it may be specified by an editor command.

Great care is frequently needed because full source file backup (e.g., automatic creation of a suitably-named bulk-resident backup file containing the unedited source code) is rarely provided with small computer operating systems. The better editors require the user to terminate the edit session with an editor command to CREATE "FILNAM" if the source code is new and the file did not previously exist. Alternately, if "FILNAM" had been specified for SI, the CREATE will fail; an editor REPLACE command must be used instead. Though the correct code is always REPLACEd, the user must make certain he has not inadvertently deleted a large code segment before terminating the edit session.

The least desirable editor implementations output source code generated in an edit session to WS, leaving the original input source code untouched. Again, we must make sure the output code is correct. However, we must now issue a monitor REPLACE command specifying both SI and SO. This can be hazardous during development of application programs whose source code resides in multiple-named disk files because the wrong source code might—and all too frequently does—get REPLACEd!

In passing, we should note that on many systems the effect of a REPLACE command, no matter how issued, is not to actually replace old code with new. Rather the current name of the input source file is deleted from the appropriate disk directory and replaced by a system default name (e.g., ######); actual source code is not touched. The contents of WS are then moved to a new (hopefully available) disk area and given the name used during the edit session. Obviously in such systems, the disk directory gradually accumulates large numbers of system default name entries with an attendant accumulation of much wasted disk storage space. At some point, an operator command to PACK the disk must be issued to recover the otherwise unusable disc data storage areas. Such PACK operations are slow and, in general, should not be issued during periods of heavy system use.

12.2.3 Editing Commands

Editing commands are as varied as the number of vendor product lines available. Commands may be split into two groups: those that interface with the operating system (e.g., for code input, code output, name specification, etc.) and those that input to, or modify source code in, the editor buffers. It is this latter class of commands that has the largest impact on editor power and flexibility.

In general, every editor operates easily in an "input" mode (with or without an editor "prompt" character) where each record is stored sequentially in an editor buffer after input. The performance of various editors in the "command" mode (invariably with a prompt character)—where a command character is followed by special characters, numbers, or text strings—is an entirely different matter. In the command mode, the editor maintains a current line pointer and current line working buffer for use by certain commands. Editors use several special control characters—in addition to single character commands—for easy,

Table 12.3. Typical Nonbulk Editor Commands

Command	Description
T	Go to the top of the buffer.
B	Go to the bottom of the buffer.
P 10	Print 10 lines including the current line.
L STRING	Locate a string.
C OLD/NEW	Replace an OLD string with a NEW string on the current line.
I NEWLINE	Insert a line after the current line.
D	Delete the current line.
N	Go to the next line in the buffer.

effective interactive editing. Most editors are line-oriented and insert line numbers before each line as the core buffer is filled from the input device. When sequential line numbers are in the form 10, 20, 30, etc., line insertions are identified by a line number between the preceding and following lines. If the source code lines are numbered sequentially, e.g., 1, 2, 3, etc., then insertions between lines 2 and 3 are identified as 2.1, 2.2, etc.

Primitive editors used with smaller, limited-capability machines have a somewhat restrictive command repertoire. Also, since they do not number buffer lines, the user must keep track of the current line pointer. As a minimum, such editors should at least include the single character commands of Table 12.3 though many provide more capability. Nonetheless, such editors remain difficult to use.

Let us now turn to consideration of bulk editor command features for interline as well as intraline editing. This type of editor is usually used as a background processor in disk-based multitasking applications and generally has more memory in which to operate than does the nonbulk editor discussed previously. Table 12.4 summarizes the features that a good bulk text editor should support.

The best editors for small computer systems provide some exceptionally powerful and useful features such as the "sticky" string. Thus for Items 11 and 14 in Table 12.4, reissuing the interline edit command character without any string would cause a repeat of the previous action on the next occurrence of the old string. This feature, when combined with Items 12 and 15, can decrease editing time by a factor of 5 to 10 in many cases.

A simple illustration will perhaps further indicate some of the differences between a poor editor and a good one. In the example of Table 12.5, we assume that we are in the interline edit mode with the current line pointer at Line 50. We wish to replace the string TCRL by the string CTRL everywhere and then find the second occurrence of the string DTIR and delete that line.

Given the amount of editing required in real-time installations, time invested in careful evaluation of key utility software such as editors prior to purchase is well worthwhile.

Nonetheless, such considerations are often overlooked.

12.3 SOURCE CODE PROCESSORS

All languages at a level higher than absolute machine code can be classified as either Translators (e.g., Assemblers and FORTRAN compilers) or Interpreters (e.g., BASIC). Translators and interpreters differ chiefly in their method of processing source code. Translators accept alphanumeric input and, via one or more passes over the source code, reduce it to an equivalent sequence of (usually) relocatable machine language instructions and data. Such output must then be processed further—usually by a Linking Loader or Memory Load Builder—before execution can begin. Interpreters, on the other hand, decode or interpret and execute each line of source code as it is encountered. Thus translators process source code as a file of sequential input records, while interpreters follow the logic flow of the program. Interpreters are always much slower in execution than translators because interpreters must decode a given source line each and every time it is encountered in executing the user program.

12.3.1 Assemblers

Assemblers (cf. Chapter 11) are designed to read source text in symbolic machine language and produce an equivalent (relocatable or absolute) machine-oriented module suitable for loading into computer memory. Almost all assemblers have variations which in one way or another reflect the operating system under which they run or the hardware configuration of the host computer. In general, assemblers, as with most of the utility programs discussed in this chapter, are not real-time programs; rather they execute as part of *non*-real-time or background operation. Basically, assemblers perform a one-to-one translation of source code to relocatable or absolute object (i.e., machine) code and are the lowest level languages in which program development is at all practical. Of course, most assemblers also include "pseudo operation" codes and "macro expansion" capability. The former are simply directives to the assembler (e.g., set a number base), while

Table 12.4. Typical Bulk Editor Commands

Feature	Interline Command	Note
1	Insert new line after current line.	1
2	Insert new line before current line.	2
3	Replace current line with new line.	1
4	Delete current line.	1
5	Delete M lines starting at line N.	1
6	Transfer line M to after line N.	2
7	Transfer a range of lines to after line N.	3
8	Copy line M to after line N.	2
9	Copy a range of lines to after line N.	3
10	Merge lines M and N; delete original line N.	3
11	Locate and list first occurrence of string after current line.	1
12	Locate and list all occurrences of string after current line.	3
13	Append a string to the current line.	2
14	Replace first occurrence of old string after current line with new string.	1
15	Replace all occurrences of old string after current line with new string.	3
16	List the first, current, Nth or last line.	1
17	List the first through last lines.	1
18	List a range of lines.	2
19	Renumber lines by increment N.	4
20	Enter intraline command mode.	5
21	Copy text from/to another file after current line. Delimiters shall be line numbers or strings.	3
22	Delete next line starting at current line containing string.	3
23	Delete all lines starting at current line containing string.	3
24	Delete, starting at current line, all lines between the line containing string 1 through the line containing string 2.	3
25	Locate and list text from line containing string 1 to line containing string 2.	3

Feature	Intraline Command
1	Scan for the Nth occurrence of character.
2	Change the next N characters
3	Move intraline pointer right (or left) N characters.
4	Insert N characters before (or after) the current character.
5	Delete N characters starting with the current character.
6	Transpose the next two characters before (or after) the current character.

Notes: (1) An absolute must.

(2) Almost an absolute must.

(3) Not implemented in many cases, though exceedingly useful.

(4) Useful especially when line numbers are saved for use in subsequent edit sessions.

(5) Though exceedingly powerful, most small computer editors do not have a separate intraline edit mode; rather additional special characters are used in the interline edit mode to effect intraline edit action.

the latter cause an appropriately defined block of source code to be expanded, modified by argument parameters, and inserted wherever the macro name is used. All assemblers invariably include a permanent symbol table so that user-oriented operation code mnemonics can be converted into absolute machine code instructions.

In general a line of assembler source code appears as:

$$(\text{LABEL}) \rightarrow| \text{OPERATION} \rightarrow| (\text{OPERAND}) \rightarrow| (\text{COMMENT})$$

where the parentheses indicate an optional line component and the symbol $\rightarrow|$ designates a delimiter (usually one or more spaces or a tab character). Even when the assembly language program consists of a series of consecutive one-line statements, the assembler cannot simply input each statement, translate that statement and output a line of machine code to a *Binary Output* (BO) device (e.g., a disk, diskette, or paper tape punch). This characteristic is a direct result of the "forward-reference" problem. Thus for an unconditional branch (e.g.,

Table 12.5. The Poor vs the Good Editor

Line No.	Code	
50	TCRL	2,X´15´
60	TCRL	4,X´15´
70	DTIR	A,X´15´
80	DTIR	0,X´15´

Action	The Poor Editor	The Good Editor
Print current line:	@$,P ⟩	P$ ⟩
Replace 1st string:	@´TCRL´,´CTRL´ ⟩	STCRL$CTRL$ ⟩
Replace 2nd string:	@´TCRL´,´CTRL´ ⟩	S ⟩
Find 1st string:	@´DTIR´,P ⟩	FDTIR$ ⟩
Move to next line:	@+80 ⟩	not necessary
Find 2nd string:	@´DTIR´,P ⟩	F ⟩
Delete 2nd string:	@$,1 ⟩	DS ⟩
Total Key Strokes:	65	29

Note: ⟩ = RETURN Key

JMP) on Line m to a label on Line n (m < n), the Line m containing the JMP cannot be assembled and output until Line n containing the referenced label is encountered. Such problems may be handled by:

- Single-pass assembly that requires that all forward references be resolved by the programmer.
- Single-pass assembly that resolves forward references by building a forward-reference table that is processed only after all source code has been input.
- Multipass assembly that on the first pass builds a user symbol table containing the name and location of all labels relative to the start of the module and which uses this information on subsequent passes to resolve forward references.

The assembly process, regardless of the number of passes, requires that all user references within a single program or subprogram be placed in a "symbol table". Each entry of this table indicates the symbol, its value (location), and whether it is external to the program module being assembled, or an absolute or relative (displacement) address within that same module.   The complexity of the assembly process is essentially one of symbol table manipulation and look-up.

The simple single-pass assembler that cannot handle forward references is best suited for very small computers—especially those with no bulk storage.   The burden of satisfying forward references, starting location, and other book-keeping chores is most unpleasant for the user/programmer.   Assembly is reduced to essentially a direct one-to-one translation of the symbolic text into machine object code. The lack of versatility and the imposition of these additional tasks on the programmer makes single-pass assemblers of this type generally undesirable. They should be avoided if at all possible. The more versatile single-pass assembler saves statements containing forward references in a forward-reference table and only assembles such statements after all source code has been processed. Such one-pass assemblers tend to be more complicated and require temporary memory storage in proportion to the number of forward references present. Further, source/object listings can be confusing, since the output object code sequence can differ from that of the input source code. Nonetheless, one-pass assemblers of this type can be very useful with small computer systems whose only I/O device is a low-speed terminal that

Table 12.6. Macro vs. Subroutine Characteristics

	TECHNIQUE	
CHARACTERISTIC	MACRO	SUBROUTINE
Assembly Time Processing Overhead	Definition Storage and Macro Expansion	None
Execution Time Processing Overhead	None	Argument Handling and Return Instructions
Occurrence of Code in Module	Once for each Macro Reference	Once

includes an integral paper tape reader and punch.

Multipass assemblers, most of which use only two passes, handle the forward-reference problem in an entirely different manner. Construction of a user symbol table is the primary function of the first assembler pass over the source code. Each symbol table entry contains the symbol itself (or a portion of it), as associated numerical value, and, usually, other information as well. On the second pass, the assembler outputs an (optional) source listing, an (optional) object code listing, and, for bulk storage systems, (optional) absolute or relocatable binary code for subsequent input to a loader.

The need to repeat short, almost identical instruction sequences at several points in an assembly language module is quite common. For example, DCB code segments for asynchronous communication channels can require from 20 to 30 source lines with only minor variations in baud rate, buffer size, etc. Such coding is best handled through the use of macro instructions or "macros" rather than the obviously tedious process of keying in such repeated code segments manually. Almost all multipass assemblers for small computer systems include such macro definition and expansion capability. However, macros should not be confused with subroutines. The differences are indicated in Table 12.6.

Macros are especially useful where the object code, though almost identical, includes numerous execution-time modified scratch pad areas, buffers, pointers, status flags, and the like as in the case of device DCBs. In other cases, subroutines may be preferred, depending on code segment size, the number of subroutine arguments, and the run-time subroutine call overhead. In general, for very short repeated program segments, a macro is almost always more efficient at run time than a subroutine call.

Upon encountering a macro definition, the assembler stores the macro and associated arguments in a macro definition table. When the macro name occurs subsequently in the source code, the definition is retrieved and expanded using current arguments. Note that exapansion

Table 12.7. Conventional Assembly Coding vs Macro Usage

Conventional Source				Macro Assembler Source			
Line No.				Line No.			
1	START	LDR	A,NUM1	1	#EXCHNG	LDR	A,#1
2		LDR	B,NUM2	2		LDR	B,#2
3		STR	A,NUM2	3		STR	A,#2
4		STR	B,NUM1	4		STR	B,#1
5		LDR	A,NUM3	5		EXCHNG	NUM1,NUM2
6		LDR	B,NUM4	6		EXCHNG	NUM3,NUM4
7		STR	A,NUM4	7		DSECT	
8		STR	B,NUM3	8	NUM1	DS	X'B'
9		DSECT		9	NUM2	DS	X'C'
10	NUM1	DS	X'B'	10	NUM3	DS	X'D'
11	NUM2	DS	X'C'	11	NUM4	DS	X'E'
12	NUM3	DS	X'D'	12	END		
13	NUM4	DS	X'E'				
14		END					

(i.e., replacing the macro name and optional arguments by actual object code) occurs at assembly time not run time. Table 12.7 illustrates the difference between conventional code and macro assembler code. The program in this example simply exchanges the contents of two pairs of specified core registers (NUM1 and NUM2 or NUM3 and NUM4) using two working registers (A and B). Though both are functionally equivalent, note that each additional "exchange" using the macro assembler version would require only one extra line of source code instead of four.

As a final point, we should discuss the class of assembler directives called pseudo operations or instructions. Though such "pseudo ops" must appear in the operator field of the source code record, they do not in general cause generation of object code although memory size can be affected, e.g., as with a pseudo op that reserves a block of memory. Rather they set flags internal to the assembler itself and thus affect subsequent processing of source code. Table 12.8 summarizes just a few typical pseudo ops. Note the use of a special character ($) to identify this special class of instruction. Actual options are almost as varied as the number of vendor product lines.

12.3.2 Compilers

Compilers in many respects are much like assemblers in that they accept source code records in some higher-level language (e.g., FORTRAN, PL/I, PASCAL), make one or more passes over the source code, and generate relocatable object code modules suitable for input to a linking loader or memory load builder.

Many small computer compilers also accept in-line assembly code. This feature can be especially useful during development of special-purpose device drivers, since part of the required code can be written at the compiler level with special features (e.g., Input/Output Transfer instructions) handled at the assembler level. The ability to generate reentrant code can be an extremely important consideration in real-time systems. Reentrant code can be interrupted during execution, used by another (higher-priority) task, and restored to its previous state prior to resumption of (lower-priority) task execution. Thus a single copy of the reentrant routine can be built into the resident monitor and subsequently used by several unrelated tasks rather than building a separate copy of the routine into each program or task that

Table 12.8. Some Typical Pseudo Instructions

Assembler Directive	Action
►| $ABS	Output an absolute binary tape preceded by a binary loader.
►| $ASCII (string)	Convert character string to internal binary.
►| $BLOCK (expression)	Reserve a block of memory.
►| $BYTE (value)	Reserve a byte of memory and assign value.
►| $END	End of source code input.
►| $GLOBL (name list)	Identify procedures external to current module.
►| $LOC (abs location)	Set or reset the program counter.
►| $OCT or $HEX or $DEC	Set the radix (number base) as 8, 16 or 10 respectively.
►| $REPT (count, inc)	Repeat next instruction "count" times incrementing absolute value by "inc".
►| $TITLE	Set up a label to appear at the top of each page of output.

references the routine. Depending on the routines, this can result in a large savings in roll time, since many roll modules will be much smaller. In some cases, such user-developed shared code is not counted as part of the task memory load. This feature can be important as the task memory maximum is usually only 28 KW (28,000 words) to 32 KW (sometimes only 16 KW!) in 16-bit machines.

A few compilers can also generate recursive code (i.e., a routine can call itself). This feature is usually of minor importance, however. Some small computer compilers even generate a pure procedure code section for code that is not modified during execution (e.g., instructions) as well as a separate data code section for program variables, registers, and other items that are normally modified during execution. When such code is properly processed to run in separate (not necessarily contiguous) areas of memory, system throughput is increased because the pure procedure code section does not ever need to be rolled out (since it never changes), only rolled in.

Most compilers perform three functional steps: source code scan, source code parse, and object code generation. The scanner handles lexical analysis (i.e., spelling and punctuation). "Tokens" output by the scanner are passed to the parser which handles the syntactic analysis (i.e., grammar). Only after parsing is complete, can relocatable object code be produced by the object code generator. Obviously, as the complexity of acceptable source code increases, the compilation process can become quite involved. Indeed, compilers can be evaluated in terms of the allowable complexity of source code as well as the diagnostic aids provided. Simple compilers accept expressions of limited complexity, prohibit use of certain useful language structures (e.g., an implied "DO" in a FORTRAN DATA statement), require strict adherence to a statement-ordering hierarchy, and generate diagnostics of limited utility (e.g., "statement in error"). This type of compiler may or may not output assembly code equivalent to the input source and a symbol table, both of which can prove most useful under the right circumstances. Such compilers are usually provided with small computer systems having limited memory (less

than 8K words or 16K bytes) although they (or only slightly more powerful versions) are sometimes found in much larger systems. The headaches involved in program development using simple, limited-capability compilers should be obvious. Fortunately, modern multitasking real-time systems offer much more sophisticated compilers—and a selection of them at that—which provide most of the useful features (e.g., in-line assembly code) discussed earlier. In addition, complex source statement structure is permitted, and extensive diagnostics are provided. Detailed symbolic reference tables and cross-reference maps are (optionally) produced as well.

The degree of optimization of object code output is yet another measure of compiler effectiveness. While optimizing compilers invariably require more CPU time during compilation, the object code generated can be quite efficient. Indeed, some vendors claim that the code produced by their optimizing compiler is as efficient from the standpoint of main memory and execution time required as code produced by an experienced assembly language programmer.

In general, optimizing compilers employ several techniques to produce efficient object code. Among them are:

- Proper hardware utilization
- Redundant operation suppression
- Subexpression result retention
- Computation replacement by constant
- Equivalent operation conversion
- Branch minimization

Nonoptimizing compilers translate each line of source code to its equivalent object code without concern for other source code lines, whereas optimizing compilers retain expressions and subexpression results for possible later use. This is but one of many aspects of the optimization process during compilation. Table 12.9 illustrates some typical output from two small computer compilers—one that optimizes object code and one that doesn't. Note that the optimizing compiler does not recompute the value of the expression "J + (K - L) * J" on the second encounter; rather it reloads a working

Table 12.9. Comparison of Small-Computer Compilers

```
    Non-Optimizing Compiler Output                        Optimizing Compiler Output

C                                              C
C       REDUNDANT SUB-EXPRESSION,              C       REDUNDANT SUB-EXPRESSION
C                                              C
        I = J + (K - L) * J                            I = J + (K - L) * J
            LAC    K                                        LDA    D+0006      K
            JMS*   .AY                                      LDR    C,D+0007    L
            LAC    L                                        SUB    A,C
            JMS*   .AD                                      LDR    C,D+0005    J
            LAC    J                                        JSR    *$+0001     F$M00
            TAD    J                                        ADD    A,C
            DAC    I                                        STA    D+0008      #AIA
        K = J + (K - L) * J                                STA    D+0004      I
            LAC    K                                    K = J + (K - L) * J
            JMS*   .AY                                      LDA    D+0008      #AIA
            LAC    L                                        STA    D+0006      K
            JMS*   .AD                                             •
            LAC    J                                             •
            TAD    J                                             •
            DAC    K
                •
                •
                •
C                                              C
C       IF STATEMENT                           C       IF STATEMENT
C                                              C
        IF (K - J) 10,11,10                            IF (K - J) 10,11,10
            LAC    K                                        LDA    D+0006      K
            JMS*   .AY                                      LDR    C,D+0005    J
            LAC    J                                        SUB    A,C
            SPA                                             SKN    $+0001      )10
            JMP    .10         11               J = 1
            SNA                                            RTN    A,Z
            JMP    .11                                     STA    D+0005      J
            JMP    .10         10        CONTINUE
11      J = 1                                                 •
            LAC    (000001                                    •
            DAC    J                                          •
10      CONTINUE
            •
            •
            •
```

register (LDA D + 0008) with the numerical value saved earlier in a temporary memory location (#AIA) and then stores (STA D + 0006) the value in the proper location. The nonoptimizing compiler, however, generates essentially identical code for the two statements; only the final stores (DAC I and DAC K) are different. The execution time differences for complicated double-precision calculations are obviously appreciable. Also consider the treatment of the simple arithmetic IF statement in Table 12.9. The nonoptimizing compiler generates multiple conditional branches, whereas the optimizing compiler recognizes that only the zero condition need be checked. As yet another example, adding the number one to a variable requires the average compiler to generate code to: load a working register, load 1 into another working register, add

the registers, and store the result. The optimizing compiler on the other hand generates a single instruction to increment the variable value by 1. Optimizing compilers also replace structures such as $R**2$ by $R*R$, $R*1$ or $R**1$ by R, $R**0$ by 1, and so on.

Our comments apply to all compilers (e.g., FORTRAN, PL/I, PASCAL, etc.) available on small computer systems. Indeed, since program development is usually a small portion of the real-time system work mix, object code optimization should be a key concern in evaluating competing utility software options.

12.3.3 Interpreters

Interpreters, unlike assemblers and compilers, do not convert source code records in a higher-level language into object code at the machine language level. Although the source code is input to the interpreter, it treats the programmer's statements something like "data", interpreting them as a sequence of standardized operations and calling a set of routines within the interpreter itself to carry out these operations. Certain important features of interpreter-based languages derive from this characteristic manner of implementation; for example, many users are aware of the primary disadvantages of interpreter languages in time-sharing or batch applications, viz., the relative slowness of execution of interpreted programs compared to compiled or assembled programs. Relatively few users recognize that this feature can be unimportant in real-time applications and that, on the other hand, interpreted code can be much easier to debug than compiled code, particularly in a small-scale application.

An interpreter may be designed for any high-level language including those that ordinarily are compiled; however, some languages most often are found in the interpreter form: BASIC and APL are the most important examples. BASIC largely was developed as a language that would be simple to learn and use and has been adapted widely to time-sharing applications where its debugging advantages can be exploited usefully; APL, by contrast, is a very complex language that permits dynamic storage allocation during the execution phase and hence cannot be handled easily by a compiler.

12.3.4 High-Level Source Language Comparison

We indicated in the previous section that a compile operation ordinarily consists of three steps: source code scan, source code parse, and object code generation. After compiling and linking a source program, the resulting object module can be loaded and executed. The interpreter performs the first two of these steps also but only on a single source code statement at a time. It then initiates the necessary calls to routines within the interpreter to carry out the required operations; hence execution is carried out indirectly—i.e., object code (which would directly execute) is never generated. Each source code statement must be re-

interpreted each time it is encountered when looping through a program, and for this reason interpreted programs invariably run slower than compiled and linked programs.

There are, however, several techniques that can be used in designing interpreters to speed up the execution of an interpreted program; the most important of these is to "pre-scan" the source code as it is entered, converting the original code to an internal representation that is more compact and more easily interpreted. For example, almost all BASIC operations are coded by the programmer as two- or three-letter mnemonics (LET, FOR, ON, etc.). These would be reduced to a single letter or number in obtaining the final internal code that represents the user's source program.

Two important features of interpreter systems are that the internal representation can be modified easily if the programmer inserts additional source code statements or modifies existing ones, and that the source program can be reconstructed from the internal representation at any time. Both of these features greatly enhance the interactive and debugging capabilities of interpreters, since they permit the programmer to make modifications to the program without having to recompile. For this reason the pre-scan step cannot be a complete translation or interpretation operation, and the interpreter must remain to carry out the final interpretation during the "execution" phase.

The fact that the interpreter must be capable of executing all of the operations in the source language means that all of the system library routines must be resident as well. More powerful interpreters are designed with some flexibility here; for example, the user may choose to load the interpreter without routines to handle complex arithmetic or to perform user-specified input/output formatting, etc.

Table 12.10 provides a side-by-side comparison of the characteristics of interpreters and compilers (after Lee* with modification). These comparison points were developed primarily with respect to large computer operations,

---

*Lee, John A. N., *The Anatomy of a Compiler*, 2nd Ed., Van Nostrand Reinhold Company, New York, 1974, p. 32.

Table 12.10. Comparison of Interpreters and Compilers

INTERPRETERS	COMPILERS
1. Available storage must contain at the same time the interpreter, the source text, the symbol table, <u>all</u> library routines, and the data. Hence the size of program (as measured by the number of statements in the program and the number of data elements defined) is restricted compared to that available for use with a compiler system.	1. The available storage at compile time must contain the compiler, the symbol table, and <u>one</u> statement from the source text. At execute time, the available storage must contain the compiled code, the required library routines, and the data.
2. There is always a direct relationship between the source text and the code being executed; hence there is a good relationship between detected errors and the source text, which promotes easy debugging.	2. The relationship between the source text and the code being executed can be remote. Hence the burden of error/source relations may be placed on the programmer and his knowledge of the machine and its compiler. However, modern debuggers allow run-time access to variables by name.
3. Syntactic errors detected by the interpreter can be corrected during a run (at the request of the interpreter) and do not require the whole run to be restarted. Execution errors can be reported to the programmer and source text changes made under the same controls.	3. Syntactic errors can be corrected at the instant of recognition, but the loss of the source text at execution time requires that corrections be made in the source text and recompilation of the whole text be performed.
4. Owing to the successive recompilations of the statements in the source text and the need to reference all data through the symbol table, an interpretive system is expensive to use.	4. A compiler system makes best use of the available resources of the computer system.

either batch or time-shared, and need to be reconsidered for real-time, particularly small computer, applications:

(a) Point 1, dealing with size of a user application program, is much more important with mini- and microcomputer-based systems unless the unit is equipped with bulk storage and an operating system that can use it effectively. (An interpreter with the ability to chain, swap, or overlay user program segments from high-speed bulk should be able to handle user applications programs of virtually unlimited size.)

(b) Point 4, dealing with the need to reinterpret each source code statement every time it is executed, has important ramifications for applications where high computational speed is important (large processes with many variables, very fast processes, or processes that require large amounts of calculations to be performed on acquired data). On the other hand, many real-time systems are characterized by relatively low data acquisition rates and modest computational requirements. Such systems are "idle" much of the time, and there may be no penalty (expense) involved in utilizing an interpretive system.

In closing, we should point out that with educational real-time facilities, the advantages that interpreters have in debugging programs can be quite significant depending on the quality of compiler-based real-time debugging aids available. The user should look for an interpreter that includes all of the routines necessary for process communication through standard peripheral devices and that, preferably, can support multipriority (multitask) user application programs easily and efficiently. As a further important consideration, we should note that multiuser interpreter systems are available for many small computer systems. In some systems, each user program is assigned to a single partition in memory; in other systems, individual users are rolled to or from a bulk storage unit. In either case the interpreter is responsible for time-slicing among users, and care must be taken that critical user tasks can be handled by such systems according to strict, real-time standards.

12.4 LOADERS

A loader is a system utility that, once placed in memory itself, causes other programs to be brought into memory and executed if so directed. A utility program that loads a loader or other system program after power-up is called a bootstrap loader. The bootstrap is used to bring a more versatile loader into operation. Once the bootstrap process is complete, the operating system can generally be activated. Usually, the following types of loaders are provided for all but the smallest dedicated systems:

- Bootstrap
- Absolute
- Linking
- Overlay

Bootstrap Loaders are generally less than 128 instructions in length and often only 32 (two input devices) or 64 (three input devices) instructions long. The length of the bootstrap is related to the type of device from which input is obtained. For example, a paper-tape bootstrap is usually shorter than a disk bootstrap. In order to facilitate use in small computer systems, the bootstrap program should be located in read-only memory or ROM which is activated directly from a switch on the programmer console. This is a very useful feature, especially for micro- or minicomputer systems which are not kept in continuous operation, because the manual input of even 32 instructions via console switches can rapidly prove most frustrating, especially if the system tends to "crash" frequently. The bootstrap loader usually resides at the highest memory location in the machine—minus, of course, the number of words used by the loader itself. The bootstrap loader reads and directly loads binary records starting at a specified initial location. Generally, no error checks are performed except checks for parity. At completion of the load, the bootstrap either starts execution at the entry point of the module loaded or, alternately, activates the operator console terminal and enables the keyboard for operator command input. In this manner, a series of increasingly complex programs can be loaded.

Absolute Loaders are invariably found in small computer systems where a Memory Load Builder can produce an absolute object module (as in a system generation procedure) or where the associated assembler can produce absolute object modules. As a matter of fact, this type of loader is the only loader provided (except for a ROM bootstrap) with most very small computer systems. Table 12.11 illustrates such absolute (hexadecimal) loader input for a small dedicated system equipped with only a terminal with an integral paper tape reader and punch. Many microcomputer systems use a similar loader for cassette tape absolute binary files. (Note: These loader records were output by the cross-assembler discussed in § 12.9 and correspond to the code of Table 12.17.)

After initializing the ROM resident monitor, the "L" command is given to transfer control to the absolute loader. The loader then inputs absolute machine code from an appropriate input device. (Note: This loader can also input such records at high speed, e.g., 9600 baud, over an asynchronous communications channel.

Table 12.11. Typical Absolute Loader Input for a Very Small Machine*

```
L
;18 0200 A9 7D 8D 00 17 A9 FF 8D 00 17 EA EA EA AD 20 17 99 20 02 C8 CA F0 07 20 0B36
;18 0218 1D 02 4C 00 02 60 60 00 00 50 54 90 D4 80 E0 90 F0 50 90 D0 50 F0 50 40 09C7
;00 0002 00 02 KIM
```

*   Spaces in the above records are provided for clarity only. They are not used in actual practice.

That is, the small machine can be "downloaded" by a higher-level machine.) The semicolon (;) is a loader "prompt" signifying the start of a new record; the first two characters (18 or 00) specify the number of data bytes per record or signify the last record; the next four characters (0200, 0218, or 0002) specify either the absolute starting address for this record (0200 and 0218) or the total number of records (0002) that should have been loaded. For the data records only, the next 18_{16} pairs of hexadecimal digits represent the bytes to be stored in sequential absolute memory locations. The last four digits comprise the checksum for the bytes of code in that record. (The four-digit checksum is simply the algebraic sum of all data bytes ignoring overflow.) Not all loader formats are so straightforward or so easy to describe. Absolute loaders, sometimes called block loaders, are also used in most bulk-storage-based multitasking systems. These loaders, which are invariably part of the operating system itself, accept relocatable (but linked) program modules constructed by a Memory Load Builder (cf. §12.5). In response to an executive command, the block loader determines the necessary relocation factor, loads the named program into memory with all relocatable addresses simply modified by this factor, and then transfers control to the first executable statement of the named program. In both cases, the loader is passed a starting address and a word or byte count, whereupon loading begins. Some error checking, besides parity, is usually done by this loader. Sometimes loading is broken up into blocks that have individually-stated starting addresses and lengths.

Linking Loaders are commonly found in single-application systems. These loaders load and link relocatable binary program modules produced by an assembler or compiler. All external references must of course be identified during assembly or compilation and provided in a format acceptable to the loader. Initially, the linking loader loads all program units specifically named in a console load command. Usually, the main module must be the first name encountered. After these programs have been loaded and linked, the loader automatically loads and links all unresolved external modules by accessing the user and system libraries stored on disk, paper tape, or some other medium. Note that with paper tape or magnetic tape systems, user libraries must be constructed carefully, i.e., a user routine must not reference a routine preceding it on any sequential storage medium because more than one tedious, time-consuming file scan would be required to resolve all external references. The linking loader also assigns the common data storage area. In some small systems, the common area is assigned to memory occupied by the loader itself. After the loading process is complete, i.e., the loader has built a single absolute memory load from the relocatable input routines, control is usually transferred to the initial execution address of the main program. Table 12.12 illustrates a typical load map output by a simple linking loader.

The highest available memory location (i.e., the word just below the bootstrap) is occupied by the last word of the main program. Loading then continues downward in memory in this system. The loader, which resides here in low memory just above the resident monitor and device handlers, obviously cannot load user or system routines on top of itself. If such an attempt is made, a loader error message is issued, and the loading process is terminated.

In disk-based systems using multiple program counters (e.g., separate pure procedure program and variable data code segments), a two-pass link-loading procedure may be required for proper relocation. The first pass determines the total length of the module to be loaded, satisfies external references by searching an object library, and builds a relocatable linked module on bulk. The second pass loads this module after determining and applying the required relocation offset. The final result is an absolute memory-image binary module which may be saved so as to not require the two-pass loading process but rather only a simple absolute load. This loading procedure is quite similar to a memory load build operation followed by foreground job processing or by block loading, all of which are discussed subsequently.

Overlay Loaders and Chain Loaders are in many ways similar to linking loaders. However,

Table 12.12. Load Map from a Simple Linking Loader*

Module Name	Start Address	Comment
CONTRO	17052	Mainline
WAIT	16222	
WENT	16164	
START	16002	
TRNON	15454	
TIME	15274	
WRTM	15217	User
OUTPUT	15126	Library
TASK1	15124	Routines
TASK2	15122	
TASK3	15120	
AIRDW	14616	
AOW	14413	
DIW	14277	
TIME10	14211	
FLOAT	14200	
.DA	14131	
BCDIO	11073	
.SS	11013	System
STOP	11000	Library
SPMSG	10705	Routines
FIOPS	10145	
OTSER	10051	
INTEGE	07647	
REAL	06670	
.CB	06650	

Begin
Execution
(after operator command)

* Note that, for this system, the loader places modules in memory beginning with
  high memory and proceeds downwards. In this way the area occupied by the loader
  itself can be utilized for blank COMMON.

they are appreciably more complicated because they allow application program segmentation into multiple memory load modules which occupy portions of the same memory area at load time. Various vendors use various techniques to squeeze a large (but modular) program into a small (but available) memory area or partition. For a given modular, user application program, we should also consider here the differences between "program swaps", "program chains", and "program overlays". Chains and overlays both permit much larger programs to run than would otherwise be possible in small systems with a 16K word to 32K word maximum task size. If a totally new program is called in, one that can "recall" the calling program, the process is known as "swapping". Distinct programs may be called in, each replacing the preceding one. On the other hand if a program is written in serially executable segments, each segment may call the next. This process is known as "chaining". These two concepts are shown in Figure 12.1. The vertical paths indicate chaining, while the horizontal moves shown as Levels 1, 2, and 3 illustrate swapping (one specific memory-load of machine language code for another).

Now let us consider the differences between swaps and overlays. In a swap, an absolute module is rolled out to a high-speed bulk storage device after execution, and the next module is rolled in to the same memory area completely overlaying the swapped module. After the second module terminates, it can either restore the original module or, depending on the execution path, swap with a third module, and so on. The overlay process described in detail in §12.5 is quite different. In this case, there is one main program with several support subroutines suitable for overlay. Overlays may be nested to several levels if necessary, as shown in Figure 12.2 where only routine A or B or C at Node 1 can be resident in memory at any given time. (The main program at Node 0 must of course be memory-resident). If the memory-

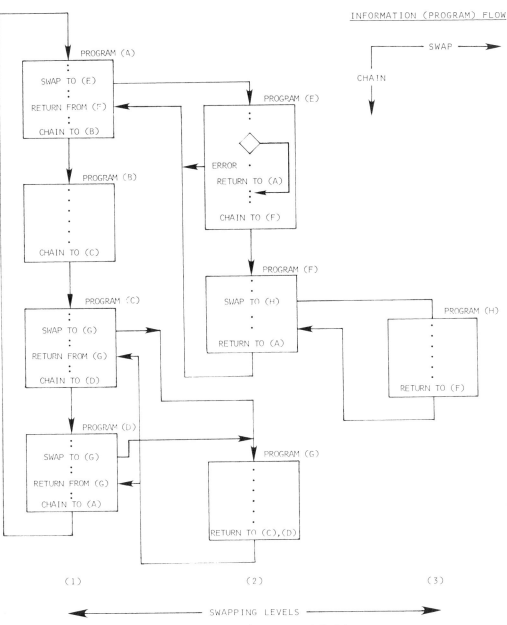

Figure 12.1. Program Swapping and Chaining

resident routine is A, then similarly only routine AA or AB at Node 2 can be memory-resident at any given time. Ideally, all branches of the tree should be of approximately the same length for best memory utilization.

12.5 MEMORY LOAD BUILDERS

At the outset we must emphasize that Memory Load Builders, with or without overlay capability, do not load modules for execution. Rather, they generate linked binary relocatable application modules for input to a block loader (§12.4) or a foreground job processor (§12.6).

Builders accept, as input, absolute or relocatable code generated by assemblers or compilers. Builder input must include a mainline routine and may include any number of user-defined subroutines. User routines can also be force-loaded if desired. This feature is espe-

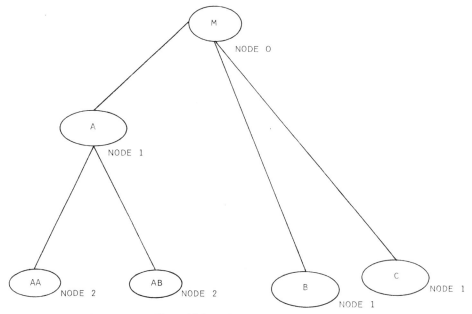

Figure 12.2. A Simple Overlay Tree

cially useful during system generation when an existing system routine must be replaced by a user routine but modification of the system library is not desired (or is not easily done). As the mainline is processed from any suitable *Binary Input* (BI) device, the builder places external references in its symbol table. Note that external references at this point usually include a mixture of user and system routine names. The builder directs its output to any suitable *Binary Output* (BO) device as memory buffer limitations require. As user subroutines are processed, the corresponding external references are added. Each routine is then linked with those that preceded it and output to BO. An (optional) Memory Load Builder map is output to any suitable listing device. If any unsatisfied externals still remain (the usual case) after all user routines have been processed, the builder initiates a search of specified user and system libraries. In bulk systems, repeated passes through the libraries occur until there are no further unsatisfied references or until no further references can be satisfied. This latter situation, of course, constitutes an error condition, and the building process is terminated. Of course, the builder also handles COMMON allocation for all modules in a particular relocatable memory load. As with loaders, great care is

needed in library preparation to avoid requiring additional passes over slow and/or sequential storage media such as paper tape.

Most builders also provide for some form of "overlay" or "chain" capability. Though there are differences (as indicated in the preceding section), we choose to concentrate on the relocatable overlay build process here. Module swapping is distinctly different, however, and is frequently handled by the foreground executive after the builder output is further processed by a foreground job processor (cf. § 12.6). The relocatable linked overlay memory load consists of the main routine, a relocatable link table to the overlay segments, and the overlay link segments themselves. The overlay segment consists of at least one external overlay component and any number of local overlay components. (An external component in this context is an overlay link that may be referenced by any higher-level routine in the memory load. Local overlay components on the other hand are routines that can only be referenced by routines within that overlay.)

Whenever an external overlay component is referenced by the routine currently executing in memory, the overlay containing the referenced component is loaded from bulk into memory by the overlay management routine unless, of course, the overlay is already in memory, in which case execution is simply transferred to that overlay segment. The overlay build process proceeds in a manner quite similar to that of a normal memory load build. Differences result from the multiple passes required and from the way overlays are linked to the main routine and to other overlays.

The builder generates an entry in the overlay link table for each overlay component because all portions of the entire overlay system cannot be in memory at one time. Each individual overlay segment as well as the main routines are quite like *non*overlay modules except that external references satisfied in the main program are not built into individual overlay segments.

Such references can be linked back to the main routine, since the foreground scheduler ensures that the main routine is in memory before loading any overlay component. The maximum amount of main memory needed by an overlaid program at any time is the sum of the memory required for the main module, the link table, and the memory required by the largest overlay component.

The right-hand side of Figure 12.3 illustrates the memory layout for an application program (identified as Task 30) with five overlays. When Task 30 is executing, the driver (i.e., the mainline), and common subroutines, and the link table are always in memory. Should a reference to an overlay component be encountered (e.g.,

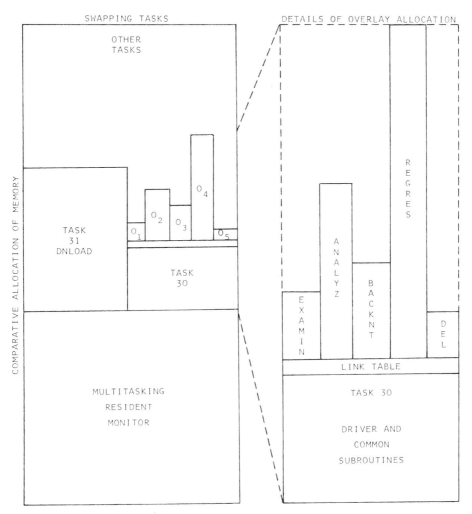

Figure 12.3. A Modular Application Using Swaps and Overlays

upon execution of a mainline call to a sub-routine such as REGRES), the foreground executive checks to see if that overlay component is in memory. If it is not, the current memory-resident overlay component is rolled out to bulk, and the required overlay component rolled in from bulk and placed just above the link table. Now as it turns out in this particular application, another job (identified as Task 31) without overlays but related to Task 30 must operate in the same memory area. The left side of Figure 12.3 illustrates the relative sizes of these concurrent tasks. Either task can issue an executive command to turn on the other task and turn itself off. The task turned off is rolled

out to bulk in its entirety and the other task rolled in from bulk. This task swap is also handled by the foreground executive. Of course, both tasks must have been converted to absolute form by a suitable Foreground Job Processor. Actual loading is handled by an Absolute Block Loader.

Table 12.13 shows a portion of the corresponding load map output by a typical memory load overlay builder. Note that the mainline and the link table require $19A2_{16}$ locations, and that the five overlay segments (called LNK1 through LNK5 in Table 12.13) require additional main memory through $1B92_{16}$, $1D63_{16}$, $1CA8_{16}$, 3096_{16}, and $1A9A_{16}$ respectively.

Table 12.13. Memory Load Overlay Builder Map

MAIN PROGRAM	NAME	LOC	NAME	LOC	NAME	LOC
PROGRAM COUNTER						
0000-0FA3-P						
0FA4-0FE3-P	KIMGO	0FB2-P				
0FE4-10A4-P	BYTCON	1064-P				
•	•	•	•	•	•	•
•	•	•	•	•	•	•
•	•	•	•	•	•	•
LINK TABLE						
PROGRAM COUNTER						
1977-19A2-P	EXAMIN	1977-P	ANALYZ	197F-P	BACKNT	1987-P
	REGRES	198F-P	DELETE	1997-P		
RESIDENT CODE MEMORY REQUIREMENTS						
0000-19A2-P						
OVERLAY - - LNK1						
PROGRAM COUNTER						
19A3-1B92-P	EXAMIN	1AA5-P				
OVERLAY MEMORY REQUIREMENTS						
19A3-1B92-P						
OVERLAY - - LNK2						
•	•	•	•	•	•	•
•	•	•	•	•	•	•
•	•	•	•	•	•	•
OVERLAY MEMORY REQUIREMENTS						
19A3-1D63-P						
OVERLAY - - LNK3						
•	•	•	•	•	•	•
•	•	•	•	•	•	•
•	•	•	•	•	•	•
OVERLAY MEMORY REQUIREMENTS						
19A3-1CA8-P						
OVERLAY - - LNK4						
PROGRAM COUNTER						
19A3-2E77-P	REGRES	273E-P				
2E78-2EBA-P	LNVECT	2E7B-P				
2EBB-2F0E-P	TIME$	2EBB-P	TIME$B	2F00-P		
2F0F-2F36-P	DATE$	2F0F-P	DATE$B	2F2B-P		
2F37-2FA7-P	ALOG	2F3B-P	ALOG10	2F37-P		
2FA8-3005-P	EXP	2FA8-P				
3006-304F-P	SQRT	3006-P				
3050-305E-P	F$V10	3050-P	F$V40	3050-P		
305F-3095-P	F$D00	305F-P				
3096-3096-P	SIM$	3096-P				
OVERLAY MEMORY REQUIREMENTS						
19A3-3096-P						
OVERLAY - - LNK5						
PROGRAM COUNTER						
19A3-1A9A-P	DELETE	1A03-P				
OVERLAY MEMORY REQUIREMENTS						
19A3-1A9A-P						
TOTAL MEMORY REQUIREMENTS						
0000-3096-P						

The maximum memory required (3096_{16} for LNK4) determines the size of the memory partition that must be available in order to execute this program. In this example, LNK4 really should be segmented further to reduce total program memory requirements.

Note that the only real difference between a Linking Loader and a Memory Load Builder (or between an Overlay Loader and a Memory Load Overlay Builder) is that the former generates absolute linked code ready for execution, whereas the latter requires further processing of its output prior to execution as a foreground task in real-time. In bulk systems, a block load is invariably required. Admittedly, the distinction between these two utilities can prove somewhat fuzzy. Sometimes a modular user application program is best implemented as a mixture of swaps and overlays as in the example discussed in this section. Rarely (if ever) can the link-up and loading process be accomplished by a single system program; rather a sequence of utility programs will be required.

12.6 FOREGROUND JOB PROCESSORS

A foreground processor accepts memory load builder output and generates an absolute memory load module tailored to the application task requirements of the foreground partition(s). Actual functions of a Foreground Job Processor (where one is provided) vary widely from system to system. Typically, this important utility allows the user to:

- Set initial task identification and priority. Task priorities can invariably be modified by an appropriate call to an executive routine.
- Set initial task state as "ready" or "suspended".
- Select the memory areas to be used by the task and its overlays if any. In some systems, memory partitions are established at system generation time; hence the task must "fit" in an available partition. Other systems allow complete freedom in selection of memory areas for task execution. In some cases, programs and data may reside in *non*contiguous areas of paged memory.

- Set the memory size required by the task program and data areas. This speeds the roll procedure, as only the actual code, rather than a whole memory area, needs to be rolled.
- Specify a task program or data area as available (or unavailable) for swapping. In this way, a high-priority task in the "ready" state is rolled into memory and locked in place until execution terminates, regardless of the priority of other tasks using that area of memory.
- Indicate secondary programs that must be memory-resident before execution of the primary program.
- Set up all bulk files (for absolute code and data) required by the task.
- Modify the resident monitor foreground table so that the task scheduler can manage the new task.
- Optionally, specify a task as a "system service routine". In this case, the task is active regardless of whether the foreground is "on" or "off".

Thus the Foreground Job Processor can allow the user great flexibility in selecting foreground job characteristics at the time a given task is created. Note that some systems require that many of these same characteristics be specified at the time the operating system is generated or installed. Others fall somewhat between the extremes of flexibility. In any case, the linked relocatable code typically output by a Memory Load Builder (§ 12.5) usually is not suitable for direct block loading under control of a task scheduler.

Background characteristics—except possibly for size and memory location (i.e., "normal memory" or "extended memory") which can be altered by executive commands—are invariably set at operating system generation time. Typically, a single background job runs concurrently with several foreground tasks though some operating systems do support multiple concurrent background tasks (e.g., concurrent editing of two different source files from two different console terminals in a multiterminal system). The background is normally used to edit, compile, load, and test modules under development or to run a developed and debugged application program. Real-time production tasks usually operate in one or more foreground memory areas.

When the functions of the Memory Load Builder and Foreground Job Processor are combined, the resulting processor is often referred to as a Linking Loader or Overlay Loader, since a single processor now accepts relocatable object modules and generates a single absolute object module suitable for loading (under control of the task scheduler) by an absolute block loader.

12.7 DEBUGGERS

On-line debuggers come in two basic versions: those that operate with interrupts "enabled", and those that do not. Obviously, the latter cannot be used during normal real-time operation of the system. As a matter of fact, this type of debugger is most useful when working on the operating system itself, since a multitasking operating system has a tendency to "crash" if key registers and memory are modified while interrupts are enabled. Here, we concentrate our attention on those debuggers that can function in an interrupt-enabled environment. (Refer to § 12.10 for a brief discussion of some typical limitations of this type of utility software.) Before continuing, however, we should emphasize that the on-line debugger is most useful for debugging assembly language modules and is of little or no help in locating FORTRAN programming errors in small systems.

In disk-based systems, the debugger can be used to modify specified words in a specified absolute disk sector location (to implement vendor "patches", usually available only in absolute binary form on cards, tapes, or floppy disks for correction of the errors that frequently are found in vendor software). The debugger also allows the user easy access to and modification of almost any memory location (data or instruction) or operating register. For example, an individual status word in a DCB can be easily modified and tested for effective-

ness without having to rebuild the operating system completely. Usually, code patches can be inserted using a combination of mnemonics (e.g., for the instruction set) and numerical data. Thus, program patches can be inserted without requiring reassembly and reloading of the program under development. This feature is especially important in nonbulk real-time systems.

Execution of the program under development is normally "monitored" through the use of debugger traps called "breakpoints". Program control returns to the debugger executive each time a breakpoint is encountered (or, in general, any time an unsolicited interrupt, e.g., CTRL T, is generated at the console terminal).

Key registers are usually dumped before the debugger issues a command prompt. Any valid debugger command may then be issued. For example, the user may wish to:

- Examine memory or registers
- Add or delete breakpoints
- Modify existing instructions or data
- Insert a code segment in some portion of available memory
- Scan a selected area of memory for a specified bit pattern
- Resume execution at the point of interruption
- Resume execution at the first executable instruction

Debuggers have one big disadvantage: their use distorts the true real-time characteristics of the module under test by appreciably slowing it down. For example, if the time allowed for an A/D conversion is too low (in systems where the A/D does not issue a "conversion complete" interrupt), debugger overhead can lead to correct operation when the debugger is present and incorrect operation otherwise.

In summary, debuggers are most useful for fixing known bugs in user or system programs in memory or on bulk or for modifying control words in modules such as a DCB. In many cases, they are of limited value in debugging application programs owing to the timing changes they cause and, for FORTRAN, the difficulty of matching the object code to source code.

12.8 OTHER UTILITIES

Here we discuss but a few of the many other utilities provided by the vendor as an adjunct to the operating system and key utilities described in earlier sections. These routines can range from marginally useful to absolutely necessary depending upon the structure of the operating system and the installation work mix. Careful evaluation of such utilities prior to selection of a small computer system can make a big difference in ease of program development and debugging as well as in overall system support.

12.8.1 Execution Monitors

Execution Monitors, as the name implies, "monitor" execution of programs under development in such a way that executing foreground tasks and/or the operating system are protected from interference due to program bugs. Such a utility program is especially important with multitasking systems that do not include hardware or software foreground memory protection capability. In general, an execution monitor does not execute CALLs to the real-time executive (e.g., WAITs, TRNONs, STARTs, special I/O calls) but simply informs the user through a dump of the routine argument list and system working registers. This feature is especially useful during development of real-time application programs that include long time delays. For example, certain sampled-data control algorithms may require task suspension for 15 minutes or longer between sampling intervals. Since WAITs are indicated as a message but otherwise ignored, logic errors can be located rapidly. Further, in modular applications, each module can be debugged separately because requests for activation of other tasks are noted but not acted upon. Error traps can be minimal (e.g., only an invalid store-to-memory request outside the background processing area of memory is detected) or extensive (e.g., in a FORTRAN program, all run-time subscripted variable store requests are checked against the DIMENSION specification). Actual features are as varied as the number of vendors providing such utility routines.

12.8.2 Map Sorters

Map Sorters operate on the load map output by a Memory Load Builder (cf. § 12.5) or, in some systems, on the load map output by a Linking Loader (§ 12.4). The map output from a builder or loader simply lists relative addresses in the order in which the corresponding routines were extracted from user or system libraries. Absolute addresses of reentrant resident monitor routines are also given. Load maps sorted alphabetically by routine name or entry point as well as load maps sorted numerically by address are frequently very useful—especially during development of large modular real-time application programs. Table 12.14 shows a portion of an alphabetic-sort load map for the modular application program discussed in § 12.5.

12.8.3 Concordance Generators

Concordance Generators are used to produce cross-reference listings of program components and are especially useful with large, modular overlayed application programs. They are also useful for building a cross-reference index for operating system routines. In general, these utilities can operate on the contents of a named bulk storage data area as well as the output of a Linking Loader or Memory Load Overlay Builder. Table 12.15 shows a small portion of a cross-reference listing for typical bulk system FORTRAN library routines. Though an example is not shown, most concordance generators can also output the inverse of Table 12.15; i.e., table entries are sorted alphebetically by calling procedure with a list of called procedures appended. Such cross-reference lists can be especially useful during debugging of application or system modules.

12.8.4 File Comparators

File Comparators are useful routines whose value is often ignored. Most such utilities can sequence and copy a file from one device to another as well as compare two files that have been read-in from two different devices. Though source (ASCII), relocatable, or absolute files can be copied and/or compared, file com-

Table 12.14. Alphabetic Load Map Segment*

TYPE	ENTRY	VALUE	TYPE	ENTRY	VALUE	TYPE	ENTRY	VALUE	TYPE	ENTRY	VALUE
S	ALOG	1C99-P	S	ALOG	1BDE-P	S	ALOG	2F3B-P	S	ALOG10	1C95-P
S	ALOG10	1BDA-P	S	ALOG10	2F37-P	L	ANALYZ	197F-P	U	ANALYZ	1AE7-P
L	BACKNT	1987-P	U	BACKNT	1A7F-P	U	BYTCON	1064-P	S	DATE$	2F0F-P
S	DATE$B	2F2B-P	S	DEL$1	009F-A	L	DELETE	1997-P	U	DELETE	1A03-P
	•	•		•	•		•	•		•	•
	•	•		•	•		•	•		•	•
	•	•		•	•		•	•		•	•
S	TIME$B	2F00-P	S	TPF$1	009E-A	S	TPF$C	00DE-A	S	TPF$CA	00DF-A
S	TPF$CR	00E0-A	S	TPF$RA	009E-A	S	TPF$RF	0080-A	S	TPN$1	009C-A
S	TPN$2	009D-A									

TOTAL MEMORY REQUIREMENTS

    0000-3096 WORDS HEX

*NOTES: Under TYPE,   S = System Routine

                          U = User Module

                          L = Overlay Link Table

           Under VALUE, A = Absolute address for reentrant resident monitor routine

                          P = Program relocatable address for system and user memory

                            load modules.

parators are especially important during backup (from bulk) of lengthy source code modules on cards or paper tape. A single character dropped in the middle of a large application program can prove most difficult to locate if the bulk copy is ever destroyed, thus necessitating restoration of module source code from the backup medium. Unfortunately, especially with paper tape reader/punch systems, the additional effort required by such file comparison is all too often considered to be not worth the trouble. Small systems using audio quality cassette tape bulk storage can make good use of the file comparators, since it is a common problem for bits to be dropped in a loading sequence.

12.8.5 Backup Processors

Backup Processors (Backup Utility Programs) differ from file comparators in that the backup processor is generally used to copy the entire current version of a bulk storage device (e.g., a permanently-mounted operating system disk) to a suitable backup medium (e.g., an equivalent removable disk or magnetic tape) though only portions (perhaps a source file data space) of the bulk device could be specified for backup if desired. Owing to the large quantity of data (typically 2.5 to 20 megabytes for a disk), paper tape or cards are never used. Especially during backup of the operating system, the

Table 12.15. Partial Cross-Reference List for FORTRAN Library Routines

CALLED PROCEDURE	CALLING PROCEDURES							
ALOG	ALOG10	CLOG	F$E11					
ATAN	ATAN2							
COS	CCOS	CEXP	CSIN					
EXP	CCOS	CEXP	CSIN	F$E11	TANH			
F$ERR	ALOG	ATAN2	DATAN2	DCOS				
	DLOG	F$ARF	F$ARU					
	F$FMT1	F$GOT	F$MABT	F$NCK	F$RED			
	F$REF	F$REU	F$TEF	F$TEU	F$WRF	F$WRT	F$WRU	SIN
	SQRT							
F$FMT1	F$FFM	FREMAT						
R$APU	APU$R	APU$S						
SIGN	ATAN							
SIN	CCOS	CEXP	COS	CSIN				
SQRT	CABS	CLOG	CSQRT	DSQRT				

copy should be verified by the backup processor prior to returning the backup pack (or tape) to storage. The importance of these processors and their proper use simply cannot be overemphasized. Attention to such details is absolutely mandatory in any real-time production environment.

12.8.6 Other Processors

Other utility programs besides those already discussed, are invariably supplied by every vendor. Such utilities may be separate memory load modules activated by a suitable command to the background or foreground executive or, in some cases, may be part of the resident monitor itself. Some other utilities that may be useful in a multitasking real-time environment are summarized in Table 12.16. Simple typical operating system commands are not shown, as they vary so widely from one system to another.

Typically, these utility programs may accept a parameter string (e.g., NAME, PAR1, ..., PARN) to set execution time characteristics. Default values are invariably supplied where possible.

12.9 CROSS-ASSEMBLERS AND CROSS-COMPILERS

A cross-assembler (or cross-compiler) executes on a (usually much larger) host computer system rather than on the target computer system itself and generates either absolute or relocatable code suitable for direct loading by the target machine. Frequently, the host and target machines are not even manufactured by the same vendor. While we emphasize cross-assemblers here, essentially all our comments pertain equally to cross-compilers. In general, though, cross-assembly is vastly more common than cross-compilation. Incidentally, assembly or

Table 12.16. Overview of Additional Utility Programs

Typical Name	Typical Function
BMEM	The size of the background processing area of memory is changed to a larger (or smaller) value and the bulk monitor is modified.
COPY	Source, relocatable, or absolute files are copied (without comparison) from one device to another.
DUMP	Hexadecimal (or octal depending on the system) contents of a bulk storage area are dumped to a hard-copy listing device.
HELP	A general system command which accepts arguments to explain the purpose and use of a supported utility.
HISTM	An execution time histogram for a named program is generated.
IDIR	A specific directoried bulk storage device or data area is initialized.
IDISK	A specified bulk storage device is initialized and a table of bad sectors is constructed.
LDIR	The contents of a bulk storage device or named data area are output. Names, bulk space occupied, and space remaining are usually provided.
LISTF	A complete listing of all background and foreground programs and their characteristics is output as is information on memory areas and status.
LPRO	Current and permanent logical unit assignments, bulk storage allocations, and system features are summarized.
PACK	A named disk or disk area is packed to recover storage area lost by previous delete and replace commands.
REPLACE	An existing source, relocatable, or absolute file is replaced by the contents of a file in a specified input device.
WHZAT	A listing of currently active (i.e. executing, ready, suspended) programs is output along with key program parameters.
WMON	The current resident monitor is written to bulk to become the new bulk monitor. In this way, many system characteristics and/or features can be altered without rebuilding the system.

compilation on a large host machine using an assembler or compiler identical to that supported by a smaller, memory-only version of the host machine does not truly constitute cross-assembly or cross-compilation.

Use of cross-assemblers for program development is a quite common practice for the simple reason that very small computers invariably are resource-bound. Most commonly, cross-assemblers are used because:

- The target computer does not have enough memory to support an assembler.
- Available I/O devices make assembly an extremely frustrating and time-consuming process.

The host computer chosen for cross-assembly usually has a wide selection of utility software as well as a reasonable peripheral configuration. Both factors, of course, greatly speed program development. Depending on the target-machine configuration, cross-assembly may also require a cross-linking loader to be effective. If the target-machine loader accepts only absolute binary records as input, then the cross-assembler can be easily designed to output such records. However, if the target computer can accept a relocatable main routine and several relocatable subroutines and then link these modules at load time, the cross-assembler must also supply needed linking information to the target-computer loader. Alternately, the host machine must support a cross-linking loader that will resolve external references and produce a linked absolute memory load which the target machine can then load.

Cross-assemblers are just host system programs written in some suitable higher-level language. The cross-assembler source language is a quite good indication of its value, though trade-offs exist. A cross-assembler written in host-machine assembly language is usually quite memory-efficient and fast. However, such an assembly language implementation limits ease of cross-assembler modification and restricts its use to a particular machine (or product line) of a single vendor. ANSI FORTRAN or some other common high-level language is a much better choice for cross-assembler implementa-

tion. Though execution is slower and more memory is required, such factors are usually a minor consideration. Cross-assemblers based on high-level languages can be easily modified and, more importantly, are portable from one host machine to another with relatively little effort. Cross-assemblers are of greatest use (indeed, we might almost say "are required") in real-time computing when the target machine has limited peripherals (e.g., many microcomputer systems) and limited memory (e.g., 4K bytes).

As a simple example, suppose we need a program to collect 1000 data points at a rate of one every 0.2 second from an 8-bit A/D converter. Typical output from a powerful FORTRAN-based cross-assembler appears in Table 12.17. Additional cross-assembler output (cf. Table 12.11) is (optionally) produced in a format suitable for direct loading by the absolute loader of the small, dedicated target machine.

Another alternative does exist, however, for very small program segments. The user may act as his own "cross-assembler" and generate the absolute machine code needed. Thus, to code Source Line 23 in Table 12.17, the user must first determine the address of the label PORTA. Since the address of PORTA is greater than 255_{10}, two machine bytes are needed for the address in this system. Hence the entire instruction takes three bytes: one for the op code and two for the address. Note that Bit 6 of the op code must be set to one to indicate that the operand address is contained in the next two bytes. Thus the op code for the LDA instruction here is AD_{16}. (In Source Line 14 the operand code for Line 23 is three bytes long, the code for Line 24 starts at Address 210_{16}. If the user makes a mistake in coding the address field at this point (i.e., only one byte for the address of PORTA is allocated), then all subsequent absolute code could be erroneous, depending upon how many additional errors exist.

Relative branch instructions are probably the most difficult to handle manually. With this type of instruction, the byte following the op code is added to the Program Counter if the condition is TRUE to determine the address of the next executable instruction. Thus, for

Table 12.17. Example of a Cross-Assembly Process

OBJECT				SOURCE CODE		
		1	;			COLLECT 1000 DATA POINTS FROM AN
		2	;			8-BIT A/D CONVERTER EVERY .2 SEC
		3	;			AND STORE DATA IN LOCATIONS
		4	;			DATA TO DATA + 999 AND THEN DUMP
		5	;			DATA TO OUTPUT DEVICE.
		6	;			
		7		*=$200:		;SET THE PROGRAM COUNTER
		8	;			
		9	;	DEFINE CONSTANTS		
		10		ON=$7D:		;PULSE A/D UP
		11		OFF=$FF:		;PULSE A/D DOWN
		12				;(i.e. NOW CONVERT)
		13	;			
200	A9	14	ADCC	LDA	#ON:	;SEND CONVERT
201	7D					
202	8D	15		STA	OUTPUT:	;COMMAND
203	00					
204	17					
205	A9	16		LDA	#OFF:	;TO
206	FF					
207	8D	17		STA	OUTPUT:	;ADC
208	00					
209	17					
		18	;			
20A	EA	19		NOP:		;WAIT
20B	EA	20		NOP:		;FOR
20C	EA	21		NOP:		;CONVERSION
		22	;			
20D	AD	23		LDA	PORTA:	;FETCH DATA FROM PORT A
20E	20					
20F	17					
210	99	24		STA,Y	DATA:	;STORE INDEXED BY Y REGISTER
211	20					
212	02					
213	C8	25		INY:		;INCREMENT STORAGE POINTER.
214	CA	26		DEX:		;DECREMENT POINT COUNTER.
215	F0	27		BEQ	DUMP:	;IF ALL READ, BRANCH & DUMP
216	07					
217	20	28		JSR	DELAY:	;DELAY FOR 0.2 SECONDS
218	1D					
219	02					
21A	4C	29		JMP	ADCC:	;CONVERT AGAIN
21B	00					
21C	02					
		30	;	DUMMY DUMP AND DELAY ROUTINE		
21D	60	31	DELAY	RTS:		;RETURN TO CALLING PROGRAM.
21E	60	32	DUMP	RTS:		;RETURN TO CALLING PROGRAM.
		33	;			
		34	;	DEFINE STORAGE		
		35		PORTA=$1720:		; LOCATION OF PORT A.
		36		OUTPUT=$1700:		; LOCATION OF OUTPUT PORT.
		37		DATA =*+1:		; START OF DATA FIELDS.
21F	0	38	.END			;TERMINATION OF PROGRAM

Source Line 27, the byte in location 216_{16} must be added to the Program Counter in order to point to location $21E_{16}$—the next executable instruction when the condition is TRUE. Note that if the dump routine is moved, then the user must recalculate this offset or displacement.

Obviously, if a program contains many relative branches and/or forward references, the user has to spend inordinate amounts of time and effort on hexadecimal or octal arithmetic.

12.10 SIMULATORS

Cross-assemblers, working in conjunction with a cross-linking loader if feasible and necessary, output absolute machine code modules acceptable to the absolute loader of the limited-configuration small computer. However, only assembly errors can be eliminated on the host computer system. Run-time errors must still be detected during actual execution on the target machine. Furthermore, even for debugging a program under control of an on-line debugger (cf. §12.7) on some small computer systems, insufficient memory is available for both the debugger and the module being debugged. The potential headaches in both cases should be obvious. During execution, a word in memory is altered that allegedly never is accessed; a process suddenly comes to a dead halt without leaving any trace of what stopped it or why;

data structures seem mysteriously "cross-wired" to point to each other instead of to the data. In such cases, the user has little recourse except to curse the so-called debugging process and to develop some dexterity at manipulating the computer system console switches. Simulators offer a way out of such small computer limitations of "no memory, no help, and virtually no hope".

The typical simulator allows I/O device input and output to be kept on the host computer's disk; that is, the simulated devices can read from and write to disk files. Thus, the input to and output from simulated devices never become physically or logically inaccessible to the simulator system. Further, the simulator debugging package resides not in simulated memory but in available host machine memory. Thus, the debugging package can be as large and complicated as necessary without affecting memory for the program being debugged. A simulator has a similar advantage over the real machine with respect to registers. The real machine can make available only those registers that it can access itself, that is, the programmable registers. The simulator, on the other hand, can make everything available—programmable registers, internal registers, even invented registers. A simulator can thus not only provide complete information about the current state of the simulated machine, but also, through judicious invention of new registers, information about past states as well.

In a properly-designed simulator, any data item of any simulated device can be examined, modified, searched, dumped, or breakpointed just as with main memory. Under simulation, disks, card readers, paper tape equipment, memory, and all other storage or I/O devices can be treated as being structurally equivalent.

The enormous advantage of a simulator with respect to access is duplicated with the problem of handling a dynamic process, i.e., of directing, altering, or interrupting a program while it is running. This seems to be by far the worst debugging headache. Small machine debugging packages offer only the most limited help with such problems. At best, relatively few breakpoints are provided, and these work only if the breakpointed location is fetched as an instruction. If the program overwrites a flagged location, the breakpoint is lost and the user is too. Further, virtually no small machine debugging package can survive in a real-time environment. Many of them will not tolerate interrupts and also tend to disrupt I/O device operations irrevocably. In a real-time system, such disruptions invariably mean that operations cannot be resumed after a debugging break but must instead by restarted.

Debugging interruptions of the simulator are totally transparent to the program being simulated. The user can interrupt execution whenever or however he chooses, poke around inside the simulator, and then, barring deliberate changes, resume simulation at exactly the same point and with the simulated machine in exactly the same state as when the interruption occurred. Nothing need be changed or lost. A simulator can provide enormously more complex control mechanisms than the real machine. For example, the simulated memory can have one or several extra bits devoted exclusively to debugging purposes. Further the simulator can detect and flag questionable machine states that the real machine normally cannot or does not detect.

A simulator is especially useful in attempting to deal with the problem of repeatability. Repeatability means the ability to reproduce on demand a particular machine state (particularly the one that caused the program to fail). Such reproduction of a particular state is simply impossible on the real machine if only because CPU and I/O device timing are subject to variation, sometimes as much as ±20%. These timing fluctuations do not occur under simulation (unless the user deliberately sets them up). An instruction always executes in the same amount of simulated time; an I/O operation always runs for the same amount of time. Further, under simulation, a particular machine state can be saved and restored as desired, whereas on the real machine at best only memory and the programmable registers can be saved and restored. Thus a well-designed simulator can fully handle the most important debugging problems, while a real-machine debugging system can only nibble at the edges of these problems. Of course, a user may not need such complex debugging tools as a simulator can offer. If the work involves writing BASIC programs on a small-machine time-sharing system, simulation is totally pointless. However, if the work is with the time-sharing system itself, a simulator might be essential.

12.11 SUMMARY

As we have seen, utility and system software can embrace an enormous range of activity and function in small computer systems. However, to paraphrase one major vendor: "Small computer systems aren't delivered—they're abandoned!" In this context, we cannot overemphasize the importance of a well-designed, well-documented, bug-free operating system and input/output subsystem for effective, rapid development and use of real-time application modules. Even in cases where the vendor provides complete source code listings for this key system software, finding and fixing major bugs (e.g., those causing frequent system crashes for no apparent reason) is a difficult, frequently impossible task for local personnel. The vendor, concentrating of course on the next sale, is usually of little or no help.

Assuming the operating system is reasonably powerful, flexible, and bug-free, then we should consider carefully the utility software provided by the vendor. For example, source file text editors can be extremely powerful and easy to use, or they can be cumbersome and inconvenient. These differences in even such a common utility program can translate into thousands of effort-hours over the life of the small computer system.

Compilers, too, can make a big difference—especially in terms of the number and complexity of real-time tasks that can be supported by a given small computer configuration. Best in this regard are (stable) optimizing compilers which can generate code approaching in efficiency that generated by experienced assembly language programmers. Loaders and Memory Load Builders are also key utility programs. Of special importance here is their ability to handle easily overlays nested to several levels. Foreground Job Processors may or may not be important, depending on overall system design philosophy. The tradeoff is simple: great

power and flexibility in assigning foreground task parameters at application program development time using this utility vs. simplicity, stability, and ease of use where such parameters are established during operating system generation with a link load then being the last step prior to task execution.

On-line debuggers of two types should be provided: those that operate with interrupts disabled (for patching certain parts of the resident monitor) and those that operate with interrupts enabled (for debugging application modules written in assembly language). Still, debuggers prove to be of little use in many installations because so much of current application software development is in FORTRAN or some other high-level language. As we have discussed, a simulator is really needed for effective debugging in many (if not most) cases.

Other support utilities (e.g., execution monitors, map sorters, concordance generators, backup programs, file comparators, execution time histogram generators, etc.) should be evaluated in the context of the projected installation work mix. The vendor user group can also be an excellent source of support utilities and should be evaluated in this context if local manpower is tight.

It almost goes without saying that cross-assemblers are an absolute must for program development on very small computer systems—particularly microcomputer systems which otherwise would have to be programmed in absolute machine code. Look for cross-assemblers written in a higher-level language that can be accessed locally rather than only through a vendor-supported time-sharing service.

Careful attention to utility and system software characteristics during evaluation of competing small computer systems prior to purchase can pay great dividends over the useful life of the system. Conversely, failure to "read-between-the-lines" of the invariably glowing vendor descriptions of their system and utility software for a given product line can lead to disaster.

12.12 SUPPLEMENTARY READING

1. Casilli, G. and Kim, W., "Microcomputer Software Development", *Digital Design*, July 1975.

2. Gear, C. W., *Computer Organization and Programming*, McGraw-Hill, New York, 1973.
3. Katzan, H., *Operating Systems*, Van Nostrand Reinhold, New York, 1973.
4. Korn, G. A., *Microprocessors and Small Digital Computer Systems for Engineers and Scientists*, McGraw-Hill, New York, 1977.
5. Korn, G. A., *Minicomputers for Engineers and Scientists*, McGraw-Hill, New York, 1973.
6. Ogdin, C., "Fundamentals of Microcomputer Systems", *Minimicro Systems*, November–December 1977.
7. Ogdin, C., "Microcomputer Support Aids", *Minimicro Systems*, February 1978.
8. Rolander, T., "How Much Software for Microcomputer Control?", *Control Engr.*, January 1978.
9. Schoeffler, J. D., *Minicomputer Real-Time Executives*, IEEE Press, New York, 1974.
10. Schoeffler, J. D. and Temple, R. H., Eds., *Minicomputers: Hardware, Software and Applications*, IEEE Press, New York, 1972.
11. Soucek, B., *Minicomputers in Data Processing and Simulation*, Wiley Interscience, New York, 1972.
12. Stiefel, M., "Venturing FORTH", *Minimicro Systems*, August 1976.
13. Weitzman, Cay, *Minicomputer Systems: Structure, Implementation and Application*, Prentice Hall, Englewood Cliffs, N.J., 1974.

12.13 EXERCISES

1. Obtain information on two or more microcomputer systems. Compare the operations required to create, edit, load, and debug a simple assembly language program. For example, create a program to add two numbers from memory and store the result.

2. From information on two or more minicomputer systems, compare the general characteristics of their FORTRAN compilers. Are they optimizing compilers? Do they accept FORTRAN 66 or FORTRAN 77 code? How many passes are required? What is the memory space required for the compilers?

3. Comment on the capabilities of the BASIC interpreter available on your system. Does it support real-time programming? Is it truly a multitasking BASIC, or are these options simulated?

4. Even if your minicomputer system has an "Automatic Program Load" switch for the bootstrap loader, key in the bootstrap loader using console switches. Why is the APL switch a very useful option?

5. What is the difference between an operating system that may require swapping and chaining and one that supports virtual memory?

6. Explain the primary problems associated with using a DEBUG utility for real-time programs. What options would you use to find a difficult run-time error in a real-time program?

7. Two vendors of home or personal computers appear to be cost-competitive, at least as far as hardware is concerned. Explain how you would evaluate their software systems in order to make a decision on which one to buy. Prepare a list of questions you would like answered by the vendors before you make your decision.

8. For a machine available to you, calculate the overhead in software and execution time for a simple subroutine call. What is the breakpoint for in-line code (i.e., macros) vs. subroutine calls for a short calculation?

9. Discuss the differences between buffered and nonbuffered I/O under a disk-based multitasking operating system, and give an example of a problem that could arise in nonbuffered systems.

10. Consider the editing example of Table 12.5 and make the changes indicated using an interactive (not full-screen) editor of choice. What is the minimum number of keystrokes required? In your opinion, what are the strengths and weaknesses of the editor used?

11. Discuss the pros and cons of single-pass vs multipass assembly or compilation.

12. Discuss the differences between recursion and reentrancy in the context of a modular high-level real-time application. Which is more important from a practical point of view and why?

13. A large modular real-time program for an industrial plant computer is to be developed in a high-level language. Assuming the system supports both BASIC and FORTRAN, which language would you prefer and why? Be specific.

14. Verify that the checksum in the first data record of Table 12.11 is 0B36 as indicated.

15. Discuss the differences between "swaps", "chains", and "overlays".

16. Guess the function of each of the op codes of Table 12.17 based on the probable meaning of each mnemonic and the comments given.

13

Real-Time Operating Systems and Multitask Programming

Joseph D. Wright
Xerox Research Centre
Mississauga, Ontario

James Wm. White
Modular Mining Systems
Tucson, Arizona

13.0 INTRODUCTION

An on-line computer system accepts input directly from the user or process that creates it and returns output directly to the user or process that requires it, regardless of where they are located. Implicit in this definition is the assumption that some remote access to the system (most frequently from keyboard-activated terminals) is required, as opposed to inputs from conventional batch-computation-oriented devices (computer-related peripherals) such as card readers or magnetic tapes which may be attached directly to the computer. Note that we exclude from this definition remote job entry terminals (RJEs) for batch computation. Included in such systems would be on-line banking systems, airline reservation systems, inventory systems, and the like. On-line systems may also include time-sharing, real-time, and multiprogramming systems, of which the last-mentioned may loosely be defined as a mode of computer operation in which various partially completed tasks or jobs are run concurrently. Most computer systems are only capable of running one job or program or task at any one instant of time. Therefore, it is important to note that concurrent operation does not imply that the tasks are running simultaneously as would be the case in a multiprocessing environment. Computer systems with parallel central processing units (CPUs) or those

with input/output peripheral processors are said to be multiprocessing systems. We are primarily concerned with multitasking and multiprogramming systems in which only one task or program is executed at a time but where the apparent operation of the system to the observer is that many tasks are operating essentially in parallel.

Many terms used here may be unfamiliar. Some of these are defined in detail in other chapters, while others find common usage among computer systems analysts. We attempt to define these terms in enough detail that the primary points of this chapter may be understood.

For purposes of this discussion, we might best divide on-line systems into the following five major categories:

- Process control systems
- Data acquisition systems
- Business-oriented information systems
- Scientific/engineering time-sharing systems
- Remote batch systems (or remote job entry)

Process control systems generally imply those in which: data are accepted from sensors (strain gauges, thermocouples, transducers, counters) that measure process variables; calculations are performed using these data; and results are output back to the process to manipulate control valves, pumps, switches, and so on. The primary concern is to perform all the data manipulations in a short period of time, frequently as fast as 100 times per second. Although control algorithms may be complex, it usually is possible to be specific about the calculation time requirements. By contrast with large inventory control systems, most real-time process control systems require relatively small information files or data bases and can frequently be developed within a minimal hardware configuration. In practice, however, it is often the case that process control computer systems are justified from the point of view that the data they collect can be used for plant scale optimization studies or for accounting purposes in manufacturing. In these examples it may be difficult to separate that part of the system which is necessary for process control from that which is used for data collection and storage.

Data acquisition systems often are very similar to process control systems with the main difference being the lack of output back to the process. We might say that data acquisition systems are "passive" with respect to the process from which data are acquired, while process control systems are "active". Many scientific applications involve very high-speed data collection rates for such things as mass spectrometers and X-ray machines, although the majority are rather slow. Data may either be processed on-line or merely collected, reduced, and formatted for subsequent analysis elsewhere.

The important difference between process control or data acquisition systems and the other categories of on-line systems is that the individual programs must normally be executed at specific times of the day or at specific time intervals. The few minutes delay one might accept in booking an airline reservation could have disastrous consequences in a critical reactor control application. Similarly the results of an entire experiment could be lost if the computer system were unable to respond owing to some external delay or through excessive system software overhead.

Apart from the many questions involved with overall system architecture and hardware specifications, many questions must be answered in the design or specification of an operating system for a particular application. We consider here a few of the hardware and software requirements which enable application programs to be developed and run effectively. Among the more important factors to consider are:

- What multitasking features are required?
- What programming languages should the system support?
- What file handling capabilities are required?
- Is the system dedicated to a simple control task or must it support program development and real-time programming applications simultaneously? (This point asks the question: Is Foreground/Background or Multiground programming support required?)
- How much memory is required?
- How many process inputs and outputs must be supported?
- What bulk storage or extended memory management system is required?

Answers to many of these questions are dependent upon the process, the programming staff, and the capital cost of hardware vs. system programming costs. Definitions of individual elements are discussed in this chapter. In this discussion we tend to use the terms "multitasking" and "multiprogramming" synonymously. Some vendors actually treat multitasking as a subset of multiprogramming; however, such a distinction is not important at the introductory level of the material presented here.

System design specifications must also cover a series of potential operating (run-time) error which might occur. The operating system should, of course, never fail. It must detect errors such as references to memory location that do not exist or that are reserved for system (as opposed to user) application use. Priority or job identifications must be checked for validity. Latency problems, wherein too many jobs or programs requiring CPU time exist must be monitored because the computer must be able to execute all critical jobs within the time constraints demanded by the process or set by the user. If these programs cannot be executed, then the system fails the fundamental definition of real-time.

13.1 OPERATING SYSTEM OVERVIEW

In configuring a real-time operating system one of the main decisions to be made is whether the system (hardware and software) is to be dedicated to a specific application such as a process control or data acquisition system, or whether the system is intended, in addition, to support extensive background programming, bulk storage facilities and high-level languages. This design decision largely determines what type of operating system will be used.

The Operating System, sometimes called a real-time "monitor" or real-time "executive", is the most complex vendor-supplied software element in any installation, especially in multi-tasking real-time applications. Various types of small computer operating systems are summarized in Table 13.1 with a short description of characteristics and approximate main memory requirements. Usually the vendor offers a choice of operating system for each machine in a given product line though not all operating

Table 13.1. Small Computer Real-Time Operating Systems

System	Key Characteristics	Approximate Memory, Words	Typical Bits/Word
Terminal Interface Monitor	Is used by only the simplest single-application dedicated machines. This type of monitor handles process interrupts and also allows the user to start a program, display or alter memory locations, set breakpoints and load or dump programs.	$2K_{10}$	8
Input/Output Monitor	Handles its own initialization and control as well as communication between itself, system programs, user programs and a simple I/O subsystem.	1-2K(a)	12-16
Keyboard Monitor	Is a single-user operating system which requires some form of bulk storage such as magnetic tapes or discs. It includes all the facilities of an I/O Monitor plus routines to accept and act on console commands as well as the ability to modify I/O assignments dynamically.	3-6K(b)	12-16
Background/Foreground Monitor	Is a dual-program executive that includes all the facilities of the Keyboard Monitor and also controls processing and I/O in a time-shared or interrupt-driven environment.	4-12K(b)	12-16
Multiprogramming Executive	Includes all the facilities of the Background/Foreground Monitor with additional capability for concurrent execution of (up to) several hundred foreground jobs.	8-80K(c)	16-32

NOTES: (a) depends on word length; normally supports few peripheral devices.

  (b) depends on word length and on the number and type of peripheral devices supported.

  (c) highly variable; depends strongly on architecture, number and type of peripherals, number and size of I/O buffers, number of foreground programs and number and type of sub-systems supported.

system types are always available or even necessary. The terminal Interface and I/O Monitors find use in dedicated service applications and multiprocessor networks, whereas the other types can be used for program development as well as such real-time production functions. Let us now consider some of the important characteristics of these system programs in more detail.

13.1.1 Terminal Interface Monitors

Terminal Interface Monitors are usually supplied in ROM (*Read Only Memory*) for the simplest dedicated small computers.   Because the monitor is implemented in ROM, the code is nonvolatile, is available at system power-up, and cannot be altered inadvertently by user programs. This type of monitor may communicate with the user via a serial full-duplex port using ASCII Code and automatically adjust to the speed of the user terminal.   Commands typed at the terminal cause this simplest of monitors to start a program, display or alter registers and memory locations, set breakpoints, and load or punch programs.   Figure 13.1 illustrates a typical memory map for a small

8-bit/word computer using a Terminal Interface Monitor.

The programs (usually generated in absolute machine code either manually or via a suitable cross-assembler on a host machine) may also be loaded via a high-speed paper tape reader or may be "downloaded" directly by a larger host computer via an asynchronous communications channel or a computer/computer communications link.   Furthermore, the larger machine can, via simple monitor commands, "dump" selected process data for further analysis. Note that this monitor occupies 2K bytes of ROM at the top of memory and may require several bytes of RAM (*Random Access Memory*) in Page 0 for monitoring program counters, registers, pointers, and accumulators.   The remaining RAM is available for user application programs though care must be exercised in using stack locations in Page 0.

13.1.2 Input/Output (I/O) Monitors

Input/Output (I/O) Monitors operate on a basic configuration machine with 8K or 16K of RAM memory with larger word length.   Usually a small bulk bootstrap ROM is located in the highest 32 (load from one of two peripheral devices) or 64 (load from one of three peripheral devices) words to provide for loading of system processors and user programs in absolute memory-image format. This monitor frequently operates in a paper tape, card, or low-speed cassette environment and provides for calling and handling of all input and output functions. An I/O Monitor represents an inexpensive and reasonable way to operate small computers dedicated to a single task. However, extensive real-time program development with frequent correction, assembly or compilation, and link-loading of code in an I/O Monitor environment is inefficient even with a high-speed paper tape reader/punch and quite impractical if the only paper tape I/O is that provided by a 10-character/sec terminal.   While several memory maps could be illustrated for the I/O Monitor (e.g., a system program map, a debugging utility map, or a chain loader map), we choose to illustrate but one in Figure 13.2–a typical memory map for an executing application program.

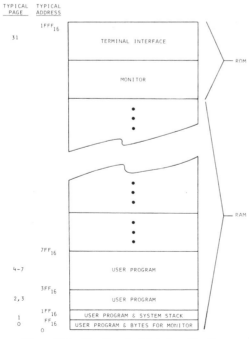

Figure 13.1. The Terminal Interface Monitor

TYPICAL
ADDRESS

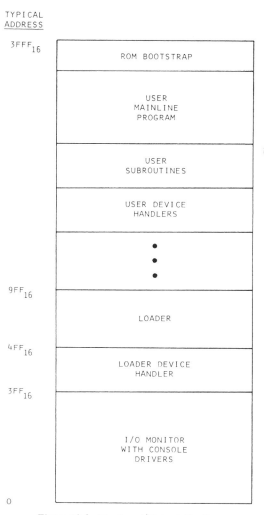

Figure 13.2. The Input/Output Monitor

storage device. For a given machine, the I/O Monitor is usually upwards-compatible with the Keyboard Monitor. Thus programs prepared in an I/O Monitor environment can be run using the Keyboard Monitor. The Keyboard Monitor responds to an interrupt when a console key is struck, reads console commands, and branches to an appropriate subroutine for executing programs in memory or loading programs from mass storage with subsequent execution. Only the portion of the executive (i.e., monitor) needed to recognize console commands is permanently stored in memory. This "resident monitor" calls the various loading and service subroutines of the executive program from the system disk or tape. However, the resident monitor in the Keyboard Monitor environment is invariably larger than the simple I/O Monitor as it must include handlers for the bulk storage devices. This usually means that less memory is available at object time to run the user's application program. Only a single application or user program can operate under the Keyboard Monitor, though the program, through overlapped I/O and processing, may appear to do several things simultaneously. The typical memory map for execution of an application program under the Keyboard Monitor is quite similar to that illustrated in Figure 13.2 for the I/O Monitor.

Invariably, memory occupied by the loader (and loader device handler if not needed by the application program) may be used for storage of certain data types after loading and during execution.

13.1.3 Keyboard Monitors

Keyboard Monitors are even more sophisticated and are usually available for systems equipped with auxiliary bulk storage devices such as tape cartridges, magnetic tapes, floppy disks, disks, or drums. This monitor allows for device-independent programming and automatic creation, modification, assembly or compilation, and loading of program files from the bulk

13.1.4 Background/Foreground Monitors

Background/Foreground Monitors are essentially extensions of Keyboard Monitors that allow concurrent, time-shared use of the small computer system by one (debugged) foreground user program and one background system or user application program. This monitor is designed to control processing and I/O operations in a real-time or time-shared environment. The foreground program has top priority at load time in selection of memory and I/O devices. At execution time it has priority on processing time and use of shared I/O. Depending on system requirements and capabilities, the foreground program could itself be a user-written "executive" capable of handling many real-time programs in a priority interrupt-driven environment. Background processing is essentially the same as processing normally performed under control of the Keyboard Monitor and could involve creation, editing, assembly or compilation, debugging, production, etc. These programs may use any facilities (memory, I/O, processing time) that are available and not simultaneously needed by the foreground job. Of course, more memory is required for the resident portion of the Background/Foreground Monitor than for the Keyboard Monitor. Indeed, even more memory is needed because two different jobs are resident in main memory simultaneously. In some implementations, background memory can be used by the foreground program through "roll-out/roll-in" to a high-speed disk, but this slows background program development and decreases precision of foreground program timing. Again, although several different memory maps could be illustrated, we show a dual-program execution memory map typical of this type of monitor in Figure 13.3. Note the memory protection boundary between the background user program and the foreground system. The desirability of such memory protection capability—which may be implemented using either software or hardware techniques though the latter are preferred—cannot be overemphasized. As a final note, an operating system of this type may allow two concurrent tasks as shown, or it may allow several tasks to use either or both memory areas under control of a task scheduler. In either case, only two distinct program memory partitions exist.

13.1.5 Multiprogramming Executives

Multiprogramming Executives are the most sophisticated of the various operating systems. Several foreground programs can reside simultaneously each in its own memory partition with access to system resources determined by a user-defined priority structure. In addition, background tasks can be executed in a separate memory partition just as under the Background/ Foreground Monitor. Some systems allow concurrent operation of several real-time background tasks as well as multiple foreground tasks, while others permit but one operating

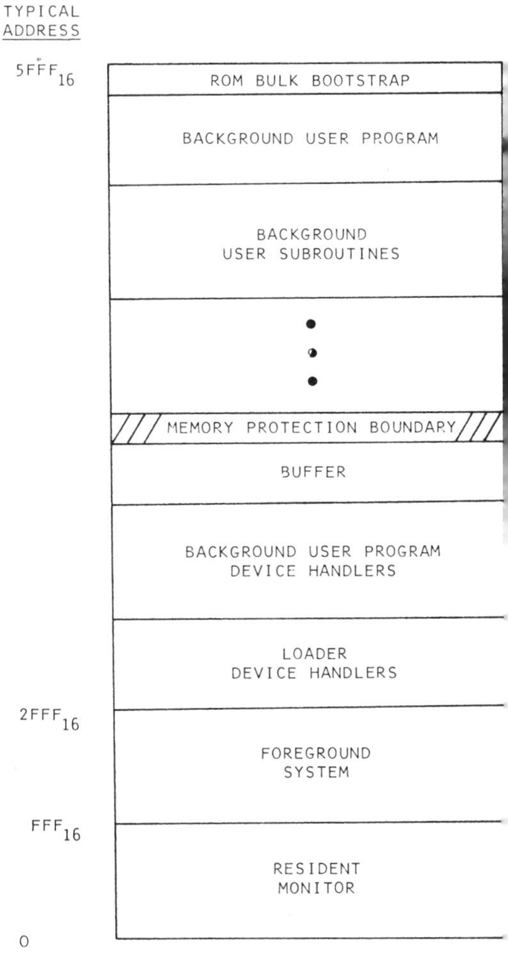

TYPICAL ADDRESS

$5FFF_{16}$

| ROM BULK BOOTSTRAP |
| BACKGROUND USER PROGRAM |
| BACKGROUND USER SUBROUTINES |
| • • • |
| MEMORY PROTECTION BOUNDARY |
| BUFFER |
| BACKGROUND USER PROGRAM DEVICE HANDLERS |
| LOADER DEVICE HANDLERS |
| FOREGROUND SYSTEM |
| RESIDENT MONITOR |

$2FFF_{16}$

FFF_{16}

0

Figure 13.3. The Background/Foreground Monitor

background task. Further, a background task may or may not operate in real-time—depending upon the particular implementation. If the system is configured with buffered I/O, then several tasks can use a given area of memory via roll-out/roll-in to a high-speed bulk storage device. Even more memory is required for a Multiprogramming Executive (MPX)—especially with buffered I/O—and much effort invariably is needed to support properly such complex operating systems, especially when diverse real-time production tasks run concurrently with tasks under development.

As with Background/Foreground Monitors we distinguish between two executives—the resident monitor and the bulk monitor. On initial system load—usually via a hardware ROM

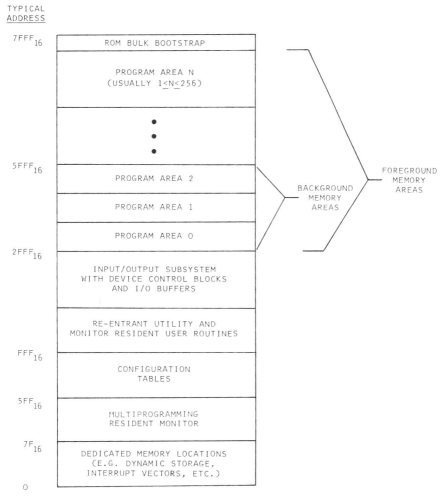

TYPICAL
ADDRESS

$7FFF_{16}$ — ROM BULK BOOTSTRAP

PROGRAM AREA N
(USUALLY $1 < N < 256$)

$5FFF_{16}$ — PROGRAM AREA 2

PROGRAM AREA 1

PROGRAM AREA 0

$2FFF_{16}$ — INPUT/OUTPUT SUBSYSTEM WITH DEVICE CONTROL BLOCKS AND I/O BUFFERS

RE-ENTRANT UTILITY AND MONITOR RESIDENT USER ROUTINES

FFF_{16} — CONFIGURATION TABLES

$5FF_{16}$ — MULTIPROGRAMMING RESIDENT MONITOR

$7F_{16}$ — DEDICATED MEMORY LOCATIONS (E.G. DYNAMIC STORAGE, INTERRUPT VECTORS, ETC.)

0

FOREGROUND MEMORY AREAS

BACKGROUND MEMORY AREAS

Figure 13.4. The Multiprogramming Executive

bootstrap loader—the bulk monitor is copied to low memory to become the resident monitor. Certain characteristics of the resident monitor (e.g., default logical unit assignments, timing parameters, etc.) can then be changed by appropriate operator console commands and the resident monitor written out to bulk to become a revised bulk monitor. Obviously, great care is required to maintain system integrity.

One of the many possible memory maps for multitasking real-time operation appears in Figure 13.4 where the "typical addresses" are highly variable from vendor to vendor. Note that foreground memory and background memory overlap in Areas 0 through 2. Thus, foreground jobs running in these areas must be structured for roll-out/roll-in to a high-speed

bulk storage device. Since at least $12K_{10}$ is usually required for background processors (such as assemblers and compilers), the time required for "swapping" a background processor and a foreground job may be as high as hundreds of milliseconds. Hence, extremely precise timing is difficult if not impossible to achieve with most operating systems. Indeed, if a foreground job in Program Area 1 is performing extensive output to a high-speed (e.g., 9600 baud or higher) peripheral device such as a CRT or a smaller dedicated computer, background service is effectively suspended until all foreground output is complete. A similar problem exists when a foreground job in Areas 0 through 2 is engaged in high-speed input from a peripheral device. Generally, low-priority computational tasks or low-speed I/O tasks should be selected to operate concurrently with the background. Tasks requiring precise timing or high-speed I/O should be selected to operate in program areas higher than 2 (in Figure 13.4) with these program areas specified as "not available for roll-out/roll-in". Tasks falling between these extremes can be selected for concurrent operation in other foreground memory areas.

While the memory map of Figure 13.4 indicates a memory maximum size of $32K_{10}$, many vendors offer memory expansion capability to 128K words or even higher. Current mapping technology allows machine configurations with several megabytes of memory, depending on computer word length. Usually, such "extended memory" is accessed using machine instructions that "deselect" a portion of lower memory in the map of Figure 13.4 and "select" an equivalent extended memory block in its place. When the operating system is configured with full extended memory management, any area or group of areas in memory may be assigned to any task though a maximum task size (e.g., 16KB, 28KB, or 32KB, depending on architecture) must usually be observed. It is noted that some systems, especially those with 32-bit architecture, allow virtual memory operation depending upon their operating systems.

Actually, the *Memory Management System* (MMS) component of a Multiprogramming Executive varies widely from vendor to vendor. In order of increasing power and capability, MMS functions can be briefly summarized as follows:

i) The system manager must lay out memory at system generation (one background and N foreground "partitions" or "areas"). A background program or foreground task must fit in its assigned partition. Partition size can only be changed by repeating the system generation procedure.

ii) As in i) above except the background area can overlap a specified range of foreground partitions.

iii) As in i) and ii) above except that the size of the background processing area can be altered dynamically (by a command to the executive) while real-time tasks are executing.

iv) As in i), ii), and iii) above except that the size of any partition can be altered dynamically while real-time tasks are executing.

v) All memory is allocated dynamically in an n-ground sense. The user need only assign a software priority level. The executive allocates task memory (sometimes in not necessarily contiguous pages) and task resources based on a (user-parameterized) scheduling heuristic. Tasks at the same software priority level can be time-sliced.

As MMS capability increases from minimum (i above) to maximum (v above), operating system memory requirements increase appreciably (e.g., from 8–30KB to 100–180KB). In any event, either hardware or software techniques should be available to minimize catastrophic interference between executing tasks or between executing tasks and the operating system.

A single computer and disk can operate under an effective multitasking executive. However, for more efficient utilization of bulk storage (disk) facilities and to maintain different sets of peripheral devices for different applications, it is sometimes convenient to operate a dual-processor, shared-disk system. In this case, one computer might be configured for efficient data processing, file handling, or software development facilities (line-printer, card readers, etc.), while the other might have a smaller array of process-related peripherals aimed toward data acquisition and process control. The processors communicate with each other via disk files in shared or separate partitions. A further extension of this concept often is used in process control applications when both processors are programmed to execute the same programs simultaneously, one system providing total backup for the other. Process I/O devices may be shared through a default switch which automatically transfers control and data links to the other processor should the primary system fail. Parallel processor systems such as these are found in critical process control applications to maintain high system reliability.

Nuclear reactor control applications provide one of the best examples of these systems.

A variation on this theme which allows greatly increased data-handling rates is provided by systems in which part of one system's main memory is dual-ported. In such cases a front-end processor, say for data acquisition, can automatically read or write to the same memory block as the primary CPU.

User commands to a Background/Foreground Monitor or to a Multiprogramming Executive—especially the latter—are extensive and vary widely from system to system. Hence, it is inappropriate to discuss such command structures here. We must realize, however, that executive commands are almost always needed to identify each of the programs and their characteristics to the executive responsible for overseeing the timely execution of these real-time jobs.

As a final comment, we should note that both Background/Foreground Monitors and Multitasking Executives are sometimes available as much smaller memory-resident monitors for use with nonbulk storage systems. In such cases, program development and debugging in a paper tape or card environment is, for all practical purposes, difficult if not impossible owing to the much greater complexity of these operating systems over the simple I/O Monitor. Thus, programs should be developed and debugged on a larger machine using the disk-based version of the corresponding real-time executive and then transferred to the smaller dedicated machine.

13.2 BUILDING BLOCKS FOR REAL-TIME PROGRAMS

Let us now consider in some detail the basic building blocks required by any real-time application program running under control of a multitasking executive.

13.2.1 Task Definition

The concept of a task is fundamental to the design of any real-time system. The task may be defined as "a logically complete program segment", or alternatively, as "the smallest

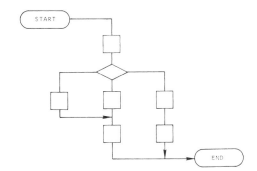

Figure 13.5a. Single-Task Environment

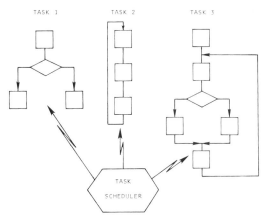

Figure 13.5b. Multitask Environment

program to which system resources (I/O devices, CPU, memory, etc.) may be allocated". The contrast between a single-task environment and a multitask one is shown clearly in Figures 13.5a and 13.5b. The single task could, in fact, have several alternative routes from START to END, but, once it is being executed, a unique path exists. For example, the single task could be a general program to calculate heat capacity of a gas as a function of composition, temperature, and pressure. Input data could include these parameters, and a logical path through the program would then be defined.

By contrast several independent tasks may exist simultaneously, each of which is logically complete and independent of the others. The three tasks shown in Figure 13.5b could be: a program to generate a process status report at the conclusion of each eight-hour shift, a program to control composition in a reactor once per second, and a program to check alarm conditions on ten critical process variables once per minute taking different action depending upon which, if any, alarm condition is set. Of course a series of logical paths may exist within each task.

A *job* may consist of a set of tasks which are intended to be executed either serially or in parallel (remembering, of course that in a multitasking, single-processor system only one task actually can have control of the CPU, i.e., be executing, at a given instant). A process control program may, for example, be comprised of all three tasks illustrated in Figure 13.5b. Because the three tasks are executing according to real-time conditions, they must be scheduled according to the time of day. Each job or task potentially running in a system must have its own priority level compared with other tasks although these need not necessarily be unique. To monitor these priorities and allocate system resources (i.e., peripheral devices, memory, or the CPU itself) to the tasks, the real-time software must include a Task Scheduler program as shown in Figure 13.5b. Simple examples illustrating these ideas are developed in Section 13.6.

13.2.2 Task States

Depending upon the particular computer and operating system being used, there are different definitions of task states as well as a different number of states which can exist. However, for our purposes, it is convenient to assume that a number of tasks exist in a multiprogramming environment, and that at any time there are four distinct task states. They are:

- EXECUTING (or running); i.e., the task has control of the CPU. The Executing Task is the highest priority task that is "ready" to run.
- READY (or active or installed), in which state one or more tasks may reside until their priority level dictates that they should have control of the CPU. In this state the task "attributes" such as its priority are completely known to the scheduler program that allocates system resources to the tasks.
- SUSPENDED (or blocked), where task execution has been halted until some real-

time event or communication from an executing task occurs. For example, a task may be waiting for input from the analog to digital converters (ADCs) before it can calculate process variable values or compute some control action to be transmitted back to the process.

- INACTIVE (or dormant), by which we mean that the task, even though potentially resident in main memory, has either not yet been introduced to the operating system (scheduler), or that it has completed an assignment and has been removed from the queue of jobs to be executed.

Figure 13.6 illustrates the various task states and the transition paths that tasks take when "moving" from one state to another. The scheduler is aware of the priority of all of the individual tasks in the system except the INACTIVE ones. Task priorities are used to determine which task has control of the CPU, since only one task may be executing at a given instant. Figure 13.6 also shows a series of general actions which cause the tasks to move, under the control of the scheduler, from one state to another. For use in subsequent examples, a set of mnemonics defined in Section 13.3 is shown parenthetically in Figure 13.6.

13.2.3 Task Data Blocks

In should be apparent from the above discussion that the system must retain certain status information for tasks in the Ready or Suspended states. Usually, task status information is retained in a *Task Data Block* (TDB). One such block is required for each task in the Ready or Suspended state. The data include task priority, identification number, status of all the active registers (accumulators, program counter, index registers), a *User Data Pointer* (UDP), and usually a link to the next-higher-priority task in the system. The UDP is basically a pointer to (address of) a user stack (memory storage array or buffer) which often is used in writing reentrant programs or subroutines. In situations where the next-higher-priority task has the same priority as a number of other tasks, a convention must be followed by the system in

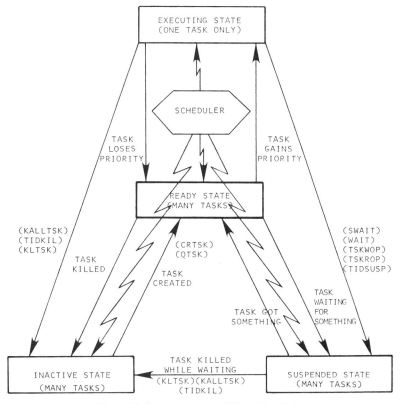

Figure 13.6. Task States and Transition Paths

scheduling the one to run next. Typically, the task that has been waiting to execute the longest has the highest priority when nominal priorities are the same. The data in the TDBs are organized and used by the scheduler program (see Section 13.3.2) within the real-time operating system.

13.3 REAL-TIME OPERATING SYSTEM STRUCTURE

Before discussing details of individual parts of a real-time operating system, we should summarize the main or general functions of this important major component of real-time software. Perhaps the key function is the handling of all the input/output (I/O) operations on peripheral devices, be they terminal, bulk storage, process I/O unit, or whatever. Basically, input/output only occurs at the request of user-written tasks or programs. The system should be capable of initiating and then handling the data transfer. In conjunction with

this, the system must be able to insert multiple requests for input/output on a particular device into a queue should that device be unavailable at the time. This potential problem can occur very easily in multitasking systems because it is quite possible for several tasks to require input from the same device at the same time. Usually there is no coordination of these requests other than by the order in which they occur as a result of scheduling of the tasks by the system task scheduler.

Efficient handling of peripheral devices by a computer depends upon the judicious use of interrupt programming. Hence, the operating system must be able to respond to hardware interrupts quickly and efficiently. The justification for interrupt handling of process I/O and of peripheral devices is illustrated in Figure 13.7. Here three tasks are illustrated in each of two systems: the first system does not use interrupts, whereas the second does. We assume that the multitasking executive allows the CPU to execute other tasks while one (or more) of

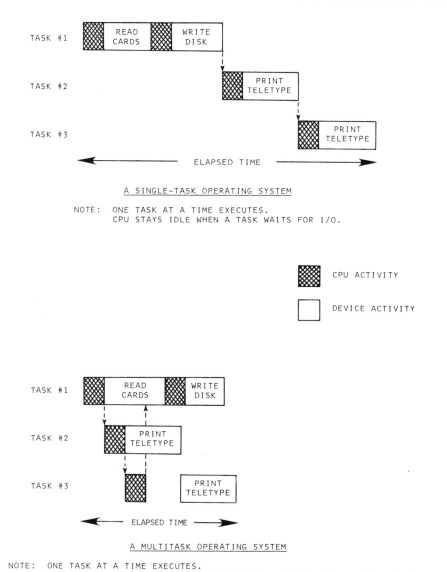

Figure 13.7. Illustration of Improved CPU Usage Using Interrupt-Driven I/O and Multitask Operating System

them is waiting for input (or output) to be handled by the operating system. Task 1 involves CPU processing, card reader input, CPU processing, and finally disk output. Task 2 requires CPU processing followed by Teletype output. Task 3 is similar to Task 2.

In the single-task operating system (System 1), Task 1 executes in its entirety before Task 2 begins. Note the large time in which the CPU is essentially idle, waiting for data transfers (as indicated by the nonhatched areas in the blocks).

By contrast, in the multitasking, interrupt-driven operating system (System 2), the time previously wasted in waiting for a device to complete an input/output operation can be used to execute other tasks. A small penalty is, of course, incurred in the time required for the operating system to monitor the status of the peripheral devices. This monitoring time, the system "overhead", is normally small in comparison with the time saved and is not even indicated in Figure 13.7.

It is interesting to view the status of the three

tasks shown in Figure 13.7 in the multitask mode of operation. Assuming that all three tasks were READY at the start, and that the priorities were high, medium, and low for Tasks 1, 2, and 3, respectively, the following would occur. Task 1 would EXECUTE until the card reader input was initiated. At this point it would be SUSPENDED (pending completion of the data transfer), and Task 2 would EXECUTE. Task 2 would become SUSPENDED when it initiated printing on the Teletype, and Task 3 would EXECUTE. Task 3 could not output to the Teletype when it finished the CPU execution because Task 2 already was using the Teletype. At this point all three tasks would be suspended with the Teletype having a queue as well. Task 1 would EXECUTE following the card reader input and begin transfer to disk, at which point it would be SUSPENDED again. Finally all tasks would become INACTIVE when their respective I/O was complete. It is important to note that the savings in elapsed time obtained by executing the three tasks in a multitasking environment result from the large difference in speed between the CPU and the peripheral devices. Most peripheral devices are so slow that their operations can be "interleaved" with an executing task with only a small fraction of time (overhead) allocated to servicing interrupts from the devices and transferring information to or from them, usually in the Programmed I/O Mode.

Perhaps the next most important function of the real-time operating system is to provide for a measure of intertask communication. Means must be provided by which messages can be passed from one task to another for such purposes as synchronization of calculations or intertask communication of "messages" or results. Facilities also should be provided to create or remove tasks in the system. Where many tasks may be required to operate at the same priority, the system should allow for time-sharing (or slicing) of available CPU time among the tasks.

The basic components of a multitask real-time system then include: (1) the operating system, including an interrupt handler or priority control program, a scheduler, a set of I/O routines, and, in some systems, memory allocation and file handling routines; and (2) the user or application-oriented tasks. Finally, for convenience in multiprogramming, a series of Macro commands is usually available which we define as system calls, i.e., calls from a task to the operating system. Common Macro commands for real-time systems include I/O, wait, memory allocation, and task linkage and communication commands. These Macro commands usually are written as a set of subroutine calls to the operating system. In this chapter we define a representative set of Macro calls. Chapters 14 and 15 present a standardized set of BASIC and FORTRAN subroutine calls which are equivalent.

As a final point, the operating system itself is comprised of a set of subroutines or system tasks, some of which are resource allocators and others of which are task managers. A graphical representation of the component parts of a complete real-time system is shown in Figure 13.8. Each circle represents a subset of the overall software. Lines between the circles indicate interactions between the groups of subroutines in each class. Two important

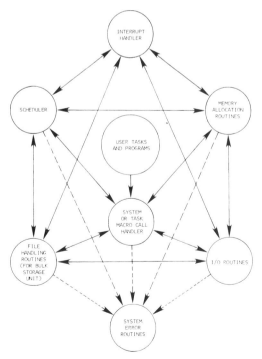

Figure 13.8. A Conceptual Overview of a Real-Time System: Component Parts and Interactions

points should be noted. The first is that the size of user programs may be relatively small compared with the size of system programs required to run them. This may be particularly true if large data buffers for user applications programs are excluded from the comparison. The second point is that the user programs can only communicate with system resources through system or task Macro calls or subroutine calls. To introduce a task to the scheduler, for example, a user program might execute a subroutine call to "create" a task.

In order to understand the interactions illustrated by Figure 13.8 let us consider some additional material on some of the main system programs required for any real-time operating system. In general, the applications programmer is not required to write system software at this level. However, a clear understanding of the principles of operation of all software is essential for development of efficient applications program code.

13.3.1 Priority Control Program

The main difference between the simple (*non-disk*) batch system and a real-time system is that in the batch system the program initiates and waits for all input/output, and each stage of the data processing is well-defined, whereas in a real-time system the program must respond to a series of random and frequently unrelated events. The basic means by which this is accomplished is through hardware or software routines which respond to and acknowledge hardware interrupts and which control internal system priorities so that requests for CPU service by I/O devices can be granted. Facilities must exist to have low-priority devices interrupted by requests from higher-priority ones, and in particular, to respond quickly to emergency alarm conditions such as powerfail interrupts.

Although a basic discussion of interrupts is presented in Chapter 11, the concepts may be sufficiently new to many readers to warrant futher discussion at the risk of repeating some of the material. In particular we are interested here in understanding the interaction of the computer hardware with the system software.

The simplest interrupt structure is one in which the computer has one hardware level of interrupt with, perhaps, a number of levels implied by the position of a single bit in a special computer register which can be loaded under program control. Each I/O device in the system is potentially tied to a different bit in this register, and each is capable of requesting service (CPU time) assuming certain conditions are fulfilled *and* the CPU has enabled hardware interrupts generally. Recall (Chapter 8) that each device in the system has a unique numeric code (or device address) so that it can be activated or identified by standard assembly or machine language level instructions. Also, each device, in general, has one or more flags (status lines or signals) which allow it to signal that it requires attention.

For this type of system, the following sequence occurs when a device requests an interrupt. The hardware automatically turns off the "master interrupt switch", stores the current contents of the working registers and the current location of program execution (program counter or PC), and branches to a specified address where the interrupt handler program resides in main memory. This program must then determine which device caused the interrupt. Two techniques are available. The first is called polling. In this mode of operation, the program sequentially scans each I/O device, in an order specified by the system programmer, checking each device flag to find which one caused the interrupt. When a raised flag is found, the interrupt program branches to a specific handler routine for the interrupting device. Alternatively, many systems allow the interrupt to be acknowledged explicitly by an instruction which directly obtains the device code (hence an address) of the interrupting device. This "address" can be used to allow direct branching to a handler program. Polling is most often used only for very high-priority interrupts, such as powerfail, or when few devices are connected to the system.

In both cases discussed here, the computer master interrupt switch is not turned on until the interrupt has been serviced. This mode of operation is inefficient and does not allow important operations such as disk transfers to take priority over slow devices such as terminals.

In order to structure a hierarchy of device priorities, the concept of masking must be introduced. The individual bit to which each device is attached can be used to disable the device from interrupting even though the master system interrupt is turned on or enabled. Thus when an intermediate-priority device causes an interrupt, the software, after determining which device is involved, can disable the device itself and all lower-priority devices, and then re-enable the system interrupts in case a higher-priority device requires service.

Some additional software is required as part of the interrupt and device priority programs. Since, in general, the CPU may be executing any program when the interrupt occurs, the software must store all critical registers which represent or uniquely define the machine status so that control may be returned to the interrupted program following completion of the interrupt service.

Some computers have an expanded set of hardware priority levels to which the peripheral devices are hardwired (connected). These automatically lockout or disable the lower levels when an interrupt occurs. Many computers are designed with an automatic hardware-operated stack which replaces much of the software described above. In these systems an interrupt automatically causes the machine status registers to be stored in a special buffer, the lower-priority devices to be masked out, and CPU service transferred directly to a handling program. At this point these systems require software for handling specific devices once again. This type of interrupt structure is called a vectored interrupt structure, and it greatly improves the speed of response to device interrupts. The entire stack can be pushed or popped (moved up or down) to store or replace one set of registers by a single CPU instruction.

Perhaps the most important point to note here, apart from the rather complex software which executes in a manner that is transparent to the user, is that the hardware priority control and interrupt handling program does not have to compete for CPU time in the sense that user or system tasks do. It has the highest system priority at all times. Once the interrupt has been identified, however, the actual servicing of the peripheral device (or data transfer) takes place as dictated by the system scheduler.

13.3.2 Scheduler Program

The scheduler program is the key to the operation of both the real-time operating system and the user tasks written to run in conjunction with it. The basic function of the scheduler is to determine which task (user task or system task) has highest priority and then to raise it to the executing state. In addition the scheduler must maintain the task Ready Queue. For each task in the Ready state the scheduler must know its priority, the status of all the active CPU registers, and the link words to the next-higher-priority Ready tasks in the queue. These have been defined (Section 13.2.3) in the task data blocks (TDBs) which the scheduler maintains and uses to restore the status of any Ready task being raised to the Executing state. An example illustrating this is discussed in Section 13.6.4.

In addition, the scheduler must know the priority of the task which is executing. Tasks may be introduced to the scheduler in three ways: by a system Macro call or equivalent assembly language Macro command to raise a task from the Inactive state to the Ready state (i.e., a user task has requested that another task be run); by a user task request for input or output of information (data); or by a request for system resources (CPU) such as one made by the interrupt handling routine when a device causes an interrupt. Tasks that are in a Suspended state have earlier been introduced to the system and are placed back in the Ready Queue when the conditions causing them to be suspended no longer exist. For example, a task that has been suspended for 15 seconds by a timing loop subroutine call is raised to the Ready state by the Scheduler when the time period has expired.

Queues of ready tasks may be randomly organized, forcing a search by the Scheduler at each scheduling time; or they may be ordered such that the highest-priority task is known immediately to the Scheduler. In the latter case, system overhead is incurred in reordering the queue each time a new task appears. The most efficient means of operation depends upon the number of tasks active in the system at any given time.

13.3.3 Real-Time Macro Commands

A complete list of all the system commands that may be useful in a real-time operating system is as long and as varied as the number of computer vendors. Each vendor has its own preferred list. Hence, it is possible here only to define, in general, the types of commands available. The system Macro commands may be generally grouped into eight categories. These are:

- Input/output commands
- Task creation and deletion commands
- Intertask communication commands
- Overlay and special queueing commands
- Clock commands
- Task identification commands
- Task/operator communication commands
- Operator/task interaction commands

Arguments (data) for each Macro command are variously passed in accumulators or memory depending upon the system and the hardware configuration. We do not discuss details of the arguments but rather the general concepts of Macro commands within each category. Equivalent BASIC subroutine calls for many of these commands are discussed in Chapter 14 and for standardized FORTRAN in Chapter 15.

Input/Output Commands effectively represent requests from a user task for data transfer to or from a specific device. They may take the form of an I/O statement or be more specific such as READ or WRITE, with arguments that provide the system with information as to device, number of pieces of data, channel numbers, coding of data, data source or destination in memory, and other information relevant to the I/O operation. The command, IOdev, where dev is a device mnemonic, is used in the examples shown in Section 13.6. With FORTRAN, dev is equivalent to the logical device number in, say, a WRITE (10, 6) statement. Table 13.2 defines the instruction and devices used in the examples in Section 13.6. The I/O commands in Table 13.2 could be used in a simple program as illustrated by Example 13.1 which reads five successive channels on the analog to digital converter beginning at channel zero (0) and stores the results in memory at starting address A.

Task Creation and Deletion Commands are used to introduce tasks to the scheduler program or change the status of a given task or tasks within the system. It is convenient but not necessary for each system task to have associated with it a specific identification number (ID). It is essential that each task have a specified software priority (P). This information is used by the scheduler in allocating system resources to the tasks. A summary of useful commands with general mnemonics is shown in Table 13.3.

Some simple examples illustrate the use of these commands. (These examples also can be developed very easily for any specific computer system and in any langauge available for multitasking.) A user function could be simply to write a message to the line printer, since the only purpose of the examples is to give an understanding of the scheduler and multitasking operation of the system. More detailed

Table 13.2. Input/Output Macros

```
INSTRUCTION:     IOdev

SYSTEM DEVICES:  dev to be replaced by one of the following

        ADC          Analog to digital converter
        DAC          Digital to analog converter
        LPT          Lineprinter
        TTI          Terminal input
        PTP          Paper tape output

Arguments:       To be shown in examples
```

Example 13.1. A Simple Input Operation

Program	Comments
.	
.	
.	
Instructions	
.	
.	
.	
IOADC	Read ADC,
5	5 channels with
0	starting channel = 0 and
A	starting address = A.
.	
.	
.	
Other Instructions	

examples are presented in Section 13.6 following a discussion of some programming and flowcharting conventions.

Assume the user program is the first task in an application. It is required to create a separate task to carry out a specific user function. Each task (of the two existing after creation) is to terminate after it prints a message on the line-printer. Note that the first user task to execute has a default value for priority when it begins. The user must change this priority if another value is desired. In Example 13.2, since the initial task had the higher priority (10), it would continue running until it finished execution. T4 would appear on the line-printer followed by T2. If the priority of the new task (2) had been set at 5, however, the output would be reversed, i.e., T2 followed by T4.

Intertask Communication Commands are used to send and receive messages between various tasks which exist under control of the operating system. Typically they communicate over software-defined channels for message transmission. Table 13.4 summarizes this class of instruction.

A situation requiring these commands could arise when two tasks must be synchronized as shown in Example 13.3.

Overlay and Special Queueing Commands are useful particularly in disk real-time operating systems to overlay selected portions of main memory with program modules stored on a bulk memory unit or for ensuring that tasks are introduced to the Scheduler at specified intervals (via the QTSK call). Macro commands are included here for completeness only. Table 13.5 lists two typical instructions.

Clock Commands are ordinarily used for timing based on the internal hardware (real-time) clock. The real-time clock is, in fact, an

Table 13.3. Task Creation and Deletion Commands

CRTSK:	Create a task with a given P and ID (identify it to the Scheduler by setting up a TDB Block in the Ready Queue)
KLTSK:	Kill a task (remove it from the Scheduler's jurisdiction) (equivalent to FORTRAN STOP for a normal program)
KALLTSK:	Kill all tasks with a specified P
STPID:	Set a task priority P and identification ID (most often from the system default value to the required value for the user's first task)
SUSTSK:	Suspend a task of given P and ID
SALLTSK:	Suspend all tasks of given P
READY:	Return to Ready status a previously suspended task of given P and ID
ALLRDY:	Return to Ready status all previously suspended tasks of a given P
Arguments:	To be shown in examples

Example 13.2. Creation of a Separate (Parallel) Task

Effective Logic Flow	Address	Program	Comments
		.	
		.	
start		instructions	
		.	
one task		.	
P = 10		STPID	Set task priority
ID = 4		10	to 10 and
		4	ID to 4.
create another task		CRTSK	Create a parallel task
		15	with priority 15
P = 15 P = 10		2	and ID 2
ID = 2 ID = 4		A	starting at address A.
		IOLPT	Print on lineprinter
write T2 write T4		2	two characters:
		T	T (for task)
		4	4 (for number 4).
		KLTSK	Kill the task.
finish finish	A	IOLPT	Print on lineprinter
		2	two characters:
		T	T (for task)
		2	2 (for number 2).
		KLTSK	Kill the task.

input device to the system, but its operation is usually restricted to providing time signals to a system software timing routine, a software clock, which executes within the operating system. The basic commands are listed in Table 13.6.

Suppose we have an application in which one task is to create a parallel task, wait for 5 seconds, send a message to the other task, print out a message, and stop. The second task is to print a message, wait for the message from the first task, print out another message, and stop. Example 13.3 illustrates how this might be programmed using the instructions defined above. For simplicity the structure from Example 13.2 is expanded. In this program the first message to be printed out would be T2, followed by T4 and finally TF. If the priorities were changed

so that Task 2 had a high priority of 5, then immediately following the transmission of "OK" from Task 4 the message TF would be printed. Thus the output would be T2, TF, and T4.

Task Identification (Task ID) Commands differ from task creation and deletion commands only in that they apply to tasks with specific ID numbers. They are summarized in Table 13.7.

Task/Operator Communications Commands are used to send and receive messages between the operator and tasks. They are most useful in real-time process-operator communication applications. Table 13.8 defines two typical commands.

Operator/Task Interaction Commands comprise a set of operator keyboard commands

Table 13.4. Intertask Communication Instructions

SEND:	Send a message to a task over a specified channel.
SWAIT:	Send a message to a task over a specified channel and wait for its receipt (here the task is suspended by the scheduler until the message has been received). This is an effective command to synchronize task scheduling and timing although variations may exist with special queueing commands.
RECEIVE:	Receive a message over a specified channel. If not yet sent wait for it. If a task was waiting for acknowledgement of receipt, send it.
CHKCHAN:	Check a specified channel, branch if no message.
Arguments:	To be shown in examples.

Example 13.3. Task Timing and Synchronization

Effective Logic Flow	Address	Program	Comments
		.	
		.	
start		instructions	
		.	
		.	
P = 10		.	
ID = 4		STPID	Set task priority
		10	to 10 and
		4	ID to 4
create another task		CRTSK	Create a parallel task
		15	with priority 15
P = 15 P = 10		2	and ID 2
ID = 2 ID = 4		A	starting at address A.
		WAIT	Wait for
		5	5 seconds.
wait 5 seconds		SEND	Send a message
		O	O
write T2		K	K
		2	to task 2.
wait for a		IOLPT	Print on lineprinter
message	send message	2	two characters:
	to task 2	T	T
write TF		4	4.
	write T4	KLTSK	Kill the task.
finish		IOLPT	Print on lineprinter
	finish	A	2	two characters:
		T	T
		2	2.
		RECEIVE	Wait for message.
		IOLPT	Print on lineprinter
		2	two characters:
		T	T
		F	F (for finished).
		KLTSK	Kill the task.

often available for direct manipulation of task status and of the Scheduler. Facilities for most of the creation, deletion, and communications Macros as defined above are usually provided.

This description is not meant to be comprehensive but only to indicate the general types of commands that may be available or desirable in a given real-time operating system.

13.3.4 Disk Files and Partitions

The primary advantages to the user of a large disk-based real-time operating system are: the improved file-handling capabilities of the system, the ability to carry out program swapping and chaining, an overlay capability to enhance program scope by judicious segmentation of routines (and for efficient input/output operations), the ability to spool output to slow peripheral devices via temporary disk files.

"File" is a name applied to any collection of information or, by extension of the concept, to any device that receives or generates information. Strictly speaking, devices cannot be files, but, insofar as they can receive (print, for example) information, they can appear to the user to be files. Disk files can be systems programs in machine language format, source or binary programs, or data. Any or all of these types of files may be available to user tasks; so a few words about their organization are

Table 13.5. Overlay and Queueing Instructions

GTSKOL: Get a task overlay from a bulk storage device (usually a disk).

QTSK: Queue a task or overlay. This is used to prepare tasks for execution at specific time intervals either by time interval from the previous execution or by time and date. As with all tasks which are introduced to the Scheduler, they are only placed in the Executing state when their priority is highest so that no absolute guarantee of execution is made.

Table 13.6. Clock and Timing Instructions

WAIT:	Suspend the executing task for a specified time interval. Note that following this time interval the task returns to the Ready state and must compete with all other Ready tasks for CPU control.
CLOCK:	Get the time and date from the system (software) clock.
SCLOCK:	Set the time and date in the system clock.
Arguments:	To be shown in examples.

warranted. Disks may be divided into a number of independent partitions (by hardware or software), each of which can be considered a separate and independent section from the operating system point of view. In each partition a set of files may reside; and for each set of files, a directory of file names is also provided. Finally, it is usually possible to link the files from one directory to another so as to provide an efficient means of accessing systems programs without duplication in various partitions of the disk. Disk file, partition layout, and directory format vary widely. Partitions have fixed boundaries, and situations can arise in which user programs require more than this fixed amount of space. Dynamic storage, a more efficient use of space, is attained by defining individual directories which can grow or shrink as the need arises. There are many very complicated algorithms used to allocate disk space, particularly on large (50–300-megabyte) disks; however detailed discussion is beyond the scope of this chapter.

An important hardware/software tradeoff should be mentioned in the context of disk files. The difference between fixed- and moving-head disks is discussed in Chapter 9, the chief difference being that a fixed-head disk usually is considerably faster than a moving-head disk, since a read/write head is provided for every track. During the design stage of a disk real-time operating system a decision must

be made as to optimum size of files (length or number of words). For example, it is possible to have files so small on a fixed-head disk that a considerable latency period develops for transfer of a large number of consecutive segments. In this context latency can be defined as the average time it takes for the system to find and transfer a randomly-specified disk file or file segment. The segments are stored at random on the disk and are linked together through software. By contrast a very large file on a moving-head disk can be transferred at high speed once it is found, provided the read head does not have to move in other than single-step increments across the disk surface. Cases exist where the fixed-head disk can transfer a file no faster than the (nominally) much slower moving-head disk. It is unlikely that the average user of a disk real-time operating system will ever actually write the operating system programs; so this decision on file size may only be relevant as a factor in purchasing the system.

13.4 SYSTEM CONFIGURATION AND GENERATION

System configuration requires the user to specify which of many possible vendor-supplied options are to be included as part of the user's own operating system. Invariably, as the number of desired options increases, the size of the execu-

Table 13.7. Task Identification Instructions

TSTATE:	Get the status of a task with a specified ID (Ready, Suspended, etc.)
TIDKIL:	Kill a task of specified ID
TIDPTY:	Change the priority of a task of specified ID from P1 to P2
TIDRDY:	Put in the Ready state a previously Suspended task of specified ID
TIDSUSP:	Suspend a task of specified ID

Table 13.8. Task/Operator Communications Commands

TSKWOP:	Send a message to the operator console.
TSKROP:	Ready a task for receiving a message from the operator console. This command frequently demands a password to be typed before the task is raised from its Suspended state back to the Ready state.

Table 13.9. A Partial Interactive Dialog for System Configuration

Question	Answer	Notes
.		
.		
.		
Use saved answer file?	NO	---
Use saved peripheral configuration?	NO	---
Save responses?	YES	---
Does processor have a switch register?	YES	---
Memory size?	128KW	16 bit words
Mapped?	YES	Required if mem > 28K
Arithmetic processing unit?	YES	---
Clock interrupts per second?	100	---
Devices?	DP	Include disk drive handlers
Devices?	IP	Include real-time subsystem handlers
.		⋮
.		Additional devices
.		⋮
Devices?	DL	Include asynch multiplexer handler
Devices?	.	Terminate device list
Data base management?	YES	Include DBM routines
Task send/receive?	YES	Allow intertask communication
Multiuser protection?	YES	Prevent user/user interference
Power fail?	YES	Include auto recovery on power fail
User written driver?	NO	No custom handlers used
Terminal driver?	FD	Make terminal driver full duplex
Number A/D converter cards?	8	8 A/D cards present
Number D/A converter cards?	4	4 D/A cards present
Any digital sense modules?	NO	---
Any latching digital outputs?	YES	Device driver determines count
Add Purdue ISA routines?	YES	Include ISA real-time FORTRAN standards
.		
.		
.		

tive increases. Options may be selected interactively during the *system gen*eration (sysgen) procedure itself (i.e., at the time that the operating system components are link-loaded to generate the system executive), or they may be specified by entering numbers and answering yes/no questions in an assembly language module which is then assembled and link-loaded with all required system object code modules provided by the vendor.

The interactive dialog approach to system generation is usually preferable to modification and subsequent assembly of operating system source code modules. This is especially so when interactive user responses can be saved in one or more disk files which can then be edited and used for operating system updates. Questions asked as system configuration and operation proceed invariably depend upon answers provided earlier in the procedure. Though actual dialog varies widely from vendor to vendor, Table 13.9 illustrates a partial (ca.

5–10% of questions actually asked) interactive sysgen session where we assume that the sysgen executive itself has already been invoked.

Table 13.10 illustrates the case where configuration options are specified within an assembly language system subroutine called a "configuration table".

The vendor usually supplies a configuration table module tailored to the CPU, memory, hardware options, and peripherals specified in the original purchase order. Invariably though, additional options or peripherals are added over the useful life of the system. For example, let's assume our system was delivered with a high-speed disk, a high-speed paper tape reader/punch, and a console terminal. Typical configuration table code appears in Table 13.11 where the REF operator designates an external subroutine, DCB designates *D*evice *C*ontrol *B*lock (cf. Section 12.2), and DC means *D*efine *C*onstant. Now suppose we add two additional peripherals—a buffered *U*niversal *A*synchronous

Table 13.10. Selection of System Configuration Options

```
*     DEFINE EXTENT OF EXTENDED MEMORY MANAGEMENT
*
*                      EXM = 0 FOR NO EXTENDED MEMORY MANAGEMENT
*                      EXM = 1 FOR FULL SYSTEM EXTENDED MEMORY MANAGEMENT
*
*
EXM         EQU     1
NEXM        EQU     24              NUMBER OF EXTENDED MEMORY MODULES
MAXAR       EQU     29              MAX NO. OF AREAS PER EXT. MEM. MOD.
STSEL       EQU     X´4000´         START ADDRESS OF SELECTABLE MEMORY
N           EQU     64              64 PROGRAMS
NPA         EQU     32              32 AREAS
NMLOG       EQU     64              64 LOGICAL UNITS IN SYSTEM
BACKPR      EQU     1               PROG 1 USES BACKGROUND LOGICAL UNIT TABLE
AUX         EQU     0               NUMBER OF AUXILIARY CLOCKS
INCTN       EQU     0               NUMBER OF DECREMENTAL COUNTERS
* MAKE THE FOLLOWING YES/NO AS DESIRED
BFIO        EQU     YES             YES FOR BUFFERED INTERFACE
USI         EQU     YES             NO INHIBITS F$USI(USER INITIALIZATION PROGRAM)
MASK3       EQU     NO              NO INHIBITS MSK$3(ALTER INTERRUPT MASK)
INTTY       EQU     YES             YES IF CTRL C INTERRUPT IS DESIRED (TY$I)
DATE        EQU     YES             NO INHIBITS CALENDAR ROUTINE (R$CAL)
TO          EQU     NO              YES FOR I/O TIME OUT
FMS         EQU     NO              YES TO INCLUDE FILE MANAGEMENT
OVLP        EQU     NO              YES FOR OVERLAPPED PROGRAM LOAD
SPOOL       EQU     NO              YES FOR SPOOLING
CNTTY       EQU     NO              YES FOR CONSOLE TELETYPE
MTS         EQU     NO              YES IF MULTITERMINAL SYSTEM
* SET HARDWARE CONFIGURATION
SCLNG       EQU     400             DISK WORDS PER SECTOR
SIMS        SET     YES             YES IF HDWR MUL/DIV IN SYSTEM
APU         SET     YES             SET YES IF HARDWARE ARITHMETIC UNIT IN SYSTEM
NDISK       EQU     2               NUMBER OF PHYSICAL DISK DCBS IN PHYS UNIT TABLE
NMDC        EQU     8               NUMBER OF DATA CHANNELS
NBUF        EQU     20              NUMBER OF BUFFERED DEVICE CONTROL BLOCKS (DCBS)
```

Table 13.11. Configuration Table Segments for Delivered System

```
•
•
•
       REF        C$DUM              SYSTEM SCRATCH FILE DCB
       REF        C$D046             SYSTEM DISK DCB
       REF        C$BTYP             SYSTEM CONSOLE DCB
       REF        C$BPRP             HI SPEED READER DCB
       REF        C$BPPP             HI SPEED PUNCH DCB

•
•
•

*PHYSICAL UNIT TABLE

*ENTRIES VARY ACCORDING TO SYSTEM I/O DEVICES

T$PUT   DC         C$DUM
        DC         C$D046
        DC         C$BTYP
        DC         C$BPRP
        DC         C$BPPP

•
•
•

*INTERRUPT MASK TABLE

E$MSK   DC         X´0809´            PAPER TAPE READER/PUNCH INTERRUPTS
        DC         X´OEOE´            DISK INTERRUPT
•
•
•
```

*R*eceiver/*T*ransmitter (UART) for a KSR-43 Teletype and a buffered UART for process I/O from (to) some ADCs (DACs). The modifications required, besides, of course, increasing the number of buffered devices by two, are illustrated in Table 13.12. Of course, since the UARTs are not "standard peripherals", assembly language subroutines for their Device Control Blocks must also be developed. This can be a nontrivial task.

In general, there are many instances where operating system source code, especially configuration table source code, must be modified as the system is expanded—especially in cases where the machine must "talk" to another (smaller or larger) machine from another vendor.

13.5 FLOWCHARTING AND PROGRAM CONVENTIONS

In order to illustrate the structure of multitasking examples unambiguously, we must define a set of conventions. The first is one of task priority: highest user priority is 1, and lowest

priority is 100 (any range of priorities may in fact be defined in an executive system). The starting point for the first user task is required as input data to the real-time operating system which normally assigns a default priority and identification number to this initial task. Hence, flowcharts begin with the first user task which should set its priority, P, and identification number, ID. Program or task execution follows through the flowchart with changes in P and ID being noted.

A set of symbols is defined in Figure 13.9 to differentiate between system and user calls (Macro commands), user code (tasks or segments of tasks), user logic symbols, subroutine calls, system tasks resulting from Macro commands (excluding actions of the scheduler), and task attribute data. The Scheduler, which operates at highest system priority (0) selects tasks of equal priority in order of their creation in the system. Therefore, a call to create a task (CRTSK) may be considered part of a task proceeding in the downwards direction with the created task branching sideways on the diagram. The Scheduler is assumed to be called *each* time an interrupt occurs or *each* time a system or

Table 13.12. Configuration Table Segments for Expanded System

```
•
•
•
          REF          C$DUM              SYSTEM SCRATCH FILE DCB
          REF          C$D046             SYSTEM DISK DCB
          REF          C$BTYP             SYSTEM CONSOLE DCB
          REF          C$BPRP             HI SPEED READER DCB
          REF          C$BPPP             HI SPEED PUNCH DCB
          REF          C$URT1             KSR I/O UART
          REF          C$URT2             SYSTEM I/O UART
•
•
•

*PHYSICAL UNIT TABLE

*ENTRIES VARY ACCORDING TO SYSTEM I/O DEVICES

T$PUT     DC           C$DUM
          DC           C$D046
          DC           C$BTYP
          DC           C$BPRP
          DC           C$BPPP
          DC           C$URT1
          DC           C$URT2
•
•
•

*INTERRUPT MASK TABLE

E$MSK     DC           X´0809´             PAPER TAPE READER/PUNCH INTERRUPTS
          DC           X´1519´             UART INTERRUPTS
          DC           X´OEOE´             DISK INTERRUPT
•
•
•
```

task Macro command is executed. At this point tasks of equal priority are rotated (round-robin scheduling), again following the order of their elevation to the Ready state. Tasks may be viewed as the minimal segments of a set of operations. Alternatively, a set of several such operations may form a larger program segment or task. We attempt to illustrate this concept in the examples following. Some of the examples are intended to highlight simple concepts such as task creation and hence are not complete in the sense that each task completes execution and is removed from the system. Other examples suspend completed tasks to show the actual operation of the system. It may be useful to keep Figure 13.6 clearly in mind as the examples are discussed. In Section 13.3.3 we did not specify which calls were System Calls and which were Task Calls. Individual systems define these differently, but here we assume I/O calls and setting the time of day are System Calls, whereas task creation, priority, communication, or waiting calls are all Task Calls.

13.6 REAL-TIME APPLICATIONS

A series of programs for typical multitask applications is shown in this section in the form of diagrams or schematics. We assume that the real-time operating system is present and executing. It is convenient to assume that some very low-priority system task which we define as a Null Task (do nothing) is always operating when no other task is available for execution. Given this situation, the Null task is interrupted only by the real-time clock (and, subsequently, the system clock routine which has the job of maintaining the time of day). Clearly, a random input from an outside event could interrupt the Null task, but we have assumed so far that no user programs have as yet been introduced to the system. In this situation, then, two system tasks exist: one does nothing, while the other, periodically (say, once every 0.1 second), adjusts a software clock following an interrupt from the real-time clock. The execution of these two tasks is transparent to the user, and they are

TASKING AND SYSTEM SYMBOLS

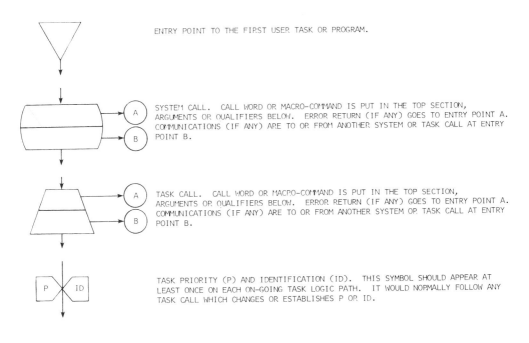

ENTRY POINT TO THE FIRST USER TASK OR PROGRAM.

SYSTEM CALL. CALL WORD OR MACRO-COMMAND IS PUT IN THE TOP SECTION, ARGUMENTS OR QUALIFIERS BELOW. ERROR RETURN (IF ANY) GOES TO ENTRY POINT A. COMMUNICATIONS (IF ANY) ARE TO OR FROM ANOTHER SYSTEM OR TASK CALL AT ENTRY POINT B.

TASK CALL. CALL WORD OR MACRO-COMMAND IS PUT IN THE TOP SECTION, ARGUMENTS OR QUALIFIERS BELOW. ERROR RETURN (IF ANY) GOES TO ENTRY POINT A. COMMUNICATIONS (IF ANY) ARE TO OR FROM ANOTHER SYSTEM OR TASK CALL AT ENTRY POINT B.

TASK PRIORITY (P) AND IDENTIFICATION (ID). THIS SYMBOL SHOULD APPEAR AT LEAST ONCE ON EACH ON-GOING TASK LOGIC PATH. IT WOULD NORMALLY FOLLOW ANY TASK CALL WHICH CHANGES OR ESTABLISHES P OR ID.

USER PROGRAM SYMBOLS

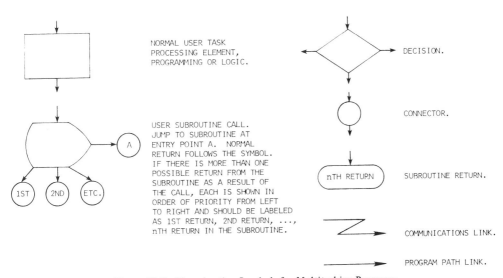

NORMAL USER TASK PROCESSING ELEMENT, PROGRAMMING OR LOGIC.

DECISION.

CONNECTOR.

USER SUBROUTINE CALL. JUMP TO SUBROUTINE AT ENTRY POINT A. NORMAL RETURN FOLLOWS THE SYMBOL. IF THERE IS MORE THAN ONE POSSIBLE RETURN FROM THE SUBROUTINE AS A RESULT OF THE CALL, EACH IS SHOWN IN ORDER OF PRIORITY FROM LEFT TO RIGHT AND SHOULD BE LABELED AS 1ST RETURN, 2ND RETURN, ..., nTH RETURN IN THE SUBROUTINE.

SUBROUTINE RETURN.

COMMUNICATIONS LINK.

PROGRAM PATH LINK.

Figure 13.9. Flowcharting Symbols for Multitasking Programs

not shown in the flow diagrams of the applications programs. These examples are intended to illustrate the structure of the applications programs rather than details of programming, and no attempt to use all of the general Macro commands or to provide definitive sets of arguments is made.

13.6.1 Process Control

The first major example concerns a simple process control application in which we wish to collect data from one or more process variables using ADC channels, calculate a set of valve positions according to a control algorithm, and

Example 13.4. Process Control

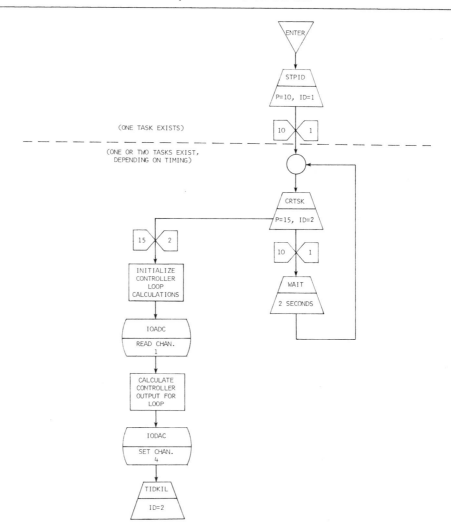

output the results to a set of DACs. This operation should be carried out periodically once every 2 seconds from the time of program initiation.

The flow chart is shown in Example 13.4. The program begins by setting the single user task priority, P, to 10 with an identification number, ID, of 1. It then executes the Macro command (CRTSK) to create a second parallel task with P = 15 and ID = 2. This task, being of lower priority, exists in a Ready State, while Task 1 continues to execute. Task 1 then initiates a wait (WAIT) loop of 2 seconds duration, suspending itself until this time is up. Task 2 is the only other task competing for CPU time so it is activated by the scheduler.

Following some initialization calculations (if any) for the control action, Task 2 requests input from the ADC through a system call (IOADC). It becomes suspended pending completion of this action. During this time the operating system activates its Null task because both Task 1 and Task 2 are suspended. Once the ADC input operation has been completed, Task 2 continues with the calculation of controller action. These results are output to the DAC, and finally Task 2 is killed, or more correctly placed in an inactive state. The Null task regains control of the CPU. Subsequently, the 2-second wait period of Task 1 expires, allowing it to be returned to the Ready State, and, since only the low-priority Null task is executing, Task 1 is immediately placed into execution. Task 1 loops back to the point where it recreates Task 2, i.e., returns it from Inactive to Ready, and the cycle is complete.

Several critical points arise in this example. First of all, it must be possible to complete execution of Task 2 within 2 seconds (less a small amount of operating system overhead time); otherwise two potential errors can occur. The first one is that on the second pass the operating system attempts to create a third task, Task 2, which still exists. Provided that the operating system monitors task ID numbers, this results in an error return from the task Macro command, with the result that the new Task 2 is not created. Alternatively, if the system does not check for unique task ID, a second parallel "Task 2" is raised to the Ready State. This cycle continues until the buffer holding TDBs is filled (there is always an upper bound on the number of tasks that can exist at any one instant) and a final system error results. In addition, it is quite likely that Task 2 was not written in reentrant code, and this would

cause loss of system integrity on subsequent passes owing to the overwriting of data being stored from the previous loop calculation.

Notice that in this example the system effectively contains two tasks (jobs). The first, Task 1, is an infinite loop involving a wait command. The second, Task 2, is a very short task which is created once every 2 seconds and is made inactive as soon as it has executed. An alternative route would have been to put the control calculations in the loop of Task 1. This would also work. However, the adjustment of the timing would require accurate determination of the time taken for the part corresponding to Task 2. This time would have to be subtracted from the 2 seconds to correct the overall loop time; hence, this approach is never recommended.

13.6.2 Timesharing

The problem illustrated by Example 13.5 is to write a program to create four equal-priority tasks which are to be executed in a round-robin

Example 13.5. Timesharing

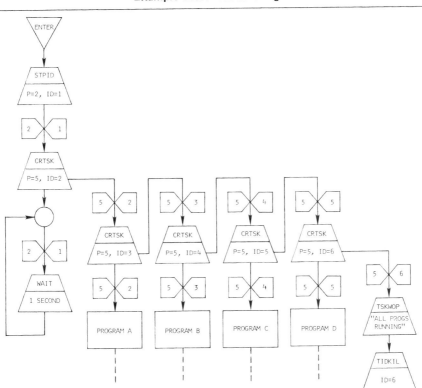

Table 13.13. Summary of Program States for the Timesharing Example

CYCLE	TASKS:	1 = EXECUTING,	0 = READY		
1	A1	BO			
2	B1	AO	CO		
3	A1	CO	BO		
4	C1	BO	AO	DO	
5	B1	AO	DO	CO	
6	A1	DO	CO	BO	
7	D1	CO	BO	AO	EO
8	C1	BO	AO	EO	DO
9	B1	AO	EO	DO	CO
10	A1	EO	DO	CO	BO
11	E1	DO	CO	BO	AO
12	D1	CO	BO	AO	

timesharing mode. Each is allowed 1 second of CPU time per "time-slice". Following the introduction of the final task, a message is to be written to tell the operator that the timesharing operation is under way.

The program begins by establishing Task 1 with priority equal to 2. Task 1 creates Task 2, but since this has a lower priority (5), Task 1 remains in the executing state until it is suspended by the 1-second wait command. Task 2 now proceeds to create Task 3. This is of the same priority as Task 2, so, by our convention, Task 2 continues with execution of Program A until the completion of the 1-second wait (Task 1). As soon as this happens, Task 1 is given priority by the scheduler but immediately suspends itself for another 1 second. Now two equal-priority Tasks, 2 and 3 exist. Again by convention, since Task 2 was last running, Task 3 is given priority. This creates Task 4, but Program B executes. Task 1 forces rescheduling again, after which Program A, which has been waiting longest, has priority. The entire operation is summarized in Table 13.13, with Task 6 being called Program E for convenience and with the numbers 1 or 0 appended for Executing or for Ready tasks, respectively. Priority is from left to right.

In Table 13.13, we assume that Program E (the message task) takes 1 second or longer to execute beginning in Cycle 11. Following Step 12 the system executes in pure timesharing mode with the tasks being executed in reverse order of creation, once per second. Of course, when the four programs have run their course, they must each be terminated, and this changes the sequencing order.

In this type of system it is quite possible to define a set of high-priority tasks to perform data logging, controller calculations, and so on with some lower-priority set of tasks that timeshare all excess time. All timesharing programs would execute at a fixed level below the priority of the time-slicing task.

13.6.3 Operator Executive and Alarm Monitoring

The structure of keyboard-oriented executive systems becomes very complicated as the level of sophistication is raised. In this relatively simple example, we wish the operator to be able to activate user tasks, i.e., raise them from the Inactive state to the Ready state by typing in a name that is checked against a status table for: (1) existence (is the name valid?), and (2) status (currently in Ready State or not?). At the same time a set of software alarm flags is to be checked once per second. If an alarm condition exists, a message is printed on the operator console. The key problem is to keep separate any operator input (which is also printed) and alarm program output; this problem illustrates the use of intertask communication commands for the purpose of synchronization of tasks. Task 1 checks operator input. Task 2 checks for alarms. Example 13.6 illustrates these ideas.

As we follow the logic of this program, Task 1 is suspended at the TSKROP Task Call pending input of the operator interrupt password. As long as no operator interrupt has been activated (password not typed), Task 2 operates in a mode wherein no message has been sent on Channel 2. Thus the receive message logic is bypassed, and only the alarm flags are checked.

Example 13.6. Operator Executive and Alarm Monitoring

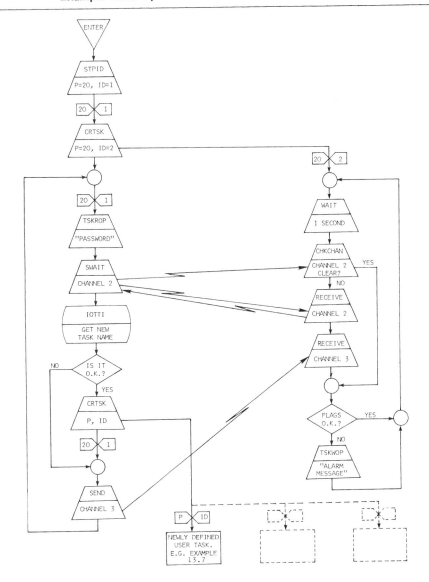

If an alarm condition is found, some appropriate message is printed. Otherwise, the task cycles once per second, checking the alarm flags.

Assuming at some point that the password has been typed by the operator, Task 1 can proceed. It sends a message on Channel 2 and waits for receipt. The reason for this action is to disable the terminal from the alarm mode long enough for a message to be typed in, free from extraneous characters, on the console printer. This action occurs because the CHKCHAN command finds Channel 2 not

clear; it acknowledges the Channel 2 Wait message and suspends itself waiting for a message on Channel 3 (to signify end of operator input to the console). Meanwhile the SWAIT command in Task 1 has received its confirmation (message received), and Task 1 proceeds. The I/O command requests terminal input, verifies it, and if the new task name is valid, a new task is introduced to the system (CRTSK) having priority P and identification ID (presumably these are supplied as arguments from a table of valid new tasks). Finally, Task 1 sends a message on Channel 3 to signify that the console is now free and to release the alarm Task 2. Task 1 cycles back to await another operator interrupt request.

The example could be improved by adding a watchdog timer loop to Task 2 between the two receive commands to prevent excessive delays in checking the alarm flags (while the operator has coffee, for example). A watchdog timer is a hardware or software unit that must be reset within a prespecified interval to prevent a branch or alteration in the program logic. It should be emphasized here that some operating systems may have specific commands to carry out these functions. We restrict the examples to use of the Macro commands defined in Section 13.3.3. In using this particular executive, the operator must be wary of specifying a new task whose priority is too high and which takes too long to execute. It could potentially disable the entire alarm monitoring system as well as the operator console keyboard.

13.6.4 Multitasking Using a Single Reentrant Program

In many real-time computer applications the same subroutine or group of subroutines is used over and over again for different logical functions. Rather than copy the subroutine many times, which could require a large amount of memory, it is good practice to develop a single routine that can be called at random by all the programs requiring it. Since most calculations involve storage of intermediate results within a subroutine, either in tables (arrays) or as single variables, reentrant routines are structured so that all of these data are stored in the calling program or in a user stack rather than in the routine itself. Only the fundamental algorithm remains in the reentrant program. It should be noted that all machine registers are stored in the task data blocks (TDBs) so that they can contain intermediate results that can be stored by the Scheduler.

Examples of reentrant programs occur in control applications (Section 13.6.1) where a single controller algorithm is used to calculate valve settings for many different control loops operating with different execution periods.

Similarly, many data-logging applications may require only one subroutine with a set of arguments to dictate specific devices, timing intervals, number of variables to be logged, and so on. This latter case is illustrated in the following example. The problem is to implement a user task to accept input from a terminal. The input specifies which device of several system devices is to be used for general-purpose data-logging activity. Selected process variables are to be logged once per time period (also specified) on the chosen device. Since the logging action involves fetching data from a table, formatting it, checking the time and then repeating the cycle, and since most of this is device-independent, a reentrant routine can be developed to carry out the logging function.

The important concept to understand in following this example is that only one program segment (for the logging function) is actually present in memory. However, as far as the scheduler is concerned, it must be made to appear to consist of as many separate logical blocks or tasks as there are independent devices in the system. This is accomplished by defining a *Device Data Block* (DDB) in which the task priority and identification number, all entries required to define the logging task (e.g., variable names, sample interval, format information), and all the temporary storage required for the logging algorithm are stored. One such block (buffer, or array) is required for each device. Assume that six output devices are present:

- TTO—terminal printer
- LPT—line-printer
- PTP—paper tape punch
- TAPE—magnetic tape
- DISK—disk
- CRT—video terminal

Recall that in the TDBs which the scheduler maintains for tasks in the Ready state, we assumed that one word of data was defined as a user data pointer (UDP) (cf. Section 13.2.3). This pointer is used by the software to link the task to its DDB as the task moves to the Executing state from the Ready state. Figure 13.10 illustrates this situation with two queues, the TDB queue and the DDB storage blocks.

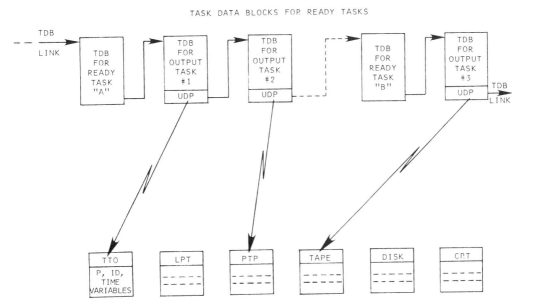

TASK DATA BLOCKS FOR READY TASKS

DEVICE DATA BLOCKS AS DEFINED BY THE USER.

Figure 13.10. Task Data Blocks and User Data Blocks for Reentrant Logging Program (Example 13.7.)

Tasks 1, 2, and 3 in the TDB queue are output logging tasks to the TTO, PTP, and TAPE, respectively. Tasks A and B in the queue are other types of tasks.

Example 13.7 illustrates the structure of a multi-tasking data logger which is activated through keyboard input. Task ID (which could, for example, be initiated as a result of an input through the operator executive in Example 13.6) accepts operator input (IOTTI) about a data-logging operation. (This includes: output device, name, logging interval, and a list of variables to be logged. The DDBs have been predefined.) The task continues, utilizing a user data pointer (UDP) which identifies the device data block, sets up the logging activity, timing, variables, etc., and then enters a loop of reentrant coding which executes the output operation. A call to get the time is made; then the data are output on the device. A second call to get the time enables a calculation of the time remaining in the logging interval, and this is used as data for a wait command. The loop is closed after the wait (WAIT), and logging continues until the task is terminated. Up to six tasks, one for each predefined logging device, can be activated using this one piece of code.

Task data blocks are defined for each one and the scheduler assumes responsibility for execution of the tasks. The TDB string as possibly seen by the scheduler has been shown in Figure 13.10. The links to the device data blocks for the logging tasks, 1, 2, . . . , m, through the user data pointer in the TDB are shown for illustration.

13.7 SUMMARY

This chapter has presented a broad picture of the many facets of real-time operating systems and multitask programming. Details on individual systems and specific examples have been avoided in an attempt to maintain generality. For specific information the reader is referred to the various vendors who can provide a wealth of information on the merits and flexibility of their own particular operating system or systems. For process control packages, particularly where small computers are involved, most of the information must come from systems houses as opposed to computer vendors, who generally prefer not to specialize in applications packages.

Example 13.7. Multitasking Data Logger

INITIALIZATION SECTION

OUTPUT ROUTINE, REENTRANT CODE

ENTER HERE
POSSIBLY USING
THE EXECUTIVE
EXAMPLE 13.6

P   ID

IOTTI
GET A DEVICE NAME
FROM OPERATOR

IS
NAME O.K.?

NO

YES

FIND DEVICE'S
INFORMATION
BLOCK. SET IT
UP FOR LOGGING
ACTIVITY

CLOCK
GET THE TIME

READ DATA
FROM DATA
TABLE. SET IT
UP FOR OUTPUT

IODEV
OUTPUT DATA TO
SELECTED DEVICE

CLOCK
GET THE
TIME AGAIN

COMPUTE DELAY
REMAINING

WAIT
COMPUTED
DELAY

13.8 SUPPLEMENTARY READING

1. Desmonde, W. H., *Real-Time Data Processing Systems: Introductory Concepts*, Prentice Hall, Englewood Cliffs, N.J., 1967.
2. Friedman, A. L. and Lees, R. A., *Real-Time Computer Systems*, Crane Russak, New York, 1977.
3. Grabbe, E. M., Ramo, S., and Wooldridge, D. E., *Handbook of Automatic Computation and Control*, 3 Vols., Wiley, New York, 1968.
4. Harrison, T. J., *Handbook of Industrial Computer Control*, Wiley Interscience, New York, 1972.
5. Korn, G. A., *Minicomputers for Engineers and Scientists*, McGraw-Hill, New York, 1973.
6. Lee, T. H., Adams, G. E., and Gaines, W. M., *Computer Process Control*, Wiley, New York, 1968.
7. Liptak, B., *Instrument Engineers Handbook*, Vol. 2, Chilton, Philadelphia, 1970.
8. Martin, J., *Programming Real-Time Computer Systems*, Prentice Hall, Englewood Cliffs, N.J., 1965.
9. Martin, J., *Design of Real-Time Computer Systems*, Prentice Hall, Englewood Cliffs, N.J., 1967.
10. Martin, J., *Design of Man Computer Dialogues*, Prentice Hall, Englewood Cliifs, N.J., 1973.

11. Rothstein, M. F., *Guide to the Design of Real-Time Systems*, Wiley, New York, 1970.
12. Savas, E. S., *Computer Control of Industrial Processes*, McGraw-Hill, New York, 1965.
13. Schomberg, M., *Real-Time Operating Systems for Minicomputers*, Crane Russak, New York, 1976.
14. Smith, C. L., *Digital Computer Process Control*, International Textbook, New York, 1972.
15. Soucek, B., *Minicomputers in Data Processing and Simulation*, Wiley Interscience, New York, 1976.
16. Stimler, P., *Real-Time Processing Systems: A Methodology for their Design and Cost Performance*, McGraw-Hill, New York, 1969.
17. Yourdon, E., *Real-Time Systems Design*, Information and Systems Press, Cambridge, Mass. 1967.
18. Yourdon, E., *Design of On-Line Computer Systems*, Prentice Hall, Englewood Cliffs, N.J., 1972.

13.9 EXERCISES

One of the most difficult concepts to learn in the transition from general programming to real-time programming is a clear appreciation and understanding of the multitasking concept;

both should be developed prior to attempting to write major application code. The most efficient way to understand how a multitasking system operates is to generate a series of elementary programming examples using console or terminal output to indicate which task is actually running. Readers are encouraged to create their own examples to completely explore the list of the tasking commands available in a system of interest to them.

1. A number of vendors now offer very attractively-priced home computer systems, most of which have operating systems supporting BASIC. Some of these claim to support real-time activities. Obtain information from several of these vendors, and evaluate whether the operating system does, indeed, support real-time operations. For example, is it possible to write multitasking programs? Summarize your results in a table comparing the systems.

2. Many computer vendors claim compatibility of their real-time disk-based operating systems with the more compact real-time operating systems offered for smaller versions of the same hardware. From information available from vendors, summarize the extent to which this is true. What are the limitations imposed on the programmer who wants to develop programs in the more powerful system to run on the small, memory-resident version?

3. Explain the differences between background/foreground disk operating systems in which real-time, multitasking features are supported and genuine multiprogramming systems which also support real-time activities. What are the limitations of the foreground/background-structured system?

4. Many real-time disk operating systems support multitasking activities. However, it is not always clear what the priority of user tasks is relative to various system activities. On a system available to you, summarize those programs or operating system subprograms that have a higher priority than the highest-priority memory-resident user real-time task. The point of this exercise is that, in some systems, it is possible that system overhead software can, or will always, have a higher priority than user programs even for utility operations. By contrast, a superior real-time operating system will allow real-time users to preempt system utilities.

5. Many real-time disk operating systems today use front-end microcomputers as preprocessors for handling communications or real-time data acquisition and transfer. An important consideration is how is this real-time information transmitted to main memory in the host computer? What is the priority of this front-end processor relative to general peripherals and system software? Also, how are the data actually transferred, i.e., in serial ASCII format, serial binary format, block mode, or others? For a system you are evaluating, what is the maximum rate at which a user-written task could ask for analog input data from a process and expect these data to be stored in memory?

6. For a typical system, what actions cause the scheduling program to reschedule tasks of the same priority. Note that many systems differ substantially in this regard. Why is this an important question? How is a series of tasks of the same priority scheduled?

7. For a system available to you and using a language of your choice, write a program that will create a parallel task of a specified priority to the mainline program. In each of the final tasks (i.e., the main program and the parallel task), include a brief console write statement indicating which task you are in. Run the program and note which task writes the message. Reverse the priorities of the parallel tasks and rerun the program.

8. Write a program to create a parallel task of a specified priority. Following task creation by the mainline program, wait for 5 seconds and then print out a message "Mainline Task". In the parallel task, print the message "Parallel Task". Following the print statement, cause each task to be deactivated. Run the program with, first, a higher priority and, then, a lower priority in the parallel task. Explain why the messages come out in the order they do.

9. Write a simple program for your system following Example 13.4.

10. Generate an example for your system to illustrate the use of task synchronization commands. The example should be simple and should include differences in priorities such that the parallel task would reach a synchronization point (i) ahead of, and (ii) following, the point at which the

mainline task would send a synchroniza-tion message. Be sure to explain what your program demonstrates.

11. Write a simple output routine that is re-entrant. The task should be able to output to at least two different devices, e.g., a console CRT and a disk-based file. To test the routine, generate a multitasking program such that it will call the routine from at least two points so that the reentrancy features of the subroutine are utilized.

12. Two tasks of roughly equal size and roughly equal resource requirements are to operate in the same memory area under control of a multitasking executive. Priority and ID assignments are 10 and 20. ($P_{20} > P_{10}$). Assume that Task 20 is memory-resident and executes a FORTRAN READ state-ment at a time when Task 10 is outputting an operator report. Explain what will happen in the context of how the Input/ Output Subsystem (cf. Chapter 12) and the associated line DCB will handle the READ request of Task 20.

13. Consider a disk-based real-time operating system running on a 32 KW, 16-bit small-computer system. The following con-siderations are pertinent:

- The system executive, as configured, supports one background (e.g., program development) job (or task) and up to 31_{10} foreground jobs (or tasks).
- Service increases as priority level in-creases (i.e., $P_1 < P_2 \ldots < P_{31}$).
- Job priority may be different from job ID.

- I/O is fully buffered.
- The user must specify an absolute memory starting address for each fore-ground job. Total job area for rolling to disk runs through the end of the last area segment that is used.
- The user must specify an absolute mem-ory starting address for a dynamic storage area for each foreground job. About $1B0_x$ words or less is usually adequate.
- Dynamic storage areas cannot be rolled to disk.
- The executive occupies locations 0_x through $2A01_x$.
- All jobs and dynamic storage areas must start at a memory boundary.
- The background uses locations $2A02_x$ through 4601_x.
- The ROM boostrap occupies locations $7FBF_x$ through $7FFF_x$.
- Foreground jobs can compete with the background for memory, but swapping overhead is high.
- Memory is segmented into logical areas 200_x words long (except for the last area right below the bootstrap) starting at $2A02_x$.

The foreground jobs summarized in the table are to be run under the executive described. Indicate where each task should begin in mem-ory, where it will end, where its dynamic storage area should be located, whether overlays should be used or not, and the priority of each job including the background. Explain your reasoning.

Task ID	Task Size w/o overlay	Task Size w/overlay	Notes
2	$2D00_x$	$1A00_x$	Little cpu Much I/O Few overlay calls
10	$C00_x$	$C00_x$	Heavy cpu Little I/O
15	2500_x	$D00_x$	Little cpu Much I/O Many overlay calls
19	$1F00_x$	$1F00_x$	Little cpu Little I/O
1	<Background>		Program Development

Part VI
Real-Time Applications Software

14

Real-Time BASIC

Duncan A. Mellichamp
Department of Chemical and Nuclear Engineering
University of California, Santa Barbara

14.0 INTRODUCTION

The BASIC Language (*B*eginners *A*ll-Purpose *S*ymbolic *I*nstruction *C*ode) was designed to give computer users a reasonably powerful high-level language suitable for mathematical analysis and, at the same time, one easy for the novice programmer to learn. In particular, the language and its interpretive mode of implementation were intended to make the iterative process of writing, running, and debugging programs as simple as possible.

Historically, BASIC and other interpreter-based languages (such as FOCAL and APL) have not been as widely used by engineers and scientists as some other compiler- or translator-based languages. FORTRAN, for one example, has become practically the universal language of scientific computing in the United States and Canada; ALGOL earlier achieved something of the same status in Europe and the Soviet Union; PASCAL presently is the language in vogue for computer scientists. The utility of an inter-preter-based language such as BASIC was strictly limited as long as most computing resources consisted of large-scale, batch-operations-oriented computing centers which, up until the late 1960s, was the case. The increasing availability of both time-sharing computing networks and single-user-oriented micro- and minicomputers has changed the picture considerably. Now, many of the same features that make BASIC a very attractive "first programming language" make BASIC with real-time extensions an excellent programming language

for a user working on a first real-time application. In fact, the interactive capabilities (user $\leftrightarrow$ computer) of the language make the handling of real-time interactions (user $\leftrightarrow$ computer $\leftrightarrow$ process) much simpler and, as a consequence, real-time programs far easier to debug on-line than programs written in assembly language or programs that have been compiled—in either case, programs intended to be run in a noninteractive mode except with the computer coupled to some process.

To see why this is so, it is necessary to spend a short amount of time discussing the important differences between compiler- and inter-preter-based programming languages. Following this, the remainder of the chapter is devoted to (1) a brief exposition or review of the BASIC language as traditionally made available to the single or time-shared user, (2) a description of an elemental set of real-time extensions to the language (subroutines) which make computer operations possible with simple processes (single task programs), and (3) a description of BASIC "dialects" which permit the user to implement full multitask programs as would be required, for example, with more complex laboratory or production processes.

Several examples have been included in this chapter to illustrate and clarify important points. Unfortunately, these examples reflect the continued lack of standardization of "Real-Time BASIC" compared to what has been accomplished with the FORTRAN language by the Instrument Society of America. It is quite possible that efforts to standardize on a real-

time version of BASIC will never reach frui-
tion.* Nevertheless, real-time versions of the
BASIC language already exist; at least one of
them is modeled somewhat along the lines of
the ISA FORTRAN specifications; and there is
no reason why the FORTRAN extensions could
not be applied to the BASIC language as well.
Those users who would make best use of an
interpreter-based real-time language—anyone
who is interested in minimizing the amount of
effort necessary to get a program operational—
in the meantime will have to push for standardization of the language.

14.1 INTERPRETER- VS. COMPILER-
BASED PROGRAMMING LANGUAGES

The important point to remember about any
programming language is that ultimately it must
reduce to a set of machine language commands,
which will be stored in memory and later executed, and to a corresponding set of memory
words, which represent program data to be used
by the machine language program commands.
The very fundamental differences between the
way a BASIC program "runs" and the way a
FORTRAN program executes depend only on
differences in how the two programs are reduced to executable machine language commands and data in memory.

Regardless of the type of programming language used, at run time the computer always
must contain the system auxiliaries necessary
for execution in addition to the program itself.
These auxiliaries include the operating system
or real-time executive program—responsible for
coordinating the operations of all peripheral devices and input/output from/to these devices—
and all system subroutines used by the program
(e.g., computational routines for multiplication,
division, floating point arithmetic, the calculation of transcendental functions, etc.).

The process of making a high-level compiler-
based program ready to load into the machine
along with these auxiliaries and then to execute
it, normally consists of a number of steps which
are shown in Figure 14.1a. First the computer
is loaded with the compiler (a type of translator), and then the FORTRAN source program
is processed through the compiler, resulting in
a relocatable machine language object program.
Several passes may be required at this stage; i.e.,
the FORTRAN source program may have to be
processed through the compiler several times.**
The computer then is loaded with the linking
loader (overwriting the compiler in the process). It then loads the relocatable object code
generated by the compiler into the computer
along with all necessary system subroutines and
links them up correctly. During all of these
steps the operating system usually is resident in
machine memory (in fact it is responsible for
carrying out all of these operations). Finally, at
run time, control of the central processor is
turned over to the machine language instructions derived from the original FORTRAN
source, and the program is executed directly.
A "map" of the machine memory at run time
is shown in Figure 14.1b. At this point even
the relocatable loader has "disappeared"; actually space occupied by the loader ordinarily is
allocated for variable storage during execution
and, hence, is not wasted.

In a computer with attached bulk storage device, many or all of the intermediate steps in
the above compilation process are "transparent"
to the user. A disk-based operating system
might be sophisticated enough to accept a
"compile-load-and-go" command along with the
FORTRAN source code, and build and delete
the intermediate files on disk until ultimately

---

*In the case of FORTRAN, economic considerations
of the major computer control users, viz. the process
industries, have been strong enough that some agreement on a first choice of language (FORTRAN) and
design of a set of uniform extensions to enable process interfacing and control has been accomplished or
is in progress (cf. Chapter 15 on Real-Time FOR-
TRAN).

**In older or smaller systems or those with more
primitively designed systems software, the compiler
may only output assembly language source code
necessitating the following further steps:
(1) The assembler is read into memory (overwriting
the compiler in the process).
(2) Assembler source code (output from the compiler) is processed through the assembler in one
or more passes resulting in relocatable object
code which can be loaded into the computer as
described above.

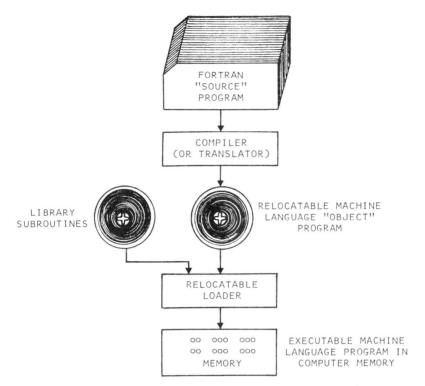

Figure 14.1a. Converting a FORTRAN Source Program to Executable Code

it has loaded machine memory and begun execution. The small-computer user working with a stand-alone operating system (i.e., paper tape or card-based input/output and without a bulk storage device such as a disk or magnetic tape unit) will have to initiate and monitor all of these operations personally.

By way of contrast, the process of making a high-level interpreter-based program ready for execution is shown in Figure 14.2a. In this case, the interpreter and all system subroutines, already in machine language form, are loaded into the computer memory, and control of the

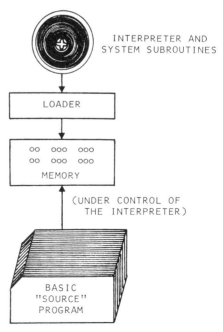

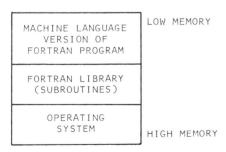

Figure 14.1b. Memory Map of a FORTRAN Program After Loading

Figure 14.2a. Loading a BASIC Program for "Running"

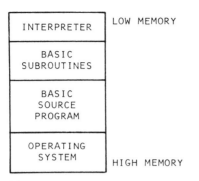

Figure 14.2b. Memory Map of a BASIC Program After Loading

CPU is given to the interpreter. Under control of the interpreter the BASIC source program is loaded into memory, usually with some encoding or packing of statements, but with no attempt to convert the source program to executable (machine language) code. After loading, the computer memory map appears as in Figure 14.2b where, it should be noted, the interpreter still is present.

Now where the FORTRAN program is reduced to executable machine code, from which it is essentially impossible to reconstruct the original source program statements, the BASIC program always exists in virtually the same format as the original source program statements. To "execute" a BASIC program, the interpreter is put into the "Run" mode via a statement typed-in at the operator terminal. At this point the interpreter (which never gives up control of the CPU) begins to treat the BASIC program statements as a form of data, interpreting them successively and branching to system subroutines as necessary to carry out the various arithmetic and logical operations that have been specified. Each BASIC program statement normally has an attached identification or statement number; hence the interpreter normally would proceed to interpret, initiate, and oversee the operations specified by each successive program statement. Branches *within* the BASIC program are carried out indirectly by causing the interpreter to branch to a nonsequential statement and begin interpretation there.

The important characteristic of computing with BASIC or any interpreter-based language

is that the user's program is *not* executed in the formal sense of the word, although the user most likely is completely unaware of the difference.

14.2 DISADVANTAGES AND ADVANTAGES OF INTERPRETER-BASED PROGRAMMING LANGUAGES

With a reasonably good understanding of how interpreter-based languages are structured for running in the computer and how this contrasts with compiler-based languages, we now are in a position to look at the pluses and minuses of each type of language, particularly at how these affect operations in real-time.

There are two major disadvantages* with interpreter-based systems and both of these potentially are important to most real-time users:

(1) The interpreter itself must be memory-resident, resulting in less main memory available for program storage. This can be a big disadvantage in a small-memory machine with no bulk storage device from which memory overlays or program chaining can be made.

(2) Because each instruction in the user's program must be interpreted before it can be carried-out, a BASIC program invariable will run slower than an equivalent FORTRAN program in the same machine. With processes requiring fast servicing or with large, complicated processes requiring a lot of computing activity, this point can represent a serious problem.

A somewhat more detailed analysis of these two points is in order before discussing potential advantages.

---

*There are a number of minor disadvantages with the BASIC language as ordinarily implemented. One is its relatively inefficient use of memory for the storage of numbers which are to be used as integers and for the storage of array variables. Another is the need to complete the interpretation and execution of a full line of code before interrupts can be recognized in a multitasking environment. This latter point is covered in detail in Section 14.5.2.

The interpreter actually is a collection of software programs which carry out a number of different functions. To begin with, it operates either in the Run mode or in the so-called Edit mode. In the Run mode it functions as an interpreting device, as a subroutine scheduler, and as a run-time monitor watching for arithmetic overflows and other operating errors. The user ordinarily changes from Run to Edit mode by typing in a single character, e.g., the "ESCAPE" Key, on the terminal. In the Edit mode the interpreter performs program inputs and modifications, permitting new program statements to be typed-in or existing statements to be changed from the terminal keyboard. Also, additional program segments can be read-in from an external input device or from bulk storage and appended or merged in memory under control of the interpreter. To perform all of these editing duties the interpreter must function much as a compiler does in parsing input statements, checking syntax and format, typing-out error messages where necessary, and encoding program statements in compressed form for storage in memory. Program listing at the terminal or on other output devices such as a line-printer is another interpreter function that can be initiated by the user. Finally, the interpreter is responsible for building (and maintaining during the editing process) all links between source program statements so that listing and executions will be consecutive even though storage in memory need not be. In this respect the interpreter must perform the linking functions that normally fall to the loader when handling executable (FORTRAN, ALGOL, PL/I, etc.) programs.

To carry out all of these functions means that the interpreter must be complex, even with the relatively small instruction set available in the standard BASIC repertoire, and this means further that computer memory must be allocated and permanently occupied by the interpreter. On the other hand, it may be the case that a BASIC program statement in slightly encoded form will require less main memory than the equivalent FORTRAN statement translated to executable machine language code. Hence there is some tradeoff available even in this area. One area where there is little potential tradeoff is in storage of system subroutines. In loading a FORTRAN program, only those subroutines specifically used by the program will be loaded; all system subroutines must be contained in memory at all times with BASIC. The user may have the option, however, of not loading some sets of subroutines he will not use, e.g., trigonometric functions or matrix operations.

Most single-user BASIC systems with real-time extensions will operate with a stand-alone operating system in a 4–8K memory system; with a disk-based operating system in 12–20K. The higher values are applicable to programs consisting of several hundred statements or involving significant variable lists and/or matrix operations.

Returning to the second disadvantage cited above, there are several other peculiarities of the BASIC language which may result in substantial penalties in execution time. Chief among these is the lack of any specific integer variable structure or arithmetic. All numbers must be represented in floating point format, usually as double words in memory. Event counters and other variables such as ADC inputs which might best be handled in integer form must be "floated" before use in a BASIC program. Conversely, output quantities for DAC channels or to binary output registers must be "fixed" before output. All of these operations incur a relatively severe time penalty unless a hardware floating point processor is available for use by the BASIC run-time arithmetic routines. Some versions of BASIC available for real-time applications permit the use of logical variables; bitwise manipulation of some variables; octal, binary, and hex representation of program constants; etc. All of these extensions assist in I/O-related (particularly binary I/O) duties. And the availability of matrix arithmetic statements in some extended versions of BASIC give computational advantages over FORTRAN, particularly when a floating-point processor is available.

Against the many significant disadvantages of an interpretive language working in a real-time environment, there are a number of major advantages. For BASIC these are:

(1) The language is easy to learn. From this point of view the limited instruction set

is no liability. The unformatted (actually automatically formatted) operations for text I/O are particularly useful in a learning situation.

(2) Programs are easy to debug. The user-interactive features are particularly useful when operating with an on-line process. Most versions of BASIC permit the user to enter the Edit mode from the Run mode at any time, to interrogate the computer via the keyboard for the values of any internal program variables, to insert modifications in the program or new variable values and re-start a run, etc. Many versions of BASIC permit the computer to be used as a calculator when in Edit mode; i.e., single line arithmetic operations may be executed immediately with results output at the terminal; and sometimes this capability extends to the execution of single-statement input/output calls such as "convert channel n of the ADC."

The combination of these two advantages makes a simple, interactive, and interpretive language such as BASIC exceptionally useful in any situation where learning will be an important element—learning the fundamentals of real-time programming, learning the operating characteristics of a new real-time computing system, learning the properties of a new or changing process, etc. The language can also be used to good advantage in any short-term (limited-time) project. These two criteria dictate that BASIC or a similar interpretive language will find its primary applications in university and research laboratories, in pilot plant and experimental operations, and in "one-time-only" applications where the costs of program development and debugging cannot be amortized over extensive periods of operation or over large numbers of similar applications.

In closing this section it should be stressed that the user with a limited real-time facility, particularly one without bulk storage equipment *or* an operating system to take advantage of it, may have to pay a large price in time required to develop and debug an applications program in FORTRAN or especially in assem-

bly language when compared to using an interpreter-based language. Many users just entering the real-time computing field have overlooked, to their later regret, the implications of the procedures depicted in Figures 14.1a and 14.2a. The BASIC user needs to load the interpreter and subroutines one time into his dedicated real-time machine and then proceeds iteratively to load his program, run it and check for "bugs", modify it, etc., until it works, then dump it into hard-copy form for program backup. Microprocessor-based personal computer systems are able to exploit these advantages even further by retaining the interpreter and associated subroutines permanently in read-only memory (firmware).

The FORTRAN user, by way of contrast, must first load the compiler, and then load (in one or two passes) his source program for compilation; second, load the relocatable loader and load the FORTRAN relocatable output plus FORTRAN library routines, including, if necessary, the run-time operating system; last, run. The inevitable result of running a real-time program—at least the first few times—will be the immediate appearance of "bugs" in the operation; the user has recourse at this point to loading all his relocatable code, library routines etc., a second time with a real-time debugging routine included to help find the source(s) of the problems. Finally, after finding the bug(s), the whole process, starting with loading the compiler, is begun again with a (hopefully) corrected version of the program. Anyone who has a limited computing facility and an even more modest application should be sure to obtain a cross-compiler and cross-assembler to use to prepare programs on a large-scale computing facility (see Chapter 12) or consider taking the interpreter route unless, of course, main memory or run-time restrictions prohibit use of an interpreter.

14.3 "BASIC" BASIC

Before beginning a discussion of several versions of Real-Time BASIC, it will be useful to review the most elementary version of single-user on-

Table 14.1. Elementary BASIC Commands

STATEMENT	USE	EXAMPLE
DATA	Delineates data table	100 DATA 4.1, 5.0, 7.3
DEF	Defines user function	110 DEF FNA(X)=SQR(X*X)
DIM	Allocates program storage	120 DIM A(5,6)
END	Delineates end of program	130 END
FOR	Starts repetitive (loop) operations	140 FOR J=0 TO 10 STEP 2
GOSUB	Transfers to subroutine	150 GOSUB 1000
GOTO	Transfers unconditionally	160 GOTO 530
IF	Checks for conditional action	170 IF J>=10 GOTO 4000
INPUT	Reads data from terminal	180 INPUT A,B,C
LET	Assigns value to variable (optional)	190 LET X=4.33
NEXT	Terminates repetitive (loop) operations	200 NEXT J
ON	Transfers conditionally on value of variable	210 ON J GOTO 500,600
PRINT	Writes to terminal	220 PRINT A,B,C
READ	Reads from "data" table	230 READ A,B,C
REM	Denotes a remark	240 REM THIS IS A COMMENT
RESTORE	Resets "data" table to first entry	250 RESTORE
RETURN	Returns from subroutine	260 RETURN
STOP	Returns to edit mode	270 STOP

time-shared BASIC.* This might be termed the original "no-frills", high-level language, as we will not deal with any extended capabilities such as matrix operations, formatted input/output, etc. Table 14.1 summarizes the elementary BASIC command structure; Table 14.2 summarizes the operators and library functions. Most of the commands are self-explanatory for anyone who has programmed a computer be-

fore; examples are given where clarification might be necessary.

Several features of the language, particularly its differences from FORTRAN, need to be described in more detail:

(1) Variable names usually are limited to a single alphabetic letter plus a single number for unsubscripted (nonarray) variables, and a single alphabetic letter for array variables.

(2) The same method is used to store out both integer and floating-point variables; hence there is no need for (specifically) integer variables and no implicit naming convention (the I,J,K,L,M,N, convention in FORTRAN).

(3) Array storage must be preallocated (via a DIM statement) if a subscript will be greater than 10. Arrays ordinarily begin with subscript zero (rather than one as in FORTRAN).

*A number of books, mostly manuals, have been written covering the BASIC language and its applications. A relatively recent paper by Lientz [1] cites a large group of references to the BASIC literature. In addition, the authors enumerate the special or unique features of several different versions of the language, most of these being designed for use on the larger, time-shared or multiuser computer installations. A good introductory text by Mullish [3] details use of the language through a number of computational examples. The computer science text by Forsythe et al. [4] is an equivalent introduction. Small-computer users are referred to manuals supplied by the individual manufacturers.

Table 14.2. BASIC Operators and Library Functions

ARITHMETIC OPERATORS	RELATORS	LIBRARY FUNCTIONS
+ Addition	= Equal	ABS(X) Absolute value of X
- Subtraction	< Less than	ATN(X) Arctangent of X
* Multiplication	<= Less than or equal to	COS(X) Cosine of X
/ Division	> Greater than	EXP(X) "e" raised to the X power
↑ Exponentiation	>= Greater than or equal to	INT(X) Greatest integer less than X
	<> Not equal to (sometimes #)	LOG(X) Natural logarithm of X
		RND(X) Random number between 0 and 1
		SGN(X) Algebraic sign of X
		SIN(X) Sine of X
		SQR(X) Square root of X
		TAN(X) Tangent of X

(4) FOR, NEXT loops (equivalent to DO, CONTINUE loops in FORTRAN) may use fractional and negative values of the loop variable (not allowed in FORTRAN).

(5) Each statement must have a statement number between 1 and 9999.

(6) Subroutines are identified and accessed by first statement number. Arguments are not required in the call, since all program variables are global.

(7) Variable values can be assigned by the LET statement, read in from the terminal using the input statement, or read from a block of data appended to the program using a READ, DATA combination.

Most other peculiarities and idiosyncracies of the language can be seen best by example and use. Perhaps two simple programs at this point will be sufficient. (See Examples 14.1 and 14.2.)

As a final set of comments the following "good programming" practices should be observed:

(1) Statement numbering is up to the user; however, it makes sense to write a program with "vacant" or unused statement numbers so that additional statements can be inserted into the program later. For example, if it is desired to modify example 14.2 to handle complex roots, this may be done by typing in a new statement 520 and inserting whatever additional statements are necessary as 522, 524, 526, etc.

(2) Unlike FORTRAN COMMENT statements, REM statements in BASIC occupy memory space continuously, usually at one or two characters (including spaces) per memory word. Good BASIC programming practice dictates the discrete use of REM statements unless plenty of memory is available.

Example 14.1. A Simple BASIC Example

```
10      REM BASIC PROGRAM TO CALCULATE SQUARES
20      REM   AND SQUARE ROOTS OF ALL NUMBERS
30      REM   FROM ZERO TO TEN INCREMENTING BY 0.1
60      PRINT
70      PRINT "NUMBER", "SQUARE", "SQUARE ROOT"
100     FOR X = 0 TO 10 STEP 0.1
110     LET Y = X↑2
120     LET Z = SQR(X)
130     PRINT X,Y,Z
140     NEXT X
9999    END
```

14.4 SINGLE-TASK REAL-TIME BASIC

Many real-time computing applications are of a relatively simple nature; i.e., the process is not complex. Perhaps only a few continuous process variables need to be measured and/or controlled; a few process status variables need to be monitored; several process input switches need

Example 14.2.   A BASIC Example with Subroutine Calls

```
  10        REM BASIC PROGRAM TO SOLVE FOR ROOTS
  20        REM    OF ANY QUADRATIC ALGEBRAIC EXPRESSION
 100        PRINT "TYPE-IN VALUES -- A,B,C, -- OF COEFFICIENTS ON REQUEST"
 150        PRINT "(A*X↑2 + B*X + C)"
 200        GOSUB 400
 220        GOSUB 500
 240        GOSUB 600
 260        GO TO 200
1000        END

 400        REM SUBROUTINE TO INPUT COEFFICIENTS
 410        PRINT "A="
 420        INPUT A
 430        PRINT "B="
 440        INPUT B
 450        PRINT "C="
 460        INPUT C
 470        RETURN

 500        REM SUBROUTINE TO FIND ROOTS
 510        LET D = B*B - 4*A*C
 520        IF D<0 RETURN
 530        LET E = SQR(D)
 540        LET X1 = (-B + E)/2*A
 550        LET X2 = (-B - E)/2*A
 560        RETURN

 600        REM SUBROUTINE TO PRINT RESULTS
 610        IF D>=0 GO TO 640
 620        PRINT "ROOTS ARE COMPLEX.  I QUIT"
 630        RETURN
 640        PRINT "X1=", X1, "X2=", X2
 650        RETURN
```

to be turned on or off. Additionally, these operations occur only infrequently, say on a time scale of seconds, and the calculations involved are not too time-consuming.

Under these circumstances an elementary interpretive language such as BASIC described in the previous section need only be augmented with several special commands (for process-oriented input and output and for accessing simple time-keeping routines driven by the computer's real-time clock) in order to run the process and log operating data. The simplest such version of BASIC is essentially single-task in nature. This means that the computer operating system is not very sophisticated and can only execute the user's program in precisely the way it was programmed. In a multitask system, any predetermined portion of a user's program (a task) can be run automatically whenever a particular external event occurs (such as when a binary input line to the computer changes state causing an interrupt, or when a certain amount of time has elapsed as evidenced by the real-time clock causing an interrupt). Further, in a multitask system, whenever two tasks might need to run at the same time, the operating system automatically would schedule the task with the higher priority. Multitask systems save the user from having to be concerned about complicated scheduling problems; however, for simple processing jobs a single-task system usually will be adequate. Since scheduling of program segments is such an important concept, we will pay some attention to these problems later in this section.

Single-tasking BASIC systems with real-time extensions are available from several manufacturers in both stand-alone and disk-based versions. In the remainder of this section we will consider in detail a system of real-time subroutines typical of those in use with a number of small-computer systems.

14.4.1   Computer/Process Interface

To begin with, no real-time system can be totally independent of either the computer's own internal architecture or the process interface configuration. As mentioned in Chapter 2, the system certainly will have a real-time clock. Beyond that the process or real-time related interface may contain any mixture of devices for inputting and outputting analog and digital signals. For purposes of all subsequent discussions

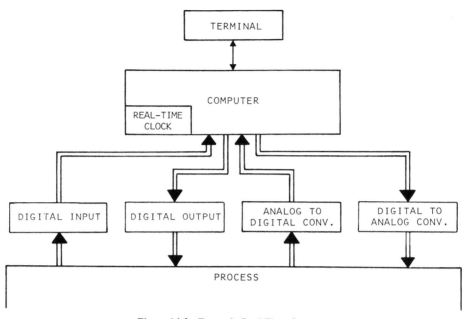

Figure 14.3. Example Real-Time System

we will assume that the computer uses a 16-bit word (an important consideration because (1) any binary I/O devices connected to the computer will utilize multiples of 16-bit registers, and (2) A/D and D/A converters ordinarily will be limited to a maximum conversion accuracy related to the computer word length). We will assume further that an ADC with 16-channel multiplexer is available along with 8 channels of latching DACs. A single 16-bit digital output register is available (TTL compatible; i.e., +5 VDC and 0 VDC represent the two states). Similarly, a 16-bit digital input register which causes an interrupt on change-of-state of any input line is available. The only device assumed to be available for user communications is a terminal. Figure 14.3 illustrates schematically how the system is structured.

14.4.2 I/O Software Drivers

Someone, either the computer manufacturer or the user, has to write the specific driver routines which operate the real-time peripherals and which allow the user access to the devices from mainline programs. For example, to output a 16-bit word to the digital output device is a very simple assignment, but the subroutine

to handle the chores of addressing the device and transferring the word into the device's own 16-bit register must exist among the routines of the operating system or library. We will assume that all such routines are present and only cite the one or two important features of such drivers as these bear on subroutine calls available to the user.

The single most important such feature concerns the use of the computer interrupt capability and what use each device makes of the computer's interrupt servicing facilities. Of the peripheral devices shown in Figure 14.3 (excluding the terminal), the digital output device and the digital to analog converters have no reason to use the interrupt structure. The driver, in each case, needs only to output one or two words of information; the devices normally process this information almost instantly, and the information should appear as a change in device output configuration (one or several changes in bit status or a change in DAC output voltage) in 10 to 50 milliseconds.

The analog to digital converter, on the other hand, might use an interrupt to signal "conversion complete". Since most ADCs on the market at the present time convert in 10 to 20 microseconds, it makes little sense to force the operating system to service such an interrupt at a cost of ∿1 millisecond of CPU overhead time. We will assume that the device driver for our example system simply waits and checks an ADC flag to make sure that conversion is completed before proceeding.

This leaves the binary input device and the real-time clock, both of which, even in our relatively primitive system, will have to use the interrupt structure at some level of operation. We will assume that the operating system allows the device service routine contained within the device driver software to store information in locations within the computer memory available to user routines written in BASIC. Hence the drivers might increment a counter at each clock interrupt ("tick") or store out useful information each time a bit in the binary input device changes status. It is the user's responsibility to access this information in a timely manner (before it is overwritten and lost), and this is the *fundamental problem* that must be considered at all times in writing a single-task program.

14.4.3 BASIC Real-Time Extensions (Subroutines)

In the next few pages we will list and describe a set of subroutines that allow the user to keep track of the passing of real time, to manipulate all process-related peripherals, to access the computer front panel switches, and to carry out noncritical (approximate) timing chores. To benefit understanding, these subroutines have been grouped according to their function.

Our single-task version of real-time BASIC will utilize a set of subroutine calls with the following format:

Statement Number	CALL	Subroutine Name (Mnemonic)	Param.* #1	Param. #2	Param. ...	Param. #8

For example, after execution of the statement

122 CALL ADC, 3, V3

representing a call to the analog to digital converter, BASIC variable V3 would have been replaced by some value representing the voltage on channel number 3 of the converter. In this case ADC is the name of a subroutine written by the user or supplied by the vendor and stored in memory along with the interpreter. The "CALL" accesses this subroutine, i.e., the ADC driver.

In this system which has rather limited flexibility, the subroutines can be written to accept up to eight parameters for the user (BASIC program) or to define values for up to eight parameters in the user program. In the above example, 3 is interpreted as the channel number, and variable V3 is assigned the converted value on return. The statement

122 CALL ADC, 3, 5

would be flagged as erroneous since the subroutine has to have some variable name (actually an address in memory) where it can store its converted result.

Routines Related to the Real-Time Clock. Table 14.3 outlines a set of routines that may be used to turn on the real-time clock, to turn it off, and to time fixed and indefinite intervals of time.

Ordinarily routine CLON would be used early in a user program to start the real-time clock and to initialize a set of clock-driven software registers. We will assume that the clock "ticks" at a frequency of 10 Hertz (10 interrupts per second) when running. In some systems a parameter may be added to specify the clock frequency when this is adjustable. Where the operating system already is real-time–oriented and the clock runs continuously, this subroutine may be used just to activate the system software timers and counters described below, or it may not be required.

Routine CLOFF ordinarily would be called at the end of a user applications program and, if

---

*From zero to eight parameters permitted.

Table 14.3. Real-Time Clock Routines

MNEMONIC	FUNCTION	PARAMETERS
CLON	Turn clock on at fixed frequency and reset all counters (or activate system timer & counter routines)	None
CLOFF	Turn clock off (or deactivate routines)	None
TMRST	Start a software timer	1) Timer no. (0-7) 2) Timer run time (no. of clock ticks)
TMRCH	Check a software timer	1) Timer no. (0-7) 2) Timer status (negative if run time not yet elapsed)
CNTRS	Reset software counters(s)	1) Counter no. (0-7, 8 = all)
CNTRD	Read count from software counter	1) Counter no. (0-7) 2) Value of count (no. of clock ticks)

required, would either turn off the clock or deactivate the clock-driven software registers.

In this system, two sets of software registers are used to keep track of the passage of real time. The "Timers" and "Counters" essentially duplicate each other's functions, but, since the programming techniques for handling such programmable clocks differ slightly, we will describe both. Two routines, callable from a BASIC program, access the eight Timers (numbered from 0 to 7) which can be considered as fixed locations in memory that accumulate clock ticks when running. Subroutine TMRST is used to start or initialize a Timer; an example of the call would be

$$980 \text{ CALL TMRST}, 3, 150$$

which would initialize Timer 3 with a count of 150 (15 seconds at 10 Hertz). In this implementation the Timers initially are set to a negative count and increment on each clock tick until zero is reached. Thus, they represent "up-counters" of a particular nature.

The second Timer subroutine, TMRCH, is available to the user to check the status of any timer from a BASIC program. For example, the statement

$$1040 \text{ CALL TMRCH}, 3, T3$$

would return a value for T3 that would be negative if the Timer was still running, zero if the

pre-set time period had elapsed. Now if a user wishes to perform some operation periodically (assume that the operation should be performed every 5 seconds and BASIC subroutine 800 has been written to do it), a set of statements to accomplish this might be arranged as in Example 14.3.

Statement 100 initializes Timer 7 the first time through. Subsequently, when the program gets to statements 300–320, it is clear that a loop has been entered that can be exited only when variable T7 becomes zero, i.e., when 5 seconds have elapsed. It is always good practice to re-initialize the count immediately, as is done on entering subroutine 800, so as not to miss any clock ticks. The program will return from the subroutine to the statement following

Example 14.3. An Elementary Timing Program

```
100   CALL TMRST, 7, 50
  .
  .
  .
300   CALL TMRCH, 7, T7
310   IF T7 < 0 GO TO 300
320   GOSUB 800
  .
  .
  .
400   GO TO 300
  .
  .
  .
800   CALL TMRST, 7, 50
  .
  .
  .
880   RETURN
  .
  .
```

statement 320 and, presumably, will reach statement 400 where it will be forced back into the wait loop. Note that it is important for the elapsed time between exiting the wait loop and returning to it, i.e., the execution time of subroutine 800 including all intermediate statements, to be less than 5 seconds for this scheme to work. More discussion on this point will be appropriate in the sequel.

Two routines, callable from a BASIC program, access the eight counters. These also can be considered to be memory locations that are incremented on each clock tick; in this case they count up to the maximum number if not reset to zero. Subroutine CNTRS can be used to reset any or all of the registers:

400 CALL CNTRS, 3

would reset Counter 3 or

450 CALL CNTRS, 8

would reset all (Counters 0-7). Subroutine CNTRD can be used to read any individual counter, for example,

600 CALL CNTRD, 3, C3

would return as variable C3, the number of clock ticks that have occurred since Counter 3 had been reset last. The software counters in this scheme can be used precisely as were the software timers; if we consider the previous example, we might have the version of Example 14.4.

Example 14.4. A Timing Program Which Uses the Software Counters

```
100        CALL CNTRS, 7
  .
  .
  .
300        CALL CNTRD, 7, C7
310        IF C7 < 50 GO TO 300
320        GOSUB 800
  .
  .
  .
400        GO TO 300
  .
  .
  .
800        CALL CNTRS, 7
  .
  .
  .
880        RETURN
```

Note, again, that good practice dictates that the count be reset immediately on exiting the wait loop. An alternative way to use a counter for time-keeping purposes would be to initialize a time variable, T7, to the desired number of clock ticks, say in a new statement 95:

95 LET T7 = 50

and then replace Statements 310 and 800 in Example 14.4 with

310 IF C7 < T7 GO TO 300
800 LET T7 = T7 + 50

Now C7 represents real time (in tenths of seconds) and T7 the time at which the periodic operation is performed (also in tenths of seconds).

One problem to watch for in this case is overflow of the software registers. For example, the subroutine CNTRD must take the contents of the appropriate software register (Counter 7) and "float" the results before returning the value as BASIC variable C7. If the software register is a single 16-bit word in memory, and the "float" routine interprets the single word as a two's complement number, then the maximum value would be +32767 (3276.7 seconds or about 50 minutes). After reaching this value the count will come back negative if no provision to handle overflow is available.

An additional point on the use of these software "clocks" concerns multiple timing operations. The programmed wait loop in the two previous examples (obtained via statements 300 and 310) results in no useful action being accomplished by the computer until after the elapsed time is up and the loop is exited. There are several ways of handling the timing of multiple operations; for example, if Timers 1, 2, 3, and 4 all are being used to time individual operations of 4, 6, 10, and 25 seconds, respectively, then the technique given in Example 14.5 could be used. Statement 1000 is the start of a routine (not subroutine) to carry out the operation with 4-second period. The routine should begin with a call to restart Timer 1 [with interval T(1)] and end with a branch to Statement 300

Example 14.5.   Timing of Multiple Operations

```
100      LET T(1) = 40
110      LET T(2) = 60
120      LET T(3) = 100
130      LET T(4) = 250
140      FOR 1 = 1 TO 4
150      CALL TMRST, I, T(I)
160      NEXT I
  .
  .
  .
300      FOR J = 1 TO 4
310      CALL TMRCH, J, C
320      IF C = 0 GO TO 400
330      NEXT J
340      GO TO 300
400      ON J GO TO 1000, 2000, 3000, 4000
  .
  .
  .
```

to begin checking the four timers again. The other routines should be similarly constructed. In this case the wait loop is more complex than in previous examples, but still serves the purpose of holding the program until the time arrives to perform some operation.

Two problems (or potential problems) involved with single-task programming methods should be apparent at this point in our discussions:

(1) A considerable amount of time is actually wasted by the computer in wait loops of the sort described.

(2) There is potential for a single routine to overrun the time interval permitted before returning to the timing loop or for multiple routines to conflict in some way such as needing to run at the same time.

There are no good solutions to these problems when using a single-tasking language. Both problems result from the inability to initiate an important routine on cue even when some other part of the program is executing. This sort of thing can be done with a multitasking language. For the moment, the user is forced

to fall back on his own resources, i.e., to time routines carefully, to make sure they execute within a reasonably small length of time, and to rewrite the routine if it is not fast enough. For example, the programming "tricks" described in Chapter 15 on FORTRAN programming all are applicable to BASIC programs. It also may be necessary to eliminate output routines (particularly on a slow terminal) storing results in arrays and outputting them on a bulk storage device or printer only at the end of the experiment. With respect to point (2), the user might utilize a set of Counters to check the timing operations. In the previous example this would only require inserting an additional statement in the initialization section

135 CALL CNTRS, 0

to reset Counter 0. Whenever the service routines (1000, 2000, 3000, 4000) were accessed, the routine could print or store out the value of the count for verification of timing. In a more complicated situation involving the introduction of additional timing loops during the course of operations, a software counter could be used with each timer. There are other ways of handling these potential problems, but the ideas all are essentially equivalent to the above.

Routines for Analog Input and Output. Only two routines are necessary to read in or read out an analog voltage if only high-level inputs are used. They are summarized in Table 14.4. As noted in the table, the first parameter refers either to the desired ADC channel or the particular DAC desired. In some systems the ADC routine might be designed to return with the BASIC variable (specified as the second parameter) set to a value of actual voltage on the input; for example,

100 CALL ADC, 9, V

Table 14.4.   Analog Input and Output Routines

MNEMONIC	FUNCTION	PARAMETERS
ADC	Convert the analog voltage on a specified channel of the ADC and furnish the result as a BASIC variable	1) Channel No. (0-15) 2) Unscaled voltage in range -32768 to +32767
DAC	Convert the value of a given BASIC variable to a voltage on a specific DAC	1) Channel No. (0-7) 2) Unscaled voltage in range -32768 to +32767

would come back with V = 5.00 if an input of 5 volts happened to be applied to ADC channel 9 at the time the call was executed. More often, V would be returned as a fraction of the full-scale voltage, i.e., some number between −1 and +1 or 0 and +1 or as a number somewhere between the largest and smallest integers expressible by the input or output device. For example, if a converter utilized an unsigned, right-justified 12-bit representation for voltages between 0 and +5 VDC, the smallest result would be 0 (zero) and the largest result would be $2^{12} - 1$ or 4095. It then would be the user's responsibility to scale the results, in this case to divide by 4095 on input or multiply a computer variable representing voltage by the same factor on output, i.e., before making a DAC call.

For purposes of example, we will assume that our system operates with inputs and outputs in the range −10.0 to +10.0 VDC and that the converters utilize left-justified, two's complement integer representation. This means that inputs and outputs on a 16-bit machine would be in the range −32768 to +32767.* If it is desired to output some BASIC variable having a voltage value, say V5 on DAC 5, an appropriate sequence of statements would be

```
.
.
340 LET V = V5 * 32768/10.
350 CALL DAC, 5, V
.
.
.
```

or

```
.
.
.
340 CALL DAC, 5, V5 * 32768/10.
.
.
.
```

---

*Actually, if the converter accuracy is less than 16 bits including sign, the upper bound is only approximate. We won't worry about minor differences here.

Similarly, to obtain the voltage on channel 7 of the ADC:

```
.
.
.
440 CALL ADC, 7, V
450 LET V7 = V*10./32768
.
.
```

Routines for the Computer Panel Switches. With small, personal or dedicated, computers it often is useful to be able to access the computer front panel switches directly from a BASIC program. The switches can then serve as a mechanism to input data to a program, to initiate some computer action, etc. Ordinarily, two different routines are useful, one that would simply read in the switch configuration as a binary (equivalently octal or hexadecimal) number and assign the value to a BASIC variable after floating it appropriately, another that could be used to test a particular switch to determine if it was set (turned on) or not set (turned off). Table 14.5 summarizes the form of two such routines.

In this case, execution of the statements

```
.
.
725 CALL SWRD, S
735 PRINT S
.
.
```

would yield a number in the range −32768 to +32767 depending on the switch configuration at the time of execution. Note that the routine might just as well have been written to treat the 16-bit input as a positive number in the range 0−65535 ($2^{16}-1$). The same result can be obtained with our routine by changing statement 735 to

735 PRINT S+32768

The status routine might be used to supply program logic from outside. For example, to

Table 14.5. Subroutines to Access the Computer Panel Switches

MNEMONIC	FUNCTION	PARAMETERS
SWRD	Read the entire set of panel switches	1) Decimal equivalent of switch settings (-32768 to +32767)

SWSTS	Check the status of a particular switch	1) Switch No. (0-15)
		2) State
		    0 = Off
		    1 = On

log data or not might be determined by

.

.

860 CALL SWSTS, 0, S
870 LET S = S + 1
880 ON S GO TO 2000, 3000

.

.

In this case, if panel switch 0 is not set, the program will branch to statement 2000, where a non-data-logging routine begins; if set, then to statement 3000 for data logging.

Note that, with a single-tasking language, it is very important that program logic be change-able via external inputs such as panel switches or binary (discrete) input lines. The terminal cannot be used directly for this purpose, since, as soon as an INPUT statement is used to obtain information from the user (operator), the program will wait until an input actually is typed-in. In the meantime, many critical timing operations may have been missed. Of course the terminal might be used to input information indirectly, i.e., only when a particular panel switch is set, it then being the user's responsibility only to set the switch during non-critical timing periods or when he is ready to type-in the necessary information quickly. An example of this approach was given in the BASIC digital clock example of Chapter 3.

Routines to Input and Output Digital Information. It is important that a real-time computer be able to read in digital (i.e., binary or discrete) information on command; for example, to read in the status of a set of input lines indicating whether valves are open or closed, switches are on or off, etc. Additionally, the

computer should be able to react to a change in status of any one of the lines; for example, to make some decision immediately if a particular sensing element detects the presence of smoke or heat. We will assume that our system has a 16-line input device that can perform these functions. It also is necessary to be able to control devices that have digital (binary) inputs; for example, to turn lights on or off, to open or close a solenoid valve, etc. Our system has a corresponding digital output device which can manipulate 16 individual lines. For applications purposes we can assume that off and on correspond to the nominal values of TTL logic, i.e., 0 and +5 VDC. Table 14.6 summarizes a set of useful binary I/O subroutines that can be called from this version of BASIC.

The digital input device can be used most simply as a passive register representing the values of the input lines, in which case the appropriate command to read the register would be

300 CALL DIRD, R

After execution, R would have the decimal value -32768 to +32767, depending on the configuration of the input lines at the moment of execution. Such a command might be useful in reading in some manual switch settings. If, for example, a single binary-coded decimal (8-4-2-1 format) switch is attached to the four least-significant bit lines (all others grounded), statement 300 above would decode the switch outputs.

For purposes of sensing input line changes it is necessary to start the device, i.e., "enable" it in such a way that a change in status of any line (from the initial configuration) will cause an interrupt in the computer. The single

Table 14.6. Routines for Digital Input and Output

MNEMONIC	FUNCTION	PARAMETERS
DIRD	Read the digital input device input line configuration	1) Decimal equivalent of the input register
DISTS	Check the status of a particular digital input line	1) No of line to check 2) Status of a line { 0 = turned off { 1 = turned on
SNSST	Start (enable) the sense function of the input device	No parameters
SNSCH	Check for change of status of the digital input lines	1) Status of device { { 1 = some line changed since last check { { 0 = no line changed 2) No. of line which changed (0-15) 3) Direction of change { 0 = Turned off { 1 = Turned on
DOWR	Write a full word to the digital output device	1) Decimal equivalent of the output word
DOOFF	Turn off (reset) a line in the digital output device	1) No. of line to turn off (0-15)
DOON	Turn on (set) a line in the digital output device	1) No. of line to turn on (0-15)

command

500 CALL SNSST

with no parameters will start this device.

Just as with the real-time clock routines it was not possible to utilize a computer interrupt directly,* we also must make use of indirect means in communicating line status changes with the sensing function of the Digital Input Device. In the case of clock interrupts (ticks), the operating system was programmed to update all of the Timer and Counter software registers. In the case of the Digital Input Device, once it is started the operating system will update three registers in memory whenever a line changes state. The user can access these registers by means of the sense check routine, SNSCH. For example, if some line has changed state since starting the device (or since the last checking operation), after executing the statement

550 CALL SNSCH, S, L, D

the variable S would be 1; and L would contain the line number that changed (0–15) and D the

direction of change (0 = Turned off, 1 = Turned on). Now in this simple version there are only three registers; hence no stacking of changes. This means that only the most recent results will be contained in the registers, and it is the user's (programmer's) responsibility to make sure that important events are not overlooked or missed while some unimportant section of code is executing. Just as with the clock, it may be necessary for critical operations involving digital inputs to provide a wait loop which is exited only on change in state of one of the lines.

A routine such as the following one might be used alone or in conjunction with a set of timing routines:

.

.

.

100 CALL SNSST

.

.

.

200 CALL SNSCH, S, L, D
210 IF S = 0 GO TO 200

.

.

.

Use of the digital output (control) device is much simpler. One routine can be used to

---

*Because of the single-task structure of this version of BASIC.

Example 14.6.   A BASIC Subroutine to Output BCD

```
7990      REM THIS ROUTINE CONVERTS A DECIMAL NUMBER
7995      REM TO BCD AND OUTPUTS TO THE CONTROL DEVICE
8000      IF N>=10000 THEN RETURN
8010      IF N<0 THEN RETURN
8020      LET N[1]=INT(N/1000)
8030      LET N=N-(N[1]*100)
8040      LET N[2]=INT(N/100)
8050      LET N=N-(N[2]*100)
8060      LET N[3]=INT(N/10)
8070      LET N[4]=N-(N[3]*10)
8080      CALL DOWR,0
8090      FOR I=1 TO 4
8100         IF N[I]<8 THEN GOTO 8130
8110         CALL DOON, 4*I-4
8120         LET N[I]=N[I]-8
8130         IF N[I]<4 THEN GOTO 8160
8140         CALL DOON,4*I-3
8150         LET N[I]=N[I]-4
8160         IF N[I]<2 THEN GOTO 8190
8170         CALL DOON,4*I-2
8180         LET N[I]=N[I]-2
8190         IF N[I]<1 THEN GOTO 8210
8200         CALL DOON,4*I-1
8210      NEXT I
8220      RETURN
```

manipulate all 16 lines directly. For example, to set-up the device with any initial decimal number in the range −32768 to +32767 the only statement required would be

220 CALL DOWR, D

where D is the number. Two other routines permit any line to be turned off directly (left off if already off), in this case bit or line 9:

230 CALL DOOFF, 9

or turned on directly (left on if already on), in this case line 12:

240 CALL DOON, 12

An example of the use of these commands is given by Example 14.6. Here, we assume that a 16-light display is to be driven by the binary output device. To output any number within range of the 16-bit display (−32768 to +32767 or 0 to 65535) as a straight binary number we

would only need to output the number using a command of the form

8000 CALL DOWR, N

or, respectively,

8000 CALL DOWR, N + 32768.

However, if we wish the display to consist of 4 digits of 8-4-2-1 binary coded decimal, the range now will be from 0 to 9999. Assuming that the four digits are numbered 1, 2, 3, 4 from left to right and the 16 lines from the control unit are numbered 0, 1, . . . , 15 also from left to right, we would have the situation shown in Figure 14.4.

The BASIC subroutine given as Example 14.6 accepts a number N and checks that it is in the acceptable range.   Statements 8020 through 8070 isolate the four individual digits N(1) . . . N(4), each of which has to be in the range 0 to 9. Statement 8080 sets the display to zero initially via the DOWR command. The remainder of the program constitutes a loop which is traversed once for each digit with the DOON command used to turn on each light (from left to right) as required to complete the display. There are, of course, other ways to program this exercise, both more or less efficient and more or less obscure. For example, instead of using the DOWR to turn all lamps off initially, the program could have been written to use only DOOFF and DOON commands to manipulate each binary line.

For user convenience, other routines can and perhaps should be supplied with a real-time version of BASIC. For example, the BCD output routine described above might be written as an assembly (machine) language program, callable as a BASIC routine. The user might then need

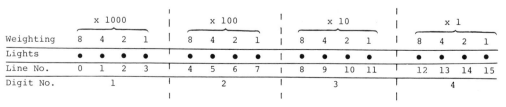

Figure 14.4. Sixteen-Bit BCD Output

only to embed in his program a statement of the form

800 CALL BCD, N

to accomplish the results of Example 14.6 much faster and with greater ease. Other useful functions which might be included:

(a) Routines to perform bit-wise manipulation of internal variables—for example, logical operations such as AND, OR, NOT, etc.

(b) A routine to hold up the program for a brief and approximate amount of time—for example, it is often the case that, in working with mechanical relays, time must be allowed for a relay to quit bouncing after being operated (a time on the order of tens of milliseconds); hence a routine to delay a program some integral number of milliseconds can be quite useful.

(c) A routine to translate binary coded information on input lines such as often is furnished by a manufacturer as the output from a laboratory instrument—for example, the voltage measured by a digital voltmeter.

The use of all of these routines should be evident at this point. Consequently, we will not discuss them further but rather take time to review the command structure already described by means of some real-time examples.

14.4.4 Single-Task Real-Time BASIC Examples

Having defined all of the real-time extensions or subroutines that we need, let us finish up this section by looking at some representative real-time exercises. As the first and simplest example we earlier considered the assignment of making a real-time computer and terminal into a "digital clock" in Chapter 3. The reader might wish to review that example (Figure 3.5 and Tables 3.2 and 3.3) at this point.

The second example, one we will discuss at some length, concerns the control of air pressure in a tank. Figure 14.5 shows the system schematically. The process is quite simple: compressed air at 20 pounds per square inch can flow into a closed tank if a solenoid valve is opened; a discharge line permits air to flow out of the tank continuously; needle valves in both inflow and discharge lines can be set to regulate how fast air flows in and out. We would like to be able to control pressure in the tank around some specified value, say 9 psig, by opening and closing the solenoid valve in the inflow line. For this purpose, the tank has been instrumented with a preset pressure-operated relay whose output turns on whenever pressure in the tank exceeds 9 psig. In addition, we would like to record tank pressure as it varies in time. Hence a pressure-to-voltage transducer whose output voltage is equal to one-half the tank pressure also is attached to the tank. In other words, when the tank pressure ranges from 0 to 20 psig, the transducer output voltage ranges from 0 to 10 VDC.

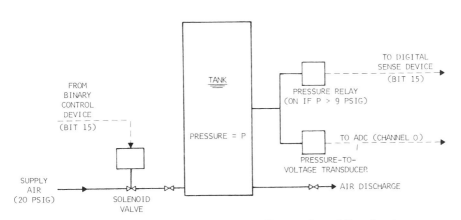

Figure 14.5. Schematic Diagram of an Air Pressure Control Experiment

Example 14.7. BASIC Program for On–Off Control of Pressure in a Tank While Logging Pressure Data

```
0010        REM SIMPLE ON/OFF CONTROL OF PRESSURE IN A TANK
0020        REM WHILE LOGGING TANK PRESSURE DATA
0030        PRINT "   TIME        PRESSURE"
0040        CALL DOWR,1
0050        LET S1=0
0060        LET T0=0
0070        LET T7=0
0080        LET F=0
0090        LET L=0
0100        LET D=0
0110        LET C=0
0120        LET V=0
0130        CALL TMRST,7,600
0140        CALL TMRST,0,10
0150        CALL CLON
0160        CALL SNSST
0170        CALL SNSCH,F,L,D
0180        IF F=1 THEN GOSUB 0500
0190        CALL TMRCH,0,T0
0200        IF T0=0 THEN GOSUB 0600
0210        CALL TMRCH,7,T7
0220        IF T7=0 THEN GOTO 0240
0230        GOTO 0170
0240        PRINT "END OF RUN"
0250        CALL CLOFF
0260        STOP

0500        REM SUBROUTINE TO CONTROL PRESSURE
0510        IF L=15 THEN GOTO 0540
0515        IF L>15 THEN GOTO 0560
0520        PRINT "SOMETHING IS AWRY"
0530        RETURN
0540        IF D=0 THEN CALL DOON,15
0550        IF D=1 THEN CALL DOOFF,15
05560       RETURN

0600        REM SUBROUTINE TO LOG DATA
0605        CALL TMRST,0,10
0610        CALL CNTRD,0,C
0620        CALL ADC,0,V
0630        LET V=10.0*V/32678
0640        LET P=2*V
0650        PRINT C/10,P
0660        RETURN
```

In order to use a real-time computer to handle this assignment, we have used Bit 15 of the Digital Output Device to operate the solenoid (Bit 15 = 1 implies valve open). Similarly, the output from the pressure relay is attached to Bit 15 of the Digital Input Device. The output from the pressure-to-voltage transducer is connected to ADC Channel 0.

A BASIC program that will startup the process and maintain operation at a pressure close to 9 psig (once there), is given as Example 14.7. Figure 14.6 gives a flowchart for the same program. The program utilizes two software timers—Timer 0, used to initiate data logging every second, and Timer 7, used to terminate the program at the end of 60 seconds.* Also, the Digital Input Device is used to determine when the single input line (Bit 15) has turned on or off. Separate subroutines for data logging

and for control of the solenoid valve have been provided. Note that the control algorithm is the simplest possible: the output to the solenoid should be the logical complement of the pressure relay state.

Assuming that implementation of the control algorithm has highest priority, data logging has next highest, and termination of the experiment has lowest priority, then the program should check the status of the Binary Input Device, Timer 0, and Timer 7 in that order as shown in the flowchart. If no operations are required, the status checking is repeated.

Whenever an input to the Input Device changes state ($F = 1$), then some control action may be warranted. The control subroutine first checks to be sure that Line 15 actually changed status ($L = 15$). If $L > 15$ then a "glitch" likely occurred (Line 15 changed state only briefly and was not latched), or if $L < 15$ then one of the other bits changed state (which is not possible in this case). Implementation of the proper solenoid control action is handled through the Digital Output Device.

---

*A second timer is used for this purpose only to illustrate the use of multiple timers. Clearly the program could be terminated when the number of accumulated points is 60.

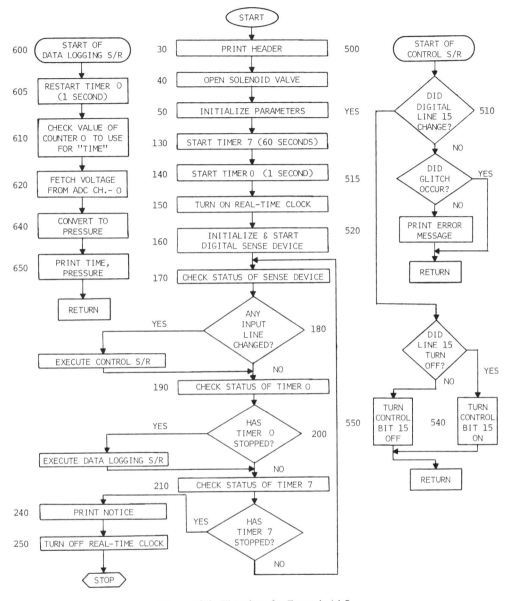

Figure 14.6. Flowchart for Example 14.7

Whenever Timer 0 is found to be stopped (T0 = 0) then data logging is implemented. Note that all of the status checking and control implementation, if required, takes much less than one period of the real-time clock (0.1 second) so it is not possible to "miss a clock tick". On the other hand, implementation of the data logging actions well may require more than 0.1 second because of the PRINT operation; hence Timer 0 should be started immedi-

ately on entering the subroutine. (This procedure is always good practice.)

Finally, when Timer 7 has stopped, (T7 = 0), the program is terminated after printing a notification.

This experimental example involves the fundamental elements of most real-time applications. Just as a batch computing job often involves three major parts in the program—initialization, implementation of the algorithm (with a heavy use of looping typical in this stage), and final calculations—so a simple, single-task, real-time program often takes the same form. In the real-time case, implementation of the algorithm usually involves "looping in time".

14.5 MULTITASK REAL-TIME BASIC

It is possible, by careful attention to the structural details of a real-time program, to handle quite complicated operating assignments with precision using only a single-task language such as the version of BASIC discussed previously. For example, it might well be possible to construct a multiloop timing stratagem to input data from or carry out control operations on several different portions of the process simultaneously. It may be necessary in this case to structure subroutine calls so that no single routine could run so long that it might interfere with some other, perhaps critical, operation. In this case the programmer may well take care of these potential problems by using some fundamental timing properties known to him in advance such as an inherent periodicity of sampling operations, a large discrepancy between intervals required for service of one part of the process as contrasted with some other part, or a large difference in the times required to execute different service routines. It might even be possible, with a precisely arranged program structure, to handle one or more random events such as alarm interrupts that conceivably could occur at any time. Here it may be necessary for the programmer to put a check for alarm status within his program where it would be accessed often enough that an alarm occurrence could not go undetected for longer than some acceptable length of time. Failing in this, it might be necessary to embed alarm status checks at many different points within the program routines so that timely detection of an alarm occurrence could be ensured.

There are two serious problems involved in this approach:

(1) The programmer has to think of *all* the possibilities in advance and structure the program to handle them.

(2) Given that all of the timing and access constraints can be met satisfactorily, any subsequent minor change in system responsibilities (for example, the addition of another routine that must be accessed periodically) likely will require that the whole structure be redesigned.

For any practical applications work these two problems are likely to be serious ones; hence some other approach clearly is desirable.

The concepts of multiple "tasks" and "multitask programs" are ones that have developed historically to handle these sorts of problems. To understand the ideas behind multitasking operations we need to review briefly several of the ideas behind the design of computer operating (executive) systems. As described in Chapters 3 and 13, one of the functions of an operating system is to service interrupts from devices connected to the computer. Most sophisticated operating systems (certainly any of those that can support a high-level language such as BASIC) have the capability to set priorities on devices capable of causing interrupts, e.g., terminals, disks, etc. Further, they are able to stack interrupts; i.e., if a slow-speed (low-priority) device such as a terminal is undergoing interrupt servicing and a high-speed (high-priority) device such as the real-time clock interrupts, then the operating system can momentarily stop servicing the low-priority device, returning to complete the job only after attending to the high-priority device.

It is conceptually simple to extend these ideas to the "user program level". Essentially what is required is to be able to organize all of the user programs into logical "tasks", each task simply being a section of code that should be executed whenever some particular event occurs. For example, one particular task would run every 5 seconds as measured by interrupts from the real-time clock, or another task would execute each time an alarm occurs as indicated by the change in status of any one of a set of digital input lines. The key idea is that the user easily can organize his program into tasks; the subdivision logically might follow the lines that are used in structuring a program into subroutines. The user also must be able to assign a priority to each task. It ordinarily is not difficult to rank tasks according to priority; for example, a task that should execute whenever an alarm occurs clearly should have higher priority than one that simply inputs data. A task that inputs data for the purpose of effecting some feedback control action should have higher priority than one that simply inputs data for record-keeping purposes, etc. Finally, it is necessary that a "task scheduler" be implemented somewhere within the system software (usually within the operating system).

The basis for multitask system operations is quite simple. Only one task (or the system programs, i.e., operating system, interpreter, etc.) can have control of the CPU at any instant. Hence it is the task scheduler's responsibility at all times to see that the highest-priority task that should be running gets control of the CPU. As each task finishes, the next-lower-priority task is given control successively until no user task needs to be run at the moment. If, at any time, something (an interrupt) occurs that would change the ranking of priorities, then the task scheduler needs to take control momentarily, decide which task needing or ready to run has highest priority, and turn over the CPU to that task. Of course, a tremendous amount of "bookkeeping" must be handled correctly by the scheduler if a multitasking system is to work correctly. If, for example, an interrupt associated with a high-priority alarm occurs while a lower-priority data-logging task is executing, then not only must the alarm task be given control of the CPU as quickly as possible, but the data-logging task which has been bypassed momentarily must be suspended in such a way that when it ultimately is regiven control of the CPU, it can resume operation with no loss of function.

A more detailed description of operating systems and multitask programming is given in Chapter 13. This very brief discussion of multitasking is given to introduce a multitask version of BASIC typical of those available for small-computer real-time systems. Since this version of BASIC is quite versatile and powerful and, also, relatively easy to learn, it can serve as a first introduction to multitasking concepts. Before discussing details of the language and some examples, however, we need to note briefly the potential differences between the way a task scheduler might be designed to function with an interpretive language such as BASIC and the implementation that would be used with a direct executing language such as assembly language or FORTRAN.

14.5.1 Multitasking Options with Interpretive Systems

The twin ideas of reentrancy and reentrant programming (cf. Chapter 13) are important ones for computing applications that involve either multiple (simultaneous) users such as in a time-sharing system or multiple tasks. In either application the users or tasks compete for use of system resources—most importantly, the central processor. As noted above, when a user or task must be bypassed to take care of another user or task, the machine state at the time of deferral must be saved so the user (task) can be restored to its former condition, subsequently. With a direct executing high-level language such as FORTRAN, the compiler has to produce what generally is called "reentrant code", and all system library routines have to be reentrant as well. What this means is that intermediate results of calculations cannot be stored within any portion of a program because, under some circumstances, this information well may be lost. The usual way to take care of this problem is to store all variables passed to and from subroutines and all intermediate results on a "run-time stack".

To accomplish the same effect with an interpretive system is not quite so straightforward. Of course, inasmuch as all of the actual computation is accomplished using library routines, these can be made reentrant. The difficulty is in the interpreter itself, which can be made reentrant only with difficulty. Consequently, a common approach is to forgo some of the advantages of immediate task or user scheduling (i.e., immediately on the occurrence of an interrupt) and permit the interpreter to complete any line of source code before invoking the scheduler. This approach considerably lightens the burdens placed on system software; however, since several tens of milliseconds may be required to complete a particular line of BASIC source code, ultrafast scheduling of critical tasks cannot be guaranteed. Furthermore, if input/output operations have been initiated within a statement, they may need to be completed before rescheduling can take place. This means that a long line of output data may hold up operations for some time, or a request to read data from a single system terminal may lock up the system completely until someone has typed-in all of the required information. Although these types of potential problems can be handled by proper design of interpreter and operating system (or by implementing automatic "spooling" of output with a bulk storage unit), it may be easier to require multiple terminals for applications requiring simultaneous input and output. Such is the case with the version of multitasking BASIC described below.

14.5.2 BASIC Command Structure

In this section we will describe a multitasking interpretive system based on a commercial version of real-time BASIC.*

Perhaps the most efficient way to introduce the language is to first list the available program statements (with emphasis on those oriented to real-time applications), to discuss some of the important programming concepts, and then to furnish some examples analogous to those described previously. Table 14.7 illustrates program statements of the following types: general program, event/time control, analog/digital input/output, and bit manipulation. Those statements dealing with other peripheral devices such as a plotter or magnetic tape unit are not given, nor are general program type statements which substantially duplicate those found in any BASIC system. The usual functions (SIN, COS, etc.) are available with the addition of OCT (to print out a variable in octal format); the usual arithmetic operators (+, -, etc.) are available with the inclusion of some logical ones (AND, OR, NOT).

---

*The author wishes to express his appreciation to the Hewlett-Packard Company for supplying the source information for this section, for answering a number of technical questions, and for "debugging" the two examples furnished in the following pages.

As general points we should first note that up to 16 individual tasks are allowed, and each task may be time- or event-scheduled and runs at a particular priority level (from 1 to 99); also, after each BASIC instruction statement is executed, the scheduler steps in and checks to see if some event has occurred (real-time clock tick, digital input line change-of-state, or operator-initiated action at the console) that might result in task rescheduling. Figure 14.7 illustrates how three separate tasks, existing together, may interleave operations, depending on how external events influence the scheduler. In this case the scheduler might return program flow to Task No. 1 if no event has occurred. Alternatively, Task No. 2 might be scheduled if it has been designated to run at a particular time, if that time has arrived (the system real-time clock has activated) *and* the task's priority is higher (in this case it is). Obviously, after a task has completed execution, the scheduler will initiate execution of the next-higher-priority waiting task.

Event/Time Control Statements. Each separate task of any program is written to look like a subroutine. In the starting portion of the program the usual initialization statements must be executed. In addition, a number of other housekeeping chores must be performed:

(1) The priority of each task must be set; e.g.,

    10   SETP(200, 10)

would set the priority of a task beginning with statement number 200 at 10.

(2) The links between a particular digital line and a task must be established; e.g.,

    100   TRAP 3 GOSUB 1000
    110   SENSE(1, 6, 0, 3)

In this case trap number 3 is associated with a task beginning with statement number 1000. This task will be scheduled whenever bit 6 of digital channel 1 turns off.

(3) The links desired between an auxiliary terminal must be established; e.g.,

Table 14.7. Additions to BASIC Instruction Set for Multitasking Software
(Supplementing Tables 14.1 and 14.2)

	INSTRUCTION	USE	FORMAT*
Logical Operators	AND OR NOT	Logical AND Logical OR Logical complement	If X=0 AND Y=1 GOTO 20 If X=0 OR Y=0 GOTO 10 If NOT(A=B) GOTO 30
Functions	SWR OCT	Status (0 or 1) of switch register bit (0-15) Print an octal variable	If SWR(6)=1 GOTO 50 PRINT OCT(X)
Event/Time Control	TRAP TIME SETP TRNON START DSABL ENABL TTYS	Link event interrupt to task Get time of day for program Set task priority Turn on task at specific time Start task after delay Disable a task Enable a task Schedule trap task from teleprinter	TRAP <u>trap#</u> GOSUB <u>numb</u> TIME(<u>time</u>) SETP(<u>task#</u>,<u>priority</u>) TRNON(<u>task#</u>,<u>time</u>) START(<u>task#</u>,<u>secs</u>) DSABL(<u>task#</u>) ENABL(<u>task#</u>) TTYS(<u>printunit#</u>,<u>trap#</u>)
Analog/Digital Input/Output	AISQV AIRDV SGAIN RGAIN AOV RDBIT RDWRD WRBIT WRWRD SENSE	Sequential mode analog input Random-scan analog input Set gain of analog channel(s) Read gain of analog channel(s) Analog output voltage Read digital input bit Read digital input word Write digital output bit Write digital output word Link contact closure event to trap	AISQV(<u>#inputs</u>,<u>firstchan#</u>, <u>resultarray</u>,<u>error</u>) AIRDV(<u>#inputs</u>,<u>chanarray</u>, <u>resultarray</u>,<u>error</u>) SGAIN(<u>chan#</u>,<u>gain</u>) RGAIN(<u>chan#</u>,<u>gain</u>) AOV(<u>#outputs</u>,<u>chanarray</u>, <u>resultarray</u>,<u>error</u>) RDBIT(<u>chan#</u>,<u>bit#</u>,<u>state</u>) RDWRD(<u>chan#</u>,<u>bit#</u>,<u>state</u>) WRBIT(<u>chan#</u>,<u>bit#</u>,<u>state</u>) WRWRD(<u>chan#</u>,<u>word</u>) SENSE(<u>chan#</u>,<u>bit#</u>,<u>state</u>, <u>trap#</u>)

Bit Manipulation	IOR
	INOT
	IEOR
	IAND
	ISHFT
	
	IBTST
	
	IBSET
	IBCLR
	ISETC
		Add two variables, bitwise
		Complement a variable, bitwise
		Exclusive OR of two variables
		Multiply two variables, bitwise
		Shift a variable ± N bit
		  positions (+ → LS, - → RS)
		Determine state of bit in
		  variable
		Set bit in variable
		Reset bit in variable
		Set variable to octal constant
			IOR(<u>var1</u>,<u>var2</u>,<u>result</u>)
			INOT(<u>var</u>,<u>result</u>)
			IEOR(<u>var1</u>,<u>var2</u>,<u>result</u>)
			IAND(<u>var1</u>,<u>var2</u>,<u>result</u>)
			ISHFT(<u>var</u>,<u>#shifts</u>,<u>result</u>)
			
			IBTST(<u>var</u>,<u>bit#</u>,<u>result</u>)
			
			IBSET(<u>var</u>,<u>bit#</u>,<u>result</u>)
			IBCLR(<u>var</u>,<u>bit#</u>,<u>result</u>)
			ISETC(<u>"octalvalue"</u>,<u>var</u>)

*Underlined quantities represent constants or variables in the user program.

```
200  TRAP 5  GOSUB 2000
210  TTYS(2, 5)
```

In this case striking any key on auxiliary terminal 2 will cause the task starting with statement number 2000 to be scheduled.

(4) Any time-dependent tasks must be scheduled; e.g.,

```
220  TRNON(1500, 101530)
```

will "turn on" a task beginning with statement number 1500 at 30 seconds past 10:15 (according to the system internal clock). Or

```
320  START(1800, 5)
```

will schedule a task beginning at statement 1800 exactly five seconds after execution of statement 320.

(5) Finally, the main program must end with an endless loop, e.g.,

```
500  GO TO 500
```

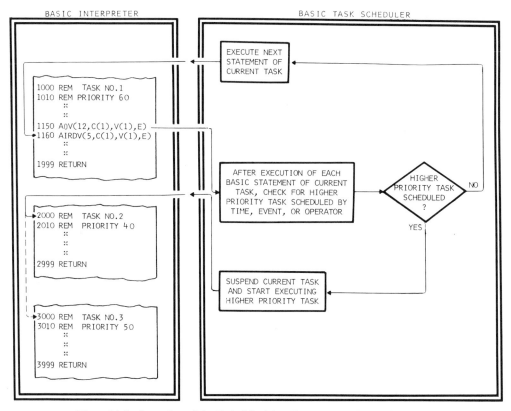

Figure 14.7. Operation of the Task Scheduler (Courtesy Hewlett-Packard Co.)

which allows the processor to continue running while all tasks are run intermittently (based on the passage of time or occurrence of external events). Note that in this system the main program executes at *lowest* priority; hence it is important to make sure that all tasks are scheduled to run only after all initialization has been completed.

Two additional statements that can be used to increase the flexibility of multitask programs are the enable and disable commands.

$$DSABL(250)$$

will remove the task beginning at statement 250 from the group of operating tasks.

$$ENABL(250)$$

will reestablish the task as a functioning program unit.

Analog/Digital Input/Output. Analog information can be read in sequentially using the AISQV command. In this case

$$200 \quad AISQV(4, 1, V(1), E)$$

will cause four ADC channels starting with channel 1 to be converted and the results stored starting with element 1 in array V. A similar statement applies to random analog input scanning.

$$300 \quad AIRDV(5, C(1), V(1), E)$$

will cause 5 ADC channels represented by the numbers stored as the first 5 elements in array C to be converted and the results stored as the corresponding elements in array V. The same sort of statement is used for outputting analog information.

$$400 \quad CALL \ AOV(6, C(1), V(1), E)$$

will cause the first 6 voltages stored in array V

to be output using the channel numbers correspondingly stored in array C.

The read bit (RDBIT), write bit (WRBIT), read word (RDWRD), and write word (WRWRD) statements are straightforward and best illustrated in the examples.

Bit Manipulation. These statements are necessary to carry out the logical algorithms inherent in most real-time applications. Note that all of these commands work with internal variables only.

14.5.3 Multitasking Examples

In this section we will return to the same examples discussed in Chapter 3 (Section 3.4.1) and in Section 14.4.4. Recall that the first example was to implement a "digital clock" which typed out the time every 15 seconds. In this case we ignore the presence of the internal system clock and program our own. Example 14.8 shows the program statements necessary to implement these functions. Note that in this version of BASIC it is necessary to have a second terminal in order to input information,

in this case to reset our internal clock. Statements 70 and 75 link up the auxiliary terminal which makes use of the task beginning at statement 1000.

One questionable point in this example is whether the time-keeping task will, in fact, re-initiate itself exactly every 15 seconds, or, because of the finite length of time required to execute statements 500 and 510,* might not "skew in time", i.e., run just slightly longer than 15 seconds. The answer to this question is based on the discrete method of keeping track of the passage of time in such a system, i.e., by "ticks" of the real-time clock. If the time required to execute the two statements is less than the real-time clock period, then there will be no skewing. Since the execution time of these statements is on the order of several milliseconds, we have made that assumption. In most cases, however, the user likely would choose to utilize the operating system software clock.

---

*Actually only statement 510 is executable, but the interpreter has to find that out.

Example 14.8.   A Multitask Version of the Digital Clock

```
    10      PRINT "THIS IS AN ACCURATE BUT RELATIVELY EXPENSIVE DIGITAL CLOCK"
    20      SETP(500,20)
    30      SETP(1000,40)
    40      LET H = 0
    50      LET M = 0
    60      LET S = -15
    65      LET N = 4
    70      TRAP N GOSUB 1000
    75      TTYS(2,N)
    80      START(500,1)
   100      GO TO 100

   500      REM TIMER TASK (PRIORITY = 20)
   510      START(500,15)
   520      LET S = S + 15
   530      IF S < 60 THEN GO TO 610
   540      LET S = 0
   550      LET M = M + 1
   560      IF M < 60 THEN GO TO 610
   570      LET M = 0
   580      LET H = H + 1
   590      IF H < 24 THEN GO TO 610
   600      LET H = 0
   610      PRINT "THE TIME IS"; H;M;S; "O'CLOCK"
   620      RETURN

  1000      REM SET-TIME TASK (PRIORITY = 40)
  1005      REM USES BREAK ON AUX. TERMINAL TO INITIATE
  1010      PRINT #2; "TYPE IN HOURS"
  1020      READ #2; H
  1030      PRINT #2; "TYPE IN MINUTES"
  1040      READ #2; M
  1050      PRINT #2; "TYPE IN SECONDS"
  1060      READ #2; S
  1070      START(500,0)
  1080      RETURN
```

Example 14.9. Multitask Control and Data Logging with the Pressure System

```
10   REM MULTITASK ON/OFF CONTROL OF PRESSURE IN A TANK
20   REM WHILE LOGGING TANK PRESSURE DATA
30   PRINT "TIME PRESSURE"
40   WRBIT(1,15,1)
50   LET T = 0
60   LET C(0) = 1
70   SETP(500,10)
80   SETP(600,20)
90   SETP(700,40)
130  START(700,61)
140  START(600,1)
160  TRAP 5 GOSUB 500
170  SENSE(1,15,5)
180  TRAP 6 GOSUB 550
190  SENSE(1,15,0,6)
200  GO TO 200

500  REM TASK CONTROL PRESSURE (PRIORITY = 10)
510  WRBIT(1,15,0)
520  RETURN
550  WRBIT(1,15,1)
560  RETURN

600  REM TASK TO LOG DATA (PRIORITY = 20)
610  START(600,5)
620  T = T + 5
630  CALL AIRDV(1, C(0), V(0), E)
640  LET P = V(0) * 2
650  PRINT T,P
660  RETURN

700  REM TASK TO STOP EXPERIMENT
710  WRBIT(1,15,0)
720  PRINT "END OF RUN"
730  STOP
740  END
```

The second example, which we treated earlier in this chapter, dealt with the on–off control of an air pressure system and with data logging of the actual pressure. In this case Example 14.9 shows the program that should do the job. An extra task (beginning with statement 700) has been added to the highest-priority control task and the middle-priority data logging task to shut down the experiment at the end of one minute. This task has been assigned a low priority (40) to reflect its relative unimportance. The programmer might have thought that shutting down the experiment on time was *most* important; in this case a higher priority, say 5, might have been assigned.

Note that two trap channels (of 16 available) have been assigned to handle the sensing of digital line 15 going both on and off. A slightly more efficient way of handling this situation using only one trap channel is shown in Example 14.10. In this case we define the bit state to be a variable and let the control task alternate the value between 0 and 1.

14.6 SUMMARY

This chapter has dealt with the use of interpretive languages for real-time applications with a specific emphasis on BASIC. The materials presented are introductory in nature and might be used to become acquainted with real-time programming techniques in other languages as well. For example, the system of calls and examples described as single-task in nature (Section 14.4) extends directly to simple applications, for example, dedicated microprocessor applications, using machine or assembly language as the program medium. The multitask concepts and techniques (Section 14.5) are largely applicable to other languages, so long as a stand-alone multitask scheduling system (monitor or executive) or a disk-based multitask executive (real-time multitask disk operating system) is available to support programs written in the language.

Example 14.10. An Alternative Way of Using the Trap

```
160  TRAP 5 GOSUB 500
170  N = 1
180  SENSE(1,15,N,5)
200  GO TO 200

500  REM TASK TO CONTROL PRESSURE
510  IF N = 0 GO TO 550
520  WRBIT(1,15,0)
530  LET N = 0
540  RETURN
550  WRBIT(1,15,1)
560  LET N = 1
570  RETURN
```

There are some fundamental differences between interpreter languages such as BASIC and compiler languages such as FORTRAN. We have dealt with the most important differences at length in the early part of this chapter and can, perhaps, sum up quite easily the single largest advantage BASIC has for real-time applications:

In a learning environment or in research applications where program development is a significant component, the ease with which BASIC programs can be written, modified, and debugged on-line gives the interpretive language a flexibility that is hard to match.

For more advanced, commercial applications the use of interpretive languages also has been significant. Chapter 16 deals with the design and application of so-called table-driven programming languages (Fill-in-the-blanks or Fill-in-the-forms software are also names frequently used) which have found rather wide application in the process industries for data logging and control projects. Such languages are, for the most part, designed around interpretive software of the sort described in this chapter.

14.7 REFERENCES AND SUPPLEMENTAL READING

1. Lientz, B. P., "A Comparative Evaluation of Versions of BASIC", *Commun. of ACM, 19* (4), 175, April 1976.
2. Hartley, P. J., *Introduction to BASIC: A Case Study Approach*, Macmillan, New York, 1976.
3. Mullish, H., *A Basic Approach to BASIC*, John Wiley & Sons, New York, 1976.
4. Forsythe, A. I., Hughes, C. E., Aiken, R. M., and Organick, E. I., *Computer Science: Programming in BASIC*, John Wiley & Sons, New York, 1976.
5. Trombetta, M., *BASIC for Students: With Applications*, Addison-Wesley, Reading, Mass., 1981.

14.8 EXERCISES

1. A physicist is trying to measure an experimental phenomenon with a real-time computer system. This particular phenomenon is characterized by a single voltage which decays in an approximately exponential manner. The physicist feels that if he arranges his sampling routine so that each sampling interval is twice the length of the previous interval, the voltage *change* between each pair of measurements will be about the same.

   The first timed interval is 1 second; the next, 2 seconds; the last, 32 seconds. Write a single-task BASIC program using software "timer" routines that will meet the required timing objectives. The voltage, between 0 and 10 VDC, is available on ADC Channel 7. Note: A phenomenon, by definition, can only be measured with much equipment and large amounts of money.

2. We wish to generate a rectangular pulse train on Bit 1 of the Digital Output Device whose duty cycle is proportional to the voltage measured on ADC Channel 3. [Duty cycle = (Ontime)/(Period of individual pulse).] The basic period of the pulse train, i.e., the period of an individual pulse, should be supplied via the terminal in milliseconds at the start of the program.

   Write a BASIC program that will generate such a pulse train, whose duty cycle is updated at the beginning of each pulse period. Use either the single- or multitask version. The program should be written so that, should the ADC voltage exceed 8 volts, pulse generation will be halted and an appropriate message printed on the system terminal.

3. Exercise 1 in Chapter 2 described a backyard irrigation system with eight solenoid-operated valves in individual watering lines and eight ground moisture sensors, one associated with each watered plot. (You may assume that sensors with digital output are used).

   Modify the digital clock examples for either single-task BASIC (Chapter 3) or multitask BASIC (Section 14.5.3) so that, instead of typing out time-of-day, the program:
   a. Checks all eight plots just after 12:00 midnight, 8:00 A.M., and 4:00 P.M.
   b. Turns on the water for 10 minutes in each plot found to be "too dry".
   Note: Since sprinklers may water adjacent plots somewhat if the wind is blowing, make sure your program determines all plots that need to be watered before commencing *any* watering operations.

4. Exercise 3 in Chapter 10 dealt with the interfacing of an optical scanner system to a

small real-time computer. Assuming that the system real-time clock operates at 10 Hz for the single-task system or 1 Hz for the multitask system, consider the following for each:

a. How would you handle the timing considerations, i.e., the length of time it takes the motors to advance, using only BASIC language I/O calls? Is this procedure foolproof?

b. How would your time compare to the minimum time for scanning an entire photograph? Would there be any incentive for writing special-purpose routines in assembly language to handle this particular interface? (Assume that user-written routines can be added to either BASIC system.)

c. Flowchart your programs.

5. Suppose you wish to monitor the response of some process to an external excitation. A sampling period of 10 seconds minimum is required. The process signal you must monitor is available through ADC Channel 12 and is known to be noisy, hence will require some input signal averaging. Write a program in BASIC that will:

a. Sample the process periodically every 10 seconds.

b. Average up to six samples, the number to be specified initially by the operator through the terminal.

c. Log the results on the terminal with appropriate sampling time.

d. Cause the rightmost eight bits of the Digital Output Device to turn on and off periodically if the monitored voltage exceeds 5 volts. (Assume that these lines are connected to a set of lights which should flash on for one second, off for one second, etc.)

6. In exercise 6 of Chapter 10 you were asked to consider the interface for a set of turbine flowmeters in a pipeline network. Assuming that you have designed an appropriate interface, draw a flowchart that defines the structure of a real-time BASIC program appropriate for measuring and logging the liquid flows in each of the pipelines. Explicitly state any assumptions that you have to make.

7. In the system depicted in Figure 14.5, a single pressure relay set to actuate at 9 psig was used. In addition a pressure-to-voltage transducer was available to supply a continuous measurement of the tank pressure.

For this problem, assume that a second pressure relay also is available such that the inputs to the Digital Input Device are as follows:

Tank Pressure, psig	Bit 14	Bit 15
$0 < P < 9$	0	0
$9 \leqslant P < 12$	0	1
$12 \leqslant P < 20$	1	1

Also assume that the pressure-to-voltage transducer has the following calibration relationship:

$$P = 2V$$

where V is the measured voltage (0 to 10 VDC).

a. Write a BASIC program that will control the tank pressure between 9 and 12 psig using only the two pressure relays and the single solenoid valve. Log the pressure on the terminal every 20 seconds using the voltage-to-pressure transducer (ADC Channel 0).

b. Repeat, controlling pressure between 9 and 12 psig and logging it every 20 seconds using only the output of the pressure-to-voltage to transducer; i.e. ignore the pressure relays.

8. A computer is interfaced to a single stirred tank water heater as shown in the figure.

Tank temperature is measured using ADC 0. Heat input to the tank is manipulated using DAC Channel 1.

The on/off control valve is connected to the Digital Output Device in such a way that it is open when Bit 15 of the output device is 1 and closed when Bit 15 is 0.

A discrete level detector is connected to the Digital Input Device so that when the level is above L2 Bit 14 is 1 (otherwise 0) and when the level is above L1 Bit 15 is (otherwise 0).

The temperature, T1, is controlled at the ADC value of 6 volts using an on-off controller to manipulate the SCR voltage. I T1 is greater than 6 volts, the SCR input voltage Q1 should be set to 0 volts; if T1 is less than 6 volts, it should be set to 10 volts.

The Flowrate F is reasonably constant but the tank level must remain between L1 an

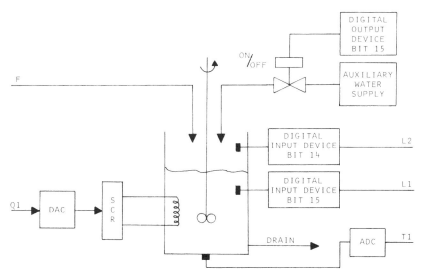

Figure for Exercise 8

L2 to keep the heating coil immersed. If the level drops below L1 the on/off valve must be turned on; the valve must be shut off when the level is greater than L2.

a. Write a BASIC program that:

i) Updates the controller (and outputs a DAC signal) every 6 seconds.

ii) Checks for tank overflow or under-flow and remedies it if necessary every 30 seconds.

iii) Records the current temperature (T1) and heat input (Q1) on a data file for future analysis every 20 seconds.

b. What problems might arise if single-tasking BASIC is used for your program?

15

Real-Time FORTRAN

James Wm. White
Modular Mining Systems
Tucson, Arizona

15.0   INTRODUCTION

FORTRAN, which stands for FORmula TRANslation, is perhaps the most widely used high-level language in batch or timesharing applications in science and engineering. This chapter is intended as an introduction to real-time FORTRAN and as such assumes that the reader is familiar with this language. For those whose FORTRAN is "rusty", we suggest review and study of suitable FORTRAN references [e.g., 3, 12] prior to reading this material.

Most small-computer systems installed today are able to support some usable FORTRAN dialect with real-time extensions. However, in the early stages of development, companies such as General Electric, Honeywell, and Westinghouse, as reported by Jarvis [2], Mensh and Diehl [4], and Roberts [8], found standard FORTRAN IV [cf. 18, 20] inadequate for real-time applications. Because FORTRAN had such influence in the scientific and engineering community and had become in effect a de facto standard, these companies as well as others developed extensions to FORTRAN IV to facilitate real-time programming in a compiler-level language. Many modern small computers support extended FORTRAN with restrictions peculiar to the particular machine. Recently, FORTRAN and other high-level languages have become increasingly popular in real-time applications because such languages are to a certain extent self-documenting, easy to maintain or modify, and, with some effort, transportable from one machine to another.

Efficiency of real-time FORTRAN program development suffers on a small computer for several reasons. Efficient compiler operation cannot be realized in less than 16 KB of main memory though a few 8 KB versions do exist. Even at the 16 KB level, small-computer FORTRAN compilers are extremely simple and do not produce extensive diagnostics as do compilers on large-computer systems. The classic debugging aid is the run-time memory dump which prints the main memory contents at a stage of computation where execution is unexpectedly halted. However, the memory dump is not practical for small machines with very limited input/output devices. In the late sixties and early seventies, most minicomputers marketed were provided with basic program development software consisting of an assembler, text editor, loader, and an on-line debugging aid [5, 11]. Program development on a minimum machine configuration of paper tape and Teletype is awkward and time-consuming even with the software aids previously mentioned. If control of a process requires the dedication of a nonnetworked small computer then on-line program development, debugging, and updating may not be practical. Such factors must be considered in configuring any small computer application where programs are to be written in FORTRAN. Note that the maximum memory size of 16-bit, small-computer systems has increased in the seventies from 64 KB to 256 KB or, in some cases, even higher (e.g., 2 MB). Since the maximum size of the single job is now usually 64 KB, the overhead associated with a good debugger is not nearly so significant. Indeed, on-line debuggers operating under control of a multitasking executive have increased appreciably in power and capability thus eliminating many of the earlier drawbacks.

As discussed in Chapter 11, assemblers are system programs that are unique to a particular computer line. Thus assembly language programs developed on one machine are in general not transferable to another machine (except within some "computer families") without extensive modification. Emulator programs [13] ease assembly language program development, but their use is usually restricted to the assembler mode of programming where execution of the application program is emulated in a time-precise, register-by-register fashion on a larger machine. Numerous cross assemblers and cross compilers exist too, of course, but these usually output binary relocatable code which must then be loaded and executed on the target machine. Assembly language is more efficient than a high-level language from the standpoint of memory requirements and execution time, but these advantages are becoming less important as small-computer CPU and memory costs and cycle times continue to drop. This is especially true in areas where memory requirements or timing considerations are not critical. Assemblers accept user-supplied source code consisting of a label field, operator field, operand field, and comments field. Mnemonics are used for each of the computer instructions, and symbolic addresses are generally allowed. Usually each line of assembly language code generates a single line of object code although numerous exceptions exist (e.g., MACRO instructions). Assembly language is the lowest (and consequently the most powerful, memory-conserving, and run-time efficient) level language at which program development is practical. Compilers, such as FORTRAN, are higher-level system programs that accept user-supplied source code consisting of a statement number, source statement, and optional identification label. Each line of FORTRAN source code usually generates from 2 to 20 lines of machine code which is invariably less efficient than assembly language from the standpoint of memory utilization and execution time. There may also be additional FORTRAN overhead factors due to extra system routines which must be memory-resident during FORTRAN execution.

Perhaps an example can clarify these points.

A simple FORTRAN-callable subroutine to compute N factorial (N!) was coded in assembly language and assembled; then coded in FORTRAN and compiled on a small minicomputer. The output of the assembler and the compiler are not shown. Assuming that both coded examples are as efficient as possible for this particular computer, FORTRAN uses 47_8 words, whereas assembly language uses 33_8. The difference in this example amounts to almost a 50% overhead for FORTRAN convenience. (Note: These figures are extremely variable and may range anywhere from 10% to 100%.) In addition, the FORTRAN subroutine requires the integer arithmetic package (116_8 locations), whereas the assembly language routine does not.

Each statement of code, *regardless* of type, is estimated to cost from \$20 to \$25 for (commercial) design, development, testing, debugging, and documentation. Since the cost of extra memory has dropped to very low levels, the extra overhead associated with FORTRAN is easy to justify. Further, FORTRAN is a "standard language" now augmented with real-time standards via the ISA S61 series of real-time standard calls [21–23] and hence is somewhat "portable" from machine to machine. This makes benchmarking for comparison much easier. Finally, consider that every application program must be supported and that this is very much harder to do with assembly language than FORTRAN. The best, most conservative approach for real-time applications—except possibly in an OEM market situation—is to insist on ANSI (*A*merican *N*ational *S*tandards *I*nstitute) standard FORTRAN capability as part of the system specification. Any special functions, e.g., real-time computer process interfacing, should be implemented as FORTRAN-callable handlers by the vendor as part of the specification.

Interpreters, such as BASIC and FOCAL, are higher-level languages similar to FORTRAN in that they are easy to learn and use. Furthermore, they are especially convenient in limited-configuration installations because program development, testing, debugging, and execution can all be accomplished with this single system program—as opposed to the editors, compilers, loaders, and library routines needed by assemblers and compilers. There are two very important disadvantages, however, which should force the user through a careful evaluation sequence before deciding to develop an application system in interpretive code. Since an "interpreter" usually interprets or decodes each user source statement each time it is encountered, an "interpreted" program runs very much slower than either compiled or assembled code. Also, the interpreter and all function subroutines that might conceivably be needed remain memory-resident all the time.

15.1 FORTRAN STANDARDS

In 1966, ANSI published standards for two then-popular FORTRAN dialects—FORTRAN II [17] and FORTRAN IV [18]. Over a decade passed before the ANSI standards were revised to reflect many commonly-implemented vendor extensions to FORTRAN IV. The new ANSI 78 FORTRAN standard [cf. 3] is designated "FORTRAN VII" by some vendors and in this text. Where applicable, the syntax of FORTRAN VII is used in the examples of this chapter. In order to facilitate portability, FORTRAN IV or FORTRAN VII are strongly recommended, with suitable real-time extensions e.g., the ISA S61 series, for all real-time FORTRAN applications. Note, however, that some time will pass before a majority of small-computer system vendors support the new FORTRAN standard. Actually, the older ANSI 66 FORTRAN standard has proved to be a mixed blessing. On the positive side, about 60% of minicomputer FORTRAN compilers support full ANSI 66, and about 80% contain at most one restriction to the standard [7]. Hence in one sense the standard is important: as long as users code in the ANSI 66 subset, FORTRAN dialect variations do not pose a major portability problem. On the negative side, most small-computer FORTRAN dialects go considerably beyond the standard for a number of reasons and in many ways—especially in real-time applications. The ISA S61 series of real-time calls and the new ANSI 78 standard should eliminate many current differences, but some portability problems are bound to remain.

Standardization of real-time FORTRAN extensions has been proposed by the FORTRAN Committee on Standardization of Industrial Computer Languages [24–33]. The Instrument Society of America (ISA) has accepted some of these extentions [21] and others in draft form [22, 23]. Several small-computer manufacturers now offer ISA standard executive routines and peripheral device handlers, and one engineering company has developed extensive programs to test the conformity of these routines to the published ISA standards. A demonstration of program portability and discussions of encountered problems is also available [16].

The ISA standard is a direct outgrowth of the International Purdue Workshop on Industrial Computer Systems whose goal was to make the definition, justification, hardware and software design, procurement, programming, installation, commissioning, operation, and maintenance of industrial computer systems more efficient and economical through the development of standards and/or guidelines on an international basis.

The following guidelines were used in the development of the ISA standards:

(1) The standards should cover features commonly used by existing industrial computer systems.
(2) The standards should be easy to implement for most vendors.
(3) The standards should follow the syntax and intent of ANSI FORTRAN as defined by the International Standards Organization [20].

The advantages of developing all code in ANSI Standard FORTRAN with ISA Real-Time Standard extensions cannot be overemphasized. The discussion of external procedure reference (i.e., in-line functions and subroutine calls) for use in industrial or laboratory real-time computing systems follows the standards already issued by ISA. The ISA specified external procedure references, rather than changes to FORTRAN syntax, primarily for ease of implementation. This does not imply that this is the only way to provide these features, nor does it exclude the possibility or desirability that th language will develop through syntax changes so that these and other related features will be included. Though most small-computer manufacturers still offer their own real-time extensions, including syntactic variations in FORTRAN, an increasing number of small computers do support at least the ISA S61.1 [21] standard as issued. Most users who wish to adopt these standards and whose FORTRAN compiler does not accept in-line assembly code can develop their own real-time I/O libraries with the aid of their FORTRAN compiler and a little knowledge of assembly language. Revision, approval, and adoption of random file handlers and multitasking handlers, with subsequent acceptance by the manufacturers, is highly probable in the future.

Some small-computer compilers offer numerous non-real-time extensions to ANSI Standard FORTRAN; others prohibit or restrict use of certain ANSI features. Furthermore, integer, real, double precision, and complex constants are permitted different maximum values depending upon system structure and layout. For maximum portability, integer constants should be restricted to ±32767 and real constants to $10^{\pm 38}$. However, even these limits exceed the single precision range allowed by several compilers [7]. In general there are many more extensions than restrictions. Of course, such extensions are only occasionally really needed for any given application, while the restrictions essentially constitute an ANSI

subset. Use of nonstandard FORTRAN extensions during program design and development greatly reduces program portability and thus should be avoided. Admittedly, any restrictions have to be observed, but the resulting programs are invariably upwards-compatible with other compilers. Users contemplating extensive FORTRAN program development or revision should consider the possibility of fully or partially automating their software production using large-scale host machines [6, 9, 14].

15.2 REAL-TIME PROCEDURES

Let us now consider in some detail the various FORTRAN "procedures", i.e., subroutines and functions, conceivably needed by any real-time system employed in data acquisition, equipment monitoring, or process control. Of course, it is unlikely that all of these procedures—or even a majority of them—will appear in any given real-time application program. Bear in mind that the information presented here is but typical of calls and function references actually used in practice. In general, each vendor specifies names and argument lists peculiar to its particular product line. Here we follow the procedural specifications recommended by ISA, though any consistent set would serve just as well.

From a functional point of view, these real-time FORTRAN-callable procedures might best be grouped as shown in Table 15.1. If

Table 15.1. Real-Time FORTRAN Procedure Groups

Group	Function
Executive Interface Routines	Provide a first-level interface between the calling program and the resident monitor for scheduling real-time application programs
Process Input/Output Routines	Provide two-way communication -- both analog and discrete -- between the equipment or process and the calling real-time FORTRAN routine
Bit String Functions	Allow bit-by-bit manipulation of values stored as integer variables
Random File Handlers	Provide for secured storage and retrieval of real-time data on any random access bulk storage device
Task Management Routines	Provide for complex intercommunication between applications tasks and the real-time executive including process interrupt and application task connection

FORTRAN-callable routines for the procedures of Table 15.1 do not exist in the FORTRAN run-time library, then the routines might best be coded directly in assembly language, assembled, debugged, and added to the system library. However, this usually requires intimate knowledge of both the operating system and assembly language programming. For the experienced professional programmer or systems analyst, this is definitely the preferred route. Virtually every computer family uses a different assembly language and operating system; hence it is extremely difficult to discuss a general procedure in any detail. However, we can say that for specialized process/computer communication, the process I/O service routine must handle receiving or sending of data, define any special Input–Output Transfer (IOT) instructions (in a programmed I/O environment), set up buffers for storing incoming or outgoing data (in systems with buffered I/O), and specify I/O data type. In an interrupt-driven environment the handlers must also be reentrant. In sophisticated multitasking environments where three separate assembly routines might be required (e.g., a Device Control Block to set device status flags and characteristics, an Interrupt Handler to recognize device interrupts, and a Device Driver to actually handle I/O), the vendor usually supplies skeletal code as part of the documentation for an "Input/Output Subsystem". Under these conditions, the vendor's guidelines must be carefully followed during handler development to ensure conformance with established system protocol.

There is another alternative to direct assembly language coding, however, which is especially useful in simple I/O monitor environments. A FORTRAN routine to handle error traps, floating point variable scaling (e.g., for input from A/D or output to D/A converters where the internal FORTRAN decimal value is converted from/to the correct image octal value input or output), subroutine linkage, buffer allocation, etc., can be written with dummy statements for any IOT instructions. The FORTRAN routine can then be compiled and the assembly language output subsequently edited to include the IOTs. The routine can then be assembled, debugged, and added to the FORTRAN run-time library. The code so generated is of course not as efficient in terms of execution time or memory requirements, but then much less development time is required. Further, if the FORTRAN compiler permits in-line assembler code—as many do—then the IOTs may be inserted and the editing and assembly steps bypassed.

Let us turn next to a detailed discussion of various real-time FORTRAN features and illustrate their use by means of simple examples.

15.2.1 Executive Interface Routines

The routines discussed here and in following subsections are intended for use in either a single-user or a multitasking environment. Date and time routines are also included for completeness. These "first-level" interface routines provide the capability to control operation of programs within the system. Through use of the external procedures discussed here one may start, stop, or delay the execution of one or more application programs and access current time and date information. The argument "MODE" used throughout this discussion is equal to or greater than two in value for all instances in which the request is not accepted by the resident monitor. Individual implementations may specify unique values of MODE within the allowable range to designate the specific reason for which the request was rejected.

Starting a Program Immediately or After a Specified Time Delay—A call to the subroutine START initiates the execution of the designated program after the expiration of the specified time delay. The actual time delay obtainable in a specific industrial computer system is subject to the resolution of that system's real time clock. Execution of the designated program commences at the program's first executable statement. The form of this call is

CALL START (IPROG, ITIME,
IUNIT, MODE)

IPROG specifies the program to be executed. The argument is either:

a. An integer constant or an integer variable; or

b. An integer array name; or

c. A procedure name.

The processor must define which one of the above three forms is acceptable.

ITIME   specifies the minimum length of time, in units as specified by IUNIT, to delay before executing the program. If the value of ITIME is zero or negative, the requested program is run as soon as permissible. This argument is an integer constant or variable.

IUNIT   specifies the units of time as follows:

0—Basic counts of the system's real-time clock

1—Milliseconds

2—Seconds

3—Minutes

This argument is also an integer constant or variable.

MODE   is set on return to the calling program to indicate disposition of the request as follows:

1—Request accepted

2 or greater—Request not accepted.

This argument must be an integer variable or integer array element.

As we see, the ISA standard is a permissive standard in that it does not prescribe how the executive is to respond to procedure references, nor does it describe how the information is passed to the executive routine. In particular, the argument "IPROG" in CALL START has three forms: an integer constant or variable, an integer array name, or a program name, only one of which is permissible for use with any particular compiler. This restriction is a necessary consequence of the requirements of the FORTRAN language that any argument in an external procedure reference be of a defined type; integer constants or variables, integer array names, and program names are different types.

Examples of the use of the argument IPROG as an integer constant or variable are:

DATA J/9/

CALL START (7, 0, 0, M)

CALL START (J, 0, 0, M)

An example of the use of the argument IPROG as an integer array name is:

INTEGER XYZ

DIMENSION XYZ(3)

DATA XYZ(1), XYZ(2), XYZ(3)/10, 20, 25/

CALL START (XYZ, 0, 0, M)

An example of the use of the argument IPROG as a procedure name is:

CALL START (TASKN, 0, 0, M)

Interchangeability of programs between different processors is reduced by permitting three possible types for this argument, but it is premature to standardize for full interchangeability in this area.

Thus, with a FORTRAN dialect that permits use of a PROGRAM header card, a program whose only function on execution is to queue TASK20 to start 60 seconds later and TASK25 to start 8 hours (i.e., 480 minutes) later could be coded as shown in Example 15.1. Of course, Tasks 20 and 25 must have been previously compiled with the object code built into an absolute memory image format which includes all support routines not part of the resident monitor, and then properly identified to the operating system using appropriate executive commands. Further, program QUE1 must have valid I/O access to Logical Unit 5. Note that in a multitasking environment, a CALL START with ITIME = 0 simply places the task to be started in the "ready" queue where it must compete with all other "ready" tasks for system resources, i.e., CPU processor time. If ITIME > 0, then the task to be "START"ed

Example 15.1. Use of Procedure START to Queue Tasks

```
      PROGRAM QUE1
      ITSK20=20
      ITSK25=25
      LUN=5
      CALL START (ITSK20,60,2,MODE)
      IF(MODE.NE.1)WRITE(LUN,100)ITSK20
      CALL START (ITSK25,480,3,MODE)
      IF(MODE.NE.1)WRITE(LUN,100)ITSK25
      STOP
100   FORMAT(1X,16HSTART ERROR TASK,I3/)
      STOP
      END
```

is placed in the "suspended" queue under control of the task scheduler.

Starting a Program at a Specified Time— Execution of a reference to the subroutine TRNON causes the designated program to be executed at a specified time of day. Execution of the designated program commences at the program's first executable statement. The form of this call is:

CALL TRNON (IPROG, IDAYTM, MODE)*

IPROG   specifies the program to be executed as defined for subroutine START.

IDAYTM   designates an array whose first three elements contain the absolute time of day at which the specified program is to be executed. These elements are as follows:
First element—Hours (0 to 23)
Second element—Minutes (0 to 59)
Third element—Seconds (0 to 59)
This argument is an integer array name.

MODE   is defined as in START.

---

*In subsequent routine descriptions, definitions of arguments identical to those cited previously are not repeated; only a brief reference is given.

Let us repeat our previous example except this time assume that TASK20 is to be "turned on" at 11:15:00 A.M. and TASK 25 is to be "turned on" at 3:45:00 P.M. (15:45:00 on a 24 hour clock). These functions might be coded as shown in Example 15.2. Again, Tasks 20 and 25 must have been previously processed by the compiler with a memory image built and properly identified to the operating system prior to execution of program QUE2. Note that if QUE2 is executed at noon, then TASK20 may be queued for execution immediately or may be queued for execution at 11:15 A.M. the following day—depending on the implementation scheme actually used for the TRNON procedure. Since the second scheme is implemented in most cases, care must be exercised when executing a task whose function is to schedule other tasks for execution. The argument MODE is examined in Example 15.1 as well as Example 15.2 to ensure that the request is accepted by the operating system. Such an examination is always advisable though statements to examine MODE are not always shown in the remaining examples in this chapter.

A simple application-oriented example of the use of the TRNON executive routine appears in Example 15.3 where we assume that PROGRAM SCHEDL itself is initially turned on by the operator.

Example 15.2. Use of Procedure TRNON to Queue Tasks

```
            PROGRAM QUE2
            INTEGER TIM20(3),TIM25(3)
            DATA ITSK20/20/,ITSK25/25/
            DATA IOPMSG/5/
            DATA TIM20/11,15,0/,TIM25/15,45,0/
      C     ***SCHEDULE TASK 20 AT 11:15:00 AM***
            CALL TRNON(ITSK20,TIM20,MODE)
            IF (MODE.EQ.1) THEN
                  WRITE (IOPMSG,100) ITSK20
            ELSE
                  WRITE (IOPMSG,110) ITSK20
            END IF
      C     ***SCHEDULE TASK 25 AT 3:45:00 PM***
            CALL TRNON(ITSK25,TIM25,MODE)
            IF (MODE.EQ.1) THEN
                  WRITE (IOPMSG,100) ITSK25
            ELSE
                  WRITE (IOPMSG,110) ITSK25
            END IF
            STOP
      100   FORMAT (1X,'***TASK ',I2,' SCHEDULED')
      110   FORMAT (1X,'***SCHEDULING ERROR:  TASK ',I2)
            END
```

Example 15.3. Use of Procedure TRNON for Report Scheduling

```
           PROGRAM SCHEDL
           CHARACTER*6 SCHPGM,SHIFT,DAILY
           DIMENSION IDAYTM(3),ITIME(3),SCHTIM(3)
           DATA IDAYTM/3*0/
           DATA SCHTIM/3,11,19/
           DATA SCHPGM/'SCHEDL'/,SHIFT/'SHIFT '/,DAILY/'DAILY '/
   C       ***FETCH CURRENT TIME***
           CALL TIME (ITIME)
   C       ***FIND NEXT SCHEDULING TIME***
           IHOUR=ITIME(1)
           DO 10 I=1,3
               IF (IHOUR.LT.SCHTIM(I)) GO TO 20
   10          CONTINUE
           I=1
   C       ***SCHEDULE NEXT SHIFT REPORT***
   20      IDAYTM(1)=SCHTIM(I)
           CALL TRNON (SHIFT,IDAYTM,MODE)
   C       ***SCHEDULE DAILY REPORT***
           IF (I.EQ.1) CALL TRNON (DAILY,IDAYTM,MODE)
   C       ***RESCHEDULE SCHEDULING TASK***
           CALL TRNON (SCHPGM,IDAYTM,MODE)
           STOP
           END
```

Delaying Continuation of a Program—Execution of a reference to the subroutine WAIT returns after a delay of a specified length of time. The form of this call is

CALL WAIT (ITIME, IUNIT, MODE)

ITIME specifies the length of time in units as specified by IUNIT to delay before returning to the calling procedure. If the value of ITIME is zero or negative, no delay occurs. Limitations of implementation must not cause the precise time to be less than requested. This argument is an integer constant or variable, and IUNIT and MODE are defined as for START.

Program Termination—Task termination is accomplished using the ANSI STOP statement. On execution of this statement, the task is placed in the "inactive" state where it remains until the executive receives another START or TRNON request. The actual implementation of these interface routines varies from manufacturer to manufacturer. For example, in the case of a single-user, single-application installation with a simple I/O monitor, a CALL WAIT may place the CPU in a wait loop until expiration of the specified time delay. Such an implementation is always undesirable as overlapped I/O and concurrent processing are not permitted. In the case of a real-time monitor or multitasking executive, a CALL WAIT effectively places the issuing task into a "suspended" state and begins execution of the highest-priority task in the "ready" state. After the specified wait time has elapsed, the task issuing the CALL WAIT is raised to the "ready" state and queued under the task scheduler to compete with all other pending tasks for control of the CPU.

Time and Date Information—The time-of-day and current date are frequently needed in real-time application programs and may be accessed using TIME and DATE routines.

The current time of day is obtained as follows:

CALL TIME (ITIME)

ITIME designates an integer array into whose first three (3) elements the absolute time-of-day is placed. The contents of these elements are as follows:
First Element—0 to 23 Hours
Second Element—0 to 59 Minutes
Third Element—0 to 59 Seconds

The current date is obtained from:

CALL DATE (IDATE)

Example 15.4. Use of Time and Date Procedures

```
      PROGRAM TIMDAT
      DIMENSION ITIME(3), IDATE(3)
      CALL TIME(ITIME)
      CALL DATE(IDATE)
      WRITE(5,100)ITIME
  100 FORMAT(1X,5HTIME-,I2,1H:,I2,1H:,I2/)
      WRITE(5,200)IDATE(2),IDATE(3),IDATE(1)
  200 FORMAT(1X,5HDATE-,I2,1H/,I2,1H/,I4/)
      STOP
      END
```

IDATE designates an integer array into whose first three (3) elements the date is placed. The contents of these elements are as follows:
First Element—AD year since zero
Second Element—Month 1 to 12
Third Element—Day 1 to 31

Some systems allow any program or task to "set" both time-of-day and date. This is a hazardous feature, as an application program could inadvertently destroy correct current time and date information. Properly speaking, setting time and date are operator responsibilities; hence the operating system should be so configured that user access to these functions is prohibited.

Use of the time and date routines, however, is quite straightforward as shown in Example 15.4. Note that an implied "DO" can be used to output time in a standard format, whereas explicit array element identification is usually required to output date information.

15.2.2 Process Input/Output Interface Routines

When real-time processing involves the input or output of information from or to a process or instrument, we must issue calls to appropriate real-time input/output subroutines. The variable names used throughout this subsection are implicitly typed according to ANSI convention. The ISA standard process I/O routines summarized in Table 15.2 allow for two forms—the basic standard and an extension. Note that the standard form in Table 15.2 is flagged by the addition of a "W" (for Wait) as the last letter of the subroutine name and is used with executive systems that suspend execution in the requesting program until the process I/O is completed.

These standard process input–output function interfaces allow access to data from specific analog and digital sensors and outputs. The execution of references to these subroutines results in return only being made to the requesting program when the requested data transfer is complete. The argument MODE is to be interrogated by the requesting program in order to determine the status of the request. An individual processor may specify unique values for MODE within the allowable range to designate the specific error conditions.

The results of the input operations and the data presented for output operations are processor-dependent. The operations described are intended to be unformatted transfers to and from the specified storage element in the processor. Therefore, the representation in the integer storage elements of the processor is generally an image of the input or output device specified. However, scaling may be performed in the device handler itself—especially those performing analog I/O—in some implementations.

The extended-form CALLs accommodate executive routines that permit continuation of

Table 15.2. Alternate Forms of ISA Standard Process I/O Routines

Function	Standard	Extension
Analog Input Sequential	AISQW	AISQ
Analog Input Random	AIRDW	AIRD
Analog Output	AOW	AO
Digital Input	DIW	DI
Digital Output Momentary	DOMW	DOM
Digital Output Latching	DOLW	DOL

Example 15.5. Use of the AIRD Procedure

```
                PROGRAM ANAIN
                DIMENSION NINCHN(2), NINVAL(2)
                DATA MAXTIM/3/,IDEL/1/
                DATA IOPMSG/5/
        C          .
        C          .
        C          .
        C       ***SET ANALOG CHANNEL NUMBERS***
                NINCHN(1)=10
                NINCHN(2)=12
                CALL AIRD (2,NINCHN,NINVAL,MODE)
        C          .
        C          .
        C          .
        C       ***CALCULATIONS NOT USING NINVAL OR MODE***
        C          .
        C          .
        C          .
        C       ***WAIT FOR CONVERSION COMPLETE***
                DO 10 I=1,MAXTIM
                    IF (MODE.GE.3) THEN
                        WRITE (IOPMSG,100)
                        STOP
                    ELSE IF (MODE.EQ.2) THEN
                        CALL WAIT(IDEL,0,MODE)
                    ELSE
                        GO TO 20
                    END IF
           10       CONTINUE
        C       ***TIME EXPIRED***
                WRITE (IOPMSG,200)
                STOP
        C       ***SCALE NINVAL***
           20   DO 30 I=1,2
           30       NINVAL(I)=FLOAT(NINVAL(I))*10000./4096.
        C          .
        C          .
        C          .
        C       ***CALCULATIONS USING NINVAL***
        C          .
        C          .
        C          .
                STOP
          100   FORMAT (1X,'***ANALOG INPUT ERROR')
          200   FORMAT (1X,'***ANALOG INPUT TIMEOUT')
                END
```

execution in the requesting program while the process input or output is being accomplished. However, the operating system must maintain association between the device data block and the calling program during execution of these subroutines. The requesting program must have provision for periodically testing the status of the request. This is accomplished by use of the argument MODE. All values returned by these extensions should not be considered as defined until such time as the parameter MODE indicates operation completed or error condition. (cf. Example 15.5.)

The system must ensure the availability of these arguments to the process input/output interface subroutine. No standard technique is specified for how this will be achieved, although most existing industrial systems have features that can be used for this purpose.

Analog Input and Output. *Analog Inputs* are obtained from an analog to digital converter. Though hardware configurations differ, analog values can be input randomly using:

CALL AIRDW(NCHAN, NINCHN, NINVAL, MODE)

NCHAN   specifies the number of analog points to input on this call and is an integer constant or variable.

NINCHN   designates a one-dimensional integer array containing terminal connection data for the analog input points. Specific information regarding this argument must be supplied by the vendor.

NINVAL   designates a one-dimensional integer array into which the analog values are stored.

Note that the order of the elements in NINVAL corresponds to the order of the elements in NINCHN.

MODE   indicates disposition of the request as follows:
             1—All data collected
             2—Operation incomplete
             3—Error conditions

Since NINCHN and NINVAL are one-dimensional integer arrays, some small-computer implementations of the handlers require that these arrays be DIMENSIONed in the calling program even if only one pin connection is used.

Suppose we have two analog field signals terminated on pins 10 and 12 of a 0 to +10 volt DC, 12-bit analog to digital converter. Our ADC resolution is one part in 2^{12} or one part in 4096_{10}. If at 10 VDC, all bits are "set" and the ADC output register is strobed (right-justified) into a 16-bit computer register, our internal unscaled representation of this analog signal is 7777_8 (assuming, of course, that the highest-order 4 bits are masked out to zero). Hence, if we wish to input these two analog signals using the (extended) ISA routine AIRD and have the data accessible as a scaled integer whose contents are VDC $*$ 1000, we might use code similar to that of Example 15.5. Note that NINVAL(I) is "floated" during the scaling calculation to avoid the possibility of an integer overflow.

Analog points may also be input sequentially through a CALL AISQW (NCHAN, NINCHN, NINVAL, MODE) where the arguments are essentially the same as those used in AIRDW except that NINCHN is now an integer constant or variable specifying the first channel to be input.

Usually, AISQW should be used where large (preferably contiguous) blocks of analog data are input in time-critical applications because ADC channel addresses are generated sequentially in the handler rather than being passed from the mainline program with subsequent look-up and random ADC channel addressing. Also, since NINCHN is an integer variable instead of an integer vector of channel connections, memory is conserved. For slow-speed or relatively small applications, AIRDW can be used.

Analog Output is handled through a call to subroutine AOW as follows:

CALL AOW (NCHAN, NOUTCH,
NOUTVL, MODE)

NCHAN   specifies the number of analog points to be output on this call and is an integer constant or variable.

NOUTCH specifies hardware or software connections and transmission information for the analog output point. Specific information relevant to construction of this argument is processor-dependent because various types of digital-to-analog systems are available. This argument is an integer array name.

NOUTVL designates a one-dimensional integer array from which the analog output values are obtained. Again, note that the order of the elements to be output from NOUTVL must correspond to the order of the elements in NOUTCH.

MODE   indicates disposition of the request as follows:
             1—All data output
             2—Operation incomplete
             3—Error conditions

A simple single-user application for *Direct Digital Control* (DDC) of two stirred tank chemical reactors in series illustrates practical use of these handlers as well as the executive routine WAIT. The example follows that of Coughanowr and Koppel [1] except that the analog controller is replaced by a simple DDC

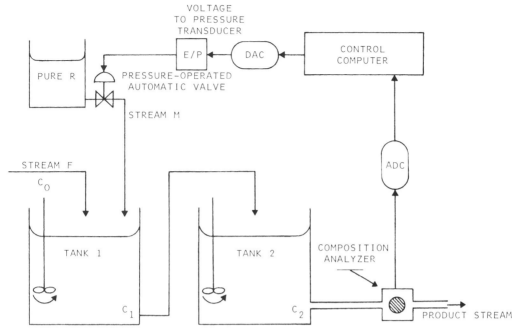

Figure 15.1. Schematic of Simple Reactor Control Problem

algorithm. A schematic of the process is shown in Figure 15.1.

A feed stream, F, containing reactant R at a concentration of c_0 (lb mols/ft^3), enters Tank 1. Reactant R decomposes in the tanks irreversibly. The purpose of the control system is to maintain the concentration c_2, of reactant R, leaving Tank 2 at a desired value despite variations of feed composition, c_0. This is accomplished by adding a stream, M, of pure R to Tank 1. The product stream of Tank 2 flows through a remotely located composition analyzer. The time required for material to flow from the tank to the analyzer is 30 seconds. Details of the development of the process describing equations and the numerical values and units for all problem variables may be obtained from the original reference [1]. Table 15.3 summarizes the correspondence between process variables and electrical signals at the process/computer interface.

Let us now consider a straightforward single-user mainline program called APPGM which implements a standard DDC position control algorithm in a simple I/O monitor environment. The application program may be coded to include one or more subroutines, but the main

program for the real-time application should be APPGM with the service or support routines nested within APPGM. One must always recall that the interface generates and accepts signals in current or voltage units, whereas the user usually prefers to work in process units within application modules. Information for such conversion must be supplied by the vendor and used during program development. Thus, if APPGM is to function in process units, the current or voltage signal supplied by the system must be converted to process units after input and the process units back-transformed to scaled current or voltage prior to output. The process control application for the example of Figure 15.1 consists of a single DDC loop with proportional-plus-integral control action. The complete code for APPGM, without provision for checking the variable MODE, changing controller constants, compensating for valve nonlinearity, outputting periodic reports, or bumpless transfer between local and remote modes, is shown in Example 15.6.

Note that NINCHN, NOUTCH, and NVALUE in APPGM are DIMENSIONed even though there is only one element needed in each. This is always advisable and may even be necessary if

Table 15.3. Process Variables and Electrical Signals for Reactor Control Problem

Process Variable	Process Range		Interface Range		Type	Scale Factor**
Exit Concentration	MIN: 0.01 MAX: 0.05	$\frac{lb\ mols}{ft^3}$	MIN: 10 MAX: 50	ma	ADC Input on Pin 51	X100
Manipulated Flowrate	MIN: 0 MAX: 20	gpm*	MIN: 10 MAX: 50	ma	DAC Output on Pin 110	X100

*Nominal for given valve trim and upstream pressure; in actual practice, program output is coded in terms of % full scale.

**a CALL AIRDW on pin 51 with a 25 ma signal returns integer 2500 (approximately) to the calling program. Note that any scaling convention can be used for convenience but actual resolution depends on the number of bits in the ADC (DAC) input (output) register. Thus an unsigned 12-bit ADC would have a maximum resolution of 1 part in 4096 not 1 part in 5000.

the target computer routines AIRDW and AOW expect these arguments to be DIMENSIONed and function improperly if they are not. Also note that while the WAIT interval is 5 seconds, the sample time may be greater by the amount of time required for computation depending on the resolution of the real-time clock. This is a consequence of the ISA requirement that the WAIT time not be less than that requested. Thus algorithm implementations of the type illustrated in Example 15.6 must be used with caution—especially when precise timing is required in the course of lengthy computations. DATA statements initialize controller gain (XKC), integral time (TI), and manual reset (XMAN) as well as the setpoint (SETPT). Pin connections are identified for the single ADC input channel 51 and the single DAC output channel 110. One value of the analyzer output is obtained by a CALL AIRDW and converted to process units. The new desired valve position in % full scale is calculated from

$$m_{n+1} = K_c\left(e_n + \frac{5}{T_i}\sum_{k=0}^{n} e_k\right) + m_R$$

where m_{n+1} is the Manipulated Variable for the n + 1st control interval (VALPOS). K_c is the Controller Gain (XKC). e_n is the Present Error (ERROR). T_i is Integral Time, sec/repeat (TI). m_R is the Manual Reset (XMAN).

The desired valve position is converted to hardware units and output by AOW whereupon execution is suspended for the control interval of 5 seconds. The cycle then repeats. This example illustrates the salient features of analog I/O though much more capability would be required in an actual process application. While not shown here, the argument MODE should always be checked to make sure that the I/O operation was successfully completed.

Figure 15.2 shows system behavior in process units when a rectangular pulse disturbance in c_0 is encountered. c_0 (lower left-hand corner of Figure 15.2) starts at 0.10, descreases to 0.05 at 50 seconds, and increases back to 0.10 at 240 seconds. Note the response to this disturbance. As expected, the concentration measured by the analyzer shows no change for 30 seconds owing to the transportation delay in the product stream line, then drops, begins to recover under control, overshoots the setpoint, and finally settles in with the long tail characteristic of a second-order response with time delay.

Note also that the manipulated variable saturates at 20 gpm as this was the maximum possible value at 100% full-scale DAC output. The output handler AOW may automatically

Example 15.6. Dedicated Application Control Program for Stirred Tank Reactor

```
          PROGRAM APPGM
C     ***DIMENSION ANALOG I/O VAR'S FOR COMPATIBILITY***
          DIMENSION NINCHN(1),NOUTCH(1),NVALUE(1)
C     ***SET CONTROLLER GAIN, RESET RATE AND MANUAL RESET***
          DATA XKC,TI,XMAN/500.,10.,46.75/
C     ***INITIALIZE SET POINT AND INTEGRAL SUM***
          DATA SETPT,SUM/0.0244,0.0/
C     ***SET OPERATOR MESSAGE LOGICAL UNIT***
          DATA IOPMSG/5/
C     ***SET ANALOG I/O CHANNEL NUMBERS***
          DATA NINCHN,NOUTCH/51,110/
C     ***FETCH EFFLUENT CONCENTRATION AS MA*100***
   10     CALL AIRDW(1,NINCHN,NVALUE,MODE)
          IF (MODE.GT.1) GO TO 20
C     ***CONVERT ADC INPUT TO LBMOLS/FT**3***
          CONC=NVALUE(1)/1.0E+05
C     ***COMPUTE CURRENT ERROR***
          ERROR=STPT-CONC
C     ***COMPUTE INTEGRAL ERROR***
          SUM=SUM+ERROR
C     ***COMPUTE NEW VALVE POSITION BY PI ALGORITHM***
          VALVE=XMAN+XKC*(ERROR+5.0*SUM/TI)
C     ***CONVERT VALVE POSITION TO DAC SIGNAL AS MA*100***
          NVALUE(1)=(0.4*VALVE+10.0)*100.0
C     ***OUTPUT TO DAC***
          CALL AOW(1,NOUTCH,NVALUE,MODE)
          IF (MODE.GT.1) GO TO 20
C     ***WAIT 5 SECONDS FOR NEXT SAMPLE***
          CALL WAIT (5,2,MODE)
          GO TO 10
C     ***ANALOG I/O ERROR, INFORM OPERATOR***
   20     WRITE (IOPMSG,100)
          STOP
  100     FORMAT (1X,'***ANALOG I/O ERROR, GO TO MANUAL!')
          END
```

limit output to full scale on an over-range request, or it may skip output and set an error flag in MODE. Obviously, action here is dependent on the particular implementation of AOW.

Digital Input and Output. Analog input and output are used for variables that cover a range of process conditions such as the analyzer output or manipulated flowrate discussed previously. Frequently, however, a real-time application requires simply that the status of contacts (i.e., pump on or off; emergency condition or not) be tested as input or that two-state outputs (i.e., power relays) be energized or not. Further, in certain applications, one may wish to output a pulse train on a single discrete line as input to an integrating amplifier in certain types of DDC (*Direct Digital Control*) systems or as input to a stepping motor in certain types of SSP (*Supervisory Set Point*) control systems. In this case, discrete I/O is

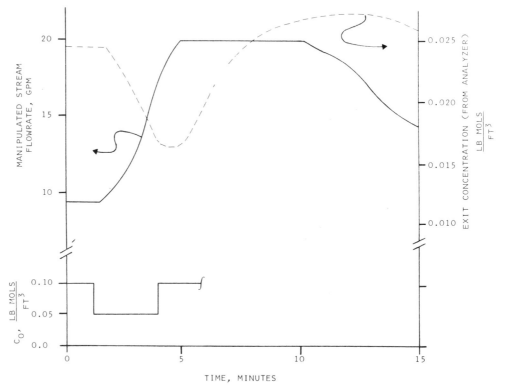

Figure 15.2. Typical Output for Reactor Control Problem

definitely preferred over analog I/O because each digital output word or register can handle multiple functions. Usually, though not always, the number of bits in the discrete I/O register corresponds to the computer word length. Thus, for a 16-bit small computer with a 16-bit discrete I/O register, 16 different functions can be accommodated for each I/O channel as compared with one for each analog I/O channel. Here, we discuss those routines adopted by ISA as standard procedures for handling discrete I/O.

Let the one-word integer be defined as the basic element of computer storage. For the purposes of industrial computing, it is necessary to view integer data as an ordered set of n bits $(a_{n-1}, a_{n-2}, \ldots a_0)$ where the set is a place-positional binary representation of an integer value.

For reasons of mathematical precision we must express the bit functions as a summation of a series. On a "normal" two's complement binary computer these functions operate in the expected manner. However, on a binary coded decimal machine, the results of these functions may not be as expected; it is not possible, for example, to reference every "bit" of a word. This is not a major problem as modern small computers are not BCD-structured.

Items such as difference in word lengths, one's complement vs. two's complement, negative numbers, and BCD vs. pure binary affect transportability. However, owing to the differences in the computers themselves, such variations are to be expected. These functions are not included to discourage the introduction of syntactic changes to FORTRAN though external procedure references are definitely preferred over such syntactic changes and will probably continue to be adopted as standards by ISA.

Digital Input—Execution of references to this subroutine causes the input of process information coded as a set of bits. A set of bits is typically organized and read as an external digital word whose length is processor-dependent. The form of this call is:

CALL DIW (IWORD, NDGCHN,
NDGVAL, MODE)

IWORD   specifies the number of external digital words to be read. This argument is an integer constant or variable.

NDGCHN specifies the hardware pin connection for each digital input register. Specific information relevant to construction of this argument is processor-dependent because various types of digital inputs are available. This argument is an integer array name or integer array element name.

NDGVAL designates an array to which the requested values are assigned. The order of the elements in NDGVAL must correspond to the order in NDGCHN. This argument must be an integer array name or integer array element name.

MODE   indicates the disposition of the request as follows:
1—All data collected
2—Operation incomplete
3 or greater—Error conditions
This argument is an integer variable or integer array element.

Momentary Digital Output—The execution of references to this subroutine causes the output of momentary digital signals. These signals consist of sets of bits typically organized as an external digital word whose length is processor-dependent. This type of output is characterized by an action that momentarily sets individual outputs when a corresponding bit in the output data word is set. These output bits are then reset after a specific time period. The form of this call is:

CALL DOMW (IWORD, NDGCHN, NDGVAL, NDGTIM, MODE)

IWORD   specifies the number of external digital words to be output. This argument is an integer constant or variable.

NDGCHN specifies the hardware pin connection for each digital output register. This argument is processor-dependent because various types of digital outputs are available. An integer array name or integer array element name must be used.

NDGVAL designates an array from which the output values are obtained. The order of the elements in NDGVAL must correspond to the order in NDGCHN. The argument must be an integer array name or integer array element name.

NDGTIM specifies the duration, measured in basic system clock counts, that all outputs are to remain set. If the processor does not allow selection of duration, this argument is ignored but must be present. This argument is an integer constant or variable.

MODE   is as defined for DIW.

Latching Digital Output—Execution of references to this subroutine causes the output of digital signals which can be latched in either the set or reset state. These signals consist of sets of bits typically organized as an external digital word whose length is processor-dependent. This type of output is characterized by an action that sets individual outputs when the corresponding bit in the internal integer word is set. The form of this call is:

CALL DOLW (IWORD, NDGCHN, NDGVAL, NDGLOK, MODE)

IWORD, NDGCHN, NDGVAL, and MODE are as defined for DOMW and

NDGLOK designates an array of words whose bit values define digital outputs *that can be changed* by the subroutine. A bit set in the NDGLOK array indicates that the digital output is changed to the state defined by the corresponding bit position in the corresponding integer array element in NDGVAL. The order of the elements in NDGVAL and NDGLOK corresponds to the order in NDGCHN. This argument is an integer array name, or an integer array element.

The three routines discussed here provide essentially all the discrete I/O capability needed in any real-time application except, perhaps, the capability of initiating execution of a program segment on charge of status of any digital input line through use of the computer interrupt system.

Example 15.7 illustrates the basic use of DOMW to drive two analog stepping motors. In this simple example, only one bit in each digital output (momentary) register was used. Since NDGVAL(2) = $4_{10} \triangleq 4_8 \triangleq 100_2$, the line connected to LSB + 2 (Bit 2) of the hardware output register 152 would receive 7 pulses of one clock count duration each. The line connected to LSB (Bit 0) of hardware output register 151 would receive 3 pulses.

In actual practice, of course, coding would be more complicated and involve use of bit string manipulation routines, since we would attempt to utilize one register fully before going on to the next. Also, each stepping motor would require two bits—one to drive the motor upscale as shown and another to drive the motor downscale.

15.2.3 Bit String Manipulation

In most applications, digital input at the FORTRAN level yields the image integer value of the register contents, i.e., a "bit string", and digital output loads the output register with the image integer value of the variable being output. Obviously, some method of bit manipulation is required. The ISA S61.1 routines which provide for such bit string manipulation are discussed here.

As with digital I/O, the bit manipulation procedures allow the programmer to view integer data as ordered sets of bits (a_{n-1}, a_{n-2}, ... a_0), where the set is a place-positional binary representation of an integer value, thus permitting interrogation and manipulation of integers on a bit-by-bit basis. The value of n depends on the processor.

The logical operations described are external functions where j and m are integer variables or constants. Operations are performed on all bits that represent the value of an integer internal to the processor. Operations are done bit-by-bit on corresponding bits; that is, the corresponding bits of the actual arguments j and m are used to generate the integer result.

Inclusive OR—The form of this external function reference is

$$IOR\,(j, m)$$

where j and m designate arguments that are logically added, bit-by-bit as follows:

j: 0 1 0 1 $\cdots$ to n bits/word
m: 0 0 1 1 $\cdots$ to n bits/word
————
Function Value: 0 1 1 1 $\cdots$ to n bits/word

Logical Product—This external function is referenced as

$$IAND\,(j, m)$$

where j and m designate arguments logically multiplied, bit-by-bit, as follows:

j: 0 1 0 1 $\cdots$ to n bits/word
m: 0 0 1 1 $\cdots$ to n bits/word
————
Function Value: 0 0 0 1 $\cdots$ to n bits/word

If we wish to access one digital input register and "mask-out" the least significant byte, then we could use the following code segment:

.

.

.

```
C  SET MASK TO ZERO LOW BYTE (LSB IS BIT 0; MSB IS BIT 15)
C  MASK BINARY IS 1,111,111,100,000,000 BUT NOTE USE
C  OF DIALECT-SPECIFIC 'B' FOR CORRESPONDING OCTAL INTEGER
   MASK = 177400B
C  INPUT ALARM REGISTER
   CALL DIW (1, NDGIN, NINVAL, MODE)
   INPUT = NINVAL(1)
C  ASSUME OCTAL INPUT IS 142017B
C  USE IAND TO MASK OUT LOW BYTE
   INPUT = IAND (INPUT, MASK)
C  FIND OCTAL RESULT AS 142000B
```

.

.

.

Example 15.7. Use of Procedure DOMW to Drive Stepping Motors Upscale

```
              PROGRAM STEP
      C       ***DEFINE NUMBER OF STEPPER MOTORS***
              PARAMETER (NMOTRS=2)
              LOGICAL DONE
              DIMENSION NDGCHN(NMOTRS),NDGVAL(NMOTRS),
             1            NOTIKS(NMOTRS),IBIT(NMOTRS)
      C       ***DEFINE CHANNEL NUMBERS***
              DATA NDGCHN/151,152/
      C       ***DEFINE BIT NUMBERS IN DIGITAL OUTPUT PORT***
              DATA IBIT/0,2/
      C            .
      C            .
      C            .
      C       ***DEFINE NUMBER OF 'TIKS' FOR MOTOR #1***
              NOTIKS(1)=7
      C            .
      C            .
      C            .
      C       ***DEFINE NUMBER OF 'TIKS' FOR MOTOR #2***
              NOTIKS(2)=3
      C       ***SET DIGITAL OUTPUTS***
              DO 10 I=1,NMOTRS
       10         NDGVAL(NMOTRS)=2**IBIT(I)
      C       ***OUTPUT STEPPER MOTOR PULSE STRINGS***
       20     DONE=.TRUE.
              DO 30 I=1,NMOTRS
                  IF (NOTIKS(I).EQ.0) THEN
                      NDGVAL(I)=0
                  ELSE
                      NOTIKS(I)=NOTIKS(I)-1
                      DONE=.FALSE.
                  END IF
       30         CONTINUE
      C       ***STOP IF ALL PULSES OUTPUT***
              IF (DONE) STOP
      C       ***OUTPUT PULSES USING DOMW***
              CALL DOMW (NMOTRS,NDGCHN,NDGVAL,1,MODE)
      C       ***WAIT ONE SECOND FOR MOTORS TO SETTLE***
              CALL WAIT (1,2,MODE)
              GO TO 20
              END
```

Logical Complement—The form of this external function reference is

$$NOT \ (j)$$

where j is logically complemented, bit-by-bit, as follows:

j: 0 1 $\cdots$ to n bits/word

Function Value: 1 0 $\cdots$ to n bits/word

Exclusive OR—This external function is referenced as

$$IEOR \ (j, m)$$

where j and m are integer variables or constants exclusively added, bit-by-bit as follows:

j: 0 1 0 1 $\cdots$ to n bits/word
m: 0 0 1 1 $\cdots$ to n bits/word

Function Value: 0 1 1 0 $\cdots$ to n bits/word

The "Exclusive Or" is especially useful if we wish to check for a change of state (of any bit) in a digital input register. For example:

The reader should note that in a sophisticated multitasking environment, such code for several input registers would be queued for execution only in response to a process interrupt caused by any one input line changing state. Nonetheless, implementations similar to that above are actually used in practice.

Shift—The shift operation is also an external function. Operations are performed on all bits that represent the value of an integer internal to the processor. A right or left shift can be specified. The form of this external function reference is

$$ISHFT \ (j, m)$$

where j designates the integer variable to be shifted, and m designates the number of positions to be shifted and the direction of the shift as follows:

$m < 0$	shift right m bits
$m > 0$	shift left m bits

```
         .
         .
         .
   C      INPUT REGISTER
  100   CALL DIW (1, NDGCHN, NDGVAL, MODE)
   C      WITH PRIOR VALUE IN NDGOLD, CHECK FOR CHANGE
          IF (IEOR (NDGVAL(1), NDGOLD). NE.0) GO TO 200
   C      WAIT TO RECHECK
          CALL WAIT (5, 2, MODE)
          GO TO 100
         .
         .
         .

   C      IF CHANGE, SAVE NEW VALUE AS OLD
  200   NDGOLD = NDGVAL(1)
   C      FIND CHANGE(S) OF STATE AND OUTPUT MSG
         .
         .
         .

   C      WAIT TO RECHECK
          CALL WAIT (5, 2, MODE)
          GO TO 100
          END
```

m = 0   no shift

Bits right-shifted out of an integer variable are lost; the variable is left-filled with zeros (and vice versa).

Routines for test, set, or clear of specified bits are also needed for proper use of the preceding routines. The external functions described here use arguments j and m as integer constants or variables.

Bit Test—This logical function tests a specified bit of an integer. The form of this external function reference is

BTEST (j, m)

where m now corresponds to the bit position with LSB defined as 0.

The result of BTEST (j, m) may be stated as follows:

If IAND (j, 2^m) = 0,
then set BTEST to .FALSE.
else set BTEST to .TRUE.

Note that since BTEST is a "logical" function

in the FORTRAN sense, it should be explicitly so typed in all routines where it is used.

Bit Set—This external function sets a specified bit of an integer varibale. The form of this function reference is

IBSET (j, m)

The result of the function reference IBSET (j, m) is

IOR (j, 2^m)

Bit Clear—This external function clears a specified bit of an integer variable. The form of this function reference is

IBCLR (j, m)

The result of the function IBCLR (j, m) is

IAND (j, NOT(2^m))

As an example of the use of the bit test and clear routines, assume that a program must "lock out" bit 15 to an output register if bit 14 in an input register is set. We could code this as follows:

```
        LOGICAL BTEST, ICHECK
        .
        .
        .
C   LET INPUT = 142000B
C   ALLOW ALL DOLW BITS TO CHANGE  STATE
        NDGLOK(1) = 177777B
C   CHECK BIT 14 OF INPUT
        ICHECK = BTEST (INPUT, 14)
C   SINCE BIT 14 OF INPUT SET, ICHECK = .TRUE.
C   CLEAR BIT 15 IN LOCK WORD
        IF (ICHECK) NDGLOK(1) = IBCLR (NDGLOK(1), 15)
C   NDGLOK(1) IS NOW 077777B
C   OUTPUT TO ALARM REGISTER
        CALL DOLW (1, NDGOUT, NOUTVL, NDGLOK, MODE)
        .
        .
        .
```

As a final example of the use of ISA S61.1 procedures, let us consider a multitasking implementation of direct digital control based on a simple "position algorithm" of the type used in Example 15.6. Here we employ

- Executive Procedures: START, TRNON, WAIT
- Analog I/O Procedures: AIRDW, AOW
- Digital I/O Procedures: DIW
- Bit String Functions: BTEST

The "process" consists of two liquid streams—manipulated flow and disturbance flow—feeding a liquid storage vessel connected to a second liquid storage vessel by a short, valved transport line. Thus our process is a second-order interacting system. Our objective is to maintain the level in the second vessel at a desired value, or setpoint, in the presence of a disturbance flow that is either zero or some constant, nonzero value. We periodically poll the position of an operator-positioned console switch via DIW and use BTEST to determine whether the process is under local analog or remote computer control. (Note: Again, in sophisticated multitasking systems equipped with a hardware priority interrupt structure, the operator console switch would generate a process interrupt when thrown. This interrupt would then queue a control mode transfer task for immediate execution.) Under local analog control, we input (via AIRDW) the current analog controller output to the process as well as the output of the level transmitter and, after proper scaling, output the controller signal (via AOW) to a DAC channel connected through an analog relay to our final control element. This procedure assures "bumpless transfer" when we go from local analog control to remote computer control.

After an indication to transfer to computer control is detected, analog I/O continues as before except that our previous scaled output via AOW is now interpreted as the manual reset in a conventional Proportional + Integral DDC position algorithm. Our DAC output for the n + 1st DDC sample interval in terms of current and previous error (setpoint minus controlled variable) is

$$m_{n+1} = K_c \left(e_n + \frac{\Delta T}{T_i} \sum_{k=0}^{n} e_k \right) + m_R$$

where the symbols and their FORTRAN variable names (cf. Table 15.4) are as follows:

m_{n+1} is the Scaled Output to the final control element, % (PCNT).

Table 15.4. Input Parameters for Example 15.8

TASK1 Prompts with	TASK1 Variable Names	Description	Units	Value of Function	Default
ADC(F)	INCHN(1)	Channel w/output of level controller	None	Available ADC Channel	None
ADC(H)	INCHN(2)	Channel w/output of level transmitter	None	Available ADC Channel	None
DAC	IOCHN(1)	Channel w/level DDC to process	None	Available DAC Channel	None
DI	INREG(1)	Register w/register to initiate DDC	None	Available DI Register	None
BIT	IBIT	Input register bit	None	Bit Position	None
SETPNT	ISET	Level setpoint*10	In	-----	240
GAIN	IGAIN	Proportional Gain*100	Pct/Ft	-----	1500
RESET	IRESET	Integral Time*10	Sec/Rep	-----	100
DDC	IDDC	Sample Interval	Sec	≥ 10	10
PRINT	IPRNT	Print Interval	Sec	$\geq$ IDDC	60

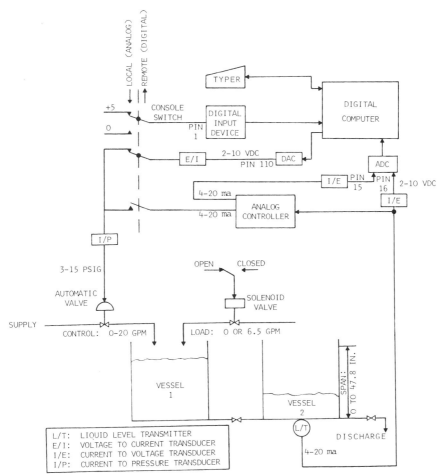

Figure 15.3. Direct Digital Control of a Liquid Level Process

K_c is the Controller Gain, % per inch (IGAIN).
e_n is the Present Error, inch (ERR).
ΔT is the DDC Sample Interval, sec (IDDC).
T_i is the Integral Time, sec/repeat (IRESET).
m_R is the Manual Reset, % (XMAN).

Schematically, our process and control options are shown in Figure 15.3, where basic scaling ranges are indicated for completeness.

Let us now develop a macro flowchart for this DDC example. Basically, we might propose two separate tasks:

- TASK1—for task initialization and report generation
- TASK2—for periodic DDC

Of course, many structures are possible; here we present a straightforward alternative especially

suited for demonstration of computer control in an instructional environment.

The flowchart of Figure 15.4 follows the conventions discussed in Chapter 13. Note that task priority is here set identical to task ID number because the first-level ISA executive routines START and TRNON do not allow separate priority and ID assignments. However, most multitasking executives generally provide default priority equal to program ID in any event; hence we need only make sure that our tasks are numbered properly. Note that some executives use a "lower ID = lower priority" scheme, whereas others use a "lower ID = higher priority" scheme. It is the latter case we assume here.

Complete source code is presented as Example 15.8. Detailed comments are included to clarify various segments of the code. Table

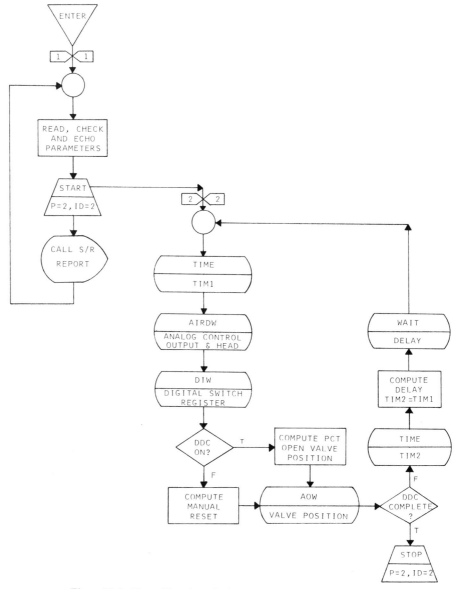

Figure 15.4. Macro Flowchart for Multitasking DDC (Example 15.8)

Example 15.8. Application Tasks for DDC Level Control

```
          PROGRAM TASK1
C        ***TASK1 INPUTS DDC PARAMETERS, QUEUES THE DDC TASK
C        ***AND OUTPUTS PERIODIC REPORTS.
          IMPLICIT INTEGER (A-Z)
          LOGICAL DDC,STATUS
          REAL HEAD,PCNT
          COMMON ANCHN(2),ANVAL(2),DACCHN(1),DACVAL(1),
     1          DDCCHN(1),DDCVAL(1),DDCBIT
          COMMON SETPNT,GAIN,RESET,DDCTIM,PRTTIM
          COMMON DDC,STATUS,HEAD,PCNT
          DATA IN,OUT/2,4/
C        ***SET INITIAL DDC STATUS***
   10     STATUS=.FALSE.
          WRITE (OUT,100)
          READ (IN,200) ADC1,ADC2,DAC1,DDC1
```

Example 15.8. Application Tasks for DDC Level Control (Cont'd.)

```
C          ***CHECK FOR ZERO ENTRY***
           IF (ADC1*ADC2*DAC1*DDC1.EQ.0) THEN
               WRITE (OUT,110)
               GO TO 10
           ELSE
C          ***SET COMMON BLOCK VALUES***
               ANCHN(1)=ADC1
               ANCHN(2)=ADC2
               DACCHN(1)=DAC1
               DDCCHN(1)=DDC1
               WRITE (OUT,200) ADC1,ADC2,DAC1,DDC1
           END IF
C          ***FETCH DDC PARAMETERS***
           WRITE (OUT,120)
           READ (IN,200) SETPNT,GAIN,RESET,DDCTIM,PRTTIM
C          ***FOR ZERO ENTRIES, SET DEFAULT VALUES***
           IF (SETPNT.EQ.0) SETPNT=240
           IF (GAIN.EQ.0) GAIN=1500
           IF (RESET.EQ.0) RESET=100
           IF (DDCTIM.EQ.0) DDCTIM=10
           IF (PRTTIM.EQ.0) PRTTIM=60
C          ***MAKE SURE THAT PRTTIM > = DDCTIM***
           IF (PRTTIM.LT.DDCTIM) PRTTIM=DDCTIM
           WRITE (OUT,200) SETPNT,GAIN,RESET,DDCTIM,PRTTIM
C          ***QUEUE DDC TASK***
           CALL START (2,1,2,MODE)
C          ***OUTPUT REPORT EVERY PRTTIM INTERVAL***
           CALL REPORT
C          ***RETURN TO FETCH NEW PARAMETERS***
           GO TO 10
    100    FORMAT (//1X,'ENTER IN N*15...',//,' ADC(F/H),DAC,DI,BIT')
    110    FORMAT (/,' ***ERROR',/)
    120    FORMAT (/,' SETPNT,GAIN,RESET,DDC,PRINT')
    200    FORMAT (5I5)
           END


           SUBROUTINE REPORT
           IMPLICIT INTEGER (A-Z)
           LOGICAL DDC,STATUS
           DIMENSION TIM1(3),TIM2(3)
           REAL HEAD,PCNT
C          ***SET DUMMY COMMON VECTOR FOR BLOCK EQUIVALENCE***
           COMMON DUMMY(13),PRTTIM,DDC,STATUS,HEAD,PCNT
           DATA IN,OUT/2,4/
C          ***DEFINE SECOND COMPUTATION STATEMENT FUNCTION***
           KSEC(HR,MIN,SEC)=3600*HR+60*MIN+SEC
C          ***OUTPUT HEADER***
           WRITE (OUT,100)
C          ***FETCH CURRENT TIME***
     10    CALL TIME (TIM1)
           WRITE (OUT,110) TIM1,PCNT,HEAD
           ***RETURN IF DDC SWITCH OFF***
           IF(STATUS.AND.NOT.DDC) RETURN
C          ***ADJUST PRINT INTERVAL FOR OUTPUT TIME***
           CALL TIME (TIM2)
           KSEC1=KSEC(TIM1(1),TIM1(2),TIM1(3))
           KSEC2=KSEC(TIM2(1),TIM2(2),TIM2(3))
           DELAY=PRTTIM-(KSEC2-KSEC1)
           CALL WAIT (DELAY,2,MODE)
           GO TO 10
    100    FORMAT (//,'HR.MN.SC   PCT OPEN   HEAD,IN')
    110    FORMAT (1X,I2,'.',I2,'.',I2,4X,F5.1,5X,F5.1)
           END
```

Example 15.8. Application Tasks for DDC Level Control (Cont'd.)

```
          PROGRAM TASK2
C     ***PROGRAM TASK2 IMPLEMENTS DDC EVERY CONTROL INTERVAL
C     ***SPECIFIED BY DDCTIM SECONDS.
          IMPLICIT INTEGER (A-Z)
          LOGICAL DDC,STATUS,BTEST
          DIMENSION TIM1(3),TIM2(3)
          REAL HEAD,PCNT,ERR,SUM,MANRST
          COMMON ANCHN(2),ANVAL(2),DACCHN(1),DACVAL(1),
      1        DDCCHN(1),DDCVAL(1),DDCBIT
          COMMON SETPNT,GAIN,RESET,DDCTIM,PRTTIM
          COMMON DDC,STATUS,HEAD,PCNT
          DATA SUM/0.0/
C     ***DEFINE SECOND COMPUTATION STATEMENT FUNCTION***
          KSEC(HR,MIN,SEC)=3600*HR+60*MIN+SEC
C     ***FETCH CONTROL START TIME***
   10     CALL TIME (TIM1)
C     ***CONVERT TO INCHES HEAD USING LINEAR CALIBRATION***
          HEAD=INVAL(2)*0.005833-10.5
C     ***FETCH CONSOLE SWITCH REGISTER***
          CALL DIW(1,DDCCHN,DDCVAL,MODE)
          DDC=BTEST(DDCVAL,DDCBIT)
          IF (DDC) THEN
C     ***IMPLEMENT DDCT IF SWITCH SET***
              STATUS=.TRUE.
C     ***COMPUTE CURRENT ERROR***
              ERR=SETPNT/10.0-HEAD
C     ***COMPUTE INTEGRAL ERROR BY RECTANGULAR INTEGRATION***
              SUM=SUM+ERR*DDCTIM/RESET
C     ***COMPUTE PERCENT OPEN VALVE POSITION FROM P/I ALGORITHM***
              PCNT=MANRST+GAIN/100.0*(ERR+SUM)
C     ***SCALE VALVE POSITION FOR OUTPUT TO DAC***
              DACVAL(1)=PCNT*80.0+2000.0
              GO TO 20
          ELSE
C     ***COMPUTE MANUAL RESET IF SWITCH IN LOCAL***
              MANRST=INVAL(1)*0.0125-25.0
              PCNT=MANRST
              DACVAL(1)=INVAL(1)
          END IF
C     ***OUTPUT VALVE POSITION TO DAC***
   20     CALL AOW(1,DACCHN,DACVAL,MODE)
C     ***DDC COMPLETE IF SWITCHED TO LOCAL***
          IF (STATUS.AND..NOT.DDC) STOP
C     ***ADJUST DDC INTERVAL FOR RUN TIME***
          CALL TIME (TIM2)
          KSEC1=KSEC(TIM1(1),TIM1(2),TIM1(3))
          KSEC2=KSEC(TIM2(1),TIM2(2),TIM2(3))
          DELAY=DDCTIM-(KSEC2-KSEC1)
C     ***WAIT ONE CONTROL INTERVAL***
          CALL WAIT(DELAY,2,MODE)
          GO TO 10
          END
```

15.4 defines key program variables and units. Remember that each task must have been previously compiled, memory load built and properly identified to the executive prior to activation of TASK1 for the first time.

This DDC example was developed using an off-line timesharing simulation system [14] and subsequently verified in a unit operations laboratory on an existing experimental apparatus. Simulator response appears in Table 15.5 where entries preceded by *** are output by the simulation system—not the application tasks—to indicate progress and simulated process status.

15.2.4 Random File Handlers

Many on-line small-computer systems provide some kind of high-speed bulk storage device such as a conventional disk or drum or the

Table 15.5. Typical Response for Multitasking DDC Example

```
*** RTS4.1 - START TASK1 EXECUTION

ENTER IN N*15...

ADC(F/H),DAC,DI,BIT
    15      16      110      1      16
    15      16      110      1      16

SETPNT,GAIN,RESET,DDC,PRINT

   240    1500     100      10      60

HR.MN.SC         PCT OPEN         HEAD,IN
13.30. 0          31.0             24.0
13.32. 0          31.0             23.9
13.34. 0          31.0             24.0
13.36. 0          31.0             24.0
13.38. 0          31.0             24.0

*** RTS4.1 - DDC ON

13.40. 0          31.0             24.0
13.42. 0          31.1             24.0
13.44. 0          31.1             24.0
13.46. 0          31.1             24.0
13.48. 0          31.1             24.0

*** RTS4.1 - LOAD REDUCED

13.50. 0          31.8             24.0
13.52. 0          66.7             23.3
13.54. 0          67.0             23.7
13.56. 0          67.1             23.9
13.58. 0          67.1             24.0
14. 0. 0          67.2             24.0
14. 2. 0          67.2             24.0
14. 4. 0          67.2             24.0
14. 6. 0          67.2             24.0
14. 8. 0          67.2             24.0
14.10. 0          67.2             24.0
14.12. 0          67.2             24.0
14.14. 0          67.2             24.0

*** RTS4.1 - LOAD INCREASED

14.16. 0          30.9             25.5
14.18. 0          31.3             24.5
14.20. 0          31.1             24.2
14.22. 0          31.1             24.1
14.24. 0          31.1             24.0
14.26. 0          31.1             24.0
14.28. 0          31.1             24.0
14.30. 0          31.1             24.0
14.32. 0          31.1             24.0
14.34. 0          31.1             24.0
14.36. 0          31.1             24.0
14.38. 0          31.1             24.0

*** RTS4.1 - DDC OFF

ENTER IN N*15...

ADC(F/H),DAC,DI,BIT
```

newer, but lower-speed, "floppy" disks; many systems only use low-speed storage devices such as cassettes or provide for no bulk storage at all. In the latter two cases, the executing jobs usually run under control of a memory-resident real-time monitor or a simple I/O monitor.

With a multitasking disk-based operating system, the disk is invariably used for data storage if only to generate hourly or shift reports. File-handling procedures provided by the manufacturer allow disk access at the FORTRAN level during program execution. Unfortunately, there seem to be as many different procedures as there are operating systems. This greatly decreases program portability. As a matter of fact, some operating systems require suballocation of disk sectors to various data spaces each with its own attributes and accessible only through its own *Device Control Block* (DCB). Such systems permit only one "OPEN" data file in any data space at any particular time (because the DCB is "busy"). This includes Background processors (e.g., editors, compilers, etc.) as well as Foreground tasks. Such a structure is quite inconvenient and seriously limits the capability of the system to handle complex multitasking applications. In many cases, a vendor-supplied File Management System can be configured with the resident monitor to provide true random file access from multiple tasks. Additional main memory requirements and system overhead are usually high, however. Standardization of procedures for handling random access files could go a long way toward alleviating problems of this type.

At the Eighth Purdue Workshop [31], standard file I/O procedures were proposed as ISA S61.2. These procedures eventually issued as ISA S61.2 draft standard in 1977 [22]. Discussion continues on the need for some of the procedures in view of the new ANSI 78 FORTRAN standards for random file I/O. In general we discriminate between files as being "sequential" or "random" and further between files that are "formatted" (i.e., read or written following a specific FORMAT statement) or "unformatted" (i.e., simply read or written as a list with no format specified). Sequential files store items of information (i.e., records) in a one-after-the-other fashion. This technique is useful in certain applications, but random files, which permit information to be read or written by specifying an arbitrary record number, provide much more powerful data storage and retrieval capability for on-line systems.

In the procedures proposed for adoption by ISA, the random files are themselves fixed-length records. Further, the files are considered to be resident in some mass memory device which is always available. The record length is defined in terms of the amount of storage occupied by one integer.

A program executing in a real-time multitasking environment is subject to problems of file integrity unless special precautions are taken in the random file access routines. For example, two tasks executing in a multitasking environment could conceivably access the same file with disastrous consequences. In order to prevent problems of this type from occurring, the ISA standard specifies access privileges associated with each task request for access to a random file.

As in the ISA FORTRAN procedures for process I/O, two forms of CALLs are suggested. The first form is used in systems that permit task continuation while file operation is in progress. Here, the argument "MODE" must be monitored, as illustrated in Example 15.5 for AIRD, to determine disposition of the request. The second form is flagged by addition of the letter "W" to the procedure name and is used with executive systems that suspend the executing task in the multiprogramming environment

or simply wait for file I/O to terminate before continuing with the executing program. The argument "MODE" should be tested in each case to ascertain the disposition of the request. Only the "W" form of these handlers is discussed here. Specifically, Table 15.6 summarizes some file features and attributes of importance in real-time industrial computer systems [33].

Basically, the FORTRAN programmer can "create" a file but not "open" it, "delete" a file from the system, "open" a file previously created, "close" a file previously opened, "modify" access privileges, "read" from the file, or "write" into it. Only argument changes are defined as they are introduced in the pertinent procedure CALLs of this section.

To Create A File:

CALL CFILW (IFILE, NUMINT, NUMREC, MODE)

IFILE   is the file number or name.
NUMINT   is the number of integers/record.
NUMREC   is the number of records/file.
MODE   designates request disposition as follows:
          1–Request successful
          ≥2–Error Condition; request failed

To Delete a File:

CALL DFILW (IFILE, MODE)

Table 15.6. Some Important Features and Attributes of Files

Included in the ISA Standard	Excluded from the ISA Standard
● Files whose content is considered as "data".	● Files whose content is not considered as "data".
● Files on fixed media only or removable media that are not removed during processing.	● Files on removable media that are removed during processing.
● Files that are external to concurrent programs.	● Files that are internal to a concurrent program.
● Creation and/or deletion of files by a concurrent program.	● Creation and/or deletion of files by a system utility or at system generation time.
● Restrictions on file access as applied to a file.	● Restrictions on file access as applied to a component of a file.
● Read and write access for random files.	● Attributes of a file used for ensuring file privacy.

To Open a File:

CALL OPENW (IO, IFILE, IPRIV, MODE)

IO   specifies the pseudo input/output unit number by which IFILE is referenced within the task.

IPRIV   designates the desired access privilege under which the task receives the file and also serves as a declaration of intended use, viz.,

     1—Read Only: The calling program can read but not write; other concurrent programs can read and write.

     2—Shared: The calling program can read or write; other programs can read or write.

     3—Exclusive Write: The calling program can read or write; other concurrent programs can only read.

     4—Exclusive All: Only the calling program can access the file.

To Close a File:

CALL CLOSEW (IO, MODE)

To Modify Privilege:

CALL MODAPW (IO, IPRIV, MODE)

To Read a File:

CALL RDRW (IO, IREC, IVAR, NCOUNT, MODE)

IREC   is the record number of the first record to be read.

IVAR   designates the first variable into which informtion is placed.

NCOUNT   specifies the number of integers to be read.

To Write a File:

CALL WRTWR (IO, IREC, IVAR, NCOUNT, MODE)

There are some obvious problems with these proposed procedures; e.g., space allocation is fixed at creation rather than allowing files to grow or contract as needs warrant and space permits. Nonetheless, these standard procedures are preferable in many cases to those normally provided by the vendor. The ISA standard does not cover all areas of file management. For example, no provision is included to ensure file privacy. However, privacy is not a significant problem in an industrial computer system because the chief intent must be to provide a common data base on plant operation for control and information purposes and where the problem of contention is acute owing to the asynchronous nature of file access.

As an example of the use of some of these procedures, let a file called PRODAT contain eight process data records of ten integers

Example 15.9. Use of ISA Random File Procedures

```
      PROGRAM MANIP
      DIMENSION IVAL(10),JVAL(10)
      INTEGER PRODAT
C     OPEN LOGICAL UNIT 10 FOR READ ACCESS ONLY
      CALL OPENW(10,PRODAT,1,MODE)
C     READ FIRST RECORD
      CALL RDRW(10,1,IVAL,10,MODE)
      .
      .
      .
C     MANIPULATE DATA
      .
      .
      .
C     CHANGE ACCESS TO EXCLUSIVE ALL
      CALL MODAPW(10,4,MODE)
      .
      .
      .
C     GET AND WRITE STATUS DATA
      .
      .
      .
      CALL WRTRW(10,8,JVAL,10,MODE)
C     CLOSE FILE
      CALL CLOSEW(10,M)
      STOP
      END

      PROGRAM STATUS
      DIMENSION KVAL(10)
      INTEGER PRODAT
C     OPEN LOGICAL UNIT 7 FOR ACCESS EXCLUSIVE ALL
C     LOOP UNTIL ACCESS GRANTED
   10 CALL OPENW(7,PRODAT,4,MODE)
      IF(MODE.EQ.1)GO TO 20
C     WAIT ONE SECOND; TRY AGAIN
      CALL WAIT(1,2,MODE)
      TO TO 10
C     READ STATUS RECORD
   20 CALL RDRW(7,8,KVAL,10,MODE)
      .
      .
      .
C     EXECUTE STATUS ACTIONS
      .
      .
      .
C     RESET STATUS RECORD
      .
      .
      .
C     WRITE CLEAR STATUS DATA
      CALL WRTRW(7,8,KVAL,10,MODE)
      CALL CLOSEW(7,MODE)
      STOP
      END
```

each. Further, let the first seven records contain process information and let the eighth record contain status information. In the code segments of Example 15.9, MANIP reads the process data in the first seven records of PRODAT and sets status flags in the eighth record; STATUS accesses the status record, performs some actions, and then resets the record. File contention procedures are used to ensure an orderly use of the status information in file PRODAT. Note the use of the variable MODE in STATUS to loop until access is granted.

15.2.5 Task Management Routines

The proposed ISA S61.3 procedures [23] for task management are presented here in summary form for completeness. These "task management routines" are in addition to START, TRNON, and WAIT as defined in ISA S61.1 and are designed for use with modern multitasking real-time computing systems. The user must always be cognizant of the complex and subtle interactions between tasks and the real-time executive if procedures of the type presented here are to be used effectively.

In industrial computer systems, tasks are independent but interrelated. The ISA standard does not address all the areas of independent task interrelationships but is concerned with those features that most commonly arise in industrial computer systems. Table 15.7 shows those features covered by the ISA standard and those excluded; however, the excluded features may affect the result of a request for cooperation between independent tasks. Such efforts are processor-dependent and outside the scope of the standard. The list of excluded features is not exhaustive.

The proposed procedures for task management and for task manipulation by events are summarized briefly in Table 15.8. Refer directly to the standard for additional information and details of use.

15.3 EFFICIENT FORTRAN CODING FOR MINICOMPUTERS

So far we have considered real-time procedures in some detail. Let us turn now to consideration of some general FORTRAN programming hints for small-computer systems. While some of the more sophisticated small-computer sys-

Table 15.7. ISA Standard S61.3

Features INCLUDED in ISA S61.3	Features EXCLUDED From ISA S61.3
• Independent but interrelated tasks.	• Dependent tasks.
• Initiation at a given time, upon event, after a period, or repetitively.	• Processor-dependent exception conditions such as overflow/underflow.
• Suspension of a task by itself.	• Task suspension by another task.
• Resumption of a task by another task.	• Suspensions caused by the processor.
• Resumption of a task by an event, at a given time, or after a period.	• Suspensions caused by the executive routine.
• Termination (by a task) of another task or itself.	• All methods of sharing variables or arrays.
• Setting/clearing/testing of event conditions.	• All methods of sharing files.
	• Sending messages between tasks.
	• Termination of tasks based on events in the executive routine.
	• All functions of the executive not involving task management.
	• Deadline scheduling of tasks.
	• Boolean operations on events.

Table 15.8. Proposed ISA Procedures for Multitasking

Description	Procedure	Notes
EVENTMARKS		
Set Eventmark	CALL POST	Set the specified eventmark to the ON condition.
Clear Eventmark	CALL CLEAR	Set the specified eventmark to the OFF condition.
Test Eventmark	CALL TESTEM	Test the current state of the specified eventmark.
RESOURCEMARKS		
Lock Resource	CALL LOCK	Lock the specified resource to the calling task.
Unlock Resource	CALL UNLOCK	Release the specified resource for use by other tasks
EXECUTIVE CALLS		
Schedule Task	CALL SKED	Schedule the task based on time of day or eventmark ON.
Cycle Task	CALL CYCLE	Schedule the task for cyclic execution.
Connect Task	CALL CON	Connect the task to a specified eventmark and execute on each transition from OFF to ON.
Cancel Task	CALL CANCEL	Cancel pending execution of a non-cyclicly-scheduled task.
Cancel Task	CALL DCYCLE	Cancel future executions of a cyclicly-scheduled task.
Cancel Task	CALL DCON	Cancel execution of an event-based task.
Delay Task	CALL DELAY	Delay task execution until specified time of day or until specified eventmark changes state from OFF to ON.
Hold Task	CALL HOLD	Suspend task execution until a specified eventmark becomes ON.

tems now offer multipass optimizing compilers, this is still the exception rather than the rule. Hence the reader should be aware of some of the simple techniques that can be used to decrease execution times and memory requirements. Alternately, FORTRAN code once developed can be processed by an off-line, source-to-source optimizing compiler [10]. Real mastery of a problem-oriented language implies the ability to describe a job so that it can be done efficiently with a minimum of unnecessary extra operations. Concern for program efficiency is considered archaic by many; however, it assumes renewed importance with the advent of the many small computers that can execute problem-oriented languages, in particular, FORTRAN. It does not take a large FORTRAN program to tax the capabilities of some of these machines. Therefore, efficiency can make the difference between being able to run such a program in a straightforward manner, or, on the other hand, having to segment the program, resort to machine language, or look for a larger computer—all of which are inconvenient.

The measures that are necessary to improve

program efficiency depend upon how optimally each type of FORTRAN statement is handled by a given compiler. In general, the compilers for the smaller machines do less well in this respect, so that more attention is required by the programmer to obtain good code. Therefore, the programmer must know the characteristics of the particular compiler that he uses. Because of the variety of compilers now available and the rate at which new ones are being introduced, it is not practical to give those characteristics here. They can be ascertained by examination of the machine-language output produced by the compiler from some benchmark programs. Ideally, such information should be provided by the computer manufacturers.

The measures discussed here save both running time and memory space where the latter normally limits the size of problems that can be run on a small computer or in a small memory partition on a larger machine. Tradeoffs between the two are noted. The points discussed may seem trivial, but it is undoubtedly true that many "professional" programmers and practially all "amateur" programmers are unaware of them. Here we present just a few typical examples.

Arithmetic Expressions and Replacement Statements—Few small-machine compilers optimize subexpressions that are redundant between two or more arithmetic replacement statements. These must invariably be optimized by the programmer as, for example, the statements.

$$X = SIN (A * B/C)$$
$$Y = COS (A * B/C)$$

should instead be written:

$$TEMP1 = A * B/C$$
$$X = SIN (TEMP1)$$
$$Y = COS (TEMP1)$$

In the above case, additional running time and memory space are required for the instructions that store the result of the redundant subexpression in the memory location assigned to the variable TEMP1. This extra time and space are outweighed, however, by the savings

resulting from elimination of the instructions required to calculate the subexpression a second time. The net savings become greater, of course, as the subexpression becomes more complicated and as it is used a greater number of times. The saving in time is especially noteworthy as floating-point operations are unduly time-consuming on small computers that do not have floating-point hardware. The memory space consumed by the temporary-storage variable TEMP1 can be used most efficiently by employing the same name wherever temporary storage is required.

Constants—Where mixed-mode arithmetic is allowed, it is best to write constants in the dominant mode of the expression to avoid needless conversions. For the expression 2 * A, many compilers store the 2 as an integer and convert it to a real number each time the expression is evaluated. With the expression in the form 2.*A (i.e., with the decimal point shown), the constant is stored as a real number and the conversion is eliminated. Some small-computer compilers do not even accept mixed mode arithmetic. Hence mixed mode usage should be avoided.

Arithmetic operations on constants should be performed by the programmer before writing an expression. In the expression 4. * A/3., the two constants are stored separately by many compilers, and the division is performed each time the expression is evaluated. It should instead be written 1.333333 * A. Such constants should be written to as many significant figures as the computer handles in its arithmetic, so that full advantage is taken of the available precision. There is no penalty for the additional digits.

Powers—In many compilers, the use of the "**" notation calls upon a special subroutine in the library. If only small whole-number powers are to be calculated, the time and space required for this subroutine can be saved by avoiding the "**" notation. Thus, for example, for X ** 2 write X * X, for X ** 3 write X * X * X, and for X ** 4 write (X * X) * (X * X) with the redundant subexpression (X * X) handled as described earlier.

Remember, too, that both time and space can be saved in evaluating polynomials if a nested computation involving only addition and multiplication is used. For example, the polynomial coded as

$$Y = X ** 4 + X ** 3 + X ** 2 + X + 1.0$$

should instead be written as

$$Y = (((X + 1.0) * X + 1.0) * X + 1.0) * X + 1.0.$$

Where the use of the "**" notation is appropriate, the mode of the exponent can make a difference, because many systems use different library subroutines for real or integer exponents of real arguments. If such a program already contains a real exponent, memory space can be saved by making whole-number exponents real, thus eliminating any need for the integer-exponent subroutine.

Other systems have a subroutine for real exponents only, and convert all integer exponents to real before calling the subroutine. There, the exponents might as well be shown as real to begin with, thus eliminating the conversion step.

Statement Numbers—Ordinarily, there is no penalty for attaching a number to a statement even when not needed for reference by another statement. However, such a penalty can arise under two special circumstances. First, some small-machine compilers are very limited in the size of programs that they can compile because of the limited memory space for the assignment tables that keep track of variables and statement numbers. Hence, elimination of unneeded statement numbers reduces the burden on the available space. Second, some compilers perform limited optimization on sequences of arithmetic replacement statements. In particular, a variable needed in one statement may not have to be fetched from memory if it is already in a register as the result of a previous statement. This optimization is done only if none of the statements has a number, indicating that the sequence is never entered in the middle. If the last statement in the sequence ends the

range of a DO, its number can be removed and attached to a CONTINUE statement following. There is never a penalty for the use of a CONTINUE statement.

Subscripted Variables—Retrieval or storage of subscripted variables always requires more work than the retrieval or storage of unsubscripted variables. This is so because the address of the variable must be calculated from the subscript combinations. In more advanced systems this arithmetic is done through indexing so no additional time or space is required that can be eliminated readily. However, less powerful compilers insert additional instructions to perform address arithmetic wherever a subscripted variable is referenced. Here, it helps to cut down on such references. Where the same subscripted variable is referenced in two or more statements, these references may be handled as redundant subexpressions, provided that the values of the subscripts do not change through the sequence. The statements

```
DO 1 I = 1, M
IF (INDEX (I, 1).EQ.INDEX (I, 2)) GO TO 1
JROW = INDEX (I, 1)
JCOLUM = INDEX (I, 2)
1   CONTINUE
```

should be optimized as

```
    DO 1 I = 1, M
    ITEMP1 = INDEX (I, 1)
    ITEMP2 = INDEX (I, 2)
    ITEMP2 = INDEX (I, 2)
    IF (ITEMP1.EQ.ITEMP2) GO TO 1
    JROW = ITEMP1
    JCOLUM = ITEMP2
1   CONTINUE
```

Input-Output Statements—The input or output of an entire array may be specified either by an implied DO loop over the array or by mention of the name of the array with no qualification. In many systems, the latter saves space by causing less code to be generated and also saves central processor time.

In many cases there is a space penalty for specifying additional input-output modes, since each mode requires its own subroutine from the library. For example, consider a program whose primary output is on a line-printer. If the programmer decides to include monitor output on the console typewriter for the convenience of the operator, the space penalty incurred may include the typewriter-output subroutine in addition to the coding for the output statements. If memory space is critical, it might be best to forgo the monitor output or take it on the line-printer.

Note that with some systems, use of an integer variable for the logical unit number in an input statement, e.g., READ (INUNIT, 10) *list*, causes the loader to load every input device handler not part of the resident monitor. In these cases, integer numbers should be used instead, e.g., READ (2, 10) *list*. If the resident monitor does include reentrant handlers for all I/O devices, then integer variables should be used with numerical values defined by replacement statements or DATA initialization statements.

Subprograms—Use of subprogram organization incurs a time and space penalty because of the linkage instructions. The space penalty is more than made up, however, if the subprogram is called from more than one place in the main program, because the coding to perform the subprogram functions need not then be repeated at each place it is needed. If, instead, a subprogram is called from only one place in the main program, it is most efficient to eliminate its separate identity as a subprogram and incorporate it directly into the main program. Where a subprogram is a function that can be executed in one replacement statement, it might best be included in the main program as an arithmetic statement function. However, programmers new to real-time computing should structure their applications as much as possible by "compartmentalizing" program functions. Subprograms are, of course, a big help in this regard. Although up to 50 machine instructions might be needed to obtain and allocate space on the run-time stack for each CALL for purposes of reentrancy, this might be worthwhile from the standpoint of overall program modularity and possible future modifi-

cation or expansion—especially for simple applications.

Attention must also be paid to the manner of linking variables between a main program and a subprogram. Argument lists or COMMON statements may be used, and each has its proper role. There is a time and space penalty associated with the use of arguments lists. The subprogram must contain coding that fetches the address of an argument from the main program and plants that address where it is accessible to those instructions in the subprogram that require it. With COMMON linkage, on the other hand, there is no penalty. In general, if a variable in the subprogram always corresponds to the same variable in the main program at every call of the subprogram, then the linkage should be through COMMON (or parameters, for an arithmetic statement function). On the other hand, where a variable in the subprogram may correspond to a different variable in the main program at different calls of the subprogram, then the linkage should be through the argument list. Of course, a reference to a FUNCTION subprogram must always have at least one argument so that the compiler can distinguish it from a reference to an ordinary variable.

There is a time penalty, and there may be a space penalty, associated with each additional reference to an argument within a subprogram. Therefore, if an unsubscripted variable is referenced more than once in a subprogram, it is usually worthwhile to use a local variable in its stead. The local variable is made equal to the argument or vice versa at the beginning or end of the subprogram, depending upon whether the argument is an input or output variable.

Some compilers handle arithmetic statement functions as if they were open subroutines or macros, repeating the instructions for the function at each place where it is called in the program. This saves a little time (by eliminating linkage) at the cost of much space. If a program on such a system is space-limited, the arithmetic statement functions should be replaced with function subprograms.

COMMON storage has an important use in addition to the linkage of variables. In some simple systems, variables declared as COMMON are assigned to the memory area that is occupied by the loader during object-program loading. This space is otherwise unavailable to the FORTRAN programmer. Therefore, a program that taxes memory can obtain some relief by use of COMMON storage, even when subprogram linkage is not in question. For best use to be made of this feature, enough arrays and unsubscripted variables should be declared COMMON to fill the loader area.

15.4 SUMMARY

Admittedly, many viable languages and techniques exist and are used for various industrial and educational real-time acquisition, equipment monitoring and process control applications. However, "supportability" and software "portability" aspects of these applications are often overlooked. This oversight has proved, in many actual cases, to be quite troublesome and expensive. The solution suggested here, i.e., application development in ANSI FORTRAN VII [19] with ISA real-time extensions, is but one possibility and really depends upon continued pressure for standardization.

For those who do decide on the FORTRAN route, the information and examples presented and discussed here should prove very useful for initial study. We have mentioned the interaction between the executive and FORTRAN tasks at several points in our discussion. Unfortunately, detailed consideration of these aspects of real-time computing are beyond the scope of this chapter. The user should always be certain that he is intimately familiar with his operating system through careful study of vendor manuals and supporting documents prior to actual implementation of a multitasking real-time application. Bear in mind, too, that attention to efficient FORTRAN coding techniques, avoidance—insofar as possible—of specialized features of a particular compiler, use of "standard" real-time procedures where they are provided, and minimization of assembly language modules except for required device handlers all contribute to increased efficiency, portability, and supportability of any real-time installation.

15.5 REFERENCES

1. Coughanowr, D. R. and Koppel, L. B., *Process Systems Analysis and Control*, pp. 123–129, McGraw-Hill, New York, 1965.
2. Jarvis, P. H., "Some Experiences with Process Control Languages", *IEEE Trans. Indust. Electron. Control Instrum.*, *15*, 54–56, 1968.
3. Katzan, Harry, *FORTRAN 77*, Van Nostrand Reinhold Co., New York, 1978.
4. Mensh, M. and Diehl, W., "Extended FORTRAN for Process Control", *IEEE Trans. Indust. Electron. Control Instrum.*, *15*, 75–79, 1968.
5. Murphy, John A., "Minicomputers", *Modern Data*, *5*, 58–72, 1971.
6. Peterson, T. W., White, J. W., and Krist, E. E., "Off-Line Development of a Minicomputer Process Control Program for an Industrial Grinding Circuit," *Trans. AIME*, *262*, 355–360, 1976.
7. Ripley, G. D. and White, J. W., "How Portable are Minicomputer FORTRAN Programs?", *Datamation*, *23* (7), 105–107, 1977.
8. Roberts, Bert C., "FORTRAN IV in a Process Control Environment", *IEEE Trans. Indust. Electron. Control Instrum.*, *15*, 61–63, 1968.
9. Ross, D. T. and Pike, J. E., "Automating Control Computer Software Production", *Control Engineering*, pp. 44–47, Oct. 1972.
10. Schneck, P. B. and Angel, E., "A FORTRAN to FORTRAN Optimizing Compiler", *Computer J.*, *16* (4), 322–330, 1973.
11. Spencer, H. W., Shepardson, H. P., and McGowan, L. M., "Small Computer Software", *IEEE Comput. Group News 3*, 15–20, 1970.
12. Sturgul, J. R. and Merchant, M. J., *Applied FORTRAN IV Programming*, Wadsworth Publ. Co., Belmont, Cal., 1976.
13. Williams, T. J. (Ed.). *Emulator, Simulator, and Translator Programs for the Small Digital Computers Necessary for Medical Laboratory Automation*, Purdue Laboratory for Applied Industrial Control, Purdue University, Lafayette, Ind., 1970.
14. White, J. W., "An Off-Line Simulation System for Development of Real-Time FORTRAN Programs", *Intl. J. Comp. Info. Sci.*, *5* (1), 59–79, 1976.
15. White, J. W. and Ripley, G. D., "A Survey and Analysis of Minicomputer FORTRAN Dialects", Internal Report, Dept. Comp. Sci., Univ. of Arizona, Apr. 1977.
16. Zobrist, D. W., Fassbender, P., Bearden, F. E., and Costrell, L., "Software Standards and CAMAC . . . a Real-Time Demonstration", *Instrum. Technol.*, pp. 33–38, Mar. 1975.
17. "American National Standard X3.10-1966, Basic FORTRAN", 1966.
18. "American National Standard X3.9-1966, ANSI Standard FORTRAN", 1966.
19. "Americal National Standard X3.9-1978, ANSI Standard FORTRAN", 1978.

20. Full FORTRAN R1539, International Standards Organization, 1972.

21. "Industrial Computer System FORTRAN Procedures for Executive Functions, Process Input/ Output and Bit Manipulation", ISA Standard S61.1, Revised Feb. 1976.

22. "Industrial Computer Systems FORTRAN Procedures for Handling Random Unformatted Files, Bit Manipulation, and Date and Time Information", ISA Draft Standard S61.2, 1977.

23. "Industrial Computer System FORTRAN Procedures for the Management of Independent Interrelated Tasks", ISA Draft Standard S61.3, 1978.

24. *Minutes, 5th Wrkshp. on Std. of Ind. Comp. Langs.*, Purdue Lab. for Appl. Ind. Contrl., Lafayette, Ind., May 1971.

25. *Mintues, 6th Wrkshp. on Std. of Ind. Comp. Langs.*, Purdue Lab. for Appl. Ind. Contrl., Lafayette, Ind., Oct. 1971.

26. *Mintues, 7th Wrkshp. on Std. of Ind. Comp. Langs.*, Purdue Lab. for Appl. Ind. Contrl., Lafayette, Ind., Apr. 1972.

27. *Minutes, 8th Wrkshp. on Std. of Ind. Comp. Langs.*, Purdue Lab. for Appl. Ind. Contrl., Lafayette, Ind., Oct. 1972.

28. *Minutes, 9th Wrkshp. on Std. of Ind. Comp. Langs.*, Purdue Lab. for Appl. Ind. Contrl., Lafayette, Ind., May 1973.

29. *Minutes, Spring Regional Mtgs., Intl. Purdue Wrkshps. on Ind. Comp. Sys.*, Mar.–Jun. 1974.

30. *Minutes, 2nd Annu. Mtg., Intl. Purdue Wrkshps. on Ind. Comp. Sys.*, Oct. 1974.

31. *Minutes, Spring Regional Mtgs., Intl. Purdue Wrkshps. on Ind. Comp. Sys.*, Mar.–July 1975.

32. *Mintues, 3rd Annu. Mtg., Intl. Purdue Wrkshps. on Ind. Comp. Sys.*, Oct. 1975.

33. *Minutes, Spring Regional Mtgs., Intl. Purdue Wrkshps. on Ind. Comp. Sys.*, Mar.–July 1976.

15.6 EXERCISES

1. Discuss the advantages and disadvantages of the standardization of FORTRAN from both the vendor and user points of view.

2. Give several (at least three) examples of improvements in ANSI 78 FORTRAN over ANSI 66 FORTRAN.

3. The ISA standard extensions for process/ computer I/O drop the letter W from the procedure name. Explain the difference between the standard calls and their extensions. Under what conditions would the extensions be useful?

4. Under what conditions would you use AISQW for analog input instead of AIRDW? Explain your answer in the context of how typical handlers would interface to the actual hardware.

5. The control law used in Examples 15.6 and 15.8 is the discrete equivalent of Proportional plus Integral analog control. By adding the term $K_c T_d \Delta e_n$ (T_d is the rate time in seconds and Δ is the backward difference operator) to the right-hand side of that control law, one obtains the discrete equivalent of the three-mode (Proportional-Integral-Derivative) analog control law.

   a. Modify the code of Example 15.6 to include this third mode of "derivative" control.

   b. Modify the code of Example 15.6 to limit explicitly output to the valve from 10% to 90% of full scale.

6. Rework the skeleton of Example 15.7 to drive stepping motors downscale as well as upscale. Use register 151 for the upscale tiks and register 152 for the downscale tiks.

7. Polling of digital inputs is sometimes used in industrial computer control systems. Why is this generally not advisable? Be explicit.

8. The liquid level in a tank that is 1.0 meter tall must be controlled in such a way that it remains between 0.4 and 0.5 meter. Liquid level may be controlled by turning a pump on or off, while level is available as an analog voltage. The liquid level transducer has the calibration relation $L = 0.15 V$ where V is the voltage generated by the transducer and L is the corresponding level in meters. The pump may be turned on by setting bit 15 of the Digital Output Register to 1 and off by resetting the bit to 0. The analog voltage corresponding to level is available on ADC channel 1. A FORTRAN program is to be written that will accomplish the required control function and, in addition, perform incidental data logging of the process response. The FORTRAN program has been structured in such a way that it will be composed of a main program, a control subroutine, and a data logging subroutine. Write the FORTRAN *control subroutine* using ISA Input and Output subroutine calls. *Don't* write the main program or the data-logging routine.

9. Modify the code of Example 15.8 to compute one-minute averages of analog inputs and store them on a random access file

(using the ISA S61.2 routines) shift after shift for six shifts (i.e., two days) in a round robin fashion (i.e., data for the 7th shift overwrite the 1st; the 8th shift overwrites the 2nd, etc.).

10. The Aspenless Ski Corporation has decided to install a new gondola for transporting skiers in a newly-developed region of the ski area. As safety is a major concern because of recent accidents suffered at other areas, a computer will be installed to monitor and control the process. Lured by the fringe benefit of free lift tickets, and hired by Aspenless because of your real-time computing background, you have the assignment of helping with the computer software. Your job is to write a table-driven software program to monitor the weight of each gondola as it is loaded and to signal the operators if the car is over-weight. The process is interfaced with the computer as follows. A voltage input through ADC Channel 0, V, gives the weight of the car. The weight in kilograms, W, can be found using the approximate linear relation:

$$W = 1000 \, V$$

A 5-volt signal should be sent out on DAC channel 0 if the car weighs more than 10,000 pounds (2.2 lb/kg) to attract the operator's attention via an alarm. The gondolas are filled at a rate of one every 2 minutes.

a. How would you structure a multitask FORTRAN program to monitor the ski lift?

b. Draw a flow chart using standard multi-tasking symbols.

16

Control-Oriented Languages (Table-Driven Software)

Cecil L. Smith
Cecil L. Smith, Inc.
Baton Rouge, Louisiana

16.0 INTRODUCTION

While the use of FORTRAN and/or BASIC to create data acquisition and control software is certainly feasible for real-time systems with a small number of process inputs, its use in larger systems is beset with a number of problems. Perhaps the major problem is that the amount of coding becomes excessive; as a result the documentation requirements become extremely critical, and a high cost results.

The above observations do not explicitly rule out the use of FORTRAN or BASIC. Instead they imply that the amount of custom software for each system must be minimized. A systems group that has successfully installed one system and is currently faced with installing a second would be wise to use as much software as possible from the first system in implementing the second.

Of course, some organizations expect to implement several systems over a time span of a few years, and in this case should consider preparing several general-purpose software routines that would be useful in all of the projected applications. Logical candidates might be process input/output routines, control algorithms, operator interface routines, etc.

Since industrial and laboratory processes differ considerably from one to the next, the writer of these routines faces a problem in trying to provide generality. Specifically, how will the eventual user of the routine (who may or may not be the writer) tailor it to a given application? Changing the coding in the routine is highly undesirable, since the inevitable result will be one or more programming errors which will have to be found and eliminated. The approach generally adopted is to write the software in as modular a form as possible, entering information specific to a given application in an associated data base (in tabular or array form). The actual execution of the software is determined from information stored in the data table; thus the term "table-driven software" frequently is applied. Several commercial vendors have developed software systems that are based on this idea and are particularly marketed for process control applications. Hence the term "control-oriented languages" also is used, data acquisition applications being naturally subsumed within the class of process control applications.

16.1 CHARACTERISTICS OF TABLE-DRIVEN SOFTWARE

A general schematic of a table-driven software routine is illustrated in Figure 16.1. As drawn, the data table is the centrex of the system, much as memory might be considered to be the centrex of modern computer systems.

In most commercial systems, three major parts of the software interface directly to the tables:

(1) The table-driven program itself
(2) The operator communications routines
(3) Any applications programs written by the user

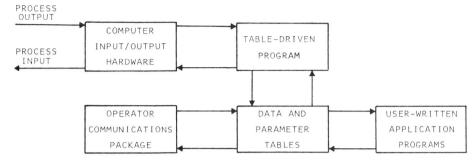

Figure 16.1. Table-Driven Software Package Information Flow

As we have mentioned, the table-driven program actually is a form of interpreter. Parameters in the table specify what actions the program is to perform and what data these actions are to be performed upon. Normally, one category of functions performed by the interpreter is that of process input/output. Other functions will retrieve data from the table, perform the specified calculations on these data, and store the results in specified locations in the tables.

The operator communications routines permit the operator to retrieve values from the data tables and, in some cases, to enter values into the tables. As operators are normally not favorably disposed toward numeric codes such as the index of an element in an array, most systems permit the user to specify a unique tag name for each data element in the tables. For convenience, this tag name is very similar to, if not identical with, the instrument or process element name with which the operator is familiar.

Although it is conceivable that some applications could be implemented utilizing only a general-purpose table-driven program, most commercial applications require some programs written by the user for the particular application. Most users prefer to write these programs in FORTRAN or BASIC, and consequently, some means must be provided for user programs to access the tables. In FORTRAN, the following possibilities exist:

(1) Put the data tables in COMMON.
(2) Provide "get" and "put" subroutine calls for accessing the data tables.

As most versions of BASIC do not provide for COMMON storage, the latter alternative is used.

From Figure 16.1, it could perhaps be implied that the data area is memory-resident. This certainly is not necessary, and in many systems much if not all of the data area is placed on disk or drum. When most small computers were limited to 32K or 64K of high-speed memory, use of the disk or drum was inevitable for large applications. However, the software is simpler and execution is faster when the tables reside in high-speed memory, and this approach will be used more often in the future as memory costs decrease.

Figure 16.1 shows all programs accessing the data area directly; however, the use of a data base manager as illustrated in Figure 16.2 has certain advantages.* Principally, it frees the programs themselves from the actual structure of the table area, and consequently the user should have more freedom in making additions, deletions, and changes to the structure of the data area. Unfortunately, some additional overhead is accrued, and generally the data base manager must be highly efficient. Of course, the concept in Figure 16.2 virtually eliminates the possibility of storing the data in FORTRAN COMMON.

---

*In this instance, by "data base" we mean the information contained in the table(s) that specifies what actions are to be taken by the table-driven programs. The "data base manager" is the single program that assumes sole responsibility for accessing information in the data base and passing it to the table-driven program, the user-written program, etc., and for keeping the contents of the tables current.

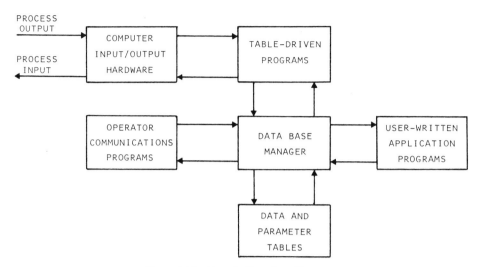

Figure 16.2. Use of a Data Base Manager

As with the most general-purpose software, the table-driven concept for process control software has disadvantages, in this case two major ones being that:

(1) A larger storage area is required.
(2) Execution speed is slower.

When minicomputer memories were limited to 32K, the larger memory requirement was a serious limitation; but this has disappeared with the newer generation of machines. The slower execution speed may necessitate a faster, and thus more expensive, computer than otherwise would be the case, but the additional cost will be more than offset by the reduction in software costs. The trend is for hardware costs to continue to decrease relative to software costs; so any method of reducing software costs will become even more important in the future.

16.2 TABLE-DRIVEN SOFTWARE FOR ANALOG INPUT CONVERSIONS

While the analog and data conversion system is only a small part of a process control software package, it is a necessary part of all such systems, and consequently is a logical candidate for implementation by means of a table-driven program. In this section the analog conversion routine will be used as an example of a table-driven program.

Suppose in our system, one storage location is reserved for the raw value read from each analog input to the system. Let these be arranged as a one-dimensional array in memory, which permits the hardware address simply to be the index of the respective location in this array. Furthermore, suppose the entire memory array is updated once each second either by a separate software routine or by input scan hardware (i.e., a free-running front end system that might utilize the computer's DMA channel).

The raw value must be converted to an engineering value before use, e.g., for display to the operator or for user-written application programs. A common approach is to convert the values on a regular interval of time and store the results in floating point in another array, often called the engineering value array or process variable array.

To keep this example as simple as possible yet still illustrate the concepts, our input conversion system will have only the following capabilities:

(1) Conversions will be made at sampling intervals based on any multiple of one second, but no provision will be made for load leveling (i.e., preventing all inputs, with, say, a 2-second scan from falling within the same 1-second processing cycle).

(2) All inputs will be linear inputs over the 2-volt to 10-volt range of a 10.24-volt A/D converter with a 12-bit resolution. That is, full range corresponds to a raw value count of 4000, and zero range corresponds to a raw value count of 800. Out-of-range conditions will be ignored.

(3) Only high alarms and low alarms will be issued for any process variable that is found to be outside prespecified limits— i.e., a message will be typed out or an alarm light turned on; no provision will be made to trigger the execution of a user-written program from an alarm.

A commercially-useful input conversion system must be far more powerful than this, but the basic structure would be the same.

In addition to the raw value array (RAWVAL) and the engineering value array (ENGVAL), the following arrays are needed:

(1) IADDR—input address, as an index in the raw value array
(2) ISAMP—sampling time, in integer seconds
(3) ICOUNT—count-down cell for control of sampling
(4) ENGRLO—low engineering range
(5) ENGRHI—high engineering range
(6) ALMHI—high alarm value
(7) ALMLO—low alarm value

Normally, all of these would contain fixed values except the ICOUNT array. Observe that for each point, 11 memory locations are required in addition to the raw value and engineering value locations if integer variables use one word of storage and real variables require two.

The flowchart for the table-driven input conversion system is illustrated in Figure 16.3. When building up the system initially, the user must specify the maximum number of analog inputs to be converted, and then dimension all of the arrays accordingly. To permit future expansions to be made to the system easily, these arrays generally are made slightly larger than the anticipated size. The flowchart uses a zero value in array IADDR to indicate that all conversions have been made. As will be shown

subsequently, additional conversions can be specified easily with this system provided space is available in the arrays.

On each execution of the input conversion routine, the contents of each location in array ICOUNT are decremented by 1 and then tested for a zero value. If nonzero, the respective conversion is not made on this execution. If zero, the contents of the respective location in ICOUNT are set back to the corresponding value in ISAMP, and the conversion to engineering value is then made for this process input.

To perform the conversion, the appropriate raw value is retrieved using the address from array IADDR and then is converted to a fraction between 0 and 1 (this is where an out-of-range test frequently would appear). Finally, the engineering value is readily calculated from the upper and lower values of the variable's range.

At the conclusion of the input conversion operations, the tests for high alarms and low alarms are made.

Even for this simple system, the arrays listed above are not quite sufficient. If the operator is permitted to inquire about engineering values, he would, at this point, have to enter the appropriate index of the variable in the engineering value array. As most operators would find this objectionable, an additional array for the tag names would be required. Since tag names usually contain about six characters, the tag name array must be dimensioned so that three words (assuming a 16-bit computer) are reserved for each tag name. For systems with a small number of inputs, a sequential search of the tag names is acceptable, but on larger systems a more systematic approach would be necessary to meet response requirements.

To build up the system initially, the user first must specify the proper dimensions for the arrays. Then he must enter the appropriate information into arrays IADDR, ISAMP, ENGRLO, ENGRHI, ALMHI, and ALMLO. This latter activity can normally be accomplished by someone who has little familiarity with programming.

To add another input conversion to the system is also quite easy, provided space is available in the arrays. Suppose the system initially

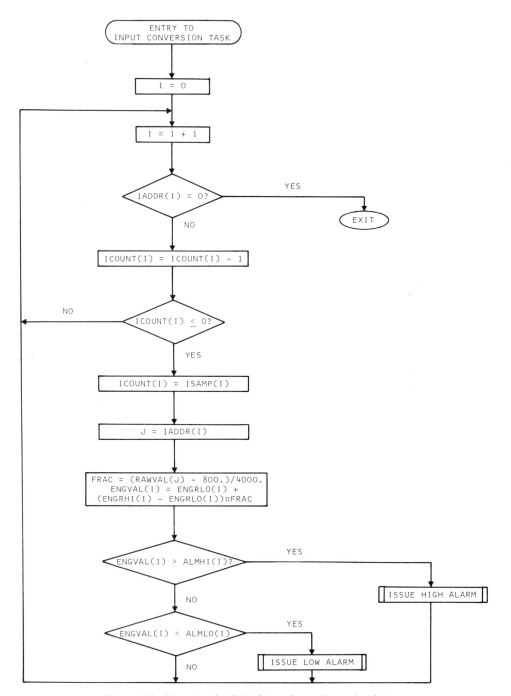

Figure 16.3. Flowchart for Table-Driven Input Conversion System

has N input conversions specified, and the user wants to specify conversion N + 1. He first enters values into the locations corresponding to ISAMP(N + 1), ENGRLO(N + 1), ENGRHI-(N + 1), ALMHI(N + 1), and ALMLO(N + 1). He should then verify that locations IADDR (N + 2) and ICOUNT (N + 1) contain zero. When the proper input address is entered into IADDR (N + 1), the new scan conversion will be completely specified to the system.

16.3 CAPABILITIES OF TABLE-DRIVEN PROCESS I/O SYSTEMS

The analog data conversion system described in the previous section is a highly simplified version of only a portion of a realistic process I/O system. Useful systems must be able to handle three types of input signals:

(1) Analog inputs
(2) Digital signals in the form of pulse trains
(3) Digital signals in the form of contact closures

In addition, special multiplexed I/O units often are needed.

Analog Signals—Common examples of analog measurements include outputs from thermocouples, pressure transmitters, flowmeters, etc. In practice, these signals must be compensated for offset and amplifier drift, smoothed or integrated, and compared to limits for alarming purposes.

To be widely applicable, the table-driven system must provide the following capabilities:

(1) *Scanning.* The time interval for the scan and/or conversion must be flexible enough to provide independent selection for each input. For industrial systems, scan intervals are normally multiples of one second. However, laboratory applications occasionally require faster scans.

(2) *Conditioning.* A variety of signal conditioning and conversion equations is required. The following capabilities are useful:

    i) Linear conversions, usually for signals coming from 4–20-ma transmitters.

    ii) Quadratic or cubic conversion equations for nonlinear calibrations.

    iii) Thermocouple conversions. Multiple types must occasionally be supported, and cold junction compensation is a must (Chapter 5).

    iv) Resistance thermometer (RTD) conversions. As the popularity of resistance temperature sensors increases, these inputs are encountered more frequently.

    v) Flow conversions. Liquid and gas flow rates often are measured indirectly, by the drop in pressure across a venturi or orifice. Flow rate usually is related to the square root of the pressure drop; hence the square root option is a must. Compensation for flowstream temperature, pressure, and specific gravity variations is desirable. Steam may be treated as a special case of flowing fluid because of its widespread use as a heating medium.

(3) *Integration or smoothing.* Some applications require digital integration of an input using the equation $y_n = y_{n-1} + kx_n$ where k is a user-specified constant, x is the input, y the output, and n and n − 1 refer to the present and previous sampled value, respectively. Although other types of smoothing are occasionally required, the most common is exponential smoothing using the equation $y_n = ky_{n-1} + (1-k)x_n$. The filter constant k is either specified directly or is computed from the filter time constant and the scan interval (cf. Chapter 19).

(4) *Alarms.* High, low, and trend alarms are most frequently encountered, although some applications require an additional level of high and low alarms, usually referred to as high-high and low-low alarms. In addition to a printed message, most really important alarms require the execution of special user-written programs. These alarms must also be applicable to computed variables as well as to direct process inputs.

In addition to the parameters required to specify the above functions, additional information frequently is required about the input hardware itself. Examples are signal ranges, input amplifier gains, etc. The actual structure of the software also is highly dependent upon the type of input hardware utilized.

Pulse Inputs—Devices such as turbine meters, tachometers, and totalizers output a train of pulses. Although the computer can be programmed to count the pulses, the software overhead is high. The most common approach is to use a pulse counter, thereby requiring that the computer only read in the value of the counting register in the pulse counter. If the register is cleared automatically each time it is read, then the total in the register is always the number of counts received since the last time the device was read.

From this reading, the computer can readily calculate the speed from a tachometer input and the flow from a flowmeter, such as a turbine meter or a vortex-shedding meter. For the latter, each pulse corresponds to a fixed volume of fluid. Thus, the number of counts received over a known interval of time can be converted readily to a volumetric flow.

Once the pulse input has been converted to engineering units, additional processing such as smoothing/integration and alarming may be required. These requirements are essentially identical to the corresponding requirements for analog inputs.

Digital Inputs—Devices such as limit-switches, contacts, or relays may be in only one of two possible states—open or closed. In most process situations, one of these states will indicate normal conditions, e.g., reactor pressure below an alarm condition. The other state will indicate a condition that calls either for some response by the system or an action by the operator.

On many computers, the digital input modules provide the main interface between the computer and the devices in the plant. Normally, each digital input module accommodates the same number of digital inputs as the number of bits in a computer word. For example, for a computer with a 16-bit word-length, the

ditigal input module would accommodate 16 digital inputs. The address of a digital input is the address of the digital input module followed by a bit address (e.g., 0 to 15) to designate the specific contact on the module.

Most systems scan all digital inputs at a preset frequency, typically once per second. The system normally looks for changes in the states of the contacts. When a change is detected, it indicates either an abnormal or a normal condition. For the abnormal condition, the message to be displayed to the operator and/or the program to be executed must be specified in the tables.

In many processes, certain conditions demand a response faster than one second. Digital signals relating to these conditions are connected to the computer's interrupt bus, which permits the processor to respond immediately rather than waiting until the next scan of the digital inputs. Programs to service these interrupts are generally supported directly by the operating system, but provision can be made for these programs to access the data base of the table-driven system.

Special Multiplexed Devices—For some applications, special devices provide input data from the process. The most common of such devices are chromatographs and pneumatic multiplexers.

As discussed in Chapter 4, chromatographs come in a variety of styles and configurations. When used to analyze a single liquid or gas stream, the inputs will be the compositions of all components in this stream. For process applications, the permissible input data rates do not permit peak integration techniques to be used; so peak picking is normally employed. In the latter approach the inputs must be sampled as the peaks occur. After all peaks have been observed, normalization and reasonableness tests are required before the results are accepted and stored in the process data base.

For some applications, calibration of the chromatograph is accomplished by processing a sample of known composition, either at regular intervals or upon demand. In some multistream chromatographs, this sample is entered essentially as one of the streams.

In many older plants, much if not all of the control hardware is pneumatic. When installing a process computer, three options are available:

(1) Replace the pneumatic instrumentation with electronic.
(2) Install pneumatic-to-current (P/I) and current-to-pneumatic (I/P) converters on the inputs and outputs.
(3) Install pneumatic multiplexers on the inputs and/or outputs.

Cost considerations strongly favor the latter approach, the major disadvantages being the low data rates which can be achieved and reliability questions. For many applications the low data rates are tolerable, and the reliability is acceptable for input devices. Consequently, input pneumatic multiplexers frequently are encountered; output pneumatic multiplexers occasionally are used.

Other multiplexed input devices, such as temperature indicators and tank gauging systems, are used occasionally as well, but the above devices are the most common.

Process Outputs—A process-control system could require four types of outputs: pulse trains, variable-width pulses, analog outputs, and digital outputs (contact closures).

The pulse-train output is a sequence of fixed-width, fixed-amplitude pulses used to drive stepper motors or to up-count or down-count the register in an external D/A converter. Such outputs are used in supervisory control to change the setpoint of a primary controller and in direct digital control (DDC) to interface to the valve station or other final control element (cf. Chapter 5). In this case the following information must be provided in the tables: the hardware address of the controller or valve station on the pulse-output system and the address of the computer/local switch indicator on the digital-input system. Some units require additional information, such as the address of the controller deviation indicator and/or computer "acknowledge light" for output of specific information to the process operator.

The variable-width pulse normally is used in DDC applications to drive valve motors. For outputs to these devices, the hardware address must be provided.

Analog outputs are generated by D/A (digital-to-analog) converters, and are used to drive analog devices such as trend recorders. Normally, only the output address is required.

Digital outputs are used to open or close field contacts such as relays for starting or stopping motors. Digital-output modules similar to the digital-input modules form the major interface between the computer and the process. The output address is composed of the digital-output module address, and a bit address within that module to specify a specific digital output.

16.4 CONTROL AND COMPUTATION FUNCTIONS

Depending upon the philosophy of the particular user, functions for control and computation may or may not be necessary. Especially for supervisory control, one philosophy is to use table-driven software only for process I/O, and implement all additional functions via user FORTRAN programs. For DDC, a Proportional-Integral-Derivative (PID) control function is generally useful (cf. Chapter 20). Most available systems provide a spectrum of additional functions.

The following is a rather abbreviated list of functions required for control purposes. Most systems currently available offer a more expanded complement.

PID Controller—This function is used to implement the three-mode (proportional, integral, derivative) control law commonly used in the process industries. The specifications must include the source of the setpoint and the source of the feedback variable. Generally, the setpoint may be specified by the operator or obtained from the data base, thereby permitting cascade or supervisory control. Other information required includes the sampling time and the tuning parameters (gain, integral or reset time, and derivative time). Some systems accommodate variations in the basic control law, including basing the proportional and/or derivative modes on the feedback variable instead of the error signal, using error-squared

($e|e|$) instead of error, and permitting only integral action to be used (floating control). A deviation alarm feature is also commonly included.

Add/Subtract—General equations of this type follow the form:

$$Y = \frac{k_1 X_1 \pm (k_2 X_2 + b)}{k_3}$$

where Y = output; X_1, X_2 = inputs; k_1, k_2 = scaling factors; k_3 = weighting factor; and b = bias. This equation can be used to add, to subtract, to bias a reading, to compute a weighted average, or to implement other similar functions.

Multiply/Divide—The generalized equations of this type permit multiplications and divisions according to:

$$Y = \frac{k X_1 X_2}{X_3}$$

where Y = block output; X_1, X_2, X_3 = block inputs; and k = gain.

By selectively specifying one or two of the three inputs as unity, one can perform multiplication, division, ratio control, multiplication by a constant, or taking the reciprocal. At least one commercial system combines multiplication and addition into a single equation, with exponentiation also included.

Lead/Lag—Difference equations are used to approximate the conventional lead-lag transfer function of the form:

$$Y(s) = k \left[\frac{\tau_1 s + 1}{\tau_2 s + 1} \right] X(s)$$

where Y = output; X = input; k = gain; τ_1 = lead time constant; and τ_2 = lag time constant. In addition to k, τ_1, and τ_2, the sample interval must be specified.

Dead Time—The delaying of signals—primarily for dynamic compensation in feedforward control systems, or for use in deadtime compensa-

tion algorithms—is described by the equations:

$$Y(t) = kX(t - \theta)$$

or

$$Y(s) = ke^{-\theta s} X(s)$$

in transfer function form where Y = output; X = input; k = gain; and θ = delay time. In addition, the sample interval must be specified.

Comparator—This function compares two inputs and sets the output according to the equation:

$$Y = \begin{cases} b_1 \text{ if } X_1 > X_2 \\ b_2 \text{ if } X_1 < X_2 \end{cases}$$

where Y = output; X_1, X_2 = inputs; and b_1, b_2 = output values. To prevent excessive chatter, i.e., output oscillation when the two inputs are nearly equal, a deadband is normally incorporated. This function can be used either to emulate the conventional on–off controller, or as a single-stage cutoff in batch-sequencing operations.

High/Low Select—For this function, the output is simply the larger (or smaller) of the two inputs.

Setpoint Ramp—The ramping of setpoint changes, at a predetermined rate, is frequently required in batch control.

Switch—This function is used to select between two inputs (or two outputs), depending upon the status of a two-state device such as an alarm input line.

16.5 FILL-IN-THE-FORMS SYSTEMS

In the prior discussion of table-driven software, very little attention was devoted to how the data are entered into the tables. One approach would be to prepare DATA statements for FORTRAN or BASIC programs, and in fact at least one commercially available system uses this method. Although the simplest to imple-

Block Number	1)	_ _ _ _ _
Scan Period (1-32767 seconds)	2)	_ _ _ _ _
Block Type	3)	<u>S C A N</u>
Is the input from another block? (Y or N) Is the input from a supervisory program (measurement supercascade)? (Y or N) Is the input from the Pulse Counter Module (PCM)? (Y or N) Only one Y is permitted. If all N's, input is from the Analog Input Module.	4)	_ _ _

If the input is from another block, enter the block number. or	5)	_ _ _ _ _
If the input is from the PCM, enter the register number. (0-59) or	5)	_ _

If the input is from the Analog Input Module, enter the following:
    Analog Input Module Type (1=contact, 2=fixed gain, 3=programmable gain,
                         4=Interspec)
	5)	_

If type 1, enter the following information:
    Multiplexer address. (0-1023)
    Gain code. (0=1V, 2=50MV, 3=10MV)
	5)	_ _ _ _
		_

If type 2, enter the following information:
    Nest address. (0-15)
    Card address. (0-13)
    Point address. (0-7)
    Gain code. (0=X_1, 1=X_2, 2=X_3, 3=X_4)
 	
      (Note: X_1-X_4 are defined at SYSGEN time)
	5)	_ _
		_ _
		_

If type 3, enter the following information:
    Nest address. (0-15)
    Card address. (0-13)
    Point address. (0-7)
    Gain code. (3=1V, 4=500MV, 5=200MV, 6=100MV, 7=50MV, 8=20MV,
              9=10MV)
    Bandwidth. (0=1KH, 1=3KH, 2=10KH, 3=100KH)
	5)	_ _
		_ _
		_
		_

If type 4, enter the following information:
    ISCM number. (1-3)
    CCM number. (1-16)
    Type of input. (M=Measurement, S=Setpoint, 0=Output)
    Point number. (1-16)
	5)	_
		_ _
		_
		_ _

Range of the input in engineering units:
    Lowest value. (-32767. to +32767.)
    Highest value. (-32767. to +99999.)
    Units. (As specified by user at System Generation.)
	6)	_ _ _ _ _ _ _
		_ _ _ _ _ _ _
		_ _ _ _

Signal conditioning index. (0-7) Thermocouple type if thermocouple input is through the Analog Input Module. (J, K, T, R) [Otherwise enter N.] Linearization polynomial index. (0-511) [For signal conditioning indexes 0 or 5 only Enter 5 only if input is from PCM.]	7)	_ _ _ _ _

Is digital integration required? (Y or N)
        If Y, enter integration multiplier K1, (1-32767)
        and integration divisor K2. (1-32767)
        If N, enter the smoothing index. (0-63)
	8)	_
		_ _ _ _ _
		_ _

Operator Console Number (1, 2, or 3)	11)	_
Process unit number (1-127;0 = none)	12)	_ _ _
Block description for alarm messages. Leading and imbedded blanks will be included.	13)	_ _
Is a supervisory program called when an alarm occurs? (Y or N) If Y, enter program call number. (0-2047)	15)	_ _ _ _ _
Should this block inhibit the passing of initialization requests? (Y or N)	18)	_
If an input fails, should this block continue control using the last good value? (Y or N)	19)	_

Figure 16.4. Example of a Data Specification Form (*Courtesy The Foxboro Company*)

ment, this procedure is somewhat tedious, and most systems have adopted other methods.

A more common approach is to require the applications engineer to enter the data first onto forms, such as the one illustrated in Figure 16.4. These forms are virtually self-explanatory; furthermore the language (terminology) utilized is closer to that of the applications engineer as compared to a computer analyst. In fact, many of the commercial software or system vendors claim that the forms can be completed by an individual with little or no background in computing.

Although not very obvious from the form in Figure 16.4, the arrangement of data on the form is such that it can be punched or otherwise entered into the system readily. These data are processed by a program that often is called the "forms processor" whose primary functions are to perform certain reasonableness checks and then store the data into the proper locations in the tables. The role of the forms processor in the overall system structure is illustrated in Figure 16.5.

Although all early fill-in-the-forms systems incorporated forms processors that operated in the batch mode, forms processors with varying degrees of interactive capabilities have appeared, and future systems probably will utilize graphics terminals. However, exclusive use of a CRT terminal as the I/O medium has the serious drawback that no hard-copy record is made of

changes to the data base. A record of what changes are made and when they are made is indispensable.

Hence an important side aspect of fill-in-the-forms systems is the degree to which they can be self-documenting. A major drawback of writing custom programs is that they must be documented somehow. Programmers do not enjoy documenting their work, and it is tempting to give the documentation effort a low priority. Chapter 18 deals in part with the general documentation problem.

16.6 OTHER IMPLEMENTATIONS OF TABLE-DRIVEN APPLICATIONS SOFTWARE

The concept of table-driven software for data acquisition and control applications is not limited to the types of systems described above. The use of a table-driven structure generally is worth considering for any user application where generality and flexibility are more important than processing efficiency.

For example, another interesting commercial use of the table-driven concept is in simple process control languages, most of which employ variations of the BASIC language. As mentioned in Chapter 14, BASIC more often is implemented as an interpreter than as a compiler. In the strictest sense an interpreter should interpret the source code directly. How-

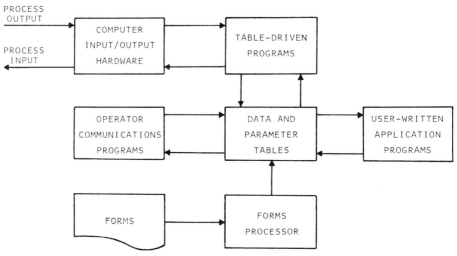

Figure 16.5. Role of the Forms Processor

ever, there are certain disadvantages to this approach:

(1) The overhead, i.e., the time required for interpreting the source code, is high.
(2) Storing the source code directly (even with unnecessary space characters deleted) consumes considerable memory.

These and other factors favor a compiler implementation, but this approach results in several undesirable features such as poor execution-time error messages.

A compromise solution to this problem is to do a partial translation, but not all the way to machine instructions as would be done by a compiler. The translator reads the source program and generates two outputs:

(1) The symbol table which is required by any compiler or interpreter.
(2) A pseudocode which specifies the actual computation or operation as called for by the source program.

Both of these must be stored for use by other parts of the system. Figure 16.6 illustrates the overall structure of the implementation.

To permit the user to edit his source program, it is necessary to be able to "inverse translate" the program. That is, the source code must be generated from the pseudocode and the symbol table. Normally some arbitrary policy for inserting space characters is assumed, and thus the inverse translation may not duplicate the source code exactly. However, the pseudocode often does contain information such as the presence of parentheses (which, by way of example, would be necessary only for the inverse translation operation).

The actual execution of the user program is carried out through the interpreter. Frequently this program element is little more than a collection of subroutine calls which corresponds to the "op-codes" comprising the pseudocode. The pseudocode, stored in table format, may be nothing more than a very high-level assembly language, i.e., a set of macroinstructions. Although not shown explicitly in Figure 16.6 for reasons of simplicity, the interpreter can utilize the symbol table to generate meaningful execution-time error messages.

This approach is especially attractive for computer configurations that do not have bulk storage. The translator, interpreter, and "inverse" translator can be relatively small, requiring as little as 10K words of memory. To keep the translator and "inverse" translator small, the source language must have a relatively simple syntax. BASIC and other special process control languages derived from BASIC meet these requirements. FORTRAN also can be translated into a pseudocode, but the size of the translator and "inverse" translator are such that mass storage support is necessary. Hence, to date these languages generally have been used only for relatively small applications, those typically without bulk storage. For

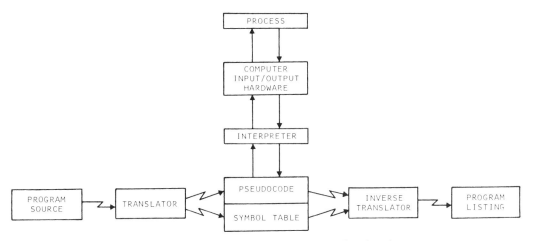

Figure 16.6. Implementation of BASIC Using Pseudocode

larger applications, the power of FORTRAN and user familiarity with the language usually outweigh the disadvantages.

16.7 SUMMARY

In this chapter we have discussed the structured or regular nature of most data acquisition and control problems, a feature that permits the efficient description of most such applications in the form of an array of data, i.e., in table format. Use of such a table—for specification of the information needed to acquire operating data from the process, to manipulate and store-out these results, and, perhaps, to utilize them to control the process as well—provides a natural application for a table-driven program. This program is structured to interpret the contents of the table, performing operations as specified by the parameters there.

Such an approach offers a number of advantages; in particular the user is required to write very little software of a specific nature, and there is relatively little documentation required other than to maintain records of the table contents as they are modified. As a result of these advantages the use of table-structured software for data acquisition and control applications has become quite widespread. Particularly for modest systems without bulk storage, these features are important. We might expect that, in the future, dedicated mini- and microcomputer systems will exploit this method of structuring their applications software or firmware routinely.

16.8 SUPPLEMENTARY READING

Smith, C. L., "Fill-in-the-Forms Computer Languages for Process Control", *Chem. Eng.*, p. 151, Mar. 1975.

Part VII
Management of Real-Time Computing Facilities

17

System Justification, Selection, and Installation

Cecil L. Smith
Cecil L. Smith, Inc.
Baton Rouge, Louisiana

17.0 INTRODUCTION

The rapidity with which developments in the computer industry have occurred has always made the justification, selection, and installation of computer systems a trying endeavor. This situation has been most evident with small-computer systems, since the typical manufacturer introduces a new (perhaps "improved" is a more appropriate word) model about every 18 months. Therefore, the person who cannot live with the fact that his computer will be "obsolete" before the installation is completed should find another occupation.

In this chapter, attention initially will be given to the various roles computers can assume in the process area, in research laboratories, and in teaching. The technical and management decisions involved in the selection and installation of computer systems are discussed subsequently.

17.1 REAL-TIME APPLICATIONS AREAS

17.1.1 Industrial Applications

Data Loggers. Perhaps the simplest function that computers can perform in the industrial area is that of data logging. As illustrated in Figure 17.1, the computer does not interact with the process in any way, but only collects data from it. In that respect, the computer's function is strictly a passive one.

With regard to production units, the usual arguments to incorporate a data logger go something like this: "From these records of the pertinent process variables, the optimum operating strategy will be obvious." Unfortunately, process units are not so simple. Furthermore, much of the data base obtained by automatic logging is redundant or uninformative, being taken at or near the normal process operating conditions.

Data loggers can serve a useful purpose in a process modeling study, but their role must clearly be supportive in nature. Modeling efforts invariably require process data, which in turn entail the careful design of process tests or experiments. Such tests do not just happen to occur during normal process operation. Furthermore, to obtain the necessary data, the normal plant instrumentation must be augmented by special instruments or by collecting samples which will be analyzed in the laboratory (off-line) at a later time.

In regard to production units, Wherry and Parsons [1] make the direct statement that data loggers cannot be justified; in fact, they even question the addition of data-logging functions to other types of computer systems. There are, of course, certain exceptions, but these usually arise from management considerations as opposed to technical considerations. For example, data loggers are frequently installed in nuclear power plants to take the data required by regulatory agencies. The Food and Drug Administration requires that drug manufacturers keep certain records for a number of years, and a data logger is one approach for insuring that these requirements are met.

In all of these considerations, the important question is, "Who will do what with the data

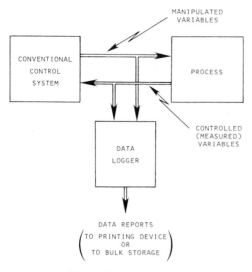

MANIPULATED
VARIABLES

CONVENTIONAL
CONTROL
SYSTEM

PROCESS

CONTROLLED
(MEASURED)
VARIABLES

DATA
LOGGER

DATA REPORTS
(TO PRINTING DEVICE
OR
TO BULK STORAGE)

Figure 17.1 Data Logger

produced?" By and large, production logs, whether taken by the operator or produced by the computer, are not used very much, if at all. The computer can be used to calculate parameters such as production rates and yields, but such data may or may not be useful. For example, if a raw material in limited supply is to be distributed between several operating plants, yield data are necessary to make this allocation. However, in most plants, the most important question is not what the yield is but instead how the yield can be improved. Simply calculating the yield does not inherently lead to yield improvements.

Data loggers are deceptively expensive in terms of machine resources required. Especially with production units, the large number of variables involved entails a large amount of disk or other bulk peripheral storage. Furthermore, the formatting of the reports involves large programs, although a number of manufacturers offer software packages that support the writing of reports. In general, computer systems for data logging in industrial units will have a large disk unit and a line printer. Occasionally a magnetic tape unit will also be included.

Direct Digital Control Systems. In existing plants, the responsibility for the primary, first-level control or regulation of the process traditionally has rested with conventional analog

control systems. In a new plant, such a control system is usually a possible alternative although there are situations (such as a batch process with multiple reactors, each capable of producing a variety of products) in which the applicability of the conventional system is questionable.

At least in the petroleum and petrochemical industry, the backbone of the conventional control system is the simple three-mode analog controller, whose mathematical algorithm is given below.

Three-Mode (Proportional-Integral-Derivative) Control Law (Ideal Form):

$$m(t) - M_R = K_c \left[e + \frac{1}{T_i} \int_0^t e(\tau)d\tau + T_d \frac{de}{dt} \right]$$

where K_c = proportional gain (1/proportional band)

T_i = integral (reset) time

T_d = derivative time

M_R = controller output value when loop is placed in automatic mode

The components of a typical control loop are illustrated in Figure 17.2. The question is, "Can a computer be used to generate a larger economic return?" Two possibilities exist:

(1) The computer might perform the basic control action better or cheaper than the analog controllers.
(2) The digital computer might be used to determine better setpoints for the analog controllers than are normally provided by the process operator.

The use of a computer in the first role is called direct digital control (DDC), and is the subject of this section. The use of a computer in the second role is normally called supervisory control, and will be discussed in the next section. Although industrial control systems are almost invariably a mixture of the two, they usually seem to be primarily one or the other.

Experience to date has been that a single digital computer, almost without exception, will be more expensive than a conventional control system. Early studies attempted to estimate

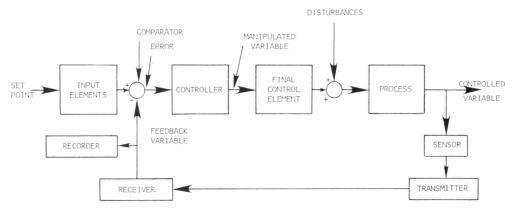

Figure 17.2. General Feedback Loop

the number of analog loops that had to be replaced in order to save enough money to pay for the computer system. Today, even with low-cost minicomputers, there does not appear to be a reasonable break-even point for a single-computer system. (With the introduction of microcomputer-based controllers which can be substituted for individual analog controllers, an economic incentive may, however, begin to appear.)

Furthermore, use of a computer system for DDC generally does not reduce personnel costs; if anything, personnel costs increase. In plants such as refineries, paper mills, etc., the number of operations personnel required on each shift is determined by considerations other than control, and the need for personnel primarily associated with the computer system simply adds to these requirements.

In effect, this says that the computer system will not be cheaper than the conventional control system. Therefore, the key is better control system performance. Unfortunately, it turns out that using a digital computer to perform the same functions as performed by analog hardware does not significantly enhance control performance, if at all [2].

Such industrial experience is typified by the system reported by Parsons et al. [3], a 120-loop DDC system in an ethylene plant, essentially representing a digital counterpart to an analog system. Although the digital system was a technical success, the economic picture was not so encouraging.

In analyzing reports such as the above, it must

be realized that the level of automation in the petroleum and petrochemical industry is relatively high. Furthermore, an ethylene plant utilizes a continuous process and as such is ideally suited for the conventional three-mode controller. In industries not so highly automated, with processes for which considerable switching or other logic is required, or with processes requiring more complex algorithms, the results would be very different.

In the petroleum and petrochemical industry, the current approach to DDC is a "hybrid" one. In essence, those loops for which the performance of the conventional control system is completely adequate are not considered; those units for which the conventional control system's performance is inadequate are considered as the most logical candidates for DDC. Using this approach, minicomputer-based systems are quite attractive.

In smaller DDC systems, the computer is usually a memory-only machine, having no disk or drum. The larger systems will include a disk or drum, with the latter being preferred because of faster access and greater reliability.*

Supervisory Control Systems. One of the difficulties in justifying DDC systems is that no economic return accrues directly from improved first-level regulation. The economic return results indirectly, from operating the process in a

---

*In this context, a fixed-head disk is considered functionally equivalent to a drum.

more profitable way, but this in turn requires good first-level control.

To illustrate, suppose a product purity specification is 98%. Suppose the purity variation under normal control is observed to be ±0.5%. Then the setpoint for the control system must be 98.5%. In essence, on the average the product is 0.5% purer than it has to be, and this excess purity results in what is known as "product give-away".

Now suppose that the control system is improved so that the purity variation is reduce to ±0.25%. If the purity setpoint is left at 98.5%, no economic benefits result. However, specifications can now be met at all times with a purity setpoint of 98.25%. This change reduces the product giveaway from 0.5% to 0.25%, yielding decreased costs and increased return. In fact, production is increased by 0.25/98.25 or about 0.25% which, depending on the scale of the process, can translate into large savings.

The basic objective of supervisory systems is to maintain the plant at or nearer to the optimum operating conditions. In many situations, this is accomplished by adjusting the setpoints to conventional analog controllers as illustrated in Figure 17.3. In other situations, this may involve scheduling critical process elements, turning equipment on and off, etc.

In this respect, three categories of plants can be identified as potential candidates for supervisory control:

(1) Plants with large throughputs, where a small percentage improvement translates into a large monetary return.

(2) Very complex plants, where the operations personnel are unable to make correct decisions consistently.

(3) Plants subject to frequent changes in raw materials, economic market position, or the like. In such plants, the operations personnel may not respond quickly enough.

Of these, the large-throughput process is usually the easiest to undertake.

The attractiveness of the large-throughput process applies both to supervisory systems and to DDC systems. For example [4], newer polyvinyl chloride plants utilize batch processes with up to 20 reactors, each having a capacity of 50,000 gallons. Assuming a 14-hour batch cycle time, 14¢/lb monomer, and 26¢/lb product, each *minute* trimmed from the batch cycle results in $125,000 in increased production over a one-year period. These results assume, of course, that the additional PVC can be sold.

Most supervisory systems entail a spectrum of programs that must be run on a periodic basis, but less frequently than DDC control calculations. The configuration almost invariably includes a disk. On-line program development is normally required, and cards or another alterna-

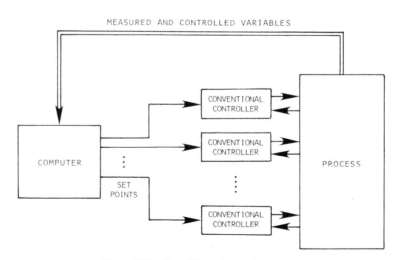

Figure 17.3. Closed-Loop Supervisory Control

tive to paper tape is often selected as the program source medium. Memory requirements usually run around 64K bytes.

17.1.2 Research Applications

By 1965, a number of companies had begun to experiment with the on-line use of computers in their research laboratories. One of the early applications was in processing data from complex instruments such as gas chromatographs. By replacing the hardwired programmers used previously by a computer, both the capability and the flexibility of the instrument were improved. The more sophisticated systems use peak integration instead of simply using peak heights to calculate compositions, and they also incorporate various peak gating techniques and other checks to insure the accuracy of the data. These techniques, often utilizing microcomputers, have been incorporated into versions of the chromatograph that now are finding use in many laboratories and plants [5].

Because of cost considerations, early applications were limited primarily to the more expensive instruments. But with the introduction of the low-cost microprocessor, the same approach can be used for relatively inexpensive sensors, especially if the number of applications is large. By providing capabilities such as self-calibration, the accuracy of results is improved, and maintenance labor is reduced.

Within research laboratories, the computer now is widely used to collect data generated by the experiments. In such efforts, a primary concern is the quality of the data. One way to screen process data is to require that material balances "close" within a specified tolerance or to use other consistency checks. Without on-line data acquisition, material balance calculations cannot be performed until samples are analyzed off-line, which is usually at least a few hours or, perhaps, several days later. If the material balance does not close, the experiment must be repeated. By performing this calculation on-line, the time required to set up the experiment for the repeat run can be saved.

The major cost in an on-line system is rarely the computer, but instead is associated with the sensors. A further complication in research operations is that material balance closure problems may be due to sensor inaccuracies. Installation of two sensors is not a satisfactory solution, since a disagreement does not indicate which is in error. By installing three sensors, the two that agree can be accepted, especially if a material balance closure is also obtained. Using this admittedly expensive approach, one company* reported an increased productivity from its research laboratories—from one run a week to four runs a day [6].

17.1.3 Instructional Applications

The low cost of computing afforded by the minicomputer has made a significant impact on the educational process. The lower-cost microprocessor promises to have an even greater impact. For example, present-day programmable calculators literally enable students to carry computers in their pockets. Coupled with higher-density and lower-cost memories, the power of these devices will increase even more. Although most are off-line devices, some can be interfaced to process sensors.

A major potential application of computers is in undergraduate laboratories, such as for analytical chemistry, unit operations, machine, and electronics laboratories. The objective is to provide the student with more data of higher quality than previously obtainable. Experiments that have been connected to computers include transport processes, analytical instruments, communications and control equipment, and larger experimental units such as distillation columns, evaporators, internal combustion engines, and the like.

Another potential area with broad possibilities for instruction is interactive graphics. Here the computer interacts directly with the student to illustrate the effect of design changes on plant or process characteristics. Substantial cost reductions in graphic video display devices are expected. Coupled with the steadily decreasing costs of computing power, graphics facilities are now coming within the purchasing

---

*Celanese Chemical Company Technical Center, Corpus Christi.

capabilities of a number of educational institutions.

17.2 SYSTEM SELECTION

In practice, system selection involves a rather complex set of procedures that vary significantly from case to case.

First, it should be determined how much application expertise is being purchased. At one extreme, the purchaser could specify the system solely in terms of process performance. In order to meet these requirements, the supplier must possess a high level of process know-how. In evaluating quotations, the purchaser must devote primary attention to the level of the bidder's process expertise, and the computer system's technical characteristics are secondary.

Although many companies would prefer to take this approach on all of their computer systems, most processes are unique to the point that no supplier has the expertise to quote. In such cases the purchaser must supply the know-how. Frequently the development of a process control strategy will require a fair amount of field experimentation with various control configurations. In this case, the purchaser must consider the computer system characteristics more closely, paying careful attention to aspects such as the facilities for on-line program development.

With regard to system selection, numerous summaries of the systems offered by various manufacturers of minicomputers have appeared [7]. For process control, to this list must be added systems offered by the traditional instrument manufacturers and system suppliers.

In selecting a system, several factors must be considered. These can be divided into four categories: computer hardware, process interface, computer software, and management aspects. Each of these will be considered in subsequent sections.

By and large, the small-computer systems offered by various manufacturers differ in performance only in relatively small degrees. Furthermore, costs are comparable and relatively low in comparison to other costs. Therefore, the expenditure of a large amount of time and effort comparing the various systems cannot be justified.

17.2.1 Hardware Considerations

In the ensuing discussion, application considerations relevant to various hardware features are briefly stated. Fundamental information concerning computer systems hardware is contained in Part IV.

Maximum Memory Size. Although minicomputers were once limited to 64K bytes or 64K words, most are now expandable to much larger memory sizes. If the cost of memory continues to decrease as expected, users will tend to have more of their software resident in memory as opposed to on disk. Thus, the trend in future systems will be to larger memory sizes.

Memory Increment. Minicomputers traditionally allow memory to be expanded in 8K increments of 16-bit words or 16K bytes. As the cost of memory continues to decrease, the minimum memory expansion is becoming 32K of 16-bit words or 64K bytes. Since the cost of memory is relatively low, having to purchase in increments this large is not disadvantageous.

Word Size (Memory). The effect of word length on effective machine speed is quite complex. However, most minicomputers have a 16-bit word length. An 18-bit word length offers little, if anything, over a 16-bit word length. A 24-bit word length machine offers substantially more capability, but the cost is also substantially higher. Users will find the 16-bit machine sufficiently powerful for all but the largest process applications, for which a 24-bit or 32-bit processor should be considered.

Cycle Time. Most minicomputers have memory cycle times below 1 microsecond.

Memory Parity. This feature is standard on some minicomputers, optional on some, and unavailable on others. There is no general agreement on the need for memory parity in process control systems. One school of thought says that any feature that would detect any error in a process control system should be included. The counter argument notes that upon detection of a parity error, the only action that can be taken is to stop the machine. It is further noted that with semiconductor memories, a single-bit error in a single word of memory is extremely unlikely, but instead the single-bit error will occur in every word within 1K of memory. In such cases, the machine will soon halt on an illegal instruction, illegal address, or other trap.

In view of these characteristics, it can be stated safely that parity has no clear-cut advantage in process control computers. Therefore, a user should not eliminate a machine from consideration simply because it does not have parity. However, since parity as an option is relatively inexpensive, he may find it more comforting to purchase it, if it is available.

Error Correcting Codes. This feature enables any single-bit memory error and most multiple-bit errors to be corrected. To accomplish this, 7 bits must be added to the 16-bit data word giving a total of 23 bits. In error-free situations, the system runs at full speed. If an error is detected, there is a degradation in memory speed due to the time required to correct the error. However, this feature permits the system to continue to function at least in some reduced-capacity state. For extremely critical process services, such a feature is attractive.

Memory Protect. In dedicated, all-memory systems, this feature is unnecessary. In systems in which on-line software development is contemplated, this feature is highly recommended.

Core or Semiconductor Memory. Most current minicomputers support either core or semiconductor memory. Semiconductor memory is presently faster and considerably cheaper than core, and the prediction is that future memories will be even faster and cheaper. No similar advances in core memories are contemplated. The volatility of semiconductor memory has been a cause for concern in process computer systems. However, a battery back-up feature for the memory is relatively inexpensive and has proved to be sufficiently reliable.

Instruction Length. Most small computers have a fixed instruction length of 16 bits or one memory word length. Some have a 32-bit instruction length, and a few have a variable instruction length. Machines with a long instruction length make up for the extra memory used by offering a more powerful instruction set. With current memory prices, the long-instruction-length machines are quite acceptable.

Addressing Modes. Small computers offer indirect, indexed, and relative addressing. Some offer only one mode, usually either indirect or indexed. Only one type of address modification is necessary. Multilevel indirect addressing is also unnecessary. If index registers are offered, hardware registers are preferred as opposed to memory locations.

Bit Manipulation. In process environments, basic hardware bit manipulation instructions are desirable but not absolutely necessary, since these operations can always be accomplished by logical instructions coupled with masks. In this respect, it should be noted that bit manipulating capability also is incorporated into high-level languages such as FORTRAN, frequently via subroutine calls. Unfortunately this approach eliminates most of the efficiency of bit manipulation instructions that function directly.

Byte Manipulation. This feature is unnecessary for process control computers unless a considerable amount of data reporting is anticipated.

Registers. A machine with multiple registers usable as accumulators is desirable.

Hardware Integer Arithmetic. The trend is for this feature to be standard. If not, it should be purchased for process control computers.

Floating-Point Hardware. This feature is not offered on some minicomputers, and it is optional on others. It is a must on very large process control computers, particularly supervisory systems, and is desirable on smaller systems.

Direct Memory Access Channel. If the system includes a disk, at least one direct memory access channel is required. If a free-running front end is used for data acquisition, it is desirable that it be attached through a direct memory access channel.

External Interrupts. Although external interrupts are an essential element in a process control computer system, the response speed requirements are relatively low. Therefore, hardware vectoring is unnecessary, the main advantage being simpler system software. If hardware vectoring is not used, interrupt response times must be estimated including the overhead of the associated software polling routines.

Power Fail-Safe. This feature is desirable on process control computers except when an uninterruptible power source is used, in which case it is unnecessary.

Automatic Restart. This feature is unnecessary when an uninterruptible power supply is used. Otherwise, it may or may not be desirable; depending on the system and the types of power interruptions experienced. For momentary losses it can effectively maintain processor performance unimpaired. However, a timer with a back-up power source is necessary so that the duration of power outage can be ascertained. The restart procedure for long power outages will be quite different than for momentary power losses.

Real-Time Clock. This is an integral feature of process control systems.

Console Printers. For industrial control systems, a heavy-duty printer is worth the extra cost. For air-conditioned environments, electrostatic or other quiet terminals operating at 30 cps are desirable.

CRT. In most industrial control systems, the CRT is being used for operator communications. Graphics capability and color displays both increase the appeal of the system to the process operator.

Paper Tape. Although paper tape has been almost the standard of the process control com-

puter industry, the trend is away from paper tape to cassette tapes or floppy disks.

Cards. A card reader is relatively inexpensive, but the additional cost of a keypunch machine should not be ignored. On-line card punches are generally too expensive, although some systems offer an interface to a keypunch machine for low-volume card punching.

Disk and Drum. The higher reliability of the drum or fixed-head disk makes it the choice for most industrial applications. In larger systems, a fixed-head and a movable-head disk are included, with the critical software routines on the fixed-head disk and the less critical on the movable-head disk. In laboratory applications, the movable-head disk is more attractive, especially if the disk cartridge is removable.

Cassette Tapes. Although a potential competitor to cards, the sequential nature of a tape and the relatively low transport speeds reduce the attractiveness of cassette tapes.

Floppy Disks. The floppy disk is rapidly replacing cards, cassette tapes, and paper tape as the primary medium for program storage. The devices are random-access and must be supported by software for cataloging entries on the disk, for packing the disk, and for editing programs.

17.2.2 Process Interface Considerations

The process/computer interface is one of the most important elements in any computer installation. Failure to structure the interface so as to input process data and to output control data to the process efficiently and accurately can lead to a considerable degradation in system performance. Chapter 10 furnishes detailed information on the components of the process interface; here we merely note the usual interface elements, and comment on their general application.

Analog Input System. Industrial systems require large numbers of analog inputs, sometimes reaching four thousand or more. The conversion accuracy of the A/D converter is typically 11 or 12 bits, with the latter being favored. Especially for laboratory applications, a 14-bit A/D converter is desirable. In fact, owing to the relatively low additional cost, it should probably be used more often in industrial systems.

Analog Output. Since most field equipment is discrete or incremental in nature, analog outputs are not used to a large extent in most process control systems. They are needed in order to record any computer-generated data on a trend recorder. Therefore, the number required is relatively small (usually six or less), and individual D/A converters are used. In laboratory and instructional systems, analog outputs generally are used to a greater extent.

Discrete Input/Output. These types of I/O signals are extensively used in process computer systems. Discrete inputs are widely used in manufacturing operations and to indicate variable "out-of-limits" conditions in process applications. Discrete outputs also are widely used in manufacturing, particularly sequential, operations and to operate a broad variety of binary and digital displays.

Pulse Inputs. Although used sparingly in process computer systems, they are attractive in applications such as speed control, flow control with a turbine meter, and the like. The accuracy of input data obtained via pulse counters is significantly better than with analog input methods for such data.

Pulse Outputs. Since most field equipment is incremental, pulse outputs form the backbone of most process computer systems. Pulse outputs are normally multiplexed to several output devices.

17.2.3 Software Considerations

As with most computer systems, software is much more difficult to evaluate than is hardware. However, software is at least as important if not more important than the hardware. As the quality of system software increases, the amount of effort and thus the expense involved in writing application software decreases.

Assemblers. A basic assembler is available for essentially all small computers, and for most minicomputers a macro-assembler is available. For most process computer systems, the bulk of the applications software is written using languages other than assembly. However, a macro-assembler is quite useful whenever any programming must be done in assembly language, for example, in writing device drivers for user-installed peripherals. Chapter 11 deals with assembly language programming.

Problem-Oriented Languages. Most process control computers support a language such as FORTRAN or BASIC. Such a language significantly reduces software development time and expense, and is very desirable in all but the smallest systems. The features supported by the language should be checked, as well as how real-time functions are implemented.

Process Control Languages. A number of manufacturers offer languages specifically developed for process control. Use of these will reduce the programming effort even below that which would be required using FORTRAN. In fact, many of them resemble FORTRAN or BASIC but contain numerous extensions to permit real-time functions to be obtained quite easily. ISA standard FORTRAN described in Chapter 15 represents the best effort to date to adapt a problem-oriented language to process control applications.

Fill-in-the-Forms Systems. The advantage with this approach for application software is that the user need not be intimately familiar with computer software; in fact, he need not even know how to program. To support these systems, the machine configuration must usually be expanded to include a disk and 64K or more words of memory. Chapter 16 gives a description of this approach to application software.

Real-Time Operating System. This software is a must for process applications. In fact, a memory-only version should be available for the smaller (nonbulk) control computers, and a disk-based version available for the larger systems. Chapter 13 covers the important features of operating systems.

On-Line Program Development. This capability is a must in larger systems. If capability does not exist to modify a program being debugged without taking all of the other control loops off control, then the entire computer system must be taken off-line in order to make changes. This drawback results in poor initial acceptance by the operations personnel, and should be avoided.

File Manager. On a disk-based system, a powerful file manager is essential. This file manager must support operations such as file creation, file deletion, copying a file, appending one file to another, etc. The file manager must maintain the file directory, and be responsible for all associated operations. A powerful source editor is also desirable if program development is contemplated.

17.2.4 Management Considerations

For many computer systems, the major factors in machine selection are in the management area. Below are a few caveats and points to be considered.

Don't Buy Serial Number 1. The purchaser of one of the first units of a new model can expect to help the manufacture locate and diagnose the inevitable system problems.

Avoid Minimal Configurations. The additional programming costs associated with such systems will more than offset any additional hardware costs.

Supplier's Financial Stability. The basic question is, "Will they be in the computer business next year?" If not, maintenance and spare parts may be hard to come by.

How Responsive is the Manufacturer? Unfortunately, the answer to this question usually depends primarily upon the local office, and secondarily upon the individuals involved. The prospective buyer can ask for the names of other customers in the area; much can be learned by contacting these people. Before purchasing the system, the prospective buyer can also devise a technical question about the system, pose it to the local representatives, and see how long it takes them to respond. Their response is unlikely to be any better after the order is placed.

Are there Other Systems of this Type at the Plant, Laboratory, or University? If so, then there exists some in-house know-how for this system. In-house programmers knowledgeable on this particular system may be available. Certain additional benefits can accrue with respect to maintenance.

Maintenance. This topic must be considered during machine selection, and not delayed until after the order has been placed. Three options are generally available: maintenance contract, on-call service, or in-house.

Superficially, the simplest approach for a user is to enter into a *maintenance contract* with the vendor. For a fixed fee, the vendor generally supplies all parts and labor. However, there are several questions that must be asked, especially if system availability is critical.

(1) Hours of Coverage: Normal maintenance contracts are for 8 hours a day, 5 days a week. Coverage on a 24-hour, 7 days a week basis is available, but for an extra charge.

(2) Local Spare Parts Inventory: To support the maintenance effort, the vendor maintains certain parts at the local office. An inquiry should be made as to where these parts are kept and what is kept. Some vendors require the user to purchase a spare parts inventory that is to be maintained on-site.

(3) Response Time. This is the time between when the vendor is called and when the serviceman begins to work on the machine. Inquire beforehand as to what can be expected.

(4) Back-up: Inquire about how long the

local personnel have to repair the machine before more experienced personnel are called in; ask on what basis these decisions are made.

On-call maintenance service implies that the manufacturer's servicemen are called out only when needed. They work on an hourly rate basis plus a charge for spare parts. Especially when some in-house capability exists to repair the simpler failures, this is likely to be less expensive than contract service. However, the vendor gives contract maintenance priority over on-call maintenance, which may be a problem when system availability is critical.

In-house maintenance is especially attractive when the user has a good in-house repair organization, when the user has several systems of the same type, or when the user's site is in a remote location. Most suppliers will recommend a spare parts inventory, but the user should inquire as to how he can obtain parts on an emergency basis. Many vendors do not permit access to local spare parts. While normal supplies can be ordered from the supplier's home office, a more responsive arrangement is needed to obtain emergency parts.

17.3 INSTALLATION

Depending upon the nature of the system and the environment in which it is to function, the installation process varies significantly.

Larger systems destined for a production environment are usually assembled at the factory and pretested before being shipped to the customer. During this period, the buyer can arrange for his people to attend training sessions conducted by the manufacturer, and even to work with the manufacturer's personnel who are assembling the system. A preshipment test is usually made with the customer's personnel in attendance.

At the installation site, several steps should be taken. Although most mini- and microcomputer CPUs will function in non-air-conditioned environments, many of the associated peripherals will not. Furthermore, the people who work with the system prefer air-conditioned space.

Thus, most sites require air conditioning. Heat load data are available from the manufacturer.

Power requirements are another consideration. Most minicomputers use relatively little power; however, it generally is not reasonable to expect that they can be plugged into an existing wall outlet. The supplier will advise as to the power requirements for their equipment. A good earth ground is usually a wise investment.

Before receipt of the system, the user should obtain adequate supplies, such as cards, paper tape, line-printer paper, etc. Storage cabinets for the various materials used in and around the computer site should be obtained.

For large systems, a raised floor is also a good investment. Electrical power is supplied beneath the floor, and the cables between units are run beneath the floor. It is also feasible to have cool air distribution from beneath the floor. This makes for very attractive and also safer working conditions because of the elimination of cables lying on the floor or strung overhead.

After the system is delivered, the manufacturer's representatives connect the various components and run their diagnostic tests. If the customer wishes to makes acceptance conditional upon certain performance criteria, these must be part of the original purchase agreement. With larger systems, the conditions for acceptance are usually negotiated between the customer and the manufacturer. With smaller systems the purchaser must be sure to determine, in advance, precisely what the manufacturer's responsibilities are. Some computer manufacturers simply ship components to the purchaser who is then expected to assemble the units, perhaps to perform some field wiring in the process, and to check out the total system to determine if it meets specifications. In the event that system performance is not satisfactory, or if a component failure subsequently occurs, it may well be difficult for the purchaser to obtain the necessary technical assistance from the manufacturer.

A further potentially troublesome and yet avoidable circumstance can occur whenever a user who is operating without a maintenance contract later expands an existing system. Computer manufacturers often make modifications to later versions of their equipment—for example, to memory units for a particular CPU—in such a way that they will not operate with earlier production units unless similar modifications are made. Sometimes they must be made at the factory. Occasionally the manufacturer's representative will not be aware that any modifications need to be made at all. In any case, the user would be well advised to inquire specifically as to the compatibility of any expansion units he plans to purchase and to obtain, preferably in writing, an understanding concerning what the conditions of acceptance will be. Users operating under a maintenance contract covering the entire system plus the new expansions will not need to be concerned with this problem.

17.4 SUMMARY

This chapter has attempted to address some of the more important questions in the justification, selection, and installation of computer systems. Owing to rapid developments, the half-life of computer system technology is rather short. Just keeping abreast of developments in the field is a problem, but somehow

one must balance the benefits promised by new features and the comfort afforded by utilizing time-proven approaches.

17.5 REFERENCES

1. Wherry, T. C. and Parsons, J. R., "Guide to Profitable Computer Control", *Hydrocarbon Processing*, *46* (4), 179-182, Apr. 1967.
2. Moore, C. F., Murrill, P. W., and Smith, C. L., "Simplifying Digital Control Dynamics for Controller Tuning and Hardware Lag Effects", *Instrument Practice, 23* (1), 45-49, Jan. 1969.
3. Parsons, J. R., Oglesby, W. E., and Smith, C. L. "Performance of a Direct Digital Control System", presented at the 25th Annual ISA Conference, Philadelphia, Oct. 26-29, 1970.
4. Kennedy, J. Patrick, "Tighter Process Design via Computer Controls", *Chemical Engineering*, pp. 54-60, Mar. 17, 1975.
5. Bobba, G. M., and Donaghey, L. F., "A Microcomputer System for Analysis and Control of Multiple Gas Chromatographs", *American Laboratory, 8*, 27, Feb. 1976.
6. Bentsen, Craig, "Computer Data Acquisition and Control for a Pilot Plant Facility Through Distributed Processing", Seminar: Department of Chemical Engineering, Louisiana State University, Oct. 21, 1977.
7. Hobbs, L. C. and McLaughlin, R. A., "Minicomputer Survey", *Datamation*, pp. 50-61, July 1974.

18

System Operations, Management, and Program Documentation

David E. Clough
Department of Chemical Engineering
University of Colorado

18.0 INTRODUCTION

In the management of large-scale computer systems, a great deal of organization has become necessary due primarily to the complexity of installations.   Real-time computing systems tend to be smaller and to be generated from the "grass roots" level, that is, from the user, rather than handed down from a computer-oriented organizational hierarchy.   As a result, the management of real-time computing systems is often poor, and in particular the lack of adequate software organization and documentation lowers the effectiveness of the systems. The goal of this chapter then is to point out management techniques that, although accepted and proven elsewhere, will improve the operation of the real-time system if they are applied judiciously. The chapter will cover maintenance of hardware, and software development and documentation.

18.1 HARDWARE MAINTENANCE

In considering the alternatives for maintenance of real-time system hardware, the system manager (who is often the primary user) must first characterize the installation's needs. It is ideal to be able to quantify the cost of the system or the cost of down time if one of its components becomes unavailable.   Then a direct comparison can be made with the cost of repair and a choice made among various repair alternatives. For example, if a small real-time computer is essential to the operation of a series of batch polymer reactors, the effect of

computer failure ("down time") can be estimated by the sum of shut-down and start-up costs plus lost product value.

In other cases it may not be possible to quantify the temporary loss of on-line computing ability. For instance, the use of a real-time computer in conjunction with experimental research might allow for rescheduling in the case of computer failure. A short delay in the availability of research results may be difficult to quantify in terms of cost.

Although there exists a spectrum of needs, it is useful to define two classifications of maintenance requirements: critical and noncritical. Critical maintenance requirements are identified by a high cost of down time. This criterion may be applied to the entire computer system; however, is better viewed in terms of system components. The mean-time-to-repair (MTTR) is a useful measure here. Noncritical requirements can tolerate a longer MTTR.

18.1.1 Alternatives for Maintenance

There are three primary sources of maintenance for real-time computer systems: the system supplier, major component manufacturer, and in-house personnel. Each has its advantages and problems. It is assumed in the following discussion that the computer system is acquired from a single vendor, often the central processing unit manufacturer, or, at least, that components are plug-in compatible. Maintaining the in-house system is discussed in a separate section below.

System Supplier. The manufacturer of the

real-time system will offer maintenance services via two modes. The first is the maintenance contract through which the customer pays a certain amount, usually monthly, according to a schedule of fees for the various major components of the system. It is difficult to provide precise estimates of these fees because they vary from one supplier to another, and the amounts are continually increasing owing to inflation of manpower costs. It should be noted that increases in maintenance costs are even more pronounced when viewed as a percentage of system hardware costs, the latter historically on the decline. As a rough estimate, however, a maintenance service contract for a small real-time system will cost several hundred dollars per month. Critical maintenance requirements may be satisfied by purchasing a higher-priority, higher-cost service contract which guarantees response within a certain period of time such as 24 or 8 hours.

The lower-priority service provided by the system supplier is simple on-call maintenance. In addition to an hourly charge (on the order of $70 at this writing) and the cost of parts, the user will usually have to pay travel expenses of the serviceman. The advantage of this arrangement is, of course, that there is expenditure only when maintenance is required. Two important disadvantages of on-call service are that costs mount up rapidly for hard-to-diagnose, open-ended repair problems, and priority is the lowest: that is, those who have maintenance contracts will be serviced first.

One trend that will continue to increase as the cost of labor increases is "board swapping". In this approach to system maintenance the serviceman simply isolates the defective component at the "board level". The customer is given a new or reconditioned and updated version of the unit in exchange for the old board at approximately 50 to 60% of the cost of a new component. The main advantages to the user are: (1) the speed of repair—the real-time system usually is down only a few hours; (2) because up-to-date hardware is substituted, the computer system is maintained in a more current state; (3) the manufacturer can have specialists perform the actual repairs at a central depot, saving the user the experience of dealing with inadequately- or poorly-trained service personnel. Against these advantages must be placed the higher cost; in many cases, the difference can be a factor of 2 to 10.

System suppliers will usually be most qualified to service equipment of their own manufacture; however, this may not be the case for major system components that the supplier obtains elsewhere and integrates into the system. Peripheral devices, such as printers, terminals, and other hard-copy equipment, and mass memory units, such as magnetic tape drives and disks or diskettes, may fall into this category. The system manufacturer will be able to check the operation of peripheral devices as integral parts of the computer system rather than as isolated units, one advantage of this maintenance service.

Component Manufacturer. Often a real-time computer system is an assemblage of large components or subsystems that the system vendor has obtained from other manufacturers. This may even be true of the central processing unit: several instrument companies market real-time computer systems in which the central processor is a slightly-disguised product of one of the major computer manufacturers.

An alternative strategy in this case is to deal directly with the component manufacturer for maintenance service. Although the overall system maintenance cost via this route may be greater, there are two possible advantages: the component manufacturer should be able to supply a more effective and rapid repair service and, second, be more aware of necessary preventive maintenance procedures. Two disadvantages are that the component supplier will usually require isolation of the particular unit from the rest of the system and that a preliminary diagnosis must be made by the user to isolate a malfunction to the particular component.

Terminals and other hard-copy devices are the best candidates for original manufacturer maintenance, since they can be operated in an off-line mode. Mass memory devices and process interface equipment are more difficult to isolate from the rest of the real-time system.

In-House Personnel. The effectiveness of maintenance by in-house personnel is highly dependent on the abilities and training of these persons. Apart from the variability, the size of many real-time systems cannot justify the employment of a full-time service person, and, if a technician is only to be involved in maintenance of the system on a part-time basis, it is even difficult to justify the proper training for that individual (this might include courses at the system manufacturer's site). In a larger organization where several computers are in use, it may be advantageous to pool maintenance resources. Not only is it more feasible to provide adequate training for more highly qualified personnel in such a situation; better test equipment can be acquired also.

In spite of the numerous pitfalls, in-house maintenance offers many advantages. It is more practical to carry out preventive maintenance and periodic check-out of components. Local employees can become familiar with the system and more aware of its maintenance history. Although effectiveness of repair service may be lower than that by the manufacturer, response to a problem is more rapid.

In considering the use of in-house personnel to provide maintenance services for a real-time system, the system manager must allow for the development of skills and familiarity with the equipment. A system will appear complex and overwhelming to a technician whose main experience is with smaller electronic components, especially if this person has no experience in elementary software.

18.1.2 Choosing Among the Alternatives

Two general recommendations are set forth here as a guide in choosing among maintenance alternatives: first, that a blend of the above choices is often best and, second, that this blend may require readjustment as the real-time system ages. Initial maintenance services are covered under a warranty agreement which typically lasts 90 days after delivery of the equipment. Shakedown of the system may require many service calls; therefore, it is imperative that the system reach some level of operation as soon as possible after delivery.

Project planning should take this as a high-priority goal.

After the warranty period, a blend of service contracts (or less preferred on-call service) dealing with both the system supplier for integral components and peripheral manufacturers, where appropriate, is recommended. As in-house personnel become familiar with the system through training courses, by looking over the shoulder of outside service persons, and on their own, they can begin to function as middlemen when a problem arises. Eventually, it may be possible to cancel the service contract and use on-call maintenance services as a last resort.

18.1.3 Maintaining the In-House System

Although there are numerous motivations for building the real-time system in-house, ease of maintenance is not one of them. Tailored design and economic constraints are perhaps foremost among the reasons for deciding on a "homemade" system. By such a system, we mean one designed around a central processing unit (although there is at least one case on record where the central computer was also designed and built in-house) that is obtained from an outside supplier, where the "homemade" components of the system consist of peripheral devices and their associated input/output controllers.

Why then is maintenance of such a system difficult? The designer, and perhaps fabricator, of the system would the ideal person to provide maintenance services but is rarely available to do so. Also, development of testing procedures and general documentation of the designs is usually inadequate and out of date. This latter condition can be ameliorated by proper attention during design, construction, and check-out of the system.

18.1.4 Documentation and Preventive Maintenance

Two documents are important to a successful maintenance program. One is a users' log in which all system activities are entered and difficulties reported. Documentation of the context of a malfunction can be invaluable as an aid to maintenance personnel. The second is a detailed maintenance log where all repair and testing procedures are entered. Such documents can be used in subsequent repair situations and to develop effective tests for periodic maintenance.

The development of a periodic maintenance schedule should occur soon after the installation of the real-time system. When activity on the system is on a daily basis, a weekly time slot for preventive maintenance will usually be sufficient. The following is a list of the activities typically scheduled periodically:

(1) Run diagnostic programs to test function of all system components.
(2) Check all power supplies for correct voltages and adjust as necessary.
(3) Check calibration of process interface for analog signal inputs and outputs.
(4) Clean terminals and other hard-copy equipment including printing mechanisms, video screens, magnetic tape guides and heads, and keyboards.
(5) Clean filters on ventilating blowers.
(6) Check ribbons on printing devices, paper supply, and paper tape.
(7) Check climatic conditions of system environment.

The diagnostic programs referred to above may take several forms. Most of these will be of the off-line variety; that is, any real-time operating system will have to be shut down for the tests to be run. More sophisticated software allows for on-line checks. The system manufacturer will have available diagnostic routines for most components; these may be available only at extra cost. The user will often develop test programs to augment the latter. In any case, periodic maintenance is documented in the appropriate log.

18.1.5 Substitution of Components

Critical maintenance requirements can sometimes be satisfied by on-the-spot substitution of a faulty component in the system by a back-up unit. Although it may be difficult to justify

a complete back-up for the entire real-time system, duplicates of certain components may be advantageous. Some that should be considered are:

(1) Central processing unit and memory modules
(2) Removable disk memory units
(3) Power supplies
(4) Hard-copy units such as printing terminals and paper tape equipment
(5) Process interface modules such as multiplexer relay arrays, D/A converter arrays, A/D converter, signal conditioning amplifiers, etc.

Apart from major components, a local supply of minor parts is always necessary for adequate system maintenance. These would include fuses, bulbs, and other simple mechanical and electronic components.

18.1.6 Development of a Maintenance Program

Managers of smaller real-time systems, especially those in the academic environment, tend to handle maintenance of their systems from the perspective of hoping that nothing will ever go wrong and, as a result, not preparing for the eventuality. The overriding theme presented in this chapter is that a maintenance program needs to be planned in advance. Decisions regarding maintenance alternatives need to be made. Provisions for adequate documentation must be taken, and a preventive maintenance program needs to be organized.

18.2 SOFTWARE DESIGN, MAINTENANCE, AND DOCUMENTATION

Software should be considered as the complete spectrum of documentation that accompanies a real-time system rather than using the restrictive definition of program coding alone. Computer systems in general have manifested an irreversible trend over the years of software gaining in importance over hardware. This trend is most easily observed in software-vs.-hardware cost figures over the last two decades and reflects both the integrated circuit revolution and inflation of software-related development costs. Software documentation, however, usually exists in a poor state when compared to the precise diagrams that describe the hardware components of a real-time system.

Maintenance of software falls into three major categories: fixing problems, making required changes, and implementing improvements. Real-time system software can be related to three types of computer programs: operating systems, utility program libraries, and applications programs. Before we consider maintenance and documentation of these types, an important underlying concept must be considered. Maintenance is facilitated by good program design, and, conversely, often made prohibitively difficult by poor design practices. Good design is accompanied by complete record-keeping. Therefore, in this section, we give emphasis to sound and proven design procedures first, then discuss software maintenance practices.

18.2.1 Program Design—Pitfalls

A primary activity of the real-time system user may well be the development of applications programs in high-level languages such as FORTRAN and BASIC. The user most often brings to these tasks only experience in programming the solution of small scientific problems on large batch or time-sharing computer systems. This section is written for such a person, and those users who have participated on teams in the development of major software will not find much enlightenment here although it may well be worthwhile to review the subject.

In programming small (small means fewer than 1000 lines of code), stand-alone, scientific

problems, one can focus immediately on the details of the solution and build the program code around that detailed core. Consider the following simple example:

A plant operator observes the level of liquid in a spherical tank via a sight gauge on the side of the tank. He requests that the sight gauge be calibrated in terms of volume of liquid in the tank. Develop a calibration table that can be used to mark the gauge in terms of volume. Figure 18.1 illustrates the situation.

The typical approach to the solution of this problem is to define the variables, H for height and V for volume, and derive or find a formula relating the two, which will involve the tank radius R. This formula is

$$V = \frac{\pi H^2}{3} (3R - H)$$

Then the algorithm for solution is built around this central relationship. A simple flowchart is shown in Figure 18.2.

Such an approach to program solution is called "bottom-up" design and becomes unsuitable under two conditions: when the complexity of the problem is high, and when the solution must fit into a larger framework. This latter characteristic is nearly always present in real-time systems software. One common result of bottom-up design is that one loses sight of the original objective. In the tank problem, the operator will not find the generated table useful because he will want to mark the sight gauge in convenient increments of volume.

Fixing-up the solution is not difficult here, but it does demonstrate another problem with

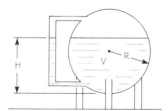

```
H: HEIGHT OF LIQUID LEVEL
   FROM BOTTOM OF TANK
R: INSIDE RADIUS OF TANK
V: VOLUME OF LIQUID IN TANK
```

Figure 18.1. Example of a Tank Volume Calibration

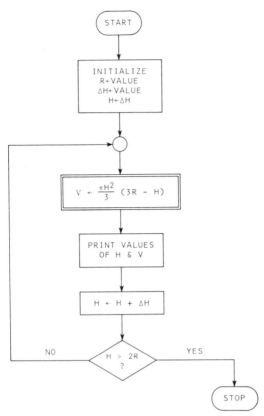

Figure 18.2. Tank Calibration Algorithm

18.2.2 Top-Down Design

The description "top-down design" implies that this reorientation represents an about-face. The design proceeds from the general to the specific in orderly steps. We will describe top-down design in detail, but first consider its application to the tank problem. Again, this elementary example does underscore some important points.

The first stage of the design process is to define the problem, which was done above. Next consider the solution of the problem in the framework of "transformation of information which must be supplied" into "results which will be used". This prompts study of what precisely the operator needs: a table that allows him to mark off volumes on the sight gauge in convenient increments of *volume*. As information that must be supplied, the tank radius is recognized along with the desired volume increment. At this point the design naturally proceeds downward to the task of transformation, that is, finding the height corresponding to a given volume. That can be conveniently considered as a module of the overall solution. Different algorithms for the transformation could be tested without affecting the other modules. In fact, the transformation module could be easily modified at a later date after the solution has been developed and used.

Note the following characteristics of top-down design as applied to this simple example:

- Only at the last stage of the design will the exact formula relating height and volume be considered.
- The design adheres to the requirements of the overall problem definition.
- Subsequent modifications are restricted to the module or level in which they naturally occur.
- Documentation of the top-down design will give a complete picture of the problem and its solution.

At this point, it is beneficial to generalize these concepts as they will apply to more complex problems, especially those solved by programs on a real-time computer system.

the bottom-up approach. We need to invert the formula to solve for H rather than V. This will require iterative approximate solution. This central change will also propagate outward into the algorithm—changing the incremented variable from H to V, reversing the table printout, and changing the termination test to "when V exceeds the total tank volume". In other words, revisions in bottom-up designs usually propagate. Another way to see this is that the flowchart in Figure 18.2 cannot be easily modified to accommodate the revised solution; it must be completely redrawn.

The fact remains that many real-time system users come from a background of bottom-up program design. Elementary scientific programming has been taught that way in the past although there are now some signs of change. A reorientation is necessary to guarantee efficient programming of real-time systems.

General Specification of the Problem. The first step in top-down design is to strive for a thorough understanding of all aspects of the problem. The problem description should be made as complete as possible, and *it must be written down.* This is a critical phase, since changes at lower levels become increasingly difficult and expensive.

Definition of Computer-Oriented Design Objectives. Consideration of factors of time and space are necessary at this point, since they will influence the structure of the problem solution. The computer job can be viewed as procedures that transform data. Sophisticated procedures can be used with accompanying reduction in the amount of data storage. These procedures will take more time to execute than their simpler counterparts which operate on an expanded data base. In other words, there is a tradeoff between speed of execution, simple procedures, and large data storage on one hand, and slow execution, complex procedures, and minimal storage on the other.

As an example of the data storage/procedure tradeoff, consider the task of sending coded information to a printing terminal. A block of computer memory serves as a buffer array containing a string of character codes to be transmitted. In one implementation, each memory word contains a single character code; so the task is to fetch the word and simply transmit it to the terminal. In another implementation, the word contains several characters which have been "packed" in. Now, the task calls for the fetching the word and "unpacking" it character by character with a sequence of transmissions to the printer. Clearly, this latter implementation economizes on memory storage yet calls for a more complex procedure which will require more time to execute.

Therefore, it is necessary to complete the specifications of the general problem by addressing the objectives of speed of execution and amount of data storage. Another pair of computer-oriented objectives also interacts with the above considerations. The amount of time to be allowed for design and implementation of the solution, that is, the project deadline, is

Figure 18.3. A Sequential Flow Structure

one. The other is how readily understandable to humans is the implemented solution to be.

Although it is rarely possible to pin down precisely computer-oriented design objectives, statements of general emphasis will influence the subsequent design steps strongly, and, again, they must be documented. It is important at this stage to address the capabilities of the real-time computer system, e.g., availability of memory storage space and computational speed, in addition to the general problem specifications which will be more process-oriented.

Problem Decomposition. The trend in program design methodology over the past decade has been toward the structured programming approach. Structured programming includes the top-down concept; however, additional aspects are important.

Algorithm design is based on modules with single entry and exit. The highest-level module naturally falls into that category, i.e., the sequential flow structure shown in Figure 18.3. This main module can be expanded using the sequential structure and two other flow schemes: decisional and conditional. The decisional structure is represented as in Figure 18.4.

Owing to its implementation in some versions of the FORTRAN, ALGOL, and PASCAL languages, the decisional structure is often called the IF/THEN/ELSE type. The conditional structure, also called the DO/WHILE type, can be depicted as in Figure 18.5. The conditional structure allows for iteration. All algorithms can be decomposed into sequential, decisional, and conditional structures.

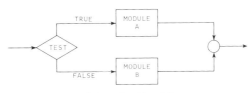

Figure 18.4. A Decisional Structure

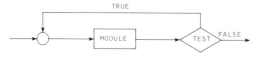

Figure 18.5. A Conditional Structure

As a problem is decomposed, some guidance is necessary in determining how much direct structure should be included in a module, and, consequently, how many details should be left to submodules. A general rule-of-thumb which has been applied widely is that one module should be embodied in no more than one page of program code (sometimes a time limit for implementation and testing of a module is used such as one week). Such conventions allow for organized distribution of workload if a programming team is involved, and provide for orderly progress toward solution of the overall problem.

As each module is designed, the structure of data representation is first considered carefully. An example of this related to real-time systems is the design of a module to update the process data base for an acquisition application. There will be a series of measurements from one or more process units and related information such as alarm limits, linearization coefficients, engineering units conversion codes, smoothing constants, etc. Several data structures are candidates to represent this information:

(1) *Individual arrays.* Separate arrays are defined for each category and elements are indexed in terms of the measurement source. This type of structure is analogous to bottom-up design.

(2) *Point-based structure.* Here, a storage block is defined for each measurement point and the internal structure of the block is designed to contain all of the information pertaining to the measurement point. Some "packing" of information may be possible with resulting economy in memory requirements.

(3) *Unit-based structure.* At this level, a storage block is allocated for each process operating unit, and a series of point-based structures fits within this block. A particular item of information will be denoted as

unit name . point tag . item name

thus implementing a top-down data structure.

All of the above represent workable schemes with associated data/procedure tradeoffs. Since the data structure chosen will affect the algorithm design for a particular module, it is important to define it first. Note that data structures usually are language-dependent.

Algorithm Representation. As procedures for modules are designed, there is a need for a consistent manner in which to document them. The flowchart has traditionally been the most common way to accomplish this. Simple examples of flowcharts have been used here to describe general algorithm structures. The drawing of flowcharts does ensure a clear algorithmic organization; however, there are several disadvantages:

- Flowcharts are inherently difficult to modify and thus tend to enforce resistance to future enhancements of the algorithm.
- Flowcharts yield an orientation toward procedures and away from data structures that may be undesirable.
- Following flowcharts back and forth from page to page via off-page connector labels can be very confusing.
- Algorithms designed in the framework of a multitasking real-time operating system are often not amenable to flowcharting, since time-coordinated and "simultaneous" tasks need to be represented (note, however, the use of multitasking flowcharting conventions described in Chapter 13, which assist in avoiding this problem).
- Flowcharts tend to reinforce "unstructured" programming.

Of course, English language descriptions can be utilized for algorithm representation, but this is an unwise choice owing to ambiguity and lack of precision. There are two additional alternatives for representations which should be considered.

(1) A "pidgin" language can be used in an

informal yet precise manner to describe algorithms. This type of representation encourages the definition of data structures prior to procedures. Referring to the previous example of a process data base, an abbreviated "pidgin" description of the data structure might be written as in Table 18.1. Then in the procedure representation, we could reference a measurement value as

reactor.P.79.rawmeas

The ease of coding such a "pidgin" programming language into a high-level language is dependent on the latter's flexibility in accepting new data types. The PASCAL language is ideal here, but it is not widely available and presently is seldom used in real-time configurations. This situation is, however, changing.

(2) Graph theory has also been applied to algorithm representation. Finite-state machine descriptions (based on graph theory) can be very useful tools in the design of real-time software in a multitasking environment.

The documentation of algorithm designs is critical to subsequent testing, installation, and maintenance of software. That a convention is chosen for such documentation and used consistently far outweighs the marginal advantages of one style over another. Algorithm designs should not be represented directly in a computer language.

Coding and Testing of Algorithms. Bottom-up program design has the deceiving characteristic of producing elements of program code very early in the development process. Serious difficulties arise in integrating the various elements later on. If top-down design is carried out exclusively, algorithm coding will be a short and smooth process. Accurate documentation of program code, especially subsequent revisions, in a working manual is recommended. If a team of programmers is involved, distribution and updating of the manual is necessary. Such a working design manual is the basis for final software documentation.

As program code is generated, an independent review by another individual will serve the dual purposes of finding errors prior to actual testing and requiring improvements in documentation

Table 18.1. Description of a Data Structure

```
type unit of column, reactor, tank;
type tag of
     begin
          type letter of T,P,F,L,A,M;
          type idno of 0..9999
     end;
type item of
     begin
          type rawmeas of -2048..2047;
          type lincode of linear,sqrt,TCJ,TCK;
          type range of
               begin
                    type lorange of real;
                    type hirange of real
               end;
          type units of degf,degc,gpm,inH₂0
     end;
```

during the generation process. It is difficult to motivate a programmer to document his or her work after the coding is complete.

Top-down design also facilitates testing and debugging of programs. Major program modules can be checked out first, independent of sub-modules, by substitution of bogus, simplified subprocedures commonly called "stubs". For example, if a real-time program is to make a control calculation but first must call a procedure that acquires a process measurement value, the control calculation procedure can be tested by supplying a dummy process value perhaps before the computer system is even interfaced to the process.

Program testing will uncover errors that cannot or should not be corrected by patching up program code. Therefore, the overall design process must allow for recycling program modules to the various development stages. As changes are incorporated, they must be noted as dated revisions and *added to* the working manual, not substituted.

Installation of New Programs. Once new software has been through the coding and testing stages, it is ready for installation. Unavailability of earlier versions of the software for similar tasks often will make necessary immediate installation; however, where the new program is a major revision of a working version, it is recommended practice to have both installed simultaneously and allow or coordinate periodic use of the new version as an advanced-stage test. In this manner the user is confronted with a direct comparison of new vs. old, and improvements will be more readily noted and accepted.

Upon successful installation of new software, subsequent modifications and their associated documentation are considered as software maintenance.

18.2.3 Maintenance of Software

The major motivation for software maintenance is to eliminate problems with the current version. These problems or "bugs" can be classified in decreasing order of urgency as

(1) Catastrophic errors that render the real-time system inoperable.

(2) Errors that degrade performance of some system elements.
(3) Major enhancements of software that is presently functional.
(4) Minor improvements to working software.
(5) Correction of errors in documentation.

Software maintenance must be recognized as an integral component of a real-time system application, and resources must be allocated to it. On large projects, i.e., when programming teams are employed, there should be a programmer (or programmers) assigned specifically to maintenance rather than the alternative approach of distributing the tasks to programmers responsible for the program modules. With regard to one-person ventures, the maintenance function should occupy a legitimate time slot on a weekly basis.

Real-time systems, which by their nature respond to random external stimuli, are notoriously difficult to analyze for errors. Error conditions often occur infrequently and unpredictably yet are serious enough to warrant attention. Multitask systems give the appearance of simultaneous execution of many program modules; therefore it is often difficult to determine which task was actually executing when a major error occurred. Careful record-keeping of the state of the system *and* its environment for a number of occurrences of an error will prove valuable. The placement of traps or breakpoints, which halt or cause major deviation in program flow when error conditions occur, is usually necessary.

18.2.4 Software Documentation

If a working design manual has been maintained during software development, it will provide the basis for formal documentation. Four components should comprise the final documentation:

General Software Description. This component is adapted from the general specifications set down at the outset of program design. What is required here is an English language description of the objectives of the software, where it fits in the overall system, what computer-oriented design objectives it satsifies, and its general capabilities and limitations.

Algorithm Descriptions. Utilizing the convention accepted during program design, the detailed algorithms of the top-down problem decomposition are documented. An explanation of the representation scheme is helpful, and the use of English language paragraph introductions to the various modules will make this information more understandable.

Program Source Code Listings and Hard Copy. Listings of the program code too often constitute the only documentation available for a computer program. Listings are, of course, necessary, and they should contain "comments" to elucidate any section of code which is not self-explanatory. Some paragraph-length comments can be adapted directly from the algorithm descriptions. All program nomenclature must be defined. Minor fixes of the program segments are often noted directly on the listings. These should be clearly marked with the date and author, and previous code should be left visible. Marking of listings is only a temporary means of documentation until fixes can be incorporated as revisions and new listings generated. An unmodified copy of program source listings should be kept separately.

Hard copy of programs in some form is necessary. Physical copies in the form of paper tape or data cards are least subject to loss; however, owing to convenience, magnetic media such as tape or disk(ette) are popular. A compromise here is to update frequently a magnetic copy of all system software and, on a much less frequent schedule, create physical copies from the magnetic media. Hard copy should be stored away from the computer system location with attention to climate and electromagnetic field environment.

User-Oriented Documentation. A reference manual and user's guide are necessary to the sucessful implementation of new software. Often only reference manuals are supplied, which adequately answer specific questions but do not serve to train the prospective user. User's guides are best written in the way many textbooks are generated, developed via efforts to educate the user with the latter's continual

evaluation. Great care must be exercised in the choice of language for user's documents. Descriptions must be precise yet should avoid jargon of the software developer. In reality, much of the useful user-oriented documentation is generated by users themselves.

18.3 SITE MANAGEMENT

In large computer installations there is much effort expended in design and operation of the computer site, i.e., the physical location of computer hardware and all documentation. Such is not often the case for smaller real-time systems, because of both their size and their physical connection to an operating process. Direct exposure to the process environment can cause rapid degradation of the computer system owing to unfavorable climatic conditions and airborne particulate matter. In this section we consider some of the factors involved in real-time computer system site layout and management.

18.3.1 Site Layout

Except under laboratory conditions where the environment is well controlled, real-time computer equipment somehow should be isolated from the process environment. If a control room or shelter is built around the system, or if it is located remotely in an enclosure, the need for control of temperature and humidity may arise. Heat dissipation from integrated circuitry and power supplies adds significantly to the cooling load of a small room.

If possible it is recommended that the real-time system site be designed with isolation of certain subsystems. Process signal termination panels should be located away from the main computer system. It is often convenient to install terminal boxes just outside the room housing the system. It is especially unwise to have process connections inside the main computer cabinets.

Although all computer equipment should be located for easy maintenance access, well-traveled traffic patterns need to be designed away from the main cabinets. If the system is large enough to involve both programmers and process personnel, its configuration should isolate the operator/engineer interface from the rest of the computer system. Programmers should not have to share input/output equipment with process personnel except on small systems.

Access to computer equipment and documentation should be controlled, and such control is facilitated by proper site layout. As mentioned above, permanent hard copy of software should be stored away from the system.

Segregation of power cables from signal cables and grounding of the system must follow recommended practice. Preparation of the site prior to delivery of the system will allow for

rapid start-up and take full advantage of the warranty period of the computer equipment.

18.3.2 Fault Procedures

In real-time system applications, planning for emergencies is frequently limited to a back-up plan in the event of catastrophic computer failure. There are many error conditions that only temporarily degrade system performance, and fault procedures need to be defined for such situations that occur frequently. For example, the flowchart in Figure 18.6 defines the response to an alarm message from the com-

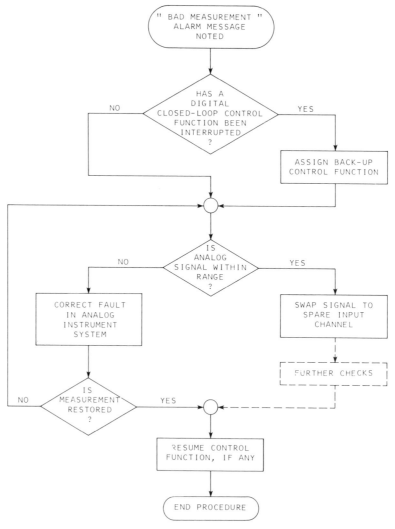

Figure 18.6. Response to a Bad Measurement Error Condition

puter, communicating that it has received a bad process measurement.

It will not be possible to design all fault procedures ahead of system installation. However, new procedures can be developed if careful records are kept of the trouble-shooting of unusual or unanticipated fault conditions. All such fault procedures should be available in a notebook organized for easy and rapid reference.

18.3.3 Documentation

The maintenance of up-to-date and orderly documentation at the real-time computer site is a continual struggle. There is a natural tendency toward degradation of documentation that can only be combatted by strong management policies such as the following:

(1) No software revisions are commissioned until all documentation is adequate.
(2) Each physical change made to the system is logged before it is carried out.
(3) Users are required to log onto the system, identifying themselves and their activity. This information is saved.
(4) A detailed maintenance log is kept that documents preventive activities.

The guiding principle throughout is that no pertinent information is to be committed only to the memory of a designer, programmer, operator, or user. Everything is written down.

18.4 SUMMARY

Alternatives for maintenance of hardware have been reviewed with emphasis on their effectiveness and cost. The need for a preventive maintenance plan has been discussed. An underlying theme to good documentation and maintenance of software, viz., program design, has been treated at length with special attention to the "top-down" design concept. Appropriate documents for software have been reviewed along with concepts of site management.

Although one of the least glamorous of real-time computing activities, after-installation site management is crucially important to successful long-term operation of a facility. Many real-time systems that were carefully planned, designed, and installed have failed to perform satisfactorily because of inattention to the principles discussed in this chapter.

18.5 SUPPLEMENTARY READING

1. Geist, R. O. and Cronin, D. J., "A Philosophy of Computer Maintenance", *Instrumentation Technology*, pp. 61–64, May 1978.
2. Harrison, Thomas J. (Ed.), *Minicomputers in Industrial Control*, Instrument Society of America, Pittsburgh, 1978.
3. Jensen, K. and Wirth, N., *PASCAL User Manual and Report*, Springer-Verlag, New York, 1975.
4. Mooney, John W., "Organized Program Maintenance", *Datamation*, 21 (2), 63–64, Feb. 1975.
5. Ogdin, Carol A., "EDN Software Design Course", *EDN*, pp. 67–200, June 1977.
6. Tausworthe, R. C., *Standardized Development of Computer Software*, Prentice-Hall, Englewood Cliffs, N.J., 1977.
7. Yourdon, Edward, *Design of On-Line Computer Systems*, Prentice-Hall, Englewood Cliffs, N.J., 1972.

18.6. EXERCISES

1. Study an existing real-time installation in terms of the criteria set forth in this chapter. Write a critical evaluation of the system.
2. A crude oil stream is split, passed through the tube-side of two heat exchangers in parallel, and recombined. The heat exchangers' shell-side fluids come from two distinct process locations and are of variable flow rate, temperature, and physical properties. Design a computer control scheme to maximize thermal energy transfer to the crude oil for steady-state operations.
3. Obtain maintenance fee schedules from two real-time system vendors and compare them.
4. Analyze several fault procedures in an existing real-time computer installation.
5. Investigate different ways of algorithm representation and data structure for batch process control (cf. Chapter 4).

Part VIII
Process Analysis, Data Acquisition, and Control Algorithms

19

Process Analysis and Description

Thomas F. Edgar
Department of Chemical Engineering
University of Texas, Austin

19.0 INTRODUCTION

The notion of a model for a process in a real-time computing application has previously been discussed in rather qualitative form in Chapter 4. The purpose of this chapter is to present introductory information on process modeling and systems analysis for real-time computer users who have little or no formal training in these subjects. While the mathematics covered here are not intended to be comprehensive, we do attempt to identify certain procedures and methods that are valuable in many real-time computing applications. In this chapter we shall limit our discussion to the subject of modeling; the use of such models for process control is reviewed in Chapter 20.

There are a number of modeling approaches used with real-time systems. While models based on the chemistry and physics of the system represent one alternative, much of the discussion in this chapter centers around so-called black-box models, namely, those developed from experimental tests of the real-time system. Models of the real-time system include not only the process but also the computer and other electronic components. Once these models have been established, we can proceed to analyze the overall system dynamics, the effect of controllers in the operating process configuration, and the stability of the system, as well as obtain other useful information.

As discussed in Chapter 4, mathematical models also provide a convenient and compact way of expressing the behavior of a process as a function of process physical parameters and process inputs. The same mathematical model can be expressed in several ways; for example, a continuous time model can be converted to a discrete time representation, or it can be transformed to a different type of independent variable altogether (e.g., Laplace transforms, z-transforms). Transform models generally feature a simplified notation, which greatly facilitates analysis of complicated systems comprised of a number of static and dynamic components. Hence mathematical modeling and the ideas presented in this chapter provide a unified procedure for analyzing the behavior of real-time systems.

One design problem commonly faced with real-time computing systems is to specify the rate at which data should be acquired. To determine an appropriate sampling rate usually requires information concerning the dynamic behavior of the process which is being monitored and/or controlled. There are some simple guidelines which can be followed in order to determine sampling rates for a given process. The evaluation of these guidelines, however, necessitates the analysis of processes in terms of the theory of linear differential equations. One of the essential elements in the interpretation of process dynamics is the notion of a dominant time constant. In Section 19.1 we show how the dominant time constant arises out of the solution to linear differential equation models of the process; in Section 19.2 several ways to calculate the dominant time constant from experimental data are presented. Common dynamic characteristics of processes are discussed in Section 19.3, and a generalization of the dominant time constant idea is given. The selection of an appropriate sampling time for a dynamic process can be based on the dominant time constant concept, and in Section 19.4 we also discuss how the sampler operates and how loss of information occurs once continuous data records are discretized in time.

Sections 19.5 and 19.6 deal with one result from the sampling operation, namely, that the computer plus process now behaves more like a discontinuous rather than a continuous system; i.e., the time variable is discretized. This allows the introduction of difference equations in place of differential equations to model the discrete time dynamic behavior. We illustrate how a first-order filter described by a continuous time dynamic model can be converted to a discrete time equation. Discrete time equations also can be readily fit to experimental data, and the procedure is illustrated for an irregular pulse input.

Probably the most popular approach for analyzing systems dynamics and control is based on transforms of the time domain equations. For continuous time systems, we can express the process dynamics in terms of Laplace transforms. Laplace transforms usually afford a reduction in mathematical complexity over the time domain approach, and in Section 19.7 we present some of the rules for transforming continuous time differential equations, manipulation of the resulting algebraic equations, and inversion back to the time domain. The utilization of block diagrams, which are of special significance for analyzing feedback control systems, is also discussed in this section. Section 19.8 presents an analogous treatment of discrete time equations followed by conversion to z-transforms. Finally Section 19.9 discusses the interrelationship of Laplace transforms to z-transforms and provides parallel treatment to the discussion in Sections 19.5 and 19.6.

While it should not be difficult to master the techniques presented in this chapter, one admonition from Chapter 4 should be repeated at this point. Mathematical modeling is not a sterile exercise but rather one that requires judgment and experience. Knowledge of the behavior of dynamic systems may be a foreign subject to many real-time users, but hopefully, through the presentation of several related examples, the reader can gain some intuition about how real-time systems behave and can learn what are some of the implications of their behavior.

19.1 ORDINARY DIFFERENTIAL EQUATION MODELS

One model dichotomy discussed in Chapter 4 is algebraic/differential, or static/dynamic. Many simple dynamic systems can be represented by the equation

$$\frac{dx}{dt} = f(x,u) \qquad (19.1)$$

where x is the output, u is the input, and t is time. The time solution to Equation (19.1) can be found by integrating Equation (19.1) for a given input, $u(t)$, and initial condition for the output, $x(0)$. By letting $(dx/dt) = 0$, the steady state model is obtained. An equilibrium value of x can be found for each selected value of u, the input, by solving the algebraic equation

$$f(x,u) = 0 \qquad (19.2)$$

An illustration of static and dynamic behavior for a stirred tank heater has been given in Chapter 4. For the agitated heating tank, depicted again in Figure 19.1, the steady state or static model is

$$FC_p T_i + W_s\lambda = FC_p T_o \qquad (19.3)$$

where F is the feed flow rate, C_p is the fluid heat capacity, T_i is the feed temperature, W_s is the steam flow rate, λ is the heat of vaporization of steam, and T_o is the tank outlet temperature. From the standpoint of a control system, ordinarily W_s is an input or manipulated variable of the process (corresponding to u in Equation 19.2), while the output or controlled variable is T_o (equivalent to x in Equation 19.2). For fixed values of F, C_p, T_i, and λ, we have a unique algebraic relationship between W_s and T_o. We define the normal operating point

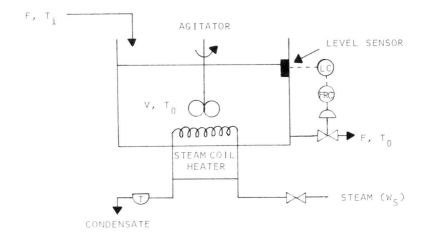

Symbols:

----	electrical or pneumatic instrument
FRC	flow recorder-controller
LC	level controller
T	steam trap

Figure 19.1. Agitated Heating Tank

for T_o as the set point; as long as the other four parameters do not change, we can set W_s at its prescribed value via Equation (19.3), and the system will remain at the set point.

However, in reality we may find that none of the four parameters is constant, but they all may vary with time, due to a variety of reasons. Once the output T_o deviates from its desired value or set point, which we shall call T_o^*, then an error occurs, which we can define as

$$x = T_o - T_o^* \qquad (19.4)$$

The above definition of the error is consistent with instrumentation practice; i.e., for maximum measurement sensitivity we subtract the steady state value, often referred to as the "d.c. component", from the actual signal. Hence rather than analyze a system in terms of absolute numbers, we almost always express the analysis in terms of error or deviation variables. This is true not only for the output variables but for the input variables as well.

The corresponding mathematical model variables also can be written in terms of deviation variables. For example, we can define the manipulated variable deviation as

$$u = W_s - W_s^* \qquad (19.5)$$

where W_s^* is the steady state value of the steam flow rate and corresponds to the solution of Equation (19.3) with T_o^* taken as the value of T_o. Therefore at the steady state (T_o^*, W_s^*), the corresponding steady state in terms of (x,u) is $(0,0)$.

The idea of deviation variables is also usefully extended to dynamic equations. Let us consider the dynamic model for the agitated heating tank illustrated and discussed in Chapter 4:

$$MC_p \frac{dT_o}{dt} = FC_p(T_i - \overline{T}) + W_s\lambda - FC_p(T_o - \overline{T}) \qquad (19.6)$$

where M is the mass of fluid in the tank (assumed to be kept constant via a level controller) and $\overline{T}$ is some arbitrary reference temperature for calculating enthalpy. Using the deviation variables defined in Equations (19.4) and (19.5), we can write this dynamic equation in terms of x and u [9]:

$$MC_p \frac{dx}{dt} = -FC_p x + \lambda u \qquad (19.7)$$

which can be rearranged to give

$$\frac{M}{F} \frac{dx}{dt} = -x + \frac{\lambda}{FC_p} u \qquad (19.8)$$

The term M/F has units of mass divided by mass per unit time, or simply time. In process dynamics nomenclature M/F is referred to as the "time constant", usually expressed as τ. The term (λ/FC_p) is called the "process gain", since it relates how much x changes at steady state to a unit change in u. This can be easily seen in Equation (19.8) by setting (dx/dt) = 0 and letting u = 1 (remember that this means a deviation in the control variable of +1). The corresponding change in x is also a deviation from T_0^*, the steady state value of the tank temperature. Denoting K as the process gain, the dynamic model, with simplified notation, can be written as

$$\tau \frac{dx}{dt} = -x + K u \qquad (19.9)$$

In a later section we shall discuss the dynamic behavior of this equation and the significance of the time constant.

Linear ordinary differential equations, of which the first-order equation derived above represents the simplest example, assume an important position in process modeling, and hence in real-time computing. Linear dynamic models also provide a theoretical means to determine the "time scale" of the process, which in turn is required to design an effective computation algorithm, e.g., for filtering process output data or for controlling the process output. A real-time algorithm must have a characteristic calculation time which is much smaller than the process time scale; this latter quantity

is usually measured in terms of a parameter called the "dominant" time constant. One way to determine the dominant time constant is to analyze the general solution to a linear ordinary differentail equation model.

The general solution to any linear differential equation is given by the sum of homogeneous and particular solutions to the problem [2]. The homogeneous solution offers a convenient way to understand the system dynamics. The homogeneous solution (x_h) to an nth-order linear differential equation with constant coefficients consists of a linear sum of n exponential functions, with each function containing a characteristic root of the system:

$$x_h(t) = c_1 \exp(p_1 t) + c_2 \exp(p_2 t) + \dots$$
$$+ c_n \exp(p_n t) \qquad (19.10)$$

The (p_i) here are assumed to be real and distinct roots, while the (c_i) are called the residues. For the agitated heating tank, which we found to be a first-order dynamic system, the homogeneous solution consists of the first term only ($c_1 \exp(p_1 t)$). For a typical physical system, which might be described by a higher-order differential equation, the roots might follow the sequence (in decreasing value): $p_1 = -1$, $p_2 = -5$, $p_3 = -20$, $p_4 = -50$, etc. By definition, $p_5 < -50$, so we disregard it and the smaller p_i, since, as shown below, exponentials with large negative arguments are very small for t > 0. The minus sign for *each* p_i indicates that $x_h(t)$ will approach zero for large values of t, a desirable operating characteristic in all systems. The negative sign of the roots insures a stable system. If the argument of any exponential function were positive, that term would grow quite large over time. This unbounded behavior is characteristic of an unstable system. Table 19.1 contains values of $\exp(p_i t)$ for various values of p_i and t.

Table 19.1. Values of exp (p_it) for Various Points in Time

	t = 0.1	t = 0.5	t = 1.0	t = 10.0
exp (-t)	0.9048	0.6065	0.3679	0.00004
exp (-5t)	.6065	.0821	.0067	0
exp (-20t)	.1304	.00004	0	0
exp (-50t)	.0067	0	0	0

It is obvious from this table that once $t >$ $-10/p_i$, that exponential term becomes quite small, and except for large values of c_i, provides a negligible contribution to $x_h(t)$. For virtually all of the time response, $\exp(-t)$ dominates the solution and is the limiting factor in Equation (19.10). Because of this fact we call $p_1 = -1$ the dominant root, and we define $\tau_1 = -1/p_1$ as the dominant time constant. For a first-order system there is only a single time constant, which obviates further analysis. If we are interested in monitoring the dynamic response of a process, the dominant time constant determines how often we need to acquire process data and how fast the computer program must be, in order to take action before the dynamics are significantly attenuated. Physical processes can have time constants from one second to several days: for example, combustion processes have a time scale less than one second, while the time constant of a petroleum reservoir can have an order of magnitude of days. This diversity demonstrates the need to characterize the process before design of a real-time computing system.

19.2 CALCULATION OF THE DOMINANT TIME CONSTANT

If a physical or chemical dynamic model is available, the dominant time constant can be found by linearizing the model (if necessary) and computing the characteristic root(s), as was done in the heating tank example. However, if physical modeling procedures are too complicated or time-consuming, the dominant time constant can also be found from experimental data. If we suppose the process can be approximated by a first-order differential equation, the dominant time constant can be obtained by fitting, graphically or numerically, the process response to some known input change. Consider an equation of the form derived earlier for the agitated heating tank where we let $\tau = \tau_1$ denote the first-order nature of the model:

$$\tau_1 \frac{dx}{dt} = -x + Ku \qquad (19.9)$$

If the system is initially at rest (i.e., $x(0) = 0$,

corresponding to u = 0) and if u, the input, is abruptly changed from zero to one at time zero, we obtain the so-called unit step response of the system.

We can obtain the step response by substituting u = 1 in Equation (19.9) and integrating the differential equation. Note that one solution (the particular solution) to Equation (19.9) is $x_p = K$ (a constant). The homogeneous solution to Equation (19.9) is

$$x_h(t) = c_1 \exp(-t/\tau_1) \qquad (19.11)$$

where c_1 is an unknown constant that must be evaluated from the initial conditions. Adding the particular and homogeneous solutions and using $x(0) = 0$, we obtain,

$$x(t) = [1.0 - \exp(t/\tau_1)] K \qquad (19.12)$$

The graphical response is shown in Figure 19.2. Note that the response of $x(t)$ reaches 63.2% of its final value (K) at $t = \tau_1$. Hence, most of the process dynamics are manifested before $t = \tau_1$. Note that the steady state response is $x(\infty) = K$, which could be found by setting $(dx/dt) = 0$ and u = 1.0.

In obtaining the dominant time constant from a step test, we compare the actual response with the true first-order response shown in Figure 19.2. For a true first-order system, the response attains 63.2% of its final steady state value when the time t is τ_1. Therefore the 63.2% response point can be a suitable approximation to the time constant. Another way to approximate τ_1 is to draw the tangent to the response at t = 0 (see Figure 19.2). For a first-order system this line intersects the steady state response $((x/K) = 1.0)$ at $t = \tau_1$.

19.3 COMMON DYNAMIC CHARACTERISTICS OF PROCESSES

Very few experimental plots of a process step response show pure first-order behavior. Since it is difficult to form a perfect step input, and the output data usually are corrupted with noise, there will be some departure from the idealized curve in Figure 19.2. Suppose the

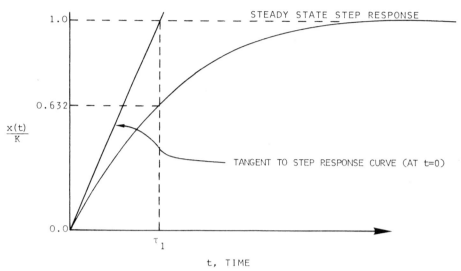

Figure 19.2. Step Response for a First-Order System and Graphical Construction for Estimating the Time Constant, τ_1

process is not first-order but is a second-order system, described by a second-order ordinary differential equation:

$$b_2 \frac{d^2 x}{dt^2} + b_1 \frac{dx}{dt} + b_0 x = a_o u \qquad (19.13)$$

Different shapes might occur for the step response, as shown in Figure 19.3. Figure 19.3(a) shows the step response for various values of the two time constants, τ_1 and τ_2. The time constants are the negative reciprocals of the characteristic roots of the system, i.e., the p_i of Equation (19.10) ($\tau_1 = (-1/p_1)$; $\tau_2 = (-1/p_2)$; the dominant τ is the larger of the two time constants). Note the appearance of the s-shaped response as the ratio of τ_2/τ_1 becomes closer to one (equal time constants). This characteristic response is referred to as "over-damped". Figure 19.3(b) depicts the step response for several cases when the two roots are imaginary (the "underdamped" case); the appearance of oscillation in the step response is due to the imaginary roots, which produce sine and cosine terms in the analytical solution. Many mechanical and electrical processes exhibit this oscillatory type of behavior. In most chemical processes, oscillation does not naturally occur. However, when a process is controlled with a feedback controller (see Chapter 20), oscillation does frequently occur; hence

the importance of covering this type of behavior.

The analysis of the step response of systems higher than second-order is often done in an approximate fashion. In practice, most high-order linear systems can be approximated by a first- or second-order model but with an initial time delay. Figure 19.4(a) shows a typical high-order process approximated by first- and second-order models, each with time delay. Note that the agreement between the true and approximate models is reasonably good.

The design of digital controllers based on an approximate first-order plus time delay model is discussed in Chapter 20. This method of model approximation/controller design is generally referred to as the process reaction curve method [3]. The technique uses a graphical fit of the actual step response* and requires the following calculations or constructions, as

---

*Note that Figure 19.4(b) shows the fractional response, defined as the deviation in the process output divided by the product of the process gain and the magnitude of the input change. This graphical construction can be made directly to evaluate the process time constant and dead time, by using the actual process output variable, so long as the intersections are made with respect to the initial and final steady state values of x.

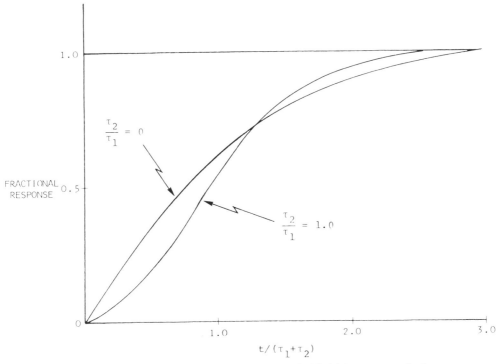

$$\frac{\tau_2}{\tau_1} = 0$$

$$\frac{\tau_2}{\tau_1} = 1.0$$

FRACTIONAL RESPONSE

$t/(\tau_1 + \tau_2)$

Figure 19.3(a). Step Response of Second-Order Model ($\tau_2 \leqslant \tau_1$; t = time)

shown in Figure 19.4(b) [7]:

(1) The process gain for the model is found by calculating the ratio of the difference between initial and final steady state values of x to the size of the step change in u.

(2) A tangent is drawn at the point of inflection of the step response; the intersection of the tangent line and the time axis (x = 0) is the time delay θ.

(3) If the tangent is extended to intersect the steady state response line (x(t) = 1), the point of intersection corresponds to time, $\theta + \tau_1$. Therefore τ_1 can be found by subtracting θ from the value of time at this intersection.

There are still other means of estimating the order of magnitude of the time constant for some physical processes; for example, the idea of residence time, defined as the sum of all process capacities divided by a flow term, has some utility. For the agitated heating tank, we showed that the time constant was equal to

M/F, where M is the mass of material in the tank, and F is the flow rate. If we placed two tanks in series with the same flow rate, we would estimate the overall time constant to be 2M/F, a larger number. The increased time constant corresponds to a more slowly responding system, which is intuitively correct. We can also use this residence time idea successfully in analyzing response times for water reservoirs and blending tanks. In systems that can be analyzed in terms of resistance (R) and capacitance (C) (e.g., power networks, flowing fluids), the time constant can be approximated as $R \cdot C$.

One implication of the physical interpretation of the time constant, as given above, is that it can change as operating conditions vary. For example, if the flow rate in the agitated heating tank increases above the design flow rate, then the time constant is reduced. Another consideration in the speed of response of the system is whether a controller is involved in the combined system, i.e., computer plus process. One purpose of any controller is to speed up the normal response of the uncontrolled process. Therefore analysis of the process alone is

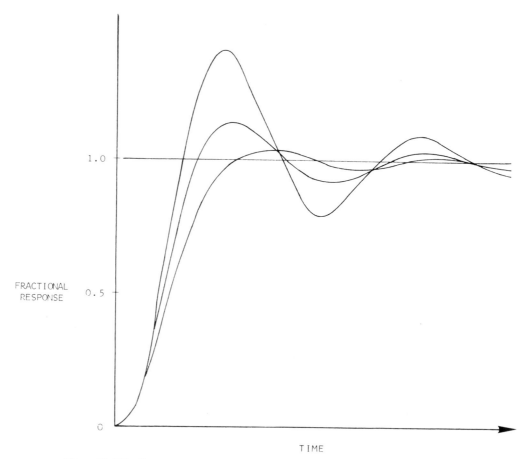

Figure 19.3(b). Step Response of Second-Order Model with Oscillation (Imaginary Roots)

often insufficient for predicting response times of the controlled process. More details on the procedure for analyzing processes under feedback control are given in Chapter 20.

19.4 SELECTION OF A SAMPLING TIME

The evaluation of the dominant time constant has direct application to the selection of a sampling time for the process of interest. In process control, we are usually concerned with influencing the process so that smooth operation is maintained. If the sampling rate is very low (sampling time or period is high), a disturbance can enter and leave the system without being observed. Fortunately, most physical and chemical processes are self-regulating and will return to steady state on their own. However, in this case we are not performing adequate process control. Hence, for the purposes of

data acquisition and control, it is important that the dynamics of a process be evaluated in selecting a sampling time.

The cost of sampling (the number of data points taken by the computer multiplied by some unit overhead cost) decreases as the sampling time increases. On the other hand, the economic cost due to lost information and control system performance deterioration increases as we increase the sampling time [7]. For a given process and for a given computer, we can empirically find the optimum value of the sampling time. However, rather than perform such an optimization, we usually select the sampling time in order to retain all the process dynamic information that we desire. If the computer is incapable of performing the necessary data acquisition activities, then we probably should expand the existing computer or obtain an additional computer.

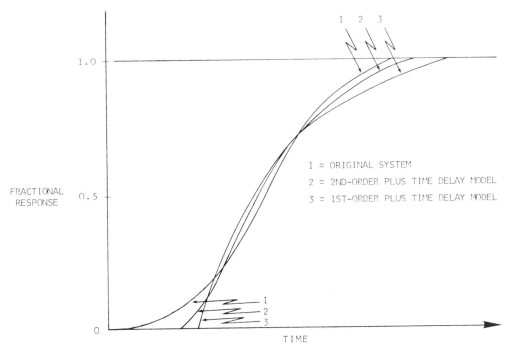

Figure 19.4(a). Step Response of Complex System and Comparison with Simple Models

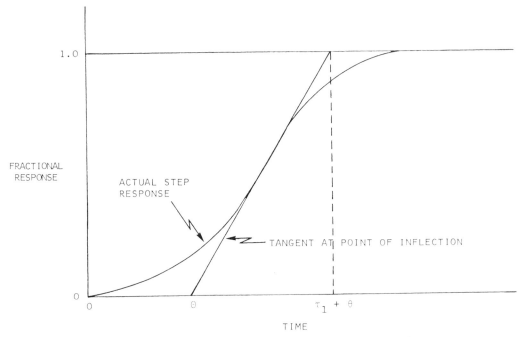

Figure 19.4(b). Graphical Construction of Process Reaction Curve, Step Response

Several textbooks on process control give guidelines for sampling rates for various types of measurements:

Type of Measurement	Sampling Time in Seconds
Flow	1
Level and pressure	5
Temperature and composition	20

It should be realized that such guidelines are arbitrary, because there are large variations from one process to another in terms of process dynamics. As an example, it would be dangerous to take a tray temperature sampling time developed empirically for a 200-tray distillation column and apply it directly to a 10-tray column.

The choice of sampling time should be related as much as possible to the operating conditions of the system. For example, if the flow or throughput to a distillation column or chemical reactor is increased, the temperature or composition response of the system will become more rapid. Therefore, as the system changes, we also need to recognize that changes in the sampling time should be made. In a practical application we would probably design the sampling rate for the worst possible case.

There are some simple cases that illustrate how information is lost in the sampling operation. In an on-line digital computing application, the computer acquires data at a uniform rate. The data are obtained from experimental equipment via the analog to digital converter (ADC, cf. Chapter 10); the digital representations can then be used as inputs in a computer program. The data acquisition step is a discrete operation; by definition, data can only be acquired at specified intervals of time. Hence there is no continuous "readout" as is the case for analog instrumentation. We can improve the sampled data approximation to the continuous function by increasing the sampling rate (and decreasing the sampling period), but, as discussed above, there can be computing equipment limitations when we need to monitor many points in the plant. Variations in the

process variables, which occur between sampling instants, represent "lost" information, since intermediate data are ignored in the internal or stored computer representation which is simply constant over the entire sampling period.

We face a similar problem when using the computer to generate an output. Smith [7] has described three types of output devices: digital to analog converters, pulse generators, and contact closures. Probably the most versatile output device is the digital to analog converter (DAC, cf. Chapter 10), which translates a digital number into an analog signal. However, this analog signal, whose digital representation is transmitted from the computer in a brief instant, must be sustained over some period of time to be of utility. This operation is performed by latching or hold circuits within the DAC which, consequently, provide a continuous value for the output over the sampling period. When this continuous value is a constant (has zero slope), we speak of the action that is obtained as that of a "zero-order hold".

Figure 19.5 illustrates how a continuous input function might look when the data are read into the computer. Obviously as the sampling interval is contracted, the input and the output from the sampler and zero-order hold become closer and closer. Again one has to balance possible errors in data acquisition with the availability of the computer for sampling operations.

A second illustration of loss of information is seen in Figure 19.6. (See, for example, Goff [4].) Suppose the signal that we are sampling is sinusoidal in nature (a), but is sampled less than once per period. It is obvious that the reconstructed signal is different from the original function. In fact, according to the Shannon Sampling Theorem [5], we need to sample this signal more than twice each period in order to recover the original sinusoidal signal; i.e., the sampling rate must be more than twice the frequency of the sine wave.

As discussed in Section 19.2, we can measure the speed of a process experimentally or, by using linear differential equation theory, find the system time constant analytically. An appropriate sampling time for the process de-

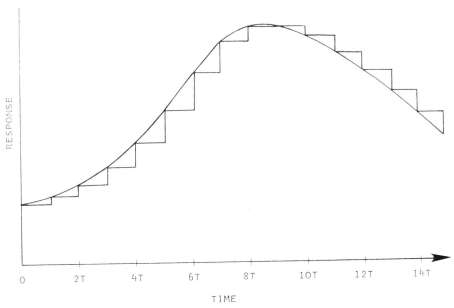

Figure 19.5. True Representation of an Arbitrary Input and Its Sampled Representation Using a Zero-Order Hold

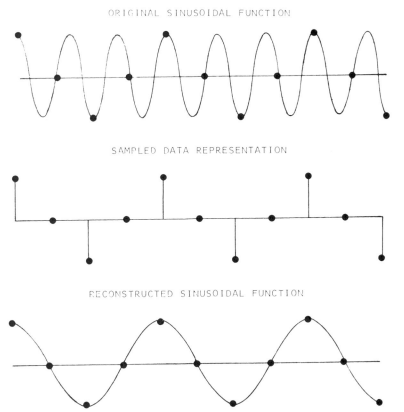

Figure 19.6. Aliasing Error Resulting from Sampling at $\frac{3}{4}$ Samples Per Cycle of Original Function

scribed by Equation (19.9) is $0.1\tau_1$. If dead time (θ) exists and is of the same order of magnitude as τ_1, we should select the sampling time to be $0.1\tau_1$ or 0.1θ, whichever is smaller. If $\theta \ll \tau_1$, we can ignore the effects of dead time in the sampling rate selection. Finally, if $\theta \gg \tau_1$ the sampling time can be as long as θ with little loss of performance. While guidelines given here are largely empirical, they are supported by more rigorous analyses [5]. Faster sampling is, in general, a waste of computing power.

One possible exception to the statements made above is with alarm systems. One should sample process measurements tied into alarm hardware fairly rapidly, for instance, on the order of every second, especially if the hazardous condition has the potential of worsening in a short amount of time. In this case we usually do not explicitly evaluate process dynamics.

In machining, discrete parts manufacturing, and assembly plants, one is faced with a somewhat different problem. One does not typically analyze the dynamics of machining processes or assembly lines per se. Here the sampling operation is concerned with the operational speeds of equipment relative to the desired resolution or accuracy in the final product.

19.5 DISCRETE-TIME MODELS

Once the computer is interfaced with a process, we must take into account the discrete-time nature of the combination of the computer plus process, and it becomes convenient to analyze the process as well as the computer using discrete-time equations. As discussed in the previous section, the computer acquires experimental data that must be translated into digital form by an analog to digital converter. The digital computer cannot directly process the analog signals without digitizing the continuous signal at discrete moments in time. In the analysis of the total system which includes the computer plus sample and hold device, it is useful to deal with discrete-time or difference equations. These equations do not provide information between sampling instants but rather only represent the input–output data at discrete moments in time. We can readily convert continuous-time models to discrete-time form, as will be shown in this section.

Some advantages of dealing with discrete-time models instead of continuous-time models are as follows:

(1) The form of the equations is such that a linear combination of previous values of inputs and outputs is used to generate the value of the output at the next sampling time. Therefore, the computational effort for integration and simulation is minimal, and discretization errors can be made negligible.

(2) Computer control and filtering algorithms can be simplified and readily implemented (see Chapter 20).

(3) Through a transformation (the so-called z-transformation), one is able to analyze rather complicated methods and systems in discrete time more easily (see Section 19.8).

(4) An empirical discrete-time model is much easier to develop on-line than its continuous-time counterpart; well-defined inputs such as a step function are not crucial to computing a difference equation model (see Section 19.6).

Let us again consider Equation (19.9) with $K = 1$; this time we interpret the equation as the equivalent of a simple RC filter used in electronic circuits. What we wish to derive is the discrete or digital version of this filter. In this development $u(t)$ is the input that is to be filtered; $x(t)$ is the filter output. For a sampled data system, $u(t)$ is considered to be held constant at u_k over the sampling interval, $kT \leqslant t \leqslant (k+1)T$ where T is the sampling interval and k is a positive integer; we can approximate Equation (19.9) by using a finite difference formula for the derivative, or

$$\tau_1 \frac{x_{k+1} - x_k}{T} = -x_k + u_k \qquad (19.14)$$

Rearranging, we write

$$x_{k+1} = (1-\alpha)x_k + \alpha u_k \qquad (19.15)$$

where $\alpha = (T/\tau_1)$. Equation (19.15) represents a first-order digital filter which predicts the new value of x based on the old values of u and x. The filter is designed to give a new value of x that is less sensitive to noise than the raw data point. Therefore we have more confidence in the filtered value for use in analysis and control. The use of the filter for noisy data is demonstrated in Chapter 20. An alternate version of Equation (19.15) can be obtained by solving Equation (19.9) analytically for a piecewise constant input; in this case we can obtain the same equation as Equation (19.15), but with $\alpha = 1 - \exp(T/\tau_1)$. The use of this expression for α in Equation (19.15) eliminates any numerical error due to discretization. If the ratio of the sampling time to time constant is 0.1, we calculate α by finite difference models (Equation 19.13) to be 0.1, while the correct value of α from the analytical formula is $1 - \exp(-0.1) = 0.0996$. If $T/\tau_1 = 1.0$, then α (approximate) = 1.0 while α (analytical) = 0.632.

More details on the analytical derivation as well as the conversion of the general nth-order linear differential equation to an nth-order linear difference equation have been provided by Cadzow and Martens [1]. Other references for the mathematics of digital or discrete-time systems include the textbooks by Kuo [6] and Tou [8].

19.6   FITTING DISCRETE-TIME EQUATIONS TO DATA

In Section 19.2, we discussed ways in which a simple continuous-time differential equation model could be developed from actual experimental data. While the graphical constructions are demonstrably easy to apply for first-order or approximately first-order systems, the analysis for second- and higher-order systems is much more cumbersome. We also require that the input change used to perturb the process be a step change, which usually necessitates isolating the piece of equipment to be modeled. In an actual operating plant, people responsible for operations generally are not enthusiastic about deliberately perturbing a process to make model calculations. On-line testing to obtain steady state and dynamic models is greatly preferred. Discrete-time models of high order can be readily developed by processing data from normal process drift and dynamic excursions, and a small computer can calculate coefficients directly from the input–output data. In the following discussion we shall demonstrate how a discrete-time model can be computed.

Consider a second-order discrete-time or difference equation,

$$x_{k+1} = a_1 x_k + a_2 x_{k-1} + b_1 u_k + b_2 u_{k-1} \tag{19.16}$$

where u_k is the input and x_k is the output at time step k. Note that Equation (19.16) is linear with respect to $a_1, a_2, b_1,$ and b_2. Since the values of x's and u's at previous times k and k-1 are known, they can be considered to be inputs, while x_{k+1} is the output predicted for the next time step. In developing discrete-time models, the parameters $a_1, a_2, b_1,$ and b_2 are treated as unknowns to be computed.

Usually we choose the unknown parameters so that the error between the measured x and predicted x (via Equation 19.16) is as small as possible. The same technique used for polynomial curve fitting with a least squares error criterion (linear regression) can be used here; the error criterion to be minimized is simply

$$J = \sum_{k=1}^{n} (x_k - \hat{x}_k)^2 \tag{19.17}$$

where the x_k are the output data points, the $\hat{x}_k$ are the corresponding predicted values, and n is the number of data points. Computer programs that calculate the unknown parameters for such a linear regression problem are readily available from a number of sources. The model can be made as complicated as desired, with any number of terms involving previous values of x and u as long as the parameters appear linearly in the equation. However, the model parameters computed are a function of the sampling time; if the sampling time is changed, then the model parameters will also change.

Consider the following set of data, which gives the response of x to an irregular (arbitrary) input pulse:

k	x_k	u_k
0	0	1.0
1	0.6	0.5
2	1.0	0.2
3	1.0	0.0
4	0.88	0.0
5	0.74	0.0
6	0.63	0.0
7	0.54	0.0

for $k < 0$, $x_k = 0$, $u_k = 0$
(at steady state)

Note that the sampling operation is such that at time zero, the output does not change instantaneously as does the input u.

In the analysis of the above data, the fact that u is an arbitrary input does not make the computation of model parameters more difficult than if we had used a step input. Secondly, the excitation of the system occurs over a limited period of time, and we do not deliberately force the system to a new steady state. This feature is attractive for a continuously operating process. We can easily program a real-time computer to perform the testing of the process as well as the data analysis.

Suppose that we desire to fit a second-order difference equation, in the form of Equation (19.15), to the input–output data. There are four inputs $(x_k, x_{k-1}, u_k, u_{k-1})$ and one output (x_{k+1}) and four unknowns (a_1, a_2, b_1, b_2). Therefore for each value of x_{k+1} generated, there must be four input values considered in the model. We then structure the data as shown in Table 19.2.

If a higher-order difference equation is desired, we merely add more columns to the table, one column for each unknown para-

meter. Using a least squares linear curve fitting computer program [7], the coefficients in Equation (19.15) are calculated to be

$$a_1 = 0.5$$
$$a_2 = 0.3$$
$$b_1 = 0.6$$
$$b_2 = 0.4$$

This model can be used in a number of ways. First it serves as a benchmark of the process dynamics. If the model parameters are computed once a month and two models are found to be quite different, we can tell that the process operation, at least in the dynamic state, has changed owing to wear in valves, fouling of heat transfer surfaces, use of a different raw material, etc. This situation implies that we may want to change the sampling rates or redesign the process controller, if one is in use.

A discrete-time model can be useful for analyzing the stability and overall dynamics of a more complicated configuration, involving process, sensors, controllers, etc. A convenient way of doing this is to express the process and other models as z-transforms, which facilitates the analysis of the total real-time system. This z-transform approach is discussed in Section 19.8 of this chapter.

19.7 LAPLACE TRANSFORMS

Anyone who has much experience in obtaining analytical solutions to linear ordinary differential equations knows that the procedure can often become quite tedious. As a way to re-

Table 19.2. Matrix for Estimating Parameters in Discrete-Time Equations

time	output	inputs			
k	x_{k+1}	x_k	x_{k-1}	u_k	u_{k-1}
k<0	0	0	0	0	0	←steady state
0	0.6	0	0	1.0	0
1	1.0	0.6	0	0.5	1.0
2	1.0	1.0	0.6	0.2	0.5
3	0.88	1.0	1.0	0.0	0.2
4	0.74	0.88	1.0	0.0	0.0
5	0.63	0.74	0.88	0.0	0.0
6	0.54	0.63	0.74	0.0	0.0

duce the effort in solving differential and difference equations, transforms are often used. At the same time, these transforms conveniently express dynamic input–output relationships.

Transformations to simplify mathematics are rather commonplace. In this section we shall introduce a transformation to convert differential equations related to time to algebraic equations related to a different variable. This approach permits an algebraic solution in terms of the transformed output variable. An inverse transformation is then used to obtain the time domain solution.

The Laplace transform is defined as follows:

$$\mathcal{L}[f(t)] = \int_0^\infty f(t)e^{-st}\, dt \qquad (19.18)$$

where $f(t)$ is some function of time. When the integration is performed, the transformed quantity becomes a function of s. Let us calculate the Laplace transform for several simple functions. For $f(t) = a_0$ (a constant),

$$\mathcal{L}(a_0) = \int_0^\infty a_0 e^{-st}\, dt = -\frac{a_0}{s} e^{-st} \Big|_0^\infty \qquad (19.19)$$

$$= 0 - \left[-\frac{a_0}{s}\right] = \frac{a_0}{s} \qquad (19.20)$$

$$\mathcal{L}(e^{-at}) = \int_0^\infty e^{-at}e^{-st}\, dt = \int_0^\infty e^{-(a+s)t}\, dt \qquad (19.21)$$

$$= \frac{1}{s+a}\left[-e^{-(a+s)t}\right]\Big|_0^\infty = \frac{1}{s+a} \qquad (19.22)$$

The transform of a derivative is more complicated,

$$\mathcal{L}\left[\frac{df}{dt}\right] = \int_0^\infty \left[\frac{df}{dt}\right]e^{-st}\, dt \qquad (19.23)$$

Integrating by parts,

$$= s\mathcal{L}(f) - f(0) = sF(s) - f(0) \qquad (19.24)$$

where $F(s)$ is a symbol for the Laplace transform of $f(t)$. If we consider a deviation variable that is zero at the initial time (time zero), then $f(0) = 0$ and $(df/dt) = sF(s)$. This greatly simplifies the manipulation of Laplace transforms.

19.7.1 Representation of Dynamic Systems with Laplace Transforms

Consider the simple first-order differential equation derived in earlier sections:

$$\tau_1 \frac{dx}{dt} = -x + Ku \qquad \begin{array}{l} \text{x: output} \\ \text{u: input} \end{array} \qquad (19.25)$$

where τ_1 is defined as the time constant of the process and K is the process gain. At time zero the system is at its steady state (equilibrium); hence $x(0) = 0$. Taking the Laplace transform of both sides of the equation, we have

$$\mathcal{L}(\tau_1 dx/dt) = \mathcal{L}(-x + Ku) \qquad (19.26)$$

$$\tau_1 \mathcal{L}(dx/dt) = \mathcal{L}(-x) + K\mathcal{L}(u) \qquad (19.27)$$

Note that τ_1 can be factored out of the transform. Equation (19.27) illustrates that the Laplace transform is a linear operator and thus satisfies the principle of superposition. Using Equation (19.24),

$$\tau_1 s\, X(s) = -X(s) + KU(s) \qquad (19.28)$$

Rearranging the equation,

$$(\tau_1 s + 1)\, X(s) = KU(s) \qquad (19.29)$$

$$X(s) = \frac{K}{\tau_1 s + 1} U(s) \qquad (19.30)$$

$$X(s) = G(s)\, U(s) \qquad (19.31)$$

where $G(s)$ is called the "transfer function". For any input $U(s)$, calculated from $u(t)$, we can find $X(s)$ by a simple multiplication. Equation (19.25) represents a first-order transfer function (or a first-order system). For a general second-order system (a second-order differential equation describes the system) and zero initial conditions,

$$b_2 \frac{d^2 x}{dt^2} + b_1 \frac{dx}{dt} + b_0 x = a_1 \frac{du}{dt} + a_0 u \quad (19.13)$$

we obtain a transfer function of the form

$$G(s) = \frac{a_0 + a_1 s}{b_0 + b_1 s + b_2 s^2} \quad (19.32)$$

where the gain of $G(s)$ is a_0/b_0. The order of the polynomial in the denominator of the transfer function corresponds to the order of the original differential equation. For a pure time delay (θ) between input and output, the transfer function numerator will be multiplied by the term, $e^{-\theta s}$.

Now let us return to the step response patterns observed earlier (Figures 19.2, 19.3, 19.4) and match the process models with appropriate transfer functions:

(1) First-order with time delay model:

$$G(s) = \frac{Ke^{-\theta s}}{1 + \tau_1 s} \quad (19.33)$$

(2) Second-order overdamped with time delay model

$$G(s) = \frac{Ke^{-\theta s}}{(1 + \tau_1 s)(1 + \tau_2 s)} \quad (19.34)$$

(3) Second-order underdamped with time delay model

$$G(s) = \frac{Ke^{-\theta s}}{1 + c_1 s + c_2 s^2} \quad (19.35)$$

(Time constants are imaginary numbers for this third type of system.)

19.7.2 Calculation of Transient Responses with Laplace Transforms

While the Laplace transform notation is compact and convenient to use, the analysis of pro-

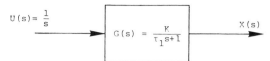

$$U(s) = \frac{1}{s}$$

$$G(s) = \frac{K}{\tau_1 s + 1}$$

$$X(s)$$

Figure 19.7. Block Digram for First-Order System

cess dynamics requires that we interpret the transform in the time domain. In order to obtain the time domain solution from a Laplace transform, we must invert the Laplace transform (a function of s) back to a function of t. A simple example patterned after Equation (19.30) is shown in Figure 19.7. In this figure $u(t)$ is a unit step input (height = 1.0) which is actually a change of 1.0 unit in the input u, since u is a deviation variable. Its transform is $1/s$. The block diagram denotes the input/output relationship. $G(s)$ operates on the input to produce $X(s)$. By multiplying, we obtain

$$X(s) = G(s) \cdot U(s) = \frac{K}{s(\tau_1 s + 1)} \quad (19.36)$$

In order to find $x(t)$, we must invert the expression in s back to a function of t. Tables of inverse Laplace transforms are generally used here such as Table 19.3.

The above table does contain the needed relationship, and the step response, $x(t)$, is $K(1 - e^{-t/\tau_1})$ as we found earlier. However, if it did not, we could decompose $X(s)$ into a sum of two fractions (a process called partial fraction decomposition)

$$\frac{K}{(\tau_1 s + 1)s} = \frac{\alpha_1}{\tau_1 s + 1} + \frac{\alpha_2}{s} \quad (19.37)$$

and solve for α_1 and α_2.

A simple procedure for obtaining α_1 and α_2 is as follows [3]:

(1) Multiply both sides of Equation (19.37) by s

$$\frac{K}{\tau_1 s + 1} = \frac{\alpha_1 s}{\tau_1 s + 1} + \alpha_2 \quad (19.38)$$

Table 19.3. Laplace Transforms for Various Time Domain Functions

f(t)	F(s)
a $\delta(t)$ (impulse)	a
a $S(t)$ (step)	a/s
at (ramp)	a/s^2
t^{n-1}	$(n-1)!/s^n$
e^{-at}	$1/(s+a)$
$\frac{1}{b-a} [e^{-at} - e^{-bt}]$	$1/(s+a)(s+b)$
te^{-at}	$1/(s+a)^2$
$e^{-at} \sin \omega t$	$\omega/((s+a)^2 + \omega^2)$
$e^{-at} \cos \omega t$	$(s+a)/((s+a)^2 + \omega^2)$
$a \sin(\omega t + \phi)$	$(a\omega \cos \phi + as \sin \phi)/(s^2 + \omega^2)$
$K(1 - e^{-t/\tau})$	$\dfrac{K}{s(\tau s + 1)}$
$\dfrac{dx}{dt}$	$s\,X(s) - x(0)$
$\dfrac{d^n x}{dt^n}$	$s^n X(s) - s^{n-1} x(0) - s^{n-2} x^1(0) - \ldots$ $- sx^{(n-2)}(0) - x^{(n-1)}(0)$

Note: $x^{(i)}$ denotes the ith time derivative of $x(t)$.

(2) Let $s = 0$

$$\alpha_2 = K$$

(3) Multiply Equation (19.37) by $\tau_1 s + 1$

$$\frac{K}{s} = \alpha_1 + \frac{\alpha_2(\tau_1 s + 1)}{s} \qquad (19.39)$$

(4) Let $s = 1/(-\tau_1)$

$$\frac{-K}{1/\tau_1} = \alpha_1 \longrightarrow \alpha_1 = -K\tau_1$$

Therefore:

$$\frac{K}{(\tau_1 s + 1)s} = \frac{-K\tau_1}{\tau_1 s + 1} + \frac{K}{s} \qquad (19.40)$$

Now we invert each term individually ($\mathcal{L}^{-1}$ signifies the inverse Laplace transform), noting

that $\mathcal{L}^{-1}$ also is a linear operator, using Table 19.3:

$$\mathcal{L}^{-1}\left[\frac{K}{(\tau_1 s + 1)s}\right] = \mathcal{L}^{-1}\left[\frac{-K\tau_1}{\tau_1 s + 1}\right] + \mathcal{L}^{-1}\left[\frac{K}{s}\right] \qquad (19.41)$$

$$\mathcal{L}^{-1}\left[\frac{-K\tau_1}{\tau_1 s + 1}\right] = \mathcal{L}^{-1}\left[\frac{-K}{s + (1/\tau_1)}\right] = -Ke^{-t/\tau_1} \qquad (19.42)$$

$$\mathcal{L}^{-1}\left[\frac{K}{s}\right] = K \qquad (19.43)$$

$$\mathcal{L}^{-1}\left[\frac{K}{(\tau_1 s + 1)s}\right] = K - Ke^{-t/\tau_1} \qquad (19.44)$$

A graph of the time domain response was shown in Figure 19.2. We can also develop the response of this first-order system subject to an impulse input, $\delta(t)$, or a sine or cosine input.

Note that these inputs are also included in Table 19.3. If G(s) is a complicated transfer function, the inversion process and partial fraction decomposition can become rather time-consuming, but in general the transform method of analysis is competitive with time domain analysis methods, especially after the user becomes experienced with the procedure.

One other useful feature of the transfer function approach is that the denominator of the transfer function can be analyzed to determine its poles or roots (pole = reciprocal of time constant). For example, if

$$G_p(s) = \frac{1}{1 + 3s + 2s^2}$$

the denominator can be factored into $(1 + 2s)$ $(1 + s)$. When the partial fraction technique is used, we see that

$$G_p(s) = \frac{\alpha_1}{1 + 2s} + \frac{\alpha_2}{1 + s} \qquad (19.45)$$

and the process has time constants of 2 and 1. However, if

$$G_p(s) = \frac{1}{1 - s - 2s^2} = \frac{1}{(1 - 2s)(1 + s)} \qquad (19.46)$$

We know that the system is unstable because it has a transient response involving $e^{t/2}$ and e^{-t}. The term $e^{t/2}$ is unbounded for large time. Therefore, the transfer function denominator gives information about stability of the system.

Other uses of the Laplace transform include determining the final $(t = \infty)$ value of a function by the final value theorem. The asymptotic value of x(t), $x(\infty)$, can be found from

$$x(\infty) = \lim_{s \to 0} (sX(s)) \qquad (19.47)$$

Consider the step response of a first-order system

$$X(s) = \frac{K}{\tau_1 s + 1} \cdot \frac{1}{s} \qquad (19.48)$$

Applying the final value theorem, we obtain

$$sX(s) = \frac{K}{\tau_1 s + 1}; \lim_{s \to 0} \frac{K}{\tau_1 s + 1} = K \qquad (19.49)$$

Hence the final value (steady state) of x(t) is found to be K without having to solve for the entire response.

It was previously mentioned that the use of a computer introduces a discontinuity at the data acquisition interface. A datum point, which is recorded by the computer at the beginning of the sampling period, is assumed to remain constant over that sampling interval. Using the Laplace transform of a pure time delay element $(e^{-\theta s}$, where θ is the time delay), we can develop a transfer function for such a sampler including the zero-order hold output circuit. This transfer function can be used to develop information on overall system dynamic behavior; its use is discussed in Sections 19.8 and 19.9. We assume that the instantaneous datum value is held constant at its measured value (f_k) over the period $kT \leqslant t \leqslant (k + 1)T$. The discretization operation yields a signal corresponding to a single rectangular pulse; Figure 19.5 shows a series of such pulses. A single rectangular pulse can be constructed by delaying the measured value (f_k) T units and subtracting the delayed measurement from the original measurement. Since the Laplace transform of a constant, f_k, is f_k/s, the transform of the rectangular pulse is

$$X(s) = \frac{f_k}{s} - e^{-Ts} \frac{f_k}{s} = (1 - e^{-Ts}) \frac{f_k}{s} \qquad (19.50)$$

If f_k is an impulse input, such as is received by the computer in data acquisition or is transmitted by the computer to a zero-order hold, the transfer function for a sampler becomes

$$H(s) = \frac{1}{s} [1 - e^{-Ts}] \qquad (19.51)$$

since the transform of an impulse of magnitude f_k is simply f_k (see Table 19.3). For more information on other properties of Laplace transforms, the reader is referred to the works of Churchill [2] and Coughanowr and Koppel [3].

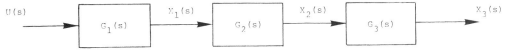

Figure 19.8. A Series of Three Transfer Functions (Blocks)

19.7.3 Information Flow (Block) Diagrams

The analysis of information flow diagrams can be greatly facilitated by the use of Laplace transforms. Block diagrams composed of transform elements aid in analyzing the interactions between input and output variables as well as in determining which subsystems can be easily obtained by decomposing a larger (more complex) system. One use of the Laplace transform in block diagram analysis was shown in Figure 19.7; viz., one merely multiplies the input $U(s)$ by the transfer function $G(s)$ to produce the output $X(s)$.

Consider a series of three processes such as is shown in Figure 19.8. $X_3(s)$ is the ultimate output response to $U(s)$. We can write therefore

$$X_1(s) = G_1(s)U(s)$$
$$X_2(s) = G_2(s)X_1(s)$$
$$X_3(s) = G_3(s)X_2(s) \qquad (19.52)$$

By successive substitution we can obtain

$$X_2(s) = G_2(s)G_1(s)U(s)$$
$$X_3(s) = G_3(s)G_2(s)G_1(s)U(s) \quad (19.53)$$

Explicit functions of s can be substituted into the above equations for each of the transform elements; such functions could be obtained by physical principles modeling (a differential equation), or by experimental testing of the process. The models G_1, G_2, and G_3 can be obtained individually, or we can obtain the overall product $G_1 \cdot G_2 \cdot G_3$. We shall discuss an example of the series rule later in this section.

Blocks also follow a rule of addition, shown in Figure 19.9.

$$X(s) = G_1(s)U_1(s) + G_2(s)U_2(s) \tag{19.54}$$

When analyzing complex block diagrams we need to invoke one of the basic laws of systems analysis; i.e., the sum of all flow variables for any junction is equal to zero. This law is exemplified by Equation (19.54); it is merely a conservation statement.

Let us consider a simple control problem related to the agitated heating tank system treated in this chapter. We shall use block diagrams to aid in analysis of the control and instrumentation system. One way of controlling the outlet temperature subject to changes in feed rate or feed temperature is to use the steady state model derived earlier as Equation (19.3). For a given tank outlet temperature, this equation allows us to calculate the necessary steam rate. This type of control is referred to as "feedforward control" [9], but is not always as effective as is desired. A more conventional control approach, one that takes into

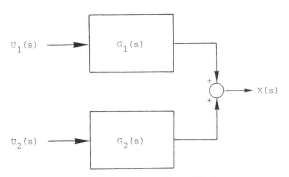

Figure 19.9. The Addition of Two Blocks

account the process dynamics, is to install a feedback controller between the outlet temperature and the steam flow rate. A feedback controller can consist of a self-contained electronic or pneumatic device, or it can be computer-based. The design of digital feedback controls is only touched on briefly here; Chapter 20 discusses this subject more fully.

The block diagram of a feedback temperature controller, shown in Figure 19.10, consists of several components. For simplicity, Figure 19.10 does not yet include a computer-based controller; we shall address that question in Section 19.8. The temperature is measured by a thermocouple; the electrical signal from the thermocouple is generated via a transmitter, $G_1(s)$. The signal is then compared with (subtracted from) the set point (desired operating temperature) to obtain the error signal. This error signal is fed to the controller, $G_2(s)$, which operates on it, ultimately setting the position of the valve. The controller is designed so that the dynamic response of the outlet temperature has desirable properties. This controller could be a digital computer interfaced with an ADC and DAC or, as in this example, an analog device. The controller output will be considered to be an electrical signal. In order for the controller to effect a change in the steam rate, we need a transducer to convert the

electronic signal to an air pressure signal, which in turn will actuate a valve to vary the steam flow rate. Hence two blocks are involved: the transducer, G_3, and the valve, G_4, since both of these components alter the magnitude and time characteristics of the input signal. The valve changes the steam rate, $U(s)$, which affects the tank outlet temperature through the process transfer function derived earlier. In this particular block diagram we show signals entering the loop in the form of a set point and disturbance changes ($R(s)$ and $D(s)$ are deviation variables). The disturbance changes affect the measured tank and outlet temperature via transfer function G_6 which can be found experimentally or analytically, just as with the process transfer function, G_5. (More details on obtaining disturbance transfer functions for the agitated heating tank have been provided by Weber [9].) Therefore the error signal, which is the input to the controller (Block G_2), can be related to the output $Y(s)$ by the following expression using the block multiplication and addition rules mentioned above:

$$Y(s) = G_2(s)G_3(s)G_4(s)G_5(s)E(s) + G_6(s)D(s)$$
$$(19.55)$$

where $E(s)$ is the transform of the error signal and $D(s)$ that of the disturbance signal. Since

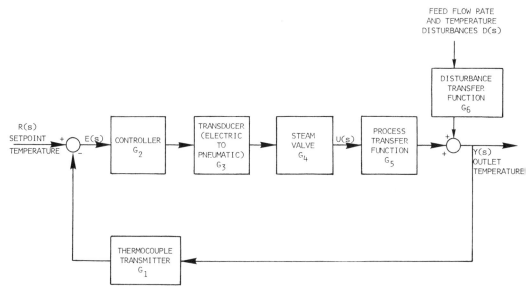

Figure 19.10. Block Diagram for a Regulator Control System, Agitated Heating Tank

Figure 19.10 represents a feedback loop ($Y(s)$ is fed back and compared with the set point), we can relate $E(s)$, the error, to $Y(s)$ by the following expression:

$$E(s) = R(s) - G_1Y(s) \qquad (19.56)$$

Suppose the set point is unchanged (its deviation $R(s)$ is zero) and $D(s)$ is nonzero; then we have a "regulator" control system. In this case, Equation (19.56) becomes

$$E(s) = -G_1Y(s) \qquad (19.57)$$

Combining Equations (19.57) and (19.55) and eliminating $E(s)$, we obtain

$$Y(s) = \left[\frac{G_6}{1 + G_1G_2G_3G_4G_5} \right] D(s) \quad (19.58)$$

where the bracketed quantity is the closed-loop transfer function. Hence under closed-loop or feedback control, the transfer function becomes much more complicated than for no control (open loop conditions). It is evident that the analysis of real control systems requires models (transfer functions) for all components in the control loop, and this may require a significant investment in component testing. However, once these models are obtained, the transfer function approach to systems analysis does allow an economy of effort.

19.8 DISCRETE-TIME TRANSFORMS

Discrete-time transforms have the same utility as continuous-time transforms in that they provide a compact mathematical description of a dynamic model. A transfer function for discrete-time processes can be defined just as for a continuous-time process. Analyses of component systems that involve sampled data operations, such as digital computers, can be performed; these transfer functions then can be manipulated and inverted to the time domain.

Figure 19.11 shows the classical discrete-time "pulse" representation of a continuous signal, $f(t)$. The computer records the discrete-time data only at the sampling instants, as controlled by the analog to digital converter. Therefore

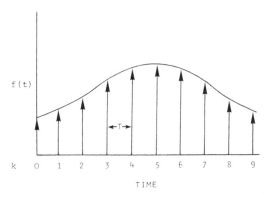

Figure 19.11. Sampled Data Pulse Representation of Continuous Signal $f(t)$

we can think of discrete-time input and output data as a series of numbers (or pulses) acquired at uniform time intervals. We let the sequence $\{f_k\}$ denote the value of the continuous curve, $f(t)$, at discrete instants of time, i.e., at $k = 0$, $k = 1$, etc. One convenient way of expressing this sequence mathematically is through the z-transform. The convention chosen in most engineering texts is that a data point f_k occurring k time steps after zero time is expressed via the transform $f_k z^{-k}$. A sequence of inputs (or outputs) $\{f_k\}$ approximating $f(t)$ can thus be expressed in terms of z-transforms as

$$F(z) = Z(f(t))$$
$$= f_0 + f_1z^{-1} + f_2z^{-2} + \cdots + f_kz^{-k}$$
$$(19.59)$$

$F(z)$ has the same significance to $f(t)$ as does the continuous time transform $F(s)$ to $f(t)$, i.e., $\mathcal{L}(f(t))$ (see Section 19.7). For example, let us consider the z-transform of a step input, where $f_k = 1$ for all k:

$$Z(1) = 1 + z^{-1} + z^{-2} + \cdots + z^{-k}$$
$$(19.60)$$

Recognizing that this infinite series is

$$Z(1) = \frac{1}{1 - z^{-1}} \qquad (19.61)$$

we then have one equivalence between s and z-transforms, namely:

$$Z\left(\frac{1}{s}\right) \equiv \frac{1}{1 - z^{-1}} \quad (19.62)$$

The shifting theorem [1] gives the z-transform of $x_{k+\varrho}$ in terms of $X(z)$:

$$Z(x_{k+\varrho}) = z^{\varrho} X(z) \text{ where } X(z) = Z(x_k) \quad (19.63)$$

Using the z-transform of difference equations, we can develop the transfer function for a dynamic model. The so-called pulse transfer function serves as a sort of shorthand notation similar to Laplace transforms. Let us consider the difference equation (19.15) presented earlier. Taking the z-transform of both sides and using the shift rule (19.63), we obtain:

$$z^1 X(z) = (1 - \alpha)X(z) + \alpha U(z) \quad (19.64)$$

Rearranging and expressing the relationship in terms of z^{-1}:

$$X(z) = \frac{\alpha z^{-1}}{1 - (1 - \alpha)z^{-1}} U(z) \quad (19.65)$$

We then define a discrete transfer function $G(z)$ as

$$\frac{X(z)}{U(z)} = G(z) = \frac{\alpha z^{-1}}{1 - (1 - \alpha)z^{-1}} \quad (19.66)$$

We can readily find the inverse transform of $X(z) = G(z)U(z)$ as a means of inverting the z-transform expression to an analogous result in the time domain. Consider a general z-transform which results from multiplying $G(z)$ and $U(z)$:

$$X(z) = \frac{a_0 + a_1 z^{-1} + a_2 z^{-2} + \cdots + a_m z^{-m}}{b_0 + b_1 z^{-1} + b_2 z^{-2} + \cdots + b_n z^{-n}} \quad (19.67)$$

If the denominator is divided into the numerator by long division, we obtain

$$X(z) = c_0 + c_1 z^{-1} + c_2 z^{-2} + \cdots \quad (19.68)$$

Referring back to Equation (19.59), we can equate the sequence $\{c_k\}$ with $\{x_k\}$, or

$$x_0 = c_0 = a_0/b_0$$
$$x_1 = c_1 = a_1/b_0 - a_0 b_1/b_0^2$$
$$x_2 = c_2 = \text{etc.}$$
$$\cdot$$
$$\cdot$$
$$\cdot$$
$$x_n = c_n$$

Therefore the simulation of a discrete-time transformed system is accomplished much more readily than for continuous transforms if only numerical results are required.

As an example, let us calculate the response of the second-order discrete-time system given in Section 19.6. First we obtain the pulse transfer function for the system. Taking the z-transform of both sides of Equation (19.16) and substituting the numerical values for a_1, a_2, b_1, and b_2, we obtain

$$zX(z) = 0.5X(z) + 0.3z^{-1}X(z) + 0.6U(z)$$
$$+ 0.4z^{-1}U(z) \quad (19.69)$$

In pulse transfer function form,

$$\frac{X(z)}{U(z)} = G(z) = \frac{0.6z^{-1} + 0.4z^{-2}}{1 - 0.5z^{-1} - 0.3z^{-2}} \quad (19.70)$$

For the input sequence $(1, 0.5, 0.2)$ given in Section (19.6), we can write the z-transform of the input as

$$U(z) = 1 + 0.5z^{-1} + 0.2z^{-2} \quad (19.71)$$

Multiplying $G(z) \cdot U(z)$, we can obtain $X(z)$:

$$X(z) = \frac{0.6z^{-1} + 0.7z^{-2} + 0.32z^{-3} + 0.08z^{-4}}{1 - 0.5z^{-1} - 0.3z^{-2}} \quad (19.72)$$

Dividing the numerator by the denominator using the lowest powers of z^{-1} as the leading coefficients, the result for Equation (19.72) becomes

$$X(z) = 0.6z^{-1} + 1.0z^{-2} + 1.0z^{-3} + .88z^{-4}$$
$$+ .74z^{-5} + .63z^{-6} + .54z^{-7} + \cdots$$
$$(19.73)$$

which is the same output sequence that we gave in Section 19.6. Similarly, if we wished to find the step response of the same system, it would only be necessary to multiply $G(z)$ by $U(z) = [1/(1 - z^{-1})]$ and then perform the division shown above to obtain the result.

Some important properties of sampled data systems can be obtained from an analysis of their z-transforms. For example, the time domain response of a first-order z-transform of the form

$$X(z) = \frac{a_0}{1 - b_0 z^{-1}} \qquad (19.74)$$

can be found by analytical long division to give the infinite series

$$X(z) = a_0(1 + b_0 z^{-1} + b_0{}^2 \cdot z^{-2}$$
$$+ \cdots + b_0{}^n z^{-n}) \qquad (19.75)$$

Therefore the sampled data response of this system is given by the sequence $\{a_0, a_0 b_0, a_0 b_0{}^2,$ etc.$\}$. We note that, if $|b_0| > 1$, the magnitude of the response grows steadily over time. We can use this characteristic to determine if a system is stable or unstable.

For systems described by second- or higher-order z-transforms, we can apply partial fraction decomposition to analyze stability in a manner analogous to that for Laplace transforms (Section 19.7.2). Let us consider a second-order transfer function:

$$G(z) = \frac{2 - 1.5z^{-1}}{1 - 2.5z^{-1} + z^{-2}} \qquad (19.76)$$

In order to perform partial fraction decomposition, we factor the denominator into a product of two terms, $(1 - 0.5z^{-1})(1 - 2z^{-1})$, and obtain

$$G(z) = \frac{1/3}{1 - 0.5z^{-1}} + \frac{5/3}{1 - 2z^{-1}} \qquad (19.77)$$

Since the second term will grow in an unbounded fashion with time (because the coefficient 2 is greater than 1), as shown in Equation (19.75), we conclude that $G(z)$ represents an unstable system.

The above discussion has presented only the most limited aspects of stability theory for discrete-time systems. For a more extensive presentation of stability analysis, the reader is urged to consult the textbooks by Kuo [6] and Tou [8].

Pulse transfer functions can also be applied to analysis of block diagrams; the discussion given for Laplace transforms in Section 19.7.3 applies also to z-transforms. For example, let us consider a simplified version of the block diagram shown in Figure 19.10. However, here we replace the continuous controller by a discrete-time computer controller and add switching elements that illustrate the sampled-data nature of the system. Figure 19.12 depicts a "hybrid" block diagram with both continuous and discrete transfer functions. The samplers are placed at the points where the output is acquired (ADC),

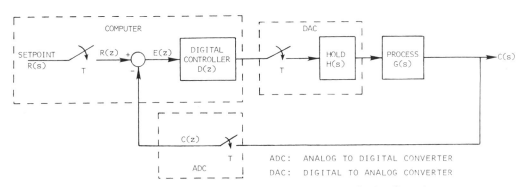

Figure 19.12. Sampled Data Block Diagram for Servomechanism Operation

where the computer controller acts upon the process (DAC), and where the set point enters the digital controller. The zero-order hold after the controller provides a piecewise constant signal from the output generated by the computer and is required to describe the latching or holding action of the DAC. This block diagram omits the measurement transfer function shown earlier in Figure 19.10, for simplicity, and it also differs from Figure 19.10 in that it illustrates a second type of process control problem, namely, servomechanism control. In this case we are concerned with changing the set point of the process from one operating point to another; we desire the process to arrive at the second operating state rather quickly. Therefore R(z), the set point input, is nonzero, and the effects of the disturbance, which were of primary interest in the regulator control problem, are not considered explicitly.

Looking at Figure 19.12, we relate the discrete output C(z) to the discrete error E(z) by the relationship

$$C(z) = HG(z)D(z)E(z) \qquad (19.78)$$

Here G is the process transfer function, H is the zero-order hold transfer function, D(z) is the discrete controller transfer function (U(z) = D(z)E(z)), and T is the sampling time or period. As discussed by Smith [7], we denote the product of the two continuous transfer functions, H(s) and G(s), in discrete time as HG(z).* This type of calculation is discussed in more detail in the next section. As a result of the feedback loop,

$$E(z) = R(z) - C(z) \qquad (19.79)$$

Therefore the closed loop transfer function relating R(z) to C(z) is given by

$$\frac{C(z)}{R(z)} = \frac{D(z)HG(z)}{1 + D(z)HG(z)} \qquad (19.80)$$

---

*Meaning that the product of H(s) and G(s) is first formed before taking the z-transform. In general, $Z(H(s)G(s)) \neq H(z)G(z)$.

19.9. INTERCONVERSION BETWEEN LAPLACE TRANSFORMS AND z-TRANSFORMS

Once a digital computer is interfaced to a continuous process, the combined closed loop process assumes a discrete nature (see Section 19.5). The key component in this transformation is the zero-order hold. The transfer function of the hold is from Equation (19.51),

$$H(s) = \frac{1 - e^{-sT}}{s} \qquad (19.81)$$

where T is the sampling time. In order to find the product of the zero-order hold with a continuous process G(s), we can use the multiplication rule to obtain the resulting transfer function; i.e.,

$$M(s) = H(s)G(s) \qquad (19.82)$$

Consider the first-order system from Section 19.2. M(s) becomes

$$\frac{1 - e^{-sT}}{s} \cdot \frac{1}{\tau_1 s + 1} \qquad (19.83)$$

In order to convert this transfer function to discrete time, we first use partial fraction decomposition

$$H(s)G(s) = \frac{1}{s(\tau_1 s + 1)} - \frac{e^{-sT}}{s(\tau_1 s + 1)} \qquad (19.84)$$

Expanding both fractions further,

$$H(s)G(s) = \frac{1}{s} - \frac{1}{s + (1/\tau_1)} - e^{-sT}\left[\frac{1}{s} - \frac{1}{s + (1/\tau_1)}\right] \qquad (19.85)$$

With this basic form, the product transfer function H(s)G(s) can be converted to z-transforms by use of Table 19.4.

Converting each term in Equation (19.85) into its equivalent z-transform, we obtain

$$Z[H(s)G(s)] = Z\left(\frac{1}{s}\right) - Z\left(\frac{1}{s + (1/\tau_1)}\right)$$

$$- Z\left[e^{-sT}\left[\frac{1}{s} - \frac{1}{s + (1/\tau_1)}\right]\right]$$

(19.86)

(19.86) becomes

$$HG(z) = \left[\frac{1}{1 - z^{-1}} - \frac{1}{1 - e^{-(T/\tau_1)}z^{-1}}\right]$$

$$- z^{-1}\left[\frac{1}{1 - z^{-1}} - \frac{1}{1 - e^{-(T/\tau_1)}z^{-1}}\right]$$

(19.87)

Since $Z[e^{-sT}F(s)] = z^{-1}Z(F(s))$,   Equation

Table 19.4. Table of z-Transforms (T = Sampling Time)

Time Function, f(t)	Laplace Transform, F(s)	z-Transform, F(z)
$a\delta(t - T)$, unit impulse	ae^{-sT}	az^{-1}
$a\delta(t)$, unit step	$\dfrac{a}{s}$	$\dfrac{a}{1-z^{-1}}$
at	$\dfrac{a}{s^2}$	$\dfrac{aT\,z^{-1}}{(1-z^{-1})^2}$
t^{n-1}	$\dfrac{(n-1)!}{s^n}$	$\lim\limits_{a \to 0}(-1)^{n-1}\dfrac{\partial^{n-1}}{\partial a^{n-1}}\left(\dfrac{1}{1-e^{-aT}z^{-1}}\right)$
e^{-at}	$\dfrac{1}{s + a}$	$\dfrac{1}{1-e^{-aT}z^{-1}}$
$\dfrac{1}{b-a}(e^{-at}-e^{-bt})$	$\dfrac{1}{(s+a)(s+b)}$	$\dfrac{1}{b-a}\left(\dfrac{1}{1-e^{-aT}z^{-1}} - \dfrac{1}{1-e^{-bT}z^{-1}}\right)$
te^{-at}	$\dfrac{1}{(s+a)^2}$	$\dfrac{T\,e^{-aT}z^{-1}}{(1-e^{-aT}z^{-1})^2}$
$e^{-at}\sin \omega t$	$\dfrac{\omega}{(s+a)^2 + \omega^2}$	$\dfrac{z^{-1}e^{-aT}\sin \omega T}{1-2z^{-1}e^{-aT}\cdot \cos \omega T + e^{-2aT}z^{-2}}$
$e^{-at}\cos \omega t$	$\dfrac{s + a}{(s+a)^2 + \omega^2}$	$\dfrac{1-z^{-1}e^{-aT}\cdot \cos \omega t}{1-2z^{-1}e^{-aT}\cdot \cos \omega T + e^{-2aT}\cdot z^{-2}}$

$$HG(z) = [1 - z^{-1}]\left[\frac{z^{-1}[1 - e^{-(T/\tau_1)}]}{[1 - z^{-1}][1 - e^{-(T/\tau_1)}z^{-1}]}\right]$$

$$(19.88)$$

Using $\alpha = 1 - e^{-T/\tau_1}$, Equation (19.88) becomes

$$HG(z) = \frac{\alpha z^{-1}}{1 - (1 - \alpha)z^{-1}}$$

$$(19.89)$$

We observe that Equations (19.89) and (19.66) are equivalent, which illustrates that the transform mathematics are consistent. Equation (19.15) gives the difference equation form of Equation (19.66), where the input variable is held at a constant value during the sampling interval, T.

Chemical processes often can be described approximately by a second-order transfer function with time delay, θ, where the process parameters are obtained empirically:

$$G(s) = \frac{Ke^{-\theta s}}{(\tau_1 s + 1)(\tau_2 s + 1)} = \frac{X(s)}{U(s)}$$

$$(19.90)$$

If we assume that $n = \theta/T$ (θ is an integer multiple of the sampling time T), and if u_k is a sampled data (zero-order hold) input, the following difference equation results:

$$x_{k+1} = a_1 x_k + a_2 x_{k-1} + b_1 u_{k-n} + b_2 u_{k-(n+1)}$$

$$(19.91)$$

where

$$a_1 = e^{-T/\tau_1} + e^{-T/\tau_2}$$

$$(19.92a)$$

$$a_2 = -e^{-T/\tau_1}e^{-T/\tau_2}$$

$$(19.92b)$$

$$b_1 = K(\tau_2 - \tau_1 + \tau_1 e^{-T/\tau_1} - \tau_2 e^{-T/\tau_2})/(\tau_2 - \tau_1)$$

$$(19.92c)$$

$$b_2 = K[(\tau_2 - \tau_1)e^{-T/\tau_1}e^{-T/\tau_2} + \tau_1 e^{-T/\tau_2} - \tau_2 e^{-T/\tau_1}]/(\tau_2 - \tau_1)$$

$$(19.92d)$$

In z-transform notation, the transfer function becomes

$$HG(z) = \frac{X(z)}{U(z)} = \frac{(b_1 z^{-1} + b_2 z^{-2})z^{-n}}{1 - a_1 z^{-1} - a_2 z^{-2}}$$

$$(19.93)$$

which is a general expression for a second-order discrete-time model with time delay. We note that, for a first-order transfer function with no time delay, $\tau_2 = 0$ and $n = 0$. Substitution of these values into Equations (19.92) and (19.93) yields the same result as given in Equations (19.88) and (19.89).

Conversely, models in the form of z-transforms can be converted to s-transforms or functions of time. For example, Equation (19.92) can be used to calculate τ_1, τ_2, and K, given that a_1, a_2, b_1, and b_2 have been found. This procedure of course would require a trial-and-error calculation [7].

Partial fraction decomposition in the z domain can also be used to convert z-transforms to the Laplace or time domain [1, 6]. Consider the response of the discrete transfer function, $G(z) = 0.5z^{-1}/(1 - 0.5z^{-1})$, to a step input, $U(z) = 1/(1 - z^{-1})$:

$$X(z) = \frac{0.5z^{-1}}{1 - 0.5z^{-1}} \cdot \frac{1}{1 - z^{-1}}$$

$$(19.94)$$

Decomposing Equation (19.94) into the sum of two fractions, we obtain

$$X(z) = \frac{1}{1 - z^{-1}} - \frac{1}{1 - 0.5z^{-1}}$$

$$(19.95)$$

Using Table 19.4 and inverting each term to the time domain (sampling time, T, equals 1.0):

$$x(t) = 1 - e^{-0.693t}$$

$$(19.96)$$

Note that this type of solution might be more desirable than a numerical response calculated by long division, as discussed in Section 19.8.

19.10. SUMMARY

In this chapter we have described some mathematical tools that are in common use in real-time computing and process control. While the material presented here is not sufficient to make the reader an expert on the subject of modeling dynamic systems, it does introduce many of the important concepts needed for analyzing process dynamics. In so doing, we have attempted to digest and simplify the rather large body of information on modeling dynamic systems. This chapter covers both continuous-time and discrete-time mathematics. The latter subject is emphasized because it is crucial to obtain an understanding of the effect of interfacing a computer, which is basically a discontinuous or discrete device, to a continuous process.

The techniques of process modeling and analysis presented here are intended to help a scientist or engineer perform the following tasks related to a real-time application:

(1) Steady state or dynamic process modeling based on physical considerations.
(2) Dynamic (black box) modeling based on experimental step testing of the process.
(3) Analysis of the resulting process model to determine appropriate sampling rates for data acquisition and/or process control.
(4) Conversion of dynamic models to transfer functions (continuous or discrete) that relate inputs and outputs dynamically. The Laplace (s) or z-transforms can be easily manipulated to analyze the dynamic effect of having various sensing and control instruments interfaced with the process. We also can convert continuous models to discrete models, depending upon the desired form of the model. Models of the overall system constructed as discussed can be used to predict or simulate the behavior of the process connected to the computer.
(5) Analysis of the effects of various controllers on the operation of the process; this question is considered in detail in Chapter 20.

19.11. REFERENCES

1. Cadzow, J. A. and Martens, H. R., *Discrete-Time and Computer Control Systems*, Prentice-Hall, Englewood Cliffs, N.J., 1970.

2. Churchill, R. V., *Operational Mathematics*, McGraw-Hill, New York, 1958.
3. Coughanowr, D. R. and Koppel, L. B., *Process Systems Analysis and Control*, McGraw-Hill, New York, 1965.
4. Goff, K. W., "Dynamics in Direct Digital Control", *ISA J.*, *13* (11), 45, 1966.
5. Koppel, L. B., *Introduction to Control Theory*, Prentice-Hall, Englewood Cliffs, N.J., 1968.
6. Kuo, B. C., *Digital Control Systems*, SRL Publishing Co., Urbana, Ill., 1977.
7. Smith, C. L., *Digital Process Control*, International Textbook, Scranton, Pa., 1972.
8. Tou, J. T., *Digital and Sampled Data Control Systems*, McGraw-Hill, New York, 1959.
9. Weber, T. W., *An Introduction to Process Dynamics and Control*, Wiley, New York, 1973.

19.12 EXERCISES

1. Using partial fraction decomposition, generate the unit step response for the process

$$G(s) = \frac{1}{(s + 2)(s + 4)(s + 6)}$$

a. Calculate the dominant time constant and time delay using the graphical method described in Section 19.2.
b. Expand the dynamic response equation in the form of Equation (19.10). Retain the term with the dominant time constant, and plot the step response for this term only. Compare with the third-order process step response.

2. Suppose the following input-output data have been obtained from a pilot plant:

k	x_k	u_k
0	0	1.0
1	0.6	1.0
2	0.5	1.0
3	0.55	1.0
4	0.525	0
5	0.317	0
6	0.210	0
7	0.137	0
8	0.089	0
9	0.058	0

Fit a second-order discrete time model to these data using linear regression.

3. Find the Laplace transform of the following differential equation:

$$\frac{d^3x}{dt^3} + 6\frac{d^2x}{dt^2} + 11\frac{dx}{dt} + 6x = \frac{du}{dt} + 4u$$

Assume all derivatives of x and u are zero at t = 0. Find the transfer function between X(s) and U(s). Using the final value theorem, find x(t) for t → ∞ and U(s) = 1/s.

4. Using partial fraction decomposition, find the unit step response for the model in Problem 3 above. The roots of the characteristic equation are 1, 2, and 3.

5. Derive the inverse transform in Table 19.3 for

$$G(s) = \frac{1}{(s + a)(s + b)}$$

6. Using Figure 19.10, derive the closed-loop transfer function between R(s) and Y(s).

7. You are asked to simulate the response, x(t), of the dynamic system,

$$G(z) = \frac{0.5z^{-1} + 0.5}{1 - 0.3z^{-1} - 0.1z^{-2}}$$

for the following rectangular pulse (k = time step)

$$0 \leqslant k \leqslant 4 \qquad u_k = 1.0$$
$$k > 4 \qquad\quad u_k = 0.0$$

a. First find the z-transform of u_k (i.e., U(z)).

b. Then compute the response using three methods:

(1) Simulation of the difference equation, G(z).

(2) Long division of the response, X(z) = G(z)U(z).

(3) Partial fraction decomposition of X(z).

8. Convert the continuous time transfer function,

$$G(s) = \frac{1}{(3s + 1)(5s + 1)}$$

to a discrete time transfer function, using the zero-order hold method of Section 19.9. Check your answer using Equation (19.92).

9. A process for which you require a discrete transfer function model has the following transfer function:

$$G(s) = \frac{e^{-2s}}{(2s + 1)(4s + 1)}$$

You have decided to sample at 0.1-second intervals. Calculate the discrete transfer function. Do not forget the sample and hold.

10. Some processes have discrete transfer functions of the form

$$G(z) = \frac{(1 - az^{-1})}{(1 + bz^{-1})}$$

For a simple case where b = 0.9, what is the effect of variations in a? Explore the step response of this process for −2.0 ⩽ a ⩽ 2.0. What are the implications for controller design?

20

Digital Computer Control and Signal Processing Algorithms

Joseph D. Wright
Xerox Research Centre
Mississauga, Ontario

Thomas F. Edgar
Department of Chemical Engineering
University of Texas, Austin

20.0 INTRODUCTION

The concepts of automatic control apply to widely different fields including applications that range from temperature control in home freezers, ovens, and the like, to the quite sophisticated guidance systems used in the aerospace industry for satellite and space module control. The objective of this introductory chapter on digital computer control is to provide an overview of some of these concepts and a discussion of the various stages a control engineer must go through in designing a control system for an industrial process. Initially we discuss a glass-forming process in order to illustrate some of the considerations involved in process analysis and control. Later sections will specifically address discrete or digital controller design, since the focus of this book is on real-time computer applications for data acquisition and control. Finally, some basic techniques for signal processing that are useful for computer data acquisition and control systems will be discussed. The ideas presented in this chapter will integrate many of the concepts of both hardware and software that have been discussed more or less separately in the preceding chapters.

20.0.1 Processes and Process Mathematical Models

In order to discuss process control it is essential to understand what is meant by a process.

Chapter 4 described a number of processes from a wide range of disciplines. The definition of a process may be quite broad. Very generally, a process is a progression to some particular end or objective through a logical and orderly sequence of events or operations. More specifically, in this chapter we shall limit the discussion to the particular chemical or physical operations necessary to convert one or more raw materials into finished products. Usually such operations are performed with the formation of one or more by-products or waste products and with quality control constraints on the desired end products.

Examples of processes that fit the above definitions are the batch reactor system shown in Figure 4.8 and the agitated heating tank system shown in Figure 4.10. The latter example is also discussed from a modeling point of view in Section 19.1.

Many process operations involve a considerable interaction of the human element with the process. At the lowest level, control can be implemented by manual sequencing of switches and valves or by manual inventory operations. However, we shall be principally concerned with applications where the human element is replaced by automatic sensors, control valves, and other mechanical or electronic devices. Furthermore, we shall emphasize processes that can be described by discrete or differential equation models such as those discussed in Chapter 19. It is not within the scope of this chapter to derive, from first principles, dynamic models that describe these processes; here we shall ordinarily either state a result or assume a representative form for the equation(s). Those readers interested in modeling techniques should refer to textbooks on process control (e.g., Refs. 2, 8, 16, and 23). It is worth noting, however, that the most common forms of models that are used for process control applications are empirical models based on first- or second-order linear-(ized) differential equations, often with an associated delay or dead time relating a process input to a process output. As such, they are not explicitly derived using a physicochemical analysis of the process.

20.0.2 Process Control Objectives

Most processes are designed from a steady state point of view, that is, by specifying raw material flow rates into the process, required amounts of energy (perhaps indirectly through calculations of the necessary fuel flow rates), and flow rates of products and by-products from the process. It is important to realize that these parameters are not necessarily constant when the process is in operation. A process seldom operates exactly at its design conditions and capacity and is subject to continuous upsets and changes

due to disturbances that tend to drive it away from equilibrium (or design values). As a result, we must manipulate certain variables (e.g., through controllers) in order to bring the process back to its desired level of operation. Disturbances may arise from changes in such variables as atmospheric conditions (temperature, relative humidity, etc.), fuel supply pressure, cooling water temperature, and the like.

This type of control (overcoming process disturbances) is called regulatory control. A different problem arises when it is required that the process move from one region of operation to another. This control problem is called servo control. The preponderance of control applications require the regulator design approach. Both problems will be discussed later.

In order to clarify some of the above concepts let us consider the process shown in Figure 20.1, namely continuous production of sheet glass. Let us specifically focus on one section of this process, the fire-polishing zone of a float bath. The desired product is a polished glass surface; the surface quality is dependent on the amount of heat it receives and the temperature of operation. Since the amount of heat transferred is a rather difficult variable to measure, but is more or less related to the furnace temperature, we choose this latter variable as the controlled or measured variable (sometimes called the output variable).

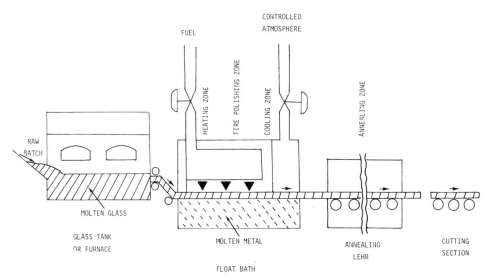

Figure 20.1. Sheet Glass Forming Process

Ordinarily this output variable is measured continuously. Suppose the furnace temperature is above or below some desired value, e.g., the design temperature of the furnace. What can we do to adjust the temperature? Clearly, we could manipulate the fuel flow rate to the burners by the opening or closing of a valve in the fuel line. The fuel flow rate is one type of input variable to the process, and is called a control variable. Another type of input variable is a disturbance variable, which is not controllable and causes the temperature to change from its desired value. For this process, several disturbance variables can be suggested, such as changes in the fuel gas supply pressure, changes in fuel gas composition (with the result that either more or less heat is produced with the same gas flow rate), and a change in the speed of passage of the glass plate through the furnace.

The desired operating point for a process is called the set point. The difference between the set point and the measured variable is the error. The most common control system configuration is called feedback control; in this design an error must be observed before a control variable can be manipulated to restore the system to the set point.* Hence, a change must occur in the controlled variable before we become aware that something is wrong. The error having been sensed, corrective action can be initiated.

The concepts of block diagrams were discussed in Chapter 19. For this simplified example, let us assume that the particular parts of the process of interest are:

(1) The control valve—for which the

Input = opening or valve stem position
Output = fuel flow rate (cubic meters/second)

(2) The burners—for which the

Input = fuel flow rate (cubic meters/second)
Output = heat (joules/second)

(3) The thermocouple—measuring furnace temperature

Input = heat (or temperature °C)
Output = millivolts (representing °C)

(4) Fuel pressure disturbance

Input = fuel pressure
Output = fuel flow rate

Assume that we have a device (called a comparator) that can subtract the measured variable from the reference or set point of the controlled variable to give the error or deviation from the desired value. The error could then be utilized in a control equation (an algorithm) to calculate how much the control valve should be opened in order to return to the desired temperature in the furnace. The above elements can all be incorporated into a block diagram, as shown in Figure 20.2 (cf. Figure 19.10). In this block diagram, we show the fuel pressure changes as a disturbance to the process; note that disturbances can cause the temperature to move from the set point if the valve position is not changed appropriately.

For the remainder of this chapter we shall assume that a dynamic model for each component of the process is available which, in this case, would include the characteristics of the valve, the burners in the furnace, and the measuring thermocouple. We shall be focusing on conventional forms of the control equation, and how values for the parameters in these equations are chosen. Furthermore, since measurements from thermocouples or other process sensors are corrupted by noise (i.e., the apparent measurement can fluctuate substantially from the true measurement), we shall examine techniques for filtering or smoothing these measurements in order to give better estimates of the measured variable. In the coverage of process control and signal processing methods, we shall emphasize the digital

---

*Sometimes it is possible to measure disturbances directly in addition to measuring the controlled variable. Assuming that we know in advance what effect such changes will have on the process, we can take corrective actions in anticipation of the changes. This approach is known as feedforward control (see Section 19.7.3). More information can be found in process control texts (e.g., Ref. 16).

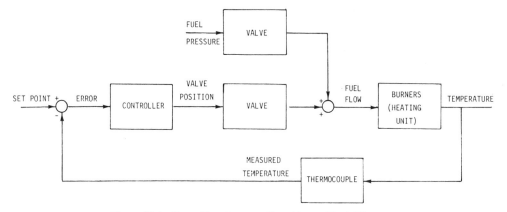

Figure 20.2. Sheet Glass Process: Closed Loop Block Diagram

techniques used currently, with initial discussion of and occasional reference to the more traditional analog techniques.

20.1 CONVENTIONAL PROCESS CONTROL ALGORITHMS

In the previous section we described conceptually a very much simplified control problem. We did not, however, provide enough details for the reader to understand what the control equation is or how it might be implemented. Let us return to the stirred tank heater discussed in Chapter 4 in order to discuss conventional control algorithms. Figure 19.1 shows this process, and Equation (20.1a) describes the dynamic model, which is derived in Chapter 19. The flow rate of steam into the tank heating coil multiplied by its latent heat, $W_s\lambda$, represents an amount of energy, q, that is used to heat the contents of the tank. Later, q will be assumed to be directly proportional to the steam valve position.

$$MC_p \frac{dT_0}{dt} = FC_p(T_i - T_0) + W_s\lambda$$

$$(20.1a)$$

or

$$\tau \frac{dT_0}{dt} = T_i - T_0 + \frac{\lambda}{FC_p} W_s \quad (20.1b)$$

Let us assume that the desired value of the tank temperature in Figure 19.1 is T_R, and that the

deviation (e) in this temperature is

$$e = T_R - T_0 \qquad (20.2)$$

where T_0 is the measured temperature from the thermocouple. (Note that we have assumed that the millivolt thermocouple signal has already been converted to temperature.) This is the information that an operator, or a controller, could use in order to make corrections in the amount of heat (steam flow) to the tank. If the temperature is too low, the error will be positive and the steam flow must be increased. However, in order to predict the necessary rate of heat addition, we must analyze the dynamic operation of the process; i.e., how fast will the temperature rise, and how should the heat input be varied? This information would be available from the differential equation (20.1) describing the process or from a dynamic model obtained from experimental (empirical) testing methods such as those described in Chapter 19. In addition, we need to select a control algorithm that relates the heat input to the tank to the error in temperature.

The control algorithm calculates the change in, or the new value of, the manipulated variable (or input) of a process in order to correct the errors in operation. A simple choice for the control algorithm can be obtained intuitively. If the error is very large, we presume that a rather large change in the input or manipulated variable is required. Alternatively, if the error is small, a small change in input variable may be appropriate. The simplest algorithm which implements this action would let the change in the control action be proportional to the error, as shown in Equation (20.3):

$$\text{Change in control action} = K_c e \quad (20.3)$$

It should be noted that this equation effectively operates in a deviation mode; that is, it calculates a controller output deviation (change) when the value of the error is nonzero. An error of zero implies that the process is operating at its desired equilibrium value. In the following analysis we assume that the models of the process are based on deviation variables (deviations from steady state values; see Chapter 19). Further details on this procedure may be found in most process control textbooks (see references).

To illustrate the effect of the proportional controller, we shall examine Equation (20.1b) but with a change in the desired set point, T_R (Equation 20.2).* The disturbance variable, T_i, is constant (zero deviation). Let

$$\frac{\lambda}{FC_p} = K_p \text{ (the process gain)}$$

and assume that the change in steam flow rate is equal to the product of the valve position deviation, $(P - P_R)$, and the valve constant (K_v),

$$W_s - \overline{W}_s = K_v(P - P_R)$$

But $(P - P_R)$ is proportional to the deviation in temperature, e:

$$P - P_R = K_c e = K_c(T_R - T_0)$$

where K_c, the controller gain, is the proportional factor. Note that if the tank temperature is less than T_R, the controller causes steam to be added. Substituting these relations into Equation (20.1b) and using the steady state relationship, $\overline{T}_0 = T_i + K_p \overline{W}_s$, where $\overline{T}_0$ is the initial operating point, we obtain

---

*Remember that T_R changes from 0 at time t = 0, since it is a deviation variable.

$$\tau \frac{dT_0}{dt} = (\overline{T}_0 - T_0) + K_p K_v K_c(T_R - T_0)$$
$$(20.4a)$$

or

$$\frac{\tau}{(1 + K_p K_v K_c)} \frac{dT_0}{dt} + T_0 = T_R \frac{K_p K_v K_c}{1 + K_p K_v K_c}$$
$$+ \frac{1}{1 + K_p K_v K_c} \cdot \overline{T}_0$$
$$(20.4b)$$

For simplicity, define

$$K = K_p K_v K_c$$

Then

$$\frac{\tau}{(1 + K)} \frac{dT_0}{dt} + T_0 = T_R \left(\frac{K}{1 + K}\right)$$
$$+ \overline{T}_0 \left(\frac{1}{1 + K}\right) \quad (20.5)$$

This is a simple first-order differential equation for which the solution to a step change in controller set point is (by Laplace transform or other methods; see Chapter 19):

$$T_0 - \overline{T}_0 = (T_R - \overline{T}_0) \frac{K}{(1 + K)} (1 - e^{-(1 + K)t/\tau})$$
$$(20.6)$$

Figure 20.3 shows a sketch of this result for various values of K (which is proportional to the controller gain, K_c). Notice that with no control (K = 0), the output remains unchanged although the reference point has changed (from 0 at t = 0 to T_R). As K is increased, the tank temperature, T_0, tracks or moves closer to T_R but never actually reaches it. The steady state error, which is called "offset", is indicated in Figure 20.3 for a value of K equal to 10. More complex controllers which can overcome this problem will be discussed later.

20.2 DIGITAL PROCESS CONTROL

Before we discuss the form of the control algorithms commonly used in industry, it is

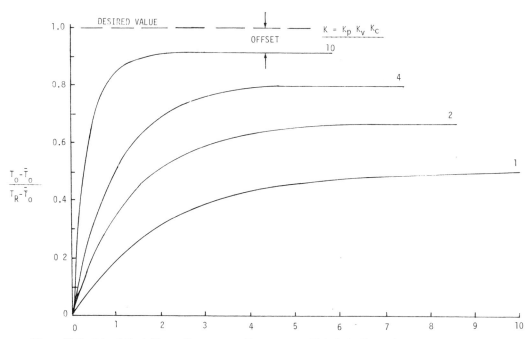

Figure 20.3. Stirred Tank Heater Temperature Response to a Unit Step Change in Controller Set Point

worth stating why we should be interested in digital (i.e., discrete) algorithms. First, the tremendous developments in computer technology are beginning to disseminate through every level of control hardware. Controllers that once were built using analog components are now being replaced by their digital equivalents. These digital controllers offer more flexibility than their analog counterparts and are easier to adjust. Large commercial computer control systems are being developed by almost all of the major control system vendors using new computer technology. Thus it is important that persons involved with real-time systems be knowledgeable about this aspect of computer applications. A second, but perhaps less well understood, reason for using digital control algorithms is that many measuring devices produce information only at discrete time intervals. One prominent application is the process chromatograph which samples a process stream periodically and, after some time delay, produces an analysis of the composition of the sample. This information can be processed by a discrete control algorithm.

A simple example of the analog proportional controller was discussed in the previous section,

and now it is useful to discuss the sequence of events that occur when a direct digital control (DDC) system is implemented using a process control computer. The algorithm is similar to that implemented by conventional analog hardware operating in continuous time, but here we shall emphasize the nature of the discrete sequence of events that occurs when conventional hardware is replaced by computer hardware.

In order to implement proportional control with a digital computer, we must specify a sampling time, T. Every T seconds the computer

(1) Samples or measures the process variable by converting a voltage representing the controlled variable using the analog to digital converter (ADC).
(2) Digitally filters the process variable measurement if necessary or desired.
(3) Subtracts the filtered or measured value of the variable from the value of the set point for the control loop (which is stored in memory) to calculate the error for that particular sample.
(4) Calculates the equivalent control output

or change in output according to some digital control algorithm.

(5) Transmits the calculated value of the output through the digital to analog converter (DAC) which is connected to the process manipulated variable.

In most instances, the measurements of the process variable enter the computer through the ADC although it is possible to use logical or BCD inputs. All the calculations are carried out using numerical algorithms within the computer. Ordinarily the numerical value of the computer output is sent back to the process through the digital to analog converter. However, an alternative to this latter step will be discussed later in this chapter.

This entire sequence of calculations takes a few milliseconds in a typical computer control system compared with several seconds or even minutes for the sample time, T. Once the current value of the output has been computed, it is maintained as a constant input to the process for the duration of the sample period. This point should be noted for simulation work as process models must be modified to take into account the constant input over the sample time as opposed to the continuously varying input from a conventional analog control scheme. Much confusion can arise if this fact is not taken into account in the modeling procedures.

In order to develop a simple example for a proportional controller operating in a discrete control environment we must first define a zero-order hold, which is equivalent to a digital to analog converter maintaining a constant output value for the duration of the sample

period. This can be represented by an equation

$$f_k(t) = f(kT), \quad kT \leqslant t < (k+1)T \quad (20.7)$$

This equation implies that the value of the function over time is equal to the value of the function at the instant of sampling. The time domain function for the zero-order hold is

$$h_o(t) = S(t) - S(t - T) \quad (20.8)$$

where $S(t)$ is a unit step function, i.e., a step increase in the input of unity magnitude. The Laplace transform of this function (see Chapter 19) is

$$H(s) = \frac{1 - e^{-sT}}{s} \quad (20.9)$$

Let us assume that the stirred tank heater is represented by a first-order transfer function with a time constant, τ, of 4 minutes. In the discrete data control system, we are sampling at one-minute intervals and are using a zero-order hold approximation. The proportional controller has a loop gain K (=$K_p K_v K_c$, see Equation 20.5). The closed loop or controlled process is shown in the block diagram in Figure 20.4, which the reader should compare with Figure 19.12. Samplers are shown on the block diagram to emphasize the discrete nature of the calculations. $R(s)$ and $R(z)$ are the analog and sampled set points, respectively, and $E(z)$ is the sampled error. $C(s)$ and $C(z)$ are the analog and sampled measurements, and the two blocks inside the dotted rectangle are a zero-order hold and the process transfer function. For simplicity we have deleted the measurement and valve

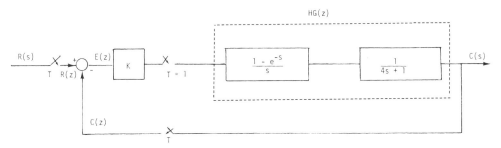

Figure 20.4. Process Block Diagram for Digital Control

blocks (or transfer functions). The proportional controller gain is K, which is the same as for analog control. Let us derive the closed loop transfer function which will represent the output of the process at the sampling instants. We introduce a unit step change in the set point (set point increases from zero to one). The transfer function for the process plus the zero-order hold is:

$$H(s)G(s) = \left(\frac{1 - e^{-s}}{s}\right)\left(\frac{1}{4s + 1}\right)$$

$$= \frac{\frac{1}{4}}{s(s + \frac{1}{4})} - \frac{\frac{1}{4}e^{-s}}{s(s + \frac{1}{4})} \quad (20.10)$$

Notice that in Equation (20.10) we have combined the two transfer functions in order to find the z-transform of the product (HG(z)) as discussed in Chapter 19.

The z-transform of this transfer function is (Equation 19.89):

$$HG(z) = \frac{(1 - e^{-T/4})z}{(z - 1)(z - e^{-T/4})} - \frac{(1 - e^{-T/4})}{(z - 1)(z - e^{-T/4})}$$

$$= \frac{(1 - e^{-1/4})}{(z - e^{-1/4})} = \frac{0.221}{(z - 0.779)} \quad (20.11)$$

since we have chosen T equal to 1. The discrete closed loop transfer function for the output is:

$$\frac{C(z)}{R(z)} = \frac{KHG(z)}{1 + KHG(z)}$$

$$= \frac{\dfrac{0.221K}{(z - 0.779)}}{1 + \dfrac{0.221K}{(z - 0.779)}} = \frac{0.221K}{z - 0.779 + 0.221K}$$

$$(20.12)$$

For a unit step change in set point, i.e., the set point increases from 0 to 1, we have

$$R(z) = \frac{z}{z - 1} \quad (20.13)$$

Multiplying Equation (20.12) by Equation (20.13) we obtain the closed loop step response (in the z-domain) of the process output

$$C(z) = \frac{0.221K}{(z - 0.779 + 0.221K)} \cdot \frac{z}{(z - 1)}$$

$$(20.14)$$

The process output in the time domain, c(nT) (note that it is only determined at discrete time intervals), may be calculated by finding the analytical expression for the inverse z-transform of Equation (20.14) or more simply by long division (Chapter 19) to find the initial response values.

Figure 20.5 shows a series of plots of the process output, c(nT), to a unit step change in set point with the controller gain ranging from zero upwards. It can be seen that for a controller gain of zero the process does not respond to the set point change. As the control loop gain is raised, the final value of the output moves increasingly toward the value of 1. As with continuous control (Figure 20.3), it only approaches 1.0 in the limit as K becomes large. As discussed by Koppel [11], the discrete system can become unstable for large values of K. The continuous process response (non-sampled) for a combined controller and process gain of 1.0 is compared with the discrete versions in Figure 20.5.

The fact that the process never reaches the new set point is a matter of concern; however, the control algorithm can be modified to eliminate this offset, as will be discussed later in this chapter.

In this section we have presented a rather simple development of the effect of a proportional controller, implemented by a digital computer system, on a first-order process model. The effect of the proportional controller in fact depends upon both the process and the value of the overall gain. This characteristic will be discussed in more detail when we talk about "tuning" the control algorithms that are typically used in practice.

20.2.1 Digital Control Algorithms

Although there is a tremendous body of discrete data control theory available today, most of the control algorithms used in practice are discrete representations of the conventional analog con-

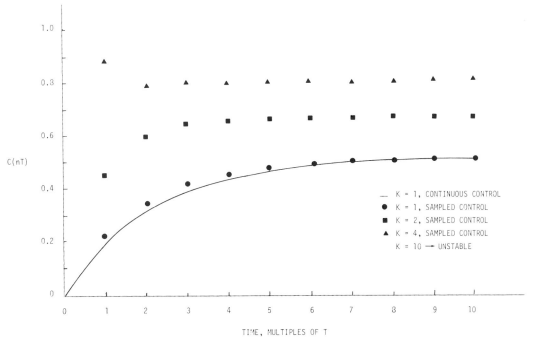

Figure 20.5. Response of the Stirred Tank Heater to a Unit Step Change in Controller Set Point

trollers. Very little discrete data control theory has been utilized either to develop these algorithms or subsequently to tune them when they are implemented with a process. The most common analog control algorithm used in practice today is the so-called three-term or three-mode controller. This controller has a term that is proportional to the error, a term that is proportional to the integral of the error, and a term that is proportional to the rate of change (derivative) of the error. The algorithm is called a proportional-integral-derivative controller or more concisely a PID controller. The general form of this controller is shown in Equation (20.15).

$$m(t) = K_c \left\{ e(t) + K_I \int_o^t e(t')dt' + K_D \frac{de}{dt} \right\}$$

(20.15)

where K_c = proportional gain
K_I = reset multiplier
K_D = derivative time
$e(t)$ = error (as a function of time)
$m(t)$ = controller output deviation

The first term in Equation (20.15) provides proportional action, as discussed earlier. The second term in the controller provides a component of the output that is equal to the integral of the error. In other words, as long as the error differs from zero, the controller output will continue to change.* The effective multiplicative constant is $K_c K_I$. Because of this property, integral action most often is used to eliminate the "offset" from the desired value of a process variable when a sustained load change or a set point change occurs. This offset is shown in Figure 20.3; without integral action, the process output never reaches the set point following the change unless the proportional gain is made unrealistically large.

The last term in the PID controller provides a component of the output when the rate of change of error is nonzero. The derivative mode thus "anticipates" the error, giving a faster process response. This mode is often not

---

*It should be noted that there are physical limits beyond which the controller output cannot change even though this algorithm potentially could require it.

used in industrial practice because it can give erroneous control actions when the process measurements are corrupted by noise. When noise is not present, however, derivative action does appreciably improve the controller performance. The multiplicative constant in Equation (20.15) is $K_c K_D$.

It is not the intention of this chapter to analyze in detail the performance of Equation (20.15) as a process control algorithm. Extensive analysis of its properties both in the time and frequency domains may be found in control texts referenced at the end of the chapter. We shall, however, discuss important features of this equation (or a discrete version of it), since it is so commonly used in practice. In addition, we shall illustrate the effects of adjusting the controller parameters on control system performance.

The digital or discrete equivalent of Equation (20.15) is given in Equation (20.16):

$$m_i = K_c \left\{ e_i + TK_I \sum_{j=1}^{i} e_j + \frac{K_D}{T} (e_i - e_{i-1}) \right\}$$

$$(20.16)$$

where T = sampling interval

e_i = error at ith sampling interval

e_{i-1} = error at previous sampling interval

The similarities between Equations (20.15) and (20.16) should be noted. The proportional term is the same. However, the integral symbol is replaced by a summation symbol, and the derivative operation is replaced by a first-order difference approximation. More accurate expressions can be used, particularly for the latter term, since numerical differentiation of process data can cause serious problems when there is appreciable process noise.

This form of PID controller is called a "position" algorithm since the computer calculates the specific value of the output at the ith sample time. In this form the computer must save the previous error and the sum of all previous errors defined by

$$S_{i-1} = \sum_{j=1}^{i-1} e_j \qquad (20.17)$$

where S_i represents the sum of all errors. When e_i is obtained, the sum S_i is updated ($S_i = S_{i-1} + e_i$) and a new output m_i is calculated from Equation (20.18):

$$m_i = K_c \left\{ 1 + \frac{K_D}{T} \right\} e_i - \left\{ \frac{K_c K_D}{T} \right\} e_{i-1}$$

$$+ \{TK_c K_I\} S_i \qquad (20.18)$$

An alternative form of the PID control algorithm is the "velocity" algorithm. In this form one calculates the change in controller output at some time, i, as opposed to the actual value of the output. One of the primary reasons for using the velocity algorithm is that rather than transmitting the value of the controller position through a DAC channel, it is frequently desirable to transmit the change in output position. This change then becomes the input to a stepper motor or an integrating amplifier which provides the final value for the output.

In a previous section we suggested that a DAC need not necessarily be used as the computer output to the process (control valve). With a velocity algorithm, pulses or single-bit outputs can be used to drive a stepper motor. The advantage of this technique is that if the computer system should fail, the integrating amplifier or the stepper motor automatically holds the control valve at the last calculated position. The form of the velocity algorithm is shown below:

$$\Delta m_i = m_i - m_{i-1}$$

$$= K_c \left(1 + TK_I + \frac{K_D}{T} \right) e_i$$

$$- K_c \left(1 + \frac{2K_D}{T} \right) e_{i-1}$$

$$+ \frac{K_c K_D}{T} e_{i-2} \qquad (20.19)$$

In this form one must save the previous errors e_{i-1} and e_{i-2} instead of S_{i-1}. In using the velocity algorithm, it is essential that the integral component (K_I) always be included. Omission of this controller mode can cause severe drift in the algorithm (i.e., very poor control), since the set point is effectively canceled out in computing both the proportional and derivative terms.

Let us return to the position algorithm. Since it is described by a difference equation (20.16), we can also express the controller by a generalized discrete transfer function:

$$D(z) = \frac{a_o + a_1 z^{-1} + a_2 z^{-2} + \cdots + a_\ell z^{-\ell}}{1 + b_1 z^{-1} + b_2 z^{-2} + \cdots + b_k z^{-k}}$$

(20.20)

We can write (in the z-domain) an expression for the current controller output as a function of the controller input, as shown below:

$$\frac{M(z)}{E(z)} = D(z)$$

(20.21)

Much has been written about various forms of this generalized digital controller; for example, adjusting parameters to obtain satisfactory control is almost impossible on an ad hoc basis. Discrete data control techniques applied to discrete transfer function models must be used. Some of these ideas are discussed in Section 20.2.3.

Let us derive the discrete PID controller shown in Equation (20.16) in the form of Equation (20.20). Rewriting Equation (20.16) as

$$m_i = K_c \left[e_i + w \sum_j e_j + d(e_i - e_{i-1}) \right]$$

(20.22)

and taking the z-transform of this relation, we obtain

$$\frac{M(z)}{E(z)} = D(z)$$

$$= K_c \left[\frac{(1 + w + d)z^2 - (1 + 2d)z + d}{z(z - 1)} \right]$$

(20.23)

Finally:

$$D(z) = K_c \left[\frac{(z - z_1)(z - z_2)}{z(z - 1)} \right]$$

(20.24)

where z_1 and z_2 are factors of the numerator polynomial that may be real or complex.

Dividing numerator and denominator by z^2, we can write

$$D(z) = \frac{K_c(1 + w + d) - K_c(1 + 2d)z^{-1} + K_c d z^{-2}}{(1 - z^{-1})}$$

(20.25)

Thus, the three-mode controller is a specific case of Equation (20.20) where:

$$\begin{aligned} a_o &= K_c(1 + w + d) \\ a_1 &= -K_c(1 + 2d) \\ a_2 &= K_c d \\ b_1 &= -1 \end{aligned}$$

(20.26)

Note that the $(1 - z^{-1})$ term in the denominator comes from the integration or summation term in Equation (20.16) and guarantees that offset can be eliminated.

Throughout the preceding discussion we have employed the conventional notation used for the PID controller in process control practice. In fact, it would be somewhat more convenient for us to write the position form of the PID controller with the gain terms separated, as shown in Equation (20.27):

$$m_i = K_c e_i + K_I T \sum_j e_j + \frac{K_D}{T} (e_i - e_{i-1})$$

(20.27)

Care should be taken when comparing this form of the PID controller with those used in the literature, particularly where general rules for adjusting the three gain terms (tuning) are discussed. Most of the literature for conventional process control uses the form shown in Equation (20.15) for controller tuning.

20.2.2 Controller Tuning

We have seen in the previous sections some typical control algorithms which are used for industrial control systems. Although the algorithms are relatively simple, implementing them in a closed loop environment requires some understanding of what the parameters do and how they are adjusted for different processes. Here we present some of the commonly accepted methods of tuning controllers without attempting to develop a fundamental understanding of the theory behind them.

First of all, the controller is used with the process because some process variable otherwise cannot be maintained at its desired value owing to changing process conditions (effects of other variables). The purpose of the controller, therefore, is to maintain the controlled variable as close as possible to the set point. It is possible to take too much corrective action in response to errors, as anyone who has learned to ride a bicycle realizes. Extensive overcorrection can make the process (that is, the bicycle) go unstable.

Figure 20.6 illustrates a number of typical responses of a process variable to changes in set point as controller parameters are varied. Figure 20.6(a) shows an oscillatory response (period of oscillation, P_u) beyond which increases in certain controller parameters would cause the "controlled" process to become unstable. Essentially what has happened is that

the system has been put in motion, and the feedback implemented by the controller is unable to damp out the oscillations. Most real industrial processes can be made to oscillate continuously using only a proportional controller with sufficiently high gain. The specific value of this gain can be calculated a priori from knowledge of the process model, using techniques discussed in any process control text.

Figure 20.6(b) shows two different responses to a set point change. One curve goes slightly past the desired new value and then slowly settles down as time increases. The other curve slowly rises to the new set point over a long time period. Figure 20.6(c) illustrates an unstable response, where the process develops increasing oscillations. In practice these could often be limited by some physical constraint such as a valve being fully opened or fully closed. In all the above cases the control is generally deemed to be less than satisfactory.

Figure 20.6(d) shows a more desirable form of response. Some overshoot or rise past the

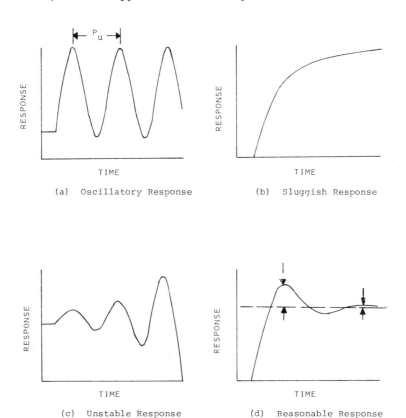

(a)   Oscillatory Response

(b)   Sluggish Response

(c)   Unstable Response

(d)   Reasonable Response

Figure 20.6. Qualitative Responses of Process Variables for Different Controller Parameter Values

desired value is permissible in order to increase the speed of the response. However, oscillations caused by this fast response should attenuate rapidly. A popular standard of performance is that the ratio of successive peak heights (maxima) as shown in Figure 20.6(d) should be about 0.25 for a reasonable response. One of the earliest tuning correlations for PID controllers (Equation 20.15) was developed by Ziegler and Nichols based on the above attenuation criterion. This method (called continuous cycling) requires that a proportional controller (K_I and K_D are set to zero in a PID controller) be implemented in the control loop and that we have the ability to change the gain K_c. Starting from an equilibrium position (i.e., where the controlled variable is constant), a particular value of the controller gain, K_c, is established. Then a small disturbance is introduced, most easily by perturbing the set point briefly. The response to this disturbance is observed. If the controlled variable response returns to an equilibrium value, then the gain is increased. The procedure is repeated until the controlled variable develops sustained oscillations as shown in Figure 20.6(a) but does not become unstable. At that point the controller gain ($K_c = K_u$) is noted, and the period of oscillation is measured (see P_u on Figure 20.6(a)) from the time response of the measured variable. Finally, the desired controller gains are chosen according to values shown in Table 20.1; these correlations were developed after extensive testing of many industrial processes. Because of their generality, they are, by nature, approximate guidelines and represent a first approximation to the optimum settings; one criticism of these settings is that they often give a response that is too oscillatory.

It should be noted that this technique for adjusting controllers was developed for continuous controllers; for discrete versions of this

controller, an extra parameter, the sample time, is included. This extra feature (i.e., sampling) effectively adds dead time into the controlled system. However, if continuous cycling is performed using a digital computer, the effect of sampling the process output is accounted for in the gains determined from the Ziegler-Nichols settings.

Cohen and Coon presented a different version of the tuning parameters which requires an empirical process model. Their method involves determination of the open loop process reaction curve as described in Chapter 19 (see Figure 19.4(b)). Values of the process gain, K_p, must be known or obtained as explained in Chapter 19. Values of the process time constant and dead time, τ and θ, are required as well. From these model parameters the controller constants can be calculated according to expressions shown in Table 20.2. A more complete discussion may be found in Murrill [19].

We can tune a controller using a simulation study provided that a reasonable process model is available. However, this process model normally will be derived from continuous ordinary differential equations; hence we face the problem of including the effect of sample time in the discrete computer control algorithm. One solution is to simulate a discrete version of the controller along with a continuous model of the process. The key here is to include the zero-order hold to account for the digital to analog converter's "holding" the controller output constant between sample periods. A second approach is to modify the continuous system model to account approximately for the sampling operations. Smith [24] has suggested that one-half the value of the sample time be added to the known process dead time in the model and that this revised dead time be used in estimating the controller parameters. This approximation works reasonably well, but

Table 20.1. Ziegler-Nichols Controller Gain Settings

Controller Type	Continuous Equation	K_c	K_I	K_D
Proportional	$m = K_c e$	$0.5\,K_u$		
Proportional-Integral	$m = K_c\left(e + K_I \int e\,dt\right)$	$0.45 K_u$	$\dfrac{1.2}{P_u}$	
Proportional-Integral-Derivative	$M = K_c\left(e + K_I \int e\,dt + K_D \dfrac{de}{dt}\right)$	$0.6\,K_u$	$\dfrac{2.0}{P_u}$	$\dfrac{P_u}{8.0}$

Table 20.2. Cohen and Coon Controller Gain Settings

Controller Type	$K_c K_p$	K_I	K_D
Proportional	$\dfrac{\tau}{\theta} + 0.33$		
Proportional-Integral	$\dfrac{0.9\tau}{\theta} + 0.082$	$\dfrac{1 + (2.2\theta/\tau)}{3.33\theta(1+0.098\theta/\tau)}$	
Proportional-Integral-	$\dfrac{1.33\tau}{\theta} + 0.25$	$\dfrac{1 + (0.6\theta/\tau)}{2.5\theta(1+0.2\theta/\tau)}$	$\dfrac{0.37\theta}{1.0.2\theta/\tau}$

once again the estimates for the controller gains are not exact and final tuning must be carried out on-line. Smith has also discussed other methods for tuning PID controllers.

20.2.3 Design of More Complex Digital Controllers

We have already seen the form of a general digital controller in Equation (20.20). This controller can have many unknown parameters depending on the order of the polynomials selected. Tuning these parameters is very difficult in practice although a number of techniques are available when an exact process model is known. Figure 20.7 shows a digital servo control loop, with a digital controller (z-domain) indicated on the figure. For practical purposes we shall assume that HG(z) includes the normal sample and hold term which would arise from the DAC output channel. The sampling switches indicate that the system is being treated in the discrete domain.

Conventional design techniques for digital controllers are based on several requirements. Typically:

(1) The system (i.e., the final closed loop system) must be physically realizable.
(2) The system should exhibit zero steady state error (for sustained disturbances of specified types).
(3) The transient response should be as fast as possible with the settling time equal to a finite number of sample periods.

Physical realizability means that we can actually implement the controller. In principle this means that if the process contains a dead time, then the controller also must have at least the equivalent dead time. If this requirement is not met, then the controller will have to predict the effect of the output of the process in future time. This is not possible for the type of control that we are considering. The second requirement, zero steady state error, is equiv-

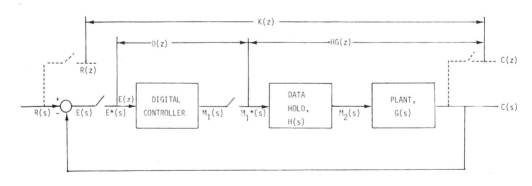

E*(s), $M_1(s)$ are sampled continuous signals

D(z), HG(z), K(z) are discrete transfer functions which define variables at sample times only

Figure 20.7. A Digital Control System

alent to introducing integral action in a conventional analog controller.

Finally the requirement for fast transient response means that within some constraints we should like set point changes to be completed as rapidly as possible; i.e., deviations of the output variable should become zero as soon as possible. It should be noted that with digital systems this condition can be met exactly only at the sampling times. The actual process output may well be varying between sample instants.

The basic technique for digital controller design assumes that we specify the character of the output response $C(z)$ for a step change in $R(z)$. Since the process transfer function is known, we can calculate the required controller transfer function, $D(z)$, such that an overall closed loop response transfer function, $K(z)$, between the process output and the set-point is realized. Equation (20.28) describes the system closed loop transfer function:

$$\frac{C(z)}{R(z)} = \frac{D(z)HG(z)}{1 + D(z)HG(z)} = K(z) \quad (20.28)$$

$K(z)$ can be any polynomial in z although it is usually a numerator polynomial with sufficient time delay to satisfy the physical realizability conditions mentioned above.

From this prespecified closed loop transfer function, we can calculate that the controller transfer function, $D(z)$, is:

$$D(z) = \frac{1}{HG(z)} \cdot \frac{K(z)}{[1 - K(z)]}$$

$$= \frac{1}{HG(z)} \cdot \frac{C(z)/R(z)}{[1 - C(z)/R(z)]}$$

$$(20.29)$$

Note that both $HG(z)$ and $K(z)$ are known a priori. The expanded form of this digital controller must match Equation (20.20) where the highest-order term of the numerator polynomial in z^{-k} must not have a positive exponent on z (otherwise the controller will not be physically realizable).*

---

*In Equation (20.20), the highest-order term is the a_0 term (the exponent on z is zero).

Typically $R(z)$ is chosen to be a step change in set point, although sometimes a ramp increase in set point is used. To illustrate the technique of digital controller design, suppose we examine the stirred tank heater process described by Equation (20.11). The transfer function can be rearranged to reflect polynomials in z^{-1} as follows:

$$HG(z) = \frac{0.221z^{-1}}{1 - 0.779z^{-1}} \quad (20.30)$$

Suppose that we wish the process to respond to a unit step change in set point with no error at the sampling instants:

$$R(z) = \frac{1}{1 - z^{-1}} \quad (20.31)$$

The error, $E(z)$, is calculated using Figure 20.7 by observing that

$$E(z) = R(z) - C(z) \quad (20.32)$$

which is combined with Equation (20.28) to obtain

$$E(z) = R(z)(1 - K(z)) \quad (20.33)$$

Using the final value theorem of the z-transform

$$\lim_{n \to \infty}\{e(nT)\} = \lim_{z \to 1}\{(1 - z^{-1})E(z)\}$$
$$(20.34)$$

Thus from Equations (20.31), (20.33), and (20.34) we require that

$$\lim_{n \to \infty}\{e(nT)\}$$
$$= \lim_{z \to 1} \left\{ (1 - z^{-1})\left(\frac{1}{1 - z^{-1}}\right)[1 - K(z)] \right\} = 0$$
$$(20.35)$$

Hence there is no steady state error if

$$K(z) = z^{-1} \quad (20.36)$$

Finally, from Equation (20.29) we calculate

$$D(z) = \frac{(1 - 0.779z^{-1})}{(0.221z^{-1})} \cdot \frac{z^{-1}}{(1 - z^{-1})} \tag{20.37}$$

or

$$D(z) = \frac{4.52 - 3.52z^{-1}}{(1 - z^{-1})} \tag{20.38}$$

The presence of the $(1 - z^{-1})$ term in the denominator comes from the integral action required to guarantee no offset. The actual performance of the controller can be calculated from Equation (20.38), since the response of the manipulated variable is of potential interest. (We have already specified the response of $c(nT)$ in designing the controller.)

A number of pitfalls and potential problems can be encountered in designing these controllers if one follows only the basic steps shown above (see Ref. 24). Ordinarily discrete controllers designed in this way often exhibit too much control action and frequently do not perform well compared to analog controllers designed using Ziegler-Nichols rules. Two main problems arise: The first is that the control action is often too severe, causing large overshoot in the controlled variable response, as illustrated in Figure 20.8. If the process transfer function is known, a simple calculation will indicate excessive control action. A second problem is often less apparent. If the de-

nominator of the controller equation, $D(z)$, as calculated from Equation (20.29), has a root in the left half plane of the z^{-1} domain on or near the unit circle, the controller output will oscillate violently from one sampling time to the next. This phenomenon is called ringing. The locations of the undesirable poles and a typical ringing controller output are shown in Figures 20.9(a) and 20.9(b), respectively.

Design techniques to eliminate these problems with digital controllers are available, but a complete discussion is beyond the scope of this chapter. One alternative to the complex design techniques has been suggested by Dahlin [3], although it does not guarantee freedom from ringing poles. He proposed that the closed loop process response, after some dead time, should resemble that of a first-order response to a step change in set point.* The Laplace transform for this desired response is given in Equation (20.39):

$$C(s) = \frac{e^{-\theta s}}{(\tau s + 1)} \cdot \frac{1}{s} \tag{20.39}$$

where θ = response dead time
        τ = response time constant

If we assume that θ can be expressed as an integral number of sample periods, NT, the discrete response for Equation (20.39) can be calculated by transform methods to be

$$C(z) = \frac{(1 - e^{-T/\tau})z^{-N-1}}{(1 - z^{-1})(1 - e^{-T/\tau}z^{-1})} \tag{20.40}$$

---

*An alternative way to develop this method is to recognize that the desired closed loop transfer function should be given by the first-order plus time delay model. However, it must be preceded by a zero-order hold to account for the fact that the output is sampled with period T. Hence:

$$\frac{C(s)}{R(s)} = {}'G(s) = \left(\frac{1 - e^{-Ts}}{s}\right)\left(\frac{e^{-\theta s}}{\tau s + 1}\right) \tag{20.43}$$

Transformation of this equation yields the identical result given in Equation (20.42). This approach is somewhat more satisfying because it directly involves the continuous transfer function we wish to obtain for this closed loop process, and it does not require any assumptions concerning process input and output.

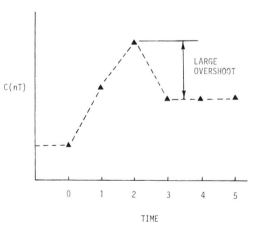

Figure 20.8. An Unsatisfactory Digital Controller Step Response

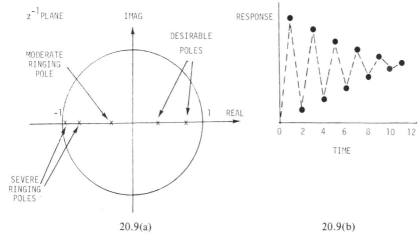

20.9(a)   20.9(b)

Figure 20.9. Controller Poles That Cause Ringing and a Controller Response That Exhibits Ringing

where N = the integer number of sample times in θ.

For a step change in set point we know that

$$R(z) = \frac{1}{1 - z^{-1}} \qquad (20.41)$$

We can then calculate the desired C/R:

$$\frac{C(z)}{R(z)} = \frac{(1 - e^{-T/\tau})z^{-N-1}}{(1 - e^{-T/\tau}z^{-1})} \qquad (20.42)$$

Substituting Equation (20.42) into (20.29) it is possible to derive a general controller for any process, G(s):

$$D(z) = \frac{(1 - e^{-T/\tau})z^{-N-1}}{1 - e^{-T/\tau}z^{-1} - (1 - e^{-T/\tau})z^{-N-1}}$$
$$\cdot \frac{1}{HG(z)} \qquad (20.44)$$

Notice that we have called the process transfer function HG(z) to emphasize that a zero-order hold is present from the DAC and should be included with the process transfer function. The closed loop time constant, τ, can be viewed as a tuning parameter while the dead time in the system is included as the largest integer multiple of the sampling period for the process. This latter point is important. Theoretically, the dead time, θ, could be as long as we choose.

It must be at least equal to the process dead time including a sample time of T for the zero-order hold. Practically, it could be somewhat longer, particularly for higher-order systems where the initial process response is slow and might be nicely approximated by a dead time term.

Let us calculate the Dahlin controller for the stirred tank heater. The pulse transfer function for the process including a zero-order hold with sampling time T = 1 and letting $\tau = 2$ is

$$D(z) = \frac{(1 - e^{-1/2})z^{-1}}{1 - e^{-1/2}z^{-1} - (1 - e^{-1/2})z^{-1}}$$
$$\cdot \frac{(1 - 0.779z^{-1})}{0.221z^{-1}} \qquad (20.45)$$

or

$$D(z) = \frac{1.37(1 - 0.779z^{-1})}{(1 - z^{-1})} \qquad (20.46)$$

Once again notice the presence of the integral term $(1 - z^{-1})$ in the denominator of the controller.

In summary, the reader should understand the following points about digital controllers. Most industrial applications of digital controllers use a digital approximation to the classical PI or PID analog controllers. In these cases the sampling rates are usually relatively high and the digital controller approximates the analog controller performance. If a process transfer function is known, a reasonable set of initial estimates for controller settings can be obtained by adding one-half of the sample time to the process dead time and then using classical continuous system techniques for controller design with the revised continuous model. The controller may also be tuned directly on-line by increasing the proportional gain until the process oscillates continuously, then applying classical settings. In this case the sampling dead time is inherently present in the actual loop.

Alternatively, a number of techniques are available for designing digital control algorithms. Most of them require knowledge of the process pulse transfer function (discrete transfer function). The single most important point to remember is that a zero-order hold must be included at the input of the process transfer function. In the Laplace domain this is represented by $(1 - e^{-Ts})/s$. The reason for this is that whenever a digital control system is implemented, the controller output will be sent to a digital to analog converter which accepts a sampled input (computer output at the sampling time) and holds it until the next output.* Most errors in calculating digital controllers arise from a failure to understand this point. In many discrete data control textbooks one sees examples of transfer function block diagrams with and without the zero-order hold. For those examples where it is omitted, the controller output is a pulse and the process inputs and outputs are truly sampled signals. Control systems using this approach, e.g., pulse-modulated systems, rarely are encountered in the process industries.

20.3 DIGITAL FILTERING OR SIGNAL PROCESSING

We have been discussing the design of digital controllers up to this point. However, another very important aspect of computer control systems involves the measurements which are

---

*Or to some other type of hold device such as a stepping motor.

brought into the computer. All signals received in a real plant can be corrupted by noise. Two types occur: process noise and measurement noise. Measurement noise can be minimized by proper selection of instrumentation and shielding of cables. On the other hand, process noise cannot be eliminated but only treated mathematically.

Analog filters have been used for many years to smooth experimental data. The electrical RC circuit will treat a noisy signal and damp out the high frequency fluctuations. The action of this filter can be described by a differential equation

$$\tau_F \frac{dy}{dt} + y(t) = x(t) \qquad (20.47)$$

where x is the input, y is the output, and τ_F is the time constant of the filter. This equation is analogous to that developed earlier for the stirred tank heater. The response of y(t) when x(t) changes by a constant (step change) has been shown in Figure 19.2. If x(t) goes from 0 to 1.0, then y(t) will be described by the equation

$$y(t) = 1.0 - e^{-t/\tau_F} \qquad (20.48)$$

Hence when x(t) changes, there is a finite time before y(t) responds equivalently. Note that when $t = \tau_F$, $y(t) = 0.632$; the 63% response can be "tuned" by adjusting τ_F so that a fast fluctuation will never be observed in the filter output. This type of filter was discussed briefly in the measurement section of Chapter 5.

The choice of filtering device depends on the time constant for the filtering operation. To remove so-called high frequency noise characteristics, i.e., when τ_F is less than 3 seconds, analog filters with resistance–capacitance components are sufficient. To filter slower dynamic signals or drifts, e.g., those requiring a filter time constant greater than 3 seconds, we must construct an analog filter system based on amplifiers. New integrated circuit technology is providing great flexibility in constructing these filters. However, an amplifier-based filter can still be quite expensive; so the digital type of filter becomes of great utility in this case.

In determining which system should be used for filtering (analog vs. digital), one can make the following observations as to the advantages or disadvantages of a given system:

(1) Digital filters can be easily tuned to fit the process and are, of course, easily modified. Analog systems require changes in actual components in order to change the filter time constant.
(2) With digital filters we must deal with an operation in which a computer samples a process and analyzes the measurements periodically. Analog filters, on the other hand, perform this operation continuously.
(3) Digital filters require more computation time and storage in the computer system; therefore, they can create natural limitations to the computing capabilities of the data acquisition and control system. Analog filters are independent units and do not interact with other computational aspects of the process.

In most practical applications it can be advantageous to use both analog and digital filters, as shown in the block diagram in Figure 20.10.

One reason for the series filter arrangement is that the sampling operation interacts with the filtering process. The filtering of high frequency noise can better be performed with analog elements (otherwise digital sampling would have to be performed much more rapidly), while the digital filter can be used to filter low frequency noise.

20.3.1 Simple Digital Filters

The simplest type of digital filter is the discrete time formulation of the RC circuit considered earlier (the first-order differential equation). It is useful to derive the first-order digital filter using the techniques we have used for pulse transfer functions. (See also Chapter 19.) Figure 20.11 shows a first-order transfer function with a zero-order hold. We sample the input but want the output to appear as though it were filtered continuously and sampled.

We can calculate the desired z-transform as in Equations (20.10) and (20.11):

$$\frac{Y(z)}{X(z)} = Z\left[\frac{(1 - e^{-Ts})}{s(\tau s + 1)}\right] \qquad (20.49)$$

or:

$$\frac{Y(z)}{X(z)} = \frac{(1 - e^{-T/\tau})z^{-1}}{(1 - e^{-T/\tau}z^{-1})} \qquad (20.50)$$

From Equation (20.50):

$$Y(z)(1 - e^{-T/\tau}z^{-1}) = (1 - e^{-T/\tau})z^{-1}X(z) \qquad (20.51)$$

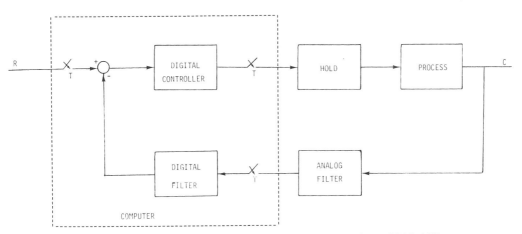

Figure 20.10. Block Diagram of a Controlled Process with Analog and Digital Filters

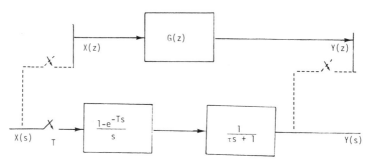

Figure 20.11. First-Order Digital Filter Design

Defining:

$$1 - e^{-T/\tau} = \alpha \quad (0 < \alpha < 1) \quad (20.52)$$

and realizing that z^{-1} in the inverse z-transform is related to a backwards shift in time we can write the difference equation:

$$y_i - (1 - \alpha)y_{i-1} = \alpha x_{i-1} \quad (20.53)$$

Finally:

$$y_i = (1 - \alpha)y_{i-1} + \alpha x_{i-1} \quad (20.54)$$

From the equations above it is seen that the filtered measurement is actually a weighted sum of the current measurement and the old filtered value. Therefore, we only need to save the last value of y in this process, called single exponential smoothing. The effect of α is as follows:

$\alpha = 1$: no smoothing (the current measurement only is used)

$\alpha = 0$: no effect from the new signal measurement

For intermediate values of α, a combination of old and new measurements is employed. Hence α can be used to tune the filter. If the input signal, x, exhibits a step change of 1.0, the output response (filtered measurement) can be written as follows

$$y_i = 1 - (1 - \alpha)^{i+1} \quad (20.55)$$

and we see the effect of α on the response y.

Another often-used filter is the double exponential filter. This type of filter is useful for data that show some drift or trends. If we wish to ignore these trends, then, in essence, we

cascade two first-order filters; i.e., we have a second filter that filters the data from the first filter. If this is translated to a digital filtering algorithm, the following equation results where y is the filtered measurement $\bar{y}$ is the doubly filtered measurement:

$$\bar{y}_i = (1 - \gamma)\bar{y}_{i-1} + \gamma y_i \quad (20.56)$$

To illustrate the effect of filtering noisy signals, a "typical" signal may be simulated. Figure 20.12 shows this signal and the responses of two different first-order filters. Notice that the more the data are filtered, the "smoother" the signal looks. However, the lower frequency peaks tend to be smoothed as well so that the filtered data signal may not represent the true signal very well even though it appears to be satisfactory.

The same example can be shown for a second-order filter. Results are given in Figure 20.13. For simplicity, two first-order filters with the same time constants were used to produce the second-order filter. It can be seen that the moderate filtering action ($\alpha = 0.5$) for the second-order filter produces a visually "better" filtered signal than the first-order example. High filtering ($\alpha = 0.2$) reduces all the high frequency noise but again distorts the original signal. It should be pointed out that the Shannon sampling theorem, i.e., that one must sample more than twice the rate of the highest frequency signal desired in the data, applies in these examples and that by sampling we have already filtered out some of the high frequency noise.

A third type of filter which occasionally is encountered, the moving average filter, essen-

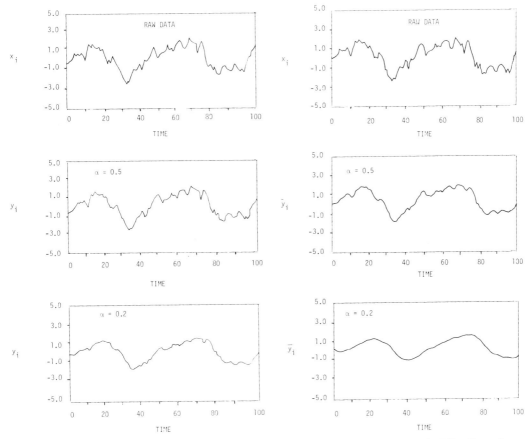

Figure 20.12. A First-Order Filter Example

Figure 20.13. A Second-Order Filter Example

tially integrates all data received, with equal weighting given to each data point recorded. This filter is not as effective for dynamic systems as the exponential filter, which usually gives more weight to the most recently measured data points. The formula for a moving average filter is

$$\bar{x}_j = \frac{1}{N} \sum_{k=0}^{N-1} x_{j-k} \qquad (20.57)$$

where $\bar{x}_j$ is the average value for a series of N readings, $x_j, x_{j-1}, \ldots, x_{j-N+1}$. Note that the time frame for averaging can be arbitrary and can be re-initialized according to the desires of the user.

20.3.2 High-Order Digital Filters

High-order filters are also commonly used in process data acquisition and control systems and often exhibit more favorable filtering characteristics than low-order filters. The study

of the frequency response characteristics of such filters can be a rather large topic. We do not cover this material here and refer the reader to textbooks on this subject (e.g., Ref. 7). As an example of a higher-order filter, let us consider the so-called three pole Butterworth low-pass filter [20]. The transfer function of this third-order filter (see Section 19.7), is:

$$G_F(s) = \frac{1}{\left(\dfrac{s}{\omega_c}\right)^3 + 2\left(\dfrac{s}{\omega_c}\right)^2 + 2\left(\dfrac{s}{\omega_c}\right) + 1} \qquad (20.58)$$

where $\omega_c = 1/\tau$ is the "cut-off" frequency measured in radians/time. The term ω_c serves as an adjustable parameter much like α in the single exponential filter. For use in a digital computer, we can convert $G_F(s)$ to a discrete form (z-transform) using the techniques discussed in Chapter 19:

$$H(z) = \frac{Cz^{-1} + Dz^{-2}}{(1 - e^{-\omega_c T}z^{-1})\left(1 - 2e^{-\omega_c T/2}\left(\cos\frac{\sqrt{3}}{2}\omega_c T\right)z^{-1} + e^{-\omega_c T}z^{-2}\right)} \quad (20.59)$$

$$C = \omega_c\left[e^{-\omega_c T} + e^{-\omega_c T/2}\left(\frac{1}{\sqrt{3}}\sin\frac{\sqrt{3}}{2}\omega_c T - \cos\frac{\sqrt{3}}{2}\omega_c T\right)\right] \quad (20.60)$$

and

$$D = \omega_c\left[e^{-\omega_c T} - e^{-3\omega_c T/2}\left(\frac{1}{\sqrt{3}}\sin\frac{\sqrt{3}}{2}\omega_c T + \cos\frac{\sqrt{3}}{2}\omega_c T\right)\right] \quad (20.61)$$

for sampling far beyond the required level for control action.

Digital and analog filters are valuable in process analysis, but they also are quite important in a real-time system where a controller is involved. The effect of a filtered vs. non-filtered measurement on the performance of the control system is important because inevitably there will be high frequency noise in the system. Filters will make the control system less sensitive to the high frequency process and instrument noise.

Once ω_c and the sampling time T are specified, C and D can be computed and the digital filter $G_F(z)$ can be implemented.

To illustrate the effect of a Butterworth filter on a noisy signal, the same signal as was used for the first- and second-order exponential filters was processed with a Butterworth filter equation. Comparisons are not exact since additional parameters have been specified. Nevertheless, one can see that the higher-order filter as shown in Figure 20.14 yields more pronounced filtering action than the simpler filters.

There is a wealth of literature available on digital signal processing, predominantly in the communications area. For this, interested readers will find that for digital signals it is probably simpler to define a second- or third-order difference equation with coefficients as adjustable parameters for filtering than it is to adapt the Butterworth or other special high-order filter. Also, it is generally good practice to filter out continuous high frequency noise using analog filters as discussed in Chapter 5 rather than with digital filters, especially high-order ones. The main reason, apart from complexity, is that one must sample at very high rates in order to get good discrete data from which to obtain a properly filtered signal. This action could easily increase computer overhead

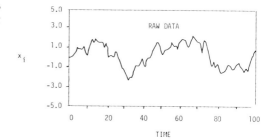

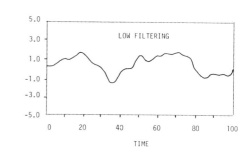

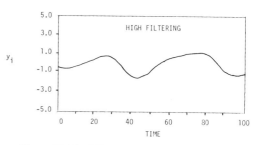

Figure 20.14. A Butterworth Filter Example

20.4 PROGRAMMING CONSIDERATIONS

In previous sections we have discussed controller design, tuning, and filtering techniques without really mentioning the importance of software in the computer control or real-time computing environment. Since this topic is discussed in several other chapters of this book, only a few specific points will be raised here.

We have largely ignored numerical errors introduced by the quantizing effect of the analog to digital converter. Typically, a continuous analog signal must be discretized into approximately 12 bits for the internal computer representation. Let us assume that an equivalent 12 bits are used for the digital to analog converter at the computer output. If a valve is to be open or shut as the DAC goes from zero to full range, then for a one-digit error, e_n, we can calculate the maximum permissible gain K_c. Limits should be put on this gain in any program, since there is no advantage in having the value of the gain larger than the maximum allowable value from a numerical point of view.

A second and perhaps more common problem arises from the summation term for the integral action. If a process maintains an error for a long period of time, it is possible that this summation can build to a very large numerical value. A well-written digital computer control program would stop implementing this summation when the value of the integral term exceeds the output range of the DAC. This limiting is commonly termed anti-reset-windup. If this precaution is not taken, it is possible that the integral term will become very large. Even though the error returns to zero or moves in the opposite direction, it will take a very long time to reduce the summation below saturation levels. Recall that the summation is normally small (but not zero) for the algorithm to maintain the controlled variable at its set point. Anti-reset-windup is commonly available on commercial analog controllers today. However, in programming a digital control system the burden is placed upon the designer of the software to include these extra features.

Finally, we have used a very simplified approximation to the derivative of the error in Equation (20.16). A better representation of this term is given in Equation (20.62):

$$\frac{de}{dt} \approx \frac{\Delta e}{T} = \frac{1}{6T}(e_i + 3e_{i-1} - 3e_{i-2} - e_{i-3})$$

(20.62)

This form should be used when the measured signal is fluctuating rapidly and when it has not been filtered (as discussed in the filtering section of this chapter).

Many other aspects of process control software should be considered in addition to the numerical problems mentioned above. It is essential for the control engineer to be able to adjust or tune the digital controllers while the plant or process is in operation. As a result, very flexible and efficient algorithms are required for parameter changes and also for set-point adjustments on-line. Refinements often include limiting the amount of change in gains in any one time step and automatic checking for the maximum and minimum values of the controller parameters (from a numerical point of view).

Another important programming consideration is that in many cases the same control structure, for example, Equation (20.27), is used for many different control loops in a computer control system. Thus major efficiencies can be gained by appropriate use of re-entrant programs for the control algorithms.

One of the problems which can arise in developing computer control software is that the theory typically (and this chapter is no exception) is developed using deviations of all input and output variables from the normal design or steady state operating conditions. Thus Equation (20.27) should actually be written

$$m_i = K_c e_i + K_I T \sum_j e_j + \frac{K_D}{T}(e_i - e_{i-1}) + \overline{m}$$

(20.63)

where $\overline{m}$ = steady state position of control valve (or controller output to the process). The additional term must also be adjustable through the parameter changing programs, since it defines the actual output of the computer system for a particular loop when the error, integral of the error, and the rate of change of the error are zero.

20.5 SUMMARY

This chapter has been intended to provide a brief overview of a number of topics in computer process control. It has been written for readers only slightly familiar with the concepts of process control, but at the same time it provides some new material not normally taught in classical process control courses at the undergraduate level. As far as sophisticated digital process control techniques are concerned, we have only been able to scratch the surface of an enormously broad and complex field. A number of standard references in this area have been included in the bibliography. Interested readers may wish to study this literature, after mastering the simple concepts.

20.6 BIBLIOGRAPHY AND REFERENCES

1. Cadzow, J. A. and Martens, H. R., *Discrete-Time and Computer Control Systems*, Prentice Hall, Englewood Cliffs, N.J., 1970.
2. Coughanowr, D. R. and Koppel, L. B., *Process Systems Analysis and Control*, McGraw-Hill, New York, 1965.
3. Dahlin, E. B., "Designing and Tuning Digital Controllers", *Instruments and Control Systems*, 4 (6), 77, 1968.
4. Douglas, J. M., *Process Dynamics and Control*, Vol. 1, *Analysis of Dynamic Systems*, Prentice Hall, Englewood Cliffs, N.J., 1972.
5. Douglas, J. M., *Process Dynamics and Control*, Vol. 2, *Control System Synthesis*, Prentice Hall, Englewood Cliffs, N.J., 1972.
6. Goff, K. W., "Dynamics in Direct Digital Control", *ISA J.*, 13 (11), 45, 1966.
7. Guilleman, E. A., *Synthesis of Passive Networks*, John Wiley and Sons, New York, 1957.
8. Harriott, P., *Process Control*, McGraw-Hill, New York, 1964.
9. Harrison, T. J., *Handbook of Industrial Control Computers*, Wiley Interscience, New York, 1972.
10. Harrison, T. J., *Minicomputers in Industrial Control*, ISA, Pittsburgh, Pa., 1978.
11. Koppel, L. B., *Introduction to Control Theory*, Prentice-Hall, Englewood Cliffs, N.J., (1968).
12. Krishnamoorthy, V. and Edgar, T. F., "Identification and Estimation of Discrete Models for a Distillation Column", *Proc. Joint Automatic Control Conference*, 880, 1977.
13. Kuo, B. C., *Discrete Data Control Systems*, Prentice Hall, Englewood Cliffs, N.J., 1970.
14. Kuo, B. C., *Digital Control Systems*, SRL Publishing Co., Urbana Ill., 1977.
15. Lindorff, D. P., *Theory of Sampled Data Systems*, John Wiley and Sons, New York, 1965.
16. Luyben, W. L., *Process Modelling, Simulation and Control for Chemical Engineers*, McGraw-Hill, New York, 1973.
17. Moore, C. F., Smith, C. L., and Murrill, P. W., "Simplifying Digital Control Dynamics for Controller Tuning and Hardware Lag Effects", *Instrument Practice*, 45, Jan. 1969.
18. Munroe, A. J., *Digital Processes for Sampled Data Systems*, John Wiley and Sons, New York, 1962.
19. Murrill, P. W., *Automatic Control of Processes*, International Textbook, Scranton, Pa., 1967.
20. Rader, C. M. and Gold, B., "Digital Filter Design Techniques in the Frequency Domain", *Proc. IEEE*, 55, 149, 1967.
21. Ragazzini, J. R. and Franklin, G. F., *Sampled Data Control Systems*, McGraw-Hill, New York, 1958.
22. Savas, E. S., *Computer Control of Industrial Processes*, McGraw-Hill, New York, 1965.
23. Shinskey, F. G., *Process Control Systems*, McGraw-Hill, New York, 1967.
24. Smith, C. L., *Digital Computer Process Control*, International Textbook, Scranton, Pa., 1972.
25. Tou, J., *Digital and Sampled Data Control Systems*, John Wiley and Sons, New York, 1962.
26. Verbruggen, H. B., Peperstraete, J. A., and Debruyn, H. P., "Digital Controllers and Digital Control Algorithms", *Quarterly Journal of Automatic Control, Journal A*, 16 (2), 53, 1975.

20.7 EXERCISES

1. Consider a home heating system where natural gas is used as the fuel for a furnace. The thermostat is an on–off device.
   a. Draw a block diagram for this control system. What elements (blocks) are static, and which exhibit dynamic behavior?
   b. Identify the inputs and outputs, including disturbance variables.
   c. How does the location of the sensor(s) affect the performance of the control system?
   d. Discuss the regulatory vs. servomechanism functions of the control system.

2. Suppose you are asked to analyze the stirred-tank heater system under proportional control (Section 20.1). Let $K_p = 1$ and $\tau = 1$.

Assume $K_c K_p K_v = K$, the loop gain, which is proportional to the controller gain.

a. If the initial deviation in temperature $(T_o - \overline{T}_o)$ is $+5°$, show the response of the temperature as the control system returns the process to near its desired steady-state operating point ($0°$ deviation or $T_o = \overline{T}_o$). Let $K = 0, 1, 5$, and 10. Show that no offset occurs for these cases. Compare the speed of response for these different values of the controller gain. Can the system be made unstable for extremely large values of K?

b. Using the procedure described in Section 20.3.1, convert the continuous time model to a discrete time model. Use a sampling time of 0.1 (since $\tau = 1$). Repeat the exercise in (a) above, using the same values of K. Show that the system can be made unstable for large values of K.

3. Consider a third-order process described by

$$G(s) = \frac{1}{(s+1)^3}.$$

a. Using $K = K_c K_v K_p$, derive the closed-loop transfer function.

b. Find the critical gain $(K = K_u)$ that causes the system to exhibit sustained oscillations. Two methods to find K_u are suggested: First, use partial fraction decomposition (see Chapter 19) and determine when instability occurs by inspecting the denominator of the closed-loop transfer function. Second, use a digital computer to integrate the third-order differential equation obtained from G(s), assuming a unit step change in controller set point. Once K_u and P_u are found from simulation, compute the settings for a PID controller.

c. Simulate the unit step response for the closed-loop system with PID controller.

4. A process is described by the differential equation model

$$150\frac{dy}{dt} + y = 4x(t - 100)$$

In this case $x(t - 100)$ indicates mathematically that the process input is delayed before it affects the process. If the sampling period is 50 time units:

a. Derive a discrete equation that approximates the continuous process model, assuming that x, the input, passes through a first-order hold.

b. Find an appropriate value for the gain of a proportional controller for this process.

c. Could you use a position or velocity control algorithm with the discrete model to find a discrete mathematical relationship describing the closed-loop system? If so, derive such a relation and explain how it could be used to "test" a particular value of controller gain for system stability.

5. A particular process responds to changes in a single input x according to the discrete equation

$$y_k = 0.8y_{k-1} + 0.6x_{k-2}$$

Derive the proportional-derivative controller settings for such a process if the sampling time is 5 seconds.

6. For the process G(z) of Problem 7 in Chapter 19, compute the closed-loop servo transfer function for proportional-only control. Using partial fraction decomposition, find the value of K_c that causes instability.

7. One form of analog controller that often is used for servomechanism positioning systems is the lead–lag network. The transfer function for this controller is:

$$G_c(s) = K_c \frac{(\tau_1 s + 1)}{(\tau_2 s + 1)}$$

The equivalent differential equation for the controller is:

$$\tau_2 \frac{dy}{dt} + y = K_c\left(\tau_1 \frac{dx}{dt} + x\right)$$

Derive a discrete version of this control algorithm based on a sampling time of T.

8. Design a Dahlin controller for a process that has a transfer function

$$G(s) = \frac{e^{-0.8s}}{(0.2s + 1)(0.4s + 1)}$$

Assume that the process is sampled at 0.2-second intervals.

9. Equation 20.43 hints at a more general derivation for Dahlin's algorithm which pro-

vides better understanding than the one originally presented. Essentially one specifies the desired closed-loop discrete transfer function and the discrete (including zero-order hold) plant transfer function. From these the required D(z) can be calculated. The technique follows the derivations of

deadbeat controllers as described, for example, in Ragazzini and Franklin [21].

Design a Dahlin algorithm for a step input using this technique. Compare your result with that obtained in Equation 20.44. Repeat the derivation for a process closed-loop response to ramp inputs.

Index